METHODS IN MOLECULAR BIOLOGY

Series Editor
John M. Walker
School of Life and Medical Sciences
University of Hertfordshire
Hatfield, Hertfordshire, AL10 9AB, UK

For further volumes:
http://www.springer.com/series/7651

Computational Biology of Non-Coding RNA

Methods and Protocols

Edited by

Xin Lai

Department of Dermatology, Erlangen University Hospital, Erlangen, Germany

Shailendra K. Gupta

Department of Systems Biology and Bioinformatics, University of Rostock, Rostock, Germany

Julio Vera

Department of Dermatology, Erlangen University Hospital, Erlangen, Germany

💥 Humana Press

Editors
Xin Lai
Department of Dermatology
Erlangen University Hospital
Erlangen, Germany

Julio Vera
Department of Dermatology
Erlangen University Hospital
Erlangen, Germany

Shailendra K. Gupta
Department of Systems Biology
and Bioinformatics
University of Rostock
Rostock, Germany

ISSN 1064-3745   ISSN 1940-6029 (electronic)
Methods in Molecular Biology
ISBN 978-1-4939-8981-2   ISBN 978-1-4939-8982-9 (eBook)
https://doi.org/10.1007/978-1-4939-8982-9

Library of Congress Control Number: 2018965598

© Springer Science+Business Media, LLC, part of Springer Nature 2019
This work is subject to copyright. All rights are reserved by the Publisher, whether the whole or part of the material is concerned, specifically the rights of translation, reprinting, reuse of illustrations, recitation, broadcasting, reproduction on microfilms or in any other physical way, and transmission or information storage and retrieval, electronic adaptation, computer software, or by similar or dissimilar methodology now known or hereafter developed.
The use of general descriptive names, registered names, trademarks, service marks, etc. in this publication does not imply, even in the absence of a specific statement, that such names are exempt from the relevant protective laws and regulations and therefore free for general use.
The publisher, the authors, and the editors are safe to assume that the advice and information in this book are believed to be true and accurate at the date of publication. Neither the publisher nor the authors or the editors give a warranty, express or implied, with respect to the material contained herein or for any errors or omissions that may have been made. The publisher remains neutral with regard to jurisdictional claims in published maps and institutional affiliations.

This Humana Press imprint is published by the registered company Springer Science+Business Media, LLC, part of Springer Nature.
The registered company address is: 233 Spring Street, New York, NY 10013, U.S.A.

Preface

Noncoding RNAs (ncRNAs) have emerged as a major class of regulatory genes for a diverse range of biological functions. ncRNAs exert their effect in the context of complex regulatory networks by targeting protein-coding genes, as well as other ncRNAs. Computational methods, such as high-throughput sequencing data analysis, biochemical network reconstruction and analysis, and kinetic modeling, have been applied to provide the functional characterization and annotation of ncRNAs. The aim of this volume is to review current knowledge, novel methods, and open challenges in understanding the regulatory role of ncRNAs in biomedicine. In particular, this volume offers a collection of state-of-the-art methods including the identification of novel ncRNAs and their targets, functional annotation, and disease association in different biological contexts.

The book is divided into five parts: (1) Overview of Disease-Specific ncRNAs, (2) Computational Methods and Workflows for ncRNA Discovery and Annotation Based on High-Throughput Sequencing Data, (3) Bioinformatics Tools and Databases for ncRNA Analyses, (4) Network-Based Methods for Characterizing ncRNA Function, and (5) Kinetic Modeling of ncRNA-Mediated Gene Regulation. In the first part, Schulte et al. discuss in detail the current knowledge on the role of regulatory microRNAs (miRNAs) and long ncRNA (lncRNA) in inflammatory and infectious diseases (Chapter 1). In Chapter 2, Logotheti, Marquardt, and Pützer discuss the cross-talk between p73 and miRNAs in realizing the cancer metastasis phenotype. This chapter also highlights translational opportunities and research challenges for the p73/miRNA cancer network in the context of cancer metastasis.

The subsequent five chapters describe experimental and computational methods, workflows that are useful for identification and functional annotation of circular RNAs (circRNAs), lncRNAs, and miRNAs. In Chapter 3, Sharma et al. describe software and workflows to identify circRNAs from RNA sequencing (RNA-seq) data as well as a step-by-step procedure to experimentally validate in silico identified circRNAs. Chapter 4 by Mathew et al. presents methods and pipelines for the discovery of novel lncRNAs from RNA-seq data of zebrafish. In Chapter 5, Wolfien et al. review and discuss state-of-the-art experimental and computational workflows for the identification and characterization of ncRNAs, and they emphasize the use of transcriptome-wide association studies, molecular network analyses, and artificial intelligence-guided predictions. They also propose a promising strategy for developing reproducible and sharable computational workflows for the study of ncRNAs using RNA-seq data. In Chapter 6, Nigita et al. review methods and resources for genome-wide detection of RNA editing in ncRNAs and discuss the features of ncRNA editing associated with their function as well as with the onset of human diseases. Finally, Demirci, Yousef, and Allmer present computational methodologies allowing the identification and validation of functional miRNA-mRNA interactions and discuss key points for investigating miRNAs as biomarkers of human diseases (Chapter 7).

The next three chapters present bioinformatics tools and databases for ncRNA analyses. In Chapter 8, Bagnacani et al. present a strategy that makes use of community effort to develop sharable and reusable computational workflows using the Galaxy platform. The concept is exemplified with a database for the identification and characterization of cooperative miRNA regulation of gene expression. In Chapter 9, Monga and Kumar provide a comprehensive review of in silico miRNA resources and databases. They present tools for miRNA annotation, miRNA-disease association and identification and characterization of miRNA variants. In Chapter 10, Maracaja-Coutinho et al. present an up-to-date review about 229 databases related with ncRNA research with particular emphasis on biomedicine, including ncRNA databases for cancer, cardiovascular, nervous systems, pathogens, and other diseases.

Chapters 11–13 highlight network-based methods for characterizing ncRNA functions. In Chapter 11, Nacher and Akutsu review the current research on network controllability methods. The authors also present a tripartite network model including ncRNA-target and protein-disease layers and analyze the network with controllability methods to identify the subset of critical control ncRNAs associated with human diseases. Piro and Marsico summarize recent developments in lncRNA function prediction and lncRNA-disease associations using methods based on network analysis and competing endogenous RNAs (ceRNAs) function prediction (Chapter 12). Paul presents a mutual information-based Maximum-Relevance Maximum-Significance algorithm to identify miRNA-mRNA regulatory modules from interaction networks in a case study on gynecologic cancer (Chapter 13).

The last four chapters of this book showcase the use of kinetic modeling for providing quantitative understanding about mechanisms of ncRNA-mediated regulation of gene expression. In Chapter 14, Bocci et al. review and discuss kinetic models that investigate the ability of miRNA-mediated feedback loops in provoking bistability in gene expression. They link this phenomenon with the switch of cellular phenotypes and the control of intrinsic and extrinsic signaling noise. In Chapter 15, Martirosyan et al. review kinetic models that describe competition and depletion of shared miRNAs by ceRNAs, showing the importance of ceRNA effect in processing gene expression noise. In Chapter 16, Tian et al. demonstrated the use of kinetic modeling to explain the emerging properties associated with miRNA-mRNA reciprocal regulation and show that such regulation can lead to bistable switches in gene expression directing cell fate decision. In the last chapter, Rabajante and Rosario discuss mathematical modeling frameworks for studying lncRNA regulation of the mammalian cell cycle. They also present a model of ordinary differential equations to simulate and analyze cell cycle regulation in response to DNA damage.

Taken together, all chapters presented in this book provide a state-of-the-art collection of computational methods and approaches that will be of value to researchers interested in the ncRNA field. Computational biology-based methods for the identification and analyses of ncRNAs in the context of biomedicine are rapidly growing with both challenges and opportunities to identify ncRNA biomarkers and to develop ncRNA therapeutics. We believe that this volume is a valuable and useful resource for addressing many of these exciting challenges in ncRNA research and hope it will be of interest to many peers. This book is also dedicated to the loss of Dr. Baltazar D. Aguda who has made numerous and important contribution to improve our understanding of ncRNA-mediated regulation of gene expression using mathematical models.

Erlangen, Germany *Xin Lai*
Rostock, Germany *Shailendra K. Gupta*
Erlangen, Germany *Julio Vera*

Contents

Preface v
Contributors ix

PART I OVERVIEW OF DISEASE-SPECIFIC ncRNAs

1 ncRNAs in Inflammatory and Infectious Diseases 3
   *Leon N. Schulte, Wilhelm Bertrams, Christina Stielow,
   and Bernd Schmeck*

2 p73-Governed miRNA Networks: Translating Bioinformatics
   Approaches to Therapeutic Solutions for Cancer Metastasis 33
   Stella Logotheti, Stephan Marquardt, and Brigitte M. Pützer

PART II COMPUTATIONAL METHODS AND WORKFLOWS FOR ncRNA
DISCOVERY AND ANNOTATION BASED ON HIGH-THROUGHPUT
SEQUENCING DATA

3 Methods for Annotation and Validation of Circular RNAs
   from RNAseq Data 55
   *Disha Sharma, Paras Sehgal, Judith Hariprakash, Sridhar Sivasubbu,
   and Vinod Scaria*

4 Methods to Study Long Noncoding RNA Expression
   and Dynamics in Zebrafish Using RNA Sequencing 77
   *Samatha Mathew, Ambily Sivadas, Paras Sehgal, Kriti Kaushik,
   Shamsudheen K. Vellarikkal, Vinod Scaria, and Sridhar Sivasubbu*

5 Workflow Development for the Functional Characterization of ncRNAs 111
   *Markus Wolfien, David Leon Brauer, Andrea Bagnacani,
   and Olaf Wolkenhauer*

6 ncRNA Editing: Functional Characterization and Computational
   Resources 133
   *Giovanni Nigita, Gioacchino P. Marceca, Luisa Tomasello,
   Rosario Distefano, Federica Calore, Dario Veneziano, Giulia Romano,
   Serge Patrick Nana-Sinkam, Mario Acunzo, and Carlo M. Croce*

7 Computational Prediction of Functional MicroRNA–mRNA
   Interactions 175
   Müşerref Duygu Saçar Demirci, Malik Yousef, and Jens Allmer

PART III BIOINFORMATICS TOOLS AND DATABASES FOR ncRNA ANALYSES

8 Tools for Understanding miRNA–mRNA Interactions for Reproducible
   RNA Analysis 199
   Andrea Bagnacani, Markus Wolfien, and Olaf Wolkenhauer

9 Computational Resources for Prediction and Analysis of Functional miRNA and Their Targetome 215
Isha Monga and Manoj Kumar

10 Noncoding RNAs Databases: Current Status and Trends 251
Vinicius Maracaja-Coutinho, Alexandre Rossi Paschoal, José Carlos Caris-Maldonado, Pedro Vinícius Borges, Almir José Ferreira, and Alan Mitchell Durham

PART IV NETWORK-BASED METHODS FOR CHARACTERIZING NCRNA FUNCTION

11 Controllability Methods for Identifying Associations Between Critical Control ncRNAs and Human Diseases 289
Jose C. Nacher and Tatsuya Akutsu

12 Network-Based Methods and Other Approaches for Predicting lncRNA Functions and Disease Associations 301
Rosario Michael Piro and Annalisa Marsico

13 Integration of miRNA and mRNA Expression Data for Understanding Etiology of Gynecologic Cancers 323
Sushmita Paul

PART V KINETIC MODELING OF NCRNA-MEDIATED GENE REGULATION

14 Quantitative Characteristic of ncRNA Regulation in Gene Regulatory Networks.. 341
Federico Bocci, Mohit Kumar Jolly, Herbert Levine, and José Nelson Onuchic

15 Kinetic Modelling of Competition and Depletion of Shared miRNAs by Competing Endogenous RNAs............................... 367
Araks Martirosyan, Marco Del Giudice, Chiara Enrico Bena, Andrea Pagnani, Carla Bosia, and Andrea De Martino

16 Modeling ncRNA-Mediated Circuits in Cell Fate Decision................. 411
Xiao-Jun Tian, Manuela Vanegas Ferro, and Hanah Goetz

17 Modeling Long ncRNA-Mediated Regulation in the Mammalian Cell Cycle.. 427
Jomar F. Rabajante and Ricardo C. H. del Rosario

Index 447

Contributors

MARIO ACUNZO • *Division of Pulmonary Diseases and Critical Care Medicine, Virginia Commonwealth University, Richmond, VA, USA*
TATSUYA AKUTSU • *Bioinformatics Center, Institute for Chemical Research, Kyoto University, Kyoto, Japan*
JENS ALLMER • *Applied Bioinformatics, Bioscience, Wageningen University & Research, Wageningen, The Netherlands*
ANDREA BAGNACANI • *Department of Systems Biology and Bioinformatics, Institute of Computer Science, University of Rostock, Rostock, Germany*
CHIARA ENRICO BENA • *DISAT, Politecnico di Torino, Turin, Italy; Italian Institute for Genomic Medicine, Turin, Italy*
WILHELM BERTRAMS • *Institute for Lung Research, Universities of Giessen and Marburg Lung Center, Member of the German Center for Lung Research (DZL), Philipps-University Marburg, Marburg, Germany*
FEDERICO BOCCI • *Center for Theoretical Biological Physics, Rice University, Houston, TX, USA; Department of Chemistry, Rice University, Houston, TX, USA*
PEDRO VINÍCIUS BORGES • *Department of Computer Science, Bioinformatics Graduation Program (PPGBIOINFO), Federal University of Technology—Paraná, Cornélio Procópio, Brazil*
CARLA BOSIA • *DISAT, Politecnico di Torino, Turin, Italy; Italian Institute for Genomic Medicine, Turin, Italy*
DAVID LEON BRAUER • *Department of Systems Biology and Bioinformatics, Institute of Computer Science, University of Rostock, Rostock, Germany*
FEDERICA CALORE • *Department of Cancer Biology and Genetics, Comprehensive Cancer Center, The Ohio State University, Columbus, OH, USA*
JOSÉ CARLOS CARIS-MALDONADO • *Advanced Center for Chronic Diseases—ACCDiS, Facultad de Ciencias Químicas y Farmacéuticas, Universidad de Chile, Santiago, Chile*
CARLO M. CROCE • *Department of Cancer Biology and Genetics, Comprehensive Cancer Center, The Ohio State University, Columbus, OH, USA*
ROSARIO DISTEFANO • *Department of Cancer Biology and Genetics, Comprehensive Cancer Center, The Ohio State University, Columbus, OH, USA*
ALAN MITCHELL DURHAM • *Computer Science Department, Instituto de Matemática e Estatística, Universidade de São Paulo, São Paulo, Brazil*
ALMIR JOSÉ FERREIRA • *Computer Science Department, Instituto de Matemática e Estatística, Universidade de São Paulo, São Paulo, Brazil*
MARCO DEL GIUDICE • *DISAT, Politecnico di Torino, Turin, Italy; Italian Institute for Genomic Medicine, Turin, Italy*
HANAH GOETZ • *School of Biological and Health Systems Engineering, Arizona State University, Tempe, AZ, USA*
JUDITH HARIPRAKASH • *G.N. Ramachandran Knowledge Center for Bioinformatics, CSIR Institute of Genomics and Integrative Biology (CSIR-IGIB), Delhi, India*
MOHIT KUMAR JOLLY • *Center for Theoretical Biological Physics, Rice University, Houston, TX, USA*

KRITI KAUSHIK • *Genomics and Molecular Medicine, CSIR Institute of Genomics and Integrative Biology (CSIR-IGIB), Delhi, India; Academy of Scientific and Innovative Research (AcSIR), New Delhi, India*

MANOJ KUMAR • *Bioinformatics Centre, Institute of Microbial Technology, Council of Scientific and Industrial Research, Chandigarh, India*

HERBERT LEVINE • *Center for Theoretical Biological Physics, Rice University, Houston, TX, USA; Department of Chemistry, Rice University, Houston, TX, USA; Department of Bioengineering, Rice University, Houston, TX, USA; Department of Physics and Astronomy, Rice University, Houston, TX, USA*

STELLA LOGOTHETI • *Institute of Experimental Gene Therapy and Cancer Research, Rostock University Medical Center, Rostock, Germany*

VINICIUS MARACAJA-COUTINHO • *Advanced Center for Chronic Diseases—ACCDiS, Facultad de Ciencias Químicas y Farmacéuticas, Universidad de Chile, Santiago, Chile*

GIOACCHINO P. MARCECA • *Department of Cancer Biology and Genetics, Comprehensive Cancer Center, The Ohio State University, Columbus, OH, USA; Department of Clinical and Experimental Medicine, University of Catania, Catania, CT, Italy*

STEPHAN MARQUARDT • *Institute of Experimental Gene Therapy and Cancer Research, Rostock University Medical Center, Rostock, Germany*

ANNALISA MARSICO • *Max-Planck-Institut für molekulare Genetik, Berlin, Germany; Institut für Informatik, Freie Universität Berlin, Berlin, Germany*

ANDREA DE MARTINO • *Soft & Living Matter Lab, CNR-NANOTEC, Rome, Italy; Italian Institute for Genomic Medicine, Turin, Italy*

ARAKS MARTIROSYAN • *Laboratory of Glia Biology, VIB-KU Leuven Center for Brain and Disease Research, Leuven, Belgium; KU Leuven Department of Neuroscience, Leuven, Belgium*

SAMATHA MATHEW • *Genomics and Molecular Medicine, CSIR Institute of Genomics and Integrative Biology (CSIR-IGIB), Delhi, India; Academy of Scientific and Innovative Research (AcSIR), New Delhi, India*

ISHA MONGA • *Bioinformatics Centre, Institute of Microbial Technology, Council of Scientific and Industrial Research, Chandigarh, India*

JOSE C. NACHER • *Faculty of Science, Department of Information Science, Toho University, Chiba, Japan*

SERGE PATRICK NANA-SINKAM • *Division of Pulmonary Diseases and Critical Care Medicine, Virginia Commonwealth University, Richmond, VA, USA*

GIOVANNI NIGITA • *Department of Cancer Biology and Genetics, Comprehensive Cancer Center, The Ohio State University, Columbus, OH, USA*

JOSÉ NELSON ONUCHIC • *Center for Theoretical Biological Physics, Rice University, Houston, TX, USA; Department of Chemistry, Rice University, Houston, TX, USA; Department of Physics and Astronomy, Rice University, Houston, TX, USA; Department of Biosciences, Rice University, Houston, TX, USA*

ANDREA PAGNANI • *DISAT, Politecnico di Torino, Turin, Italy; Italian Institute for Genomic Medicine, Turin, Italy*

ALEXANDRE ROSSI PASCHOAL • *Department of Computer Science, Bioinformatics Graduation Program (PPGBIOINFO), Federal University of Technology—Paraná, Cornélio Procópio, Brazil*

SUSHMITA PAUL • *Department of Bioscience & Bioengineering, Indian Institute of Technology Jodhpur, Jodhpur, Rajasthan, India*

ROSARIO MICHAEL PIRO • *Institut für Informatik, Freie Universität Berlin, Berlin, Germany; Institut für Medizinische Genetik und Humangenetik, Charite-Universitätsmedizin Berlin, Berlin, Germany*

BRIGITTE M. PÜTZER • *Institute of Experimental Gene Therapy and Cancer Research, Rostock University Medical Center, Rostock, Germany*

JOMAR F. RABAJANTE • *Institute of Mathematical Sciences and Physics, University of the Philippines Los Baños, Laguna, Philippines*

GIULIA ROMANO • *Division of Pulmonary Diseases and Critical Care Medicine, Virginia Commonwealth University, Richmond, VA, USA*

RICARDO C. H. DEL ROSARIO • *Stanley Center for Psychiatric Research, Broad Institute of MIT and Harvard, Cambridge, MA, USA*

MÜŞERREF DUYGU SAÇAR DEMIRCI • *Faculty of Life and Natural Sciences, Department of Bioinformatics, Abdullah Gul University, Kayseri, Turkey*

VINOD SCARIA • *G.N. Ramachandran Knowledge Center for Bioinformatics, CSIR Institute of Genomics and Integrative Biology (CSIR-IGIB), Delhi, India; Academy of Scientific and Innovative Research (AcSIR), CSIR Institute of Genomics and Integrative Biology (CSIR-IGIB), Delhi, India*

BERND SCHMECK • *Institute for Lung Research, Universities of Giessen and Marburg Lung Center, Member of the German Center for Lung Research (DZL), Philipps-University Marburg, Marburg, Germany*

LEON N. SCHULTE • *Institute for Lung Research, Universities of Giessen and Marburg Lung Center, Member of the German Center for Lung Research (DZL), Philipps-University Marburg, Marburg, Germany*

PARAS SEHGAL • *Academy of Scientific and Innovative Research (AcSIR), CSIR Institute of Genomics and Integrative Biology (CSIR-IGIB), Delhi, India; Genomics and Molecular Medicine, CSIR Institute of Genomics and Integrative Biology (CSIR-IGIB), Delhi, India*

DISHA SHARMA • *G.N. Ramachandran Knowledge Center for Bioinformatics, CSIR Institute of Genomics and Integrative Biology (CSIR-IGIB), Delhi, India; Academy of Scientific and Innovative Research (AcSIR), CSIR Institute of Genomics and Integrative Biology (CSIR-IGIB), Delhi, India*

AMBILY SIVADAS • *Academy of Scientific and Innovative Research (AcSIR), New Delhi, India; G.N. Ramachandran Knowledge Center for Bioinformatics, CSIR Institute of Genomics and Integrative Biology (CSIR-IGIB), Delhi, India*

SRIDHAR SIVASUBBU • *Academy of Scientific and Innovative Research (AcSIR), CSIR Institute of Genomics and Integrative Biology (CSIR-IGIB), Delhi, India; Genomics and Molecular Medicine, CSIR Institute of Genomics and Integrative Biology (CSIR-IGIB), Delhi, India*

CHRISTINA STIELOW • *Institute for Lung Research, Universities of Giessen and Marburg Lung Center, Member of the German Center for Lung Research (DZL), Philipps-University Marburg, Marburg, Germany*

XIAO-JUN TIAN • *School of Biological and Health Systems Engineering, Arizona State University, Tempe, AZ, USA*

LUISA TOMASELLO • *Department of Cancer Biology and Genetics, Comprehensive Cancer Center, The Ohio State University, Columbus, OH, USA*

MANUELA VANEGAS FERRO • *School of Biological and Health Systems Engineering, Arizona State University, Tempe, AZ, USA*

SHAMSUDHEEN K. VELLARIKKAL • *Genomics and Molecular Medicine, CSIR Institute of Genomics and Integrative Biology (CSIR-IGIB), Delhi, India; Academy of Scientific and Innovative Research (AcSIR), New Delhi, India*

DARIO VENEZIANO • *Department of Cancer Biology and Genetics, Comprehensive Cancer Center, The Ohio State University, Columbus, OH, USA*

MARKUS WOLFIEN • *Department of Systems Biology and Bioinformatics, Institute of Computer Science, University of Rostock, Rostock, Germany*

OLAF WOLKENHAUER • *Department of Systems Biology and Bioinformatics, Institute of Computer Science, University of Rostock, Rostock, Germany; Stellenbosch Institute for Advanced Study (STIAS), Wallenberg Research Centre, Stellenbosch University, Stellenbosch, South Africa*

MALIK YOUSEF • *Department of Community Information Systems, Zefat Academic College, Zefat, Israel*

Part I

Overview of Disease-Specific ncRNAs

Chapter 1

ncRNAs in Inflammatory and Infectious Diseases

Leon N. Schulte, Wilhelm Bertrams, Christina Stielow, and Bernd Schmeck

Abstract

Inflammatory and infectious diseases are among the main causes of morbidity and mortality worldwide. Inflammation is central to maintenance of organismal homeostasis upon infection, tissue damage, and malignancy. It occurs transiently in response to diverse stimuli (e.g., physical, radioactive, infective, pro-allergenic, or toxic), and in some cases may manifest itself in chronic diseases. To limit the potentially deleterious effects of acute or chronic inflammatory responses, complex transcriptional and posttranscriptional regulatory networks have evolved, often involving nonprotein-coding RNAs (ncRNA). MicroRNAs (miRNAs) are a class of posttranscriptional regulators that control mRNA translation and stability. Long ncRNAs (lncRNAs) are a very diverse group of transcripts >200 nt, functioning among others as scaffolds or decoys both in the nucleus and the cytoplasm. By now, it is well established that miRNAs and lncRNAs are implicated in all major cellular processes including control of cell death, proliferation, or metabolism. Extensive research over the last years furthermore revealed a fundamental role of ncRNAs in pathogen recognition and inflammatory responses. This chapter reviews and summarizes the current knowledge on regulatory ncRNA networks in infection and inflammation.

Key words miRNA, lncRNA, Infection, Inflammation, Immunity

1 Introduction

Gene expression is subjected to a vast variety of regulatory mechanisms taming transcriptional and translational noise and conferring adequate responses to cell-intrinsic and -extrinsic cues. While transcriptional and posttranscriptional regulation by protein factors has been a topic of research for a long time, the recent discovery of thousands of noncoding RNA genes in eukaryotic genomes has greatly expanded our understanding of gene expression control.

Ever since the discovery of microRNAs (miRNAs) as nonprotein-coding posttranscriptional regulators in eukaryotes in the early 1990s (*lin-4* in *Caenorhabditis elegans*, [1, 2]), the number of experimentally confirmed regulatory RNAs has continuously increased. In an attempt by the ENCODE project to map and classify the entire human transcriptome, thousands of previously unknown long noncoding RNA genes were discovered [3]. While the miRNA biogenesis and target repression pathway is well understood to date, target identification is still a challenging venture. Gene regulation by miRNAs may reach a remarkable degree of network complexity: while one miRNA species may target different mRNAs through shared binding motives, typically in the 3′ UTR [4], a single mRNA may be targeted by multiple miRNAs at the same time. MiRNA-mediated control is complemented by long noncoding RNA-mediated transcriptional and posttranscriptional regulatory mechanisms, cumulating in highly complex gene-regulatory circuitries. This chapter summarizes the current knowledge about miRNAs and lncRNAs in mammalian inflammatory responses and infection.

2 MiRNAs

2.1 MiRNAs in Inflammation

For more than a decade, the role of miRNAs in hematopoietic cells has been in the focus of biomedical research. This research direction was pioneered by the David Bartel group, which reported on the role of miRNAs in the differentiation of hematopoietic progenitor cells in 2004 [5]. Ever since, the role of miRNAs in differentiation and function of immune cells has been the subject of countless research projects (reviewed in refs. [6–8]).

The innate immune system constitutes the first line of defense against pathogenic threats. The lung, with its huge surface area, is a primary site of pathogen attack and immunogen exposure due to its extensive contact with environmental particles, allergens, and airborne pathogens. Pattern recognition receptors (PRRs) on the cell surface (e.g., Toll-like receptors, TLRs) and in the cytosol (e.g., NOD-like receptors, NLRs, or RIG-like receptors, RLRs) serve as the sensors of the innate immune system, recognizing pathogen-associated molecular patterns (PAMPs), such as microbial cell wall components or bacterial or viral nucleic acids. Ligation of PAMPs to PRRs initiates an intracellular signaling cascade that culminates in the activation of specific transcription factors such as the inflammation master regulator NFκB. Activation of target genes, encoding protein factors such as pro-inflammatory cytokines, initiates a concerted local or systemic immune response. This response needs to be tightly controlled, to ensure successful pathogen clearance while preventing overshooting inflammation, which may cause serious side effects, such as severe tissue damage [9].

David Baltimore's group was the first to link PAMP recognition to expression changes of various miRNAs. Upon stimulation of THP-1 cells (a human acute monocytic leukemia cell line) with LPS from *Escherichia coli* [10], they observed increased expression of miR-146, -155, and -132. MiR-146a was identified as an immediate early-response gene, inducible by various microbial components and pro-inflammatory mediators (e.g., LPS, flagellin, and IL-1β). This induction was shown to depend on NFκB-binding sites in the miR-146a promoter region. MiR-146a negatively regulates central PRR signal transduction components, such as IL-1 receptor-associated kinase 1 (IRAK1) and TNF receptor-associated factor 6 (TRAF6), both of which are involved in TLR-mediated NFκB activation [10]. In a recent report, Magilnick and colleagues confirmed *miR-146a* to play a crucial role in the modulation of *Traf6* expression in vivo, to achieve immune homeostasis. In *miR-146a$^{-/-}$* mice, LPS stimulation resulted in significantly elevated serum levels of IL-6 and TNF-α compared with wild-type animals [11]. In contrast, *miR-146a$^{-/-}$Traf6$^{+/-}$* mice exhibited markedly reduced IL-6 and TNF-α levels, verifying *miR-146a* as the critical regulator of the *Traf6*-NFκB signaling axis. Moreover, analysis of the activation status of peripheral T cells revealed a profound induction of these cells in *miR-146a$^{-/-}$* mice, whereas, in *miR-146a$^{-/-}$Traf6$^{+/-}$* mice, the reduction of the *Traf6* gene dose resulted in significantly reduced numbers of activated effector T cells [11]. These reports suggest miR-146a to play a central role in negative-feedback regulation of PRR-triggered inflammatory responses. This negative feedback hypothesis was confirmed for miR-146 in the A549 cell line by showing that it could attenuate the IL-1β-induced inflammatory response [12]. A particular feature of A549 cells in this context is the observed induction of miR-146b, which, unlike induction of miR-146a, was observed only in A549 cells. Targeting miRNA expression directly by synthetic miRNA mimics or antagonists revealed a significant decrease of IL-1β-induced IL-8 and RANTES secretion in A549 cells upon miR-146a overexpression and, reciprocally, an increase of IL-8 secretion upon miR-146a inhibition, albeit only at high levels of concomitant IL-1β. These data support the notion of miRNA regulation in severe inflammation [12]. Different signaling pathways control expression of the two miR-146 isoforms, which could, at least in part, account for the observed expression differences between miR-146a and miR-146b upon IL-1β stimulation. While NFκB and c-jun N-terminal kinase (JNK)-1/2 regulate miR-146a, miR-146b expression seems to be dependent on MEK-1/2 and JNK-1/2 [13].

In a follow-up study, miR-155 was found to be induced in murine bone marrow-derived macrophages (BMDMs) after treatment with a synthetic analogue of viral dsRNA (poly(I:C)) or IFN-β, a cytokine highly expressed by the host upon viral infection. IFN-γ also triggered miR-155 expression, albeit delayed when compared to poly(I:C). IFN-mediated miR-155 induction appeared to require TNF-α autocrine/paracrine signaling. In contrast, poly(I:C) exerted its influence via the JNK pathway [14]. Furthermore, Mantuano and colleagues demonstrated that blocking the endocytic cell-signaling receptor LDL receptor-related protein-1 (LRP1) rapidly activated NFκB; subsequently induced

TNF-α, IL-6, and CCL4 expression; and finally led to elevated miR-155 levels [15]. Deletion of *LRP1* in BMDMs significantly elevated miR-155 levels, which was accompanied by an increased RNA Pol II occupancy at the miR-155 promoter. Consequently, treatment of wild-type cells with LRP1 antagonists also resulted in a significant increase in miR-155 expression after a lag phase of 4–8 h. Moreover, inhibition of the nuclear translocation of NFκB prevented miR-155 induction, demonstrating the involvement of this transcription factor in miR-155 activation. In line with the delayed response of miR-155 induction, the inhibition of miR-155 1 and 8 h after LRP1 antagonist treatment, respectively, decreased TNF-α levels only at 8 h post addition of the antagonist, suggesting miR-155 to be crucial for the maintenance of the proinflammatory response to LRP1 antagonists. Upregulation of miR-155 was also observed in the RAW 264.7 macrophage cell line, accompanied by a downregulation of miR-125b. This pattern was also observed in an in vivo approach that used intraperitoneal inoculation of mice with LPS from *Salmonella enteritidis* and subsequent miRNA analysis of isolated splenocytes [16]. Similar to miR-146, miR-155 was suggested to act as a negative-feedback regulator of the TLR-NFκB axis in murine and human monocytes and macrophages. TLR feedback control by both miRNAs involves non-overlapping mRNA targets: while just miR-146 targets TRAF6 and IRAK1, only miR-155 may target the signaling component TAB2 [17]. Further support for a functional specialization of both miRNAs despite their shared role in TLR feedback control was provided by CRISPR-Cas9-mediated knockout of the miR-146 and miR-155 genes in the human U937 monocyte cell line. While deficiency of either of the two miRNAs increased the activity of the TLR4-NFkB signaling axis, only miR-155-deficient cells displayed preferential accumulation in G1 cell cycle phase upon TLR stimulation [18]. Together, the above-delineated studies suggest that miR-146 and miR-155 act as key regulators balancing the TLR-mediated inflammatory responses of monocytes and macrophages in a nonredundant manner.

Besides miR-146 and miR-155, further miRNAs were shown to be involved in TLR-induced inflammatory responses. In 2007, the Mark Lindsay group made use of a murine lung inflammation model to further elucidate in vivo miRNA kinetics. Analysis of whole-lung RNA from mice inhaling nebulized LPS versus saline revealed a significant increase in the expression of dozens of miRNAs during the course of stimulation. A subset of these was significantly increased at two time points (miR-21, -25, -27b, -100, -140, -142-3p, -181c, -187, -194, -214, and miR-224) while miR-223 was the only candidate showing a significant increase at all three time points [19]. This finding was followed up by in situ hybridization of miR-223 at 3 h after LPS exposure and a distinct increase of miR-223 expression in alveolar and bronchial epithelial cells could be confirmed. It was also established that cells migrating into the bronchioles, mostly neutrophils, showed a similar pattern of miR-223 expression. Focusing on the role of miR-223 in alveolar epithelial cells and polymorphonuclear neutrophils (PMNs) during acute lung injury, Neudecker and colleagues demonstrated an intercellular transfer of miR-223 from PMNs to epithelial cells subsequently attenuating lung inflammation [20]. Initially, profiling of human primary alveolar epithelial cells cocultured with activated human PMNs by analyzing miRNAs known as being expressed in PMNs identified a significant and time-dependent increase in miR-223 expression in epithelial cells. Moreover, coculture experiments using activated PMNs from *miR-223$^{-/y}$* mice verified the dependency of elevated epithelial miR-223 levels on miR-223 expression in PMNs. Accordingly, high levels of miR-223 could also be observed in the supernatant of activated human PMNs. Subsequent incubation with human alveolar epithelial cells increased intracellular miR-223 levels in a microvesicle-dependent manner, thereby demonstrating miR-223 being transferred via PMN-derived microvesicles. The described activation-dependent release of miR-223 from PMNs followed by its transfer to alveolar epithelial cells could also be confirmed in mice after ventilator-induced acute lung injury. Moreover, genetic studies demonstrated that the loss of miR-223 resulted in severe lung inflammation, while pulmonary overexpression of miR-223 prevented acute lung injury induced by either mechanical ventilation or *Staphylococcus aureus* infection. Searching for miR-223 target genes in pulmonary epithelial cells, whole-genome microarray analysis in miR-223-overexpressing cells identified *poly[adenosine diphosphate (ADP)-ribose] polymerase-1 (PARP1)* as the most robustly repressed gene, whose product, in addition, was demonstrated to be detrimental during ventilator-induced acute lung injury [21]. PARP1 repression turned out to be involved in the miR-223-dependent dampening of acute lung injury [20].

Besides the innate immune system, miRNAs were implicated in adaptive immunity, as well. B-cell activation as part of the germinal center response, for instance, involves negative control by miR-142 through targeting of BAFFR mRNA [22]. Furthermore, similar to the TLR response of macrophages and monocytes, in B cells, the central inflammatory signal transduction components TRAF6 and IRAK1 seem to be controlled by miR-146 [23]. Somatic hypermutation on the other hand is feedback-controlled by miR-155 through targeting of activation-induced deaminase (AID) mRNA [24]. Both miR-146 and miR-155 have well-documented functions in T-cell immunity, as well. While miR-146 functions as a negative regulator of the TCR-NFκB axis [25], miR-155 was shown to support IFNγ production by T_H1 cells [26]. Interestingly, miR-146 was reported to counterregulate IFNγ production by T cells [27]. This suggests that miR-146 and miR-155 have at least partly opposing functions in T-cell immunity. While many further roles of miRNAs in adaptive immune cell function were published [28], the findings summarized here exemplarily suggest that shared basic principles of miRNA-mediated control are realized in the innate and adaptive arm of the immune system.

The variety of inflammatory disease triggers (bacteria, viruses, allergens, irritants, toxic substances, and unknown factors), pathomechanisms (hyperproliferation, tissue destruction, etc.), or disease history (acute, chronic) renders the physiological roles of miRNAs even more diverse and complex than depicted above.

2.2 MiRNAs in Chronic Inflammation

Chronic obstructive pulmonary disease (COPD) is a widespread cause of disability and mortality, with steadily increasing patient numbers. Long-term exposure to primary or secondary tobacco smoke, or biomass combustion (e.g., by cooking in developing countries), is one of the main causes. Lately, the classification of COPD as a solitary disease has been challenged by suggesting it to rather constitute a syndrome, comprising (a) chronic, cortisone-insensitive airway inflammation with mucus hypersecretion, fixed bronchoconstriction, and airway remodeling, as well as (b) irreversible tissue destruction due to bronchiectasis and emphysema [29]. Both aspects progressively lead to failure of respiration and ventilation, worsened by a vicious cycle of innate immune dysfunction and recurrent infectious disease exacerbations [30]. Treatment of COPD by long-acting airway dilatators is merely symptomatic and requires lifelong commitment. In highly industrialized countries, long-term oxygen therapy, noninvasive mechanical ventilation, or lung transplantation is a treatment option for end-stage COPD patients, imposing a significant socio-economic burden [9].

In a rat model of environmental cigarette smoke (ECS) exposure, an observed statistically significant downregulation of 24 miR-NAs (miR-30c, -124a, -125a, -191, and members of the let-7 family, among others) in whole-lung tissue was contrasted by only one upregulated miRNA (miR-294) [31]. Computationally predicted or experimentally validated targets of the 24 downregulated miRNAs were implicated in stress response and cell proliferation and included oncogenes as well as tumor-suppressor genes. Broad downregulation of miRNAs upon ECS was confirmed by Izzotti and colleagues in mice [32]. Subsets of miRNAs found to be downregulated in mice and rats (miR-30, miR-99, and miR-125) or downregulated exclusively in rats (miR-146 and miR-223) were also identified to be downregulated in a comparative miRNA study in human bronchial epithelial cells between current smokers and volunteers with no smoking history. As already established in rodents, downregulation of a large number of miRNAs was demonstrated in the current smoker group (with miR-218, -15a, -199b, -125a/b being most notably downregulated) [33]. Targeted analysis of miR-218 expression and its target genes in vitro corroborated the observed effect. Moreover, Conickx and co-workers showed miR-218-5p to be significantly downregulated in lung tissue of smokers without airflow limitations and smokers with COPD [34]. Decreased pulmonary expression of miR-218-5p was verified in an independent validation cohort, a mouse model of COPD, as well as in primary human bronchial epithelial cells, thereby strongly correlating with airway obstruction. In situ hybridization studies in human and murine lungs revealed miR-218-5p being predominantly expressed in the bronchial airway epithelium. Perturbation experiments demonstrated a protective role of miR-218-5p in the pathogenesis of COPD in vitro as well as in vivo [34]. The downregulation of miR-218 in the context of smoking was confirmed by Davidson and co-workers in human lung squamous cell carcinoma [35].

The miRNA pattern that accompanies COPD was further elucidated by Sato and colleagues, who investigated the role of miR-146a. Exposure of rats to ECS as well as a smoking history in humans could be linked to a downregulation of miR-146a [31, 33]. This effect was recapitulated even more pronouncedly in fibroblasts from a group of healthy volunteers versus fibroblasts from COPD patients, which were closely matched for age and smoking status [36]. Failure to repair tissue, a hallmark of COPD, is partially due to increased levels of prostaglandin (PG) E_2, an inflammatory mediator known to be prominently expressed in cultured COPD fibroblasts [37]. Treatment of control and COPD fibroblasts with IL-1β and TNF-α yielded a statistically significant increase in PGE_2 levels that could be observed in control and COPD fibroblasts, notably more profoundly in the latter due to an elevated expression of the prostaglandin-endoperoxide synthase 2 (also referred to as cyclo-oxygenase-2 or COX-2). Further mechanistic analyses established miR-146a as the potential regulator of COX-2 levels in COPD, as miR-146a directly targets COX-2 mRNA and is less inducible by pro-inflammatory cytokines in fibroblasts of COPD patients than in healthy control fibroblasts [36]. Accordingly, Osei and co-workers demonstrated a significantly reduced induction of miR-146a-5p in COPD-derived primary human lung fibroblasts compared to control cells cocultured with bronchial epithelial cells, respectively [38]. Moreover, downregulation of miR-146a-5p in COPD fibroblasts was shown to depend on epithelial cell-derived IL-1α and genomic analysis identified a SNP within the primary sequence of miR-146a-5p to be highly overrepresented in fibroblasts from COPD patients. Overexpression of miR-146a-5p in a human lung fetal fibroblast cell line resulted in significantly reduced expression levels of IRAK-1, leading to a decrease in IL-8 secretion and, accordingly, confirmed miR-146a-5p as an effector of the cellular IL-1 response in human lung fibroblasts. Thus, in COPD fibroblasts, reduced induction of miR-146a-5p might trigger a more proinflammatory phenotype, thereby facilitating chronic inflammation [38]. A sputum miRNA analysis comparing smoking COPD patients, healthy current smokers, and healthy never-smokers found 27 miRNAs that were downregulated in current smokers versus never-smokers. Notably, comparison of smoking COPD patients to current smokers revealed no statistically significant miRNA expression differences [39]. In contrast, in a study primarily focusing on the effects of sex on circulating metabolites in COPD, miRNA expression analysis in cells from bronchoalveolar lavage and bronchial epithelial cells of smokers with normal lung function or COPD, respectively, Naz and colleagues identified the miR-29-3p family (miR-29a-3p, miR-29b-3p, and miR-29c-3p) as being specifically upregulated in male COPD patients [40]. In smokers without airflow obstruction, expression levels of the miR-29-3p family were not affected. In the light of the aforementioned observations concerning miR-146a induction in COPD fibroblasts by Sato and colleagues [36], it is conceivable that reduced miRNA expression, while being predominantly caused by smoking alone, might be further reduced by COPD [37]. Besides the reported downstream effects of COPD-derived deregulation of miRNA expression, Lee and co-workers identified an inverse correlation between the circulating levels of the maternally imprinted miRNA miR-485-3p in plasma samples of COPD patients and their peripheral muscle strength [41]. Furthermore, correlation strength increased with disease severity. However, there was no difference in expression levels of miR-485-3p between COPD patients and controls, implying that miR-485-3p levels in circulation were not significantly affected by COPD. As, in addition, muscle mass was not associated with expression levels of imprinted miRNAs in healthy controls, Lee and colleagues suggested that the effects of those imprinted genes only become evident as a consequence of physiological stress, like in COPD [41]. Accordingly, focusing on the role of miRNAs in the loss of skeletal muscle mass and function, Garros and co-workers identified 32 miRNAs as differentially expressed in the quadriceps of severe COPD patients compared with healthy volunteers [42]. Elevated expression levels were only found for six of these miRNAs, of which miR-542-5p and miR-542-3p showed the highest expression as well as the most significant increase, respectively. Moreover, in patients with intensive care unit-acquired weakness, an even more dramatic increase in miR-542-3p/5p expression could be observed, suggesting miR-542-3p/5p as being involved in mediating muscle atrophy in ICU patients [42].

2.3 MiRNAs in Allergic Inflammation

Asthma and allergic airway diseases, while not fatal per se, severely affect the quality of life and impose a substantial socioeconomic burden. Hallmark features of asthma are chronic airway inflammation as well as paroxysmal and reversible airway obstruction.

Potential triggers are drugs, exercise, or individual pathological history. The most relevant trigger though is the inhalation of airborne allergens. Environmental factors, even pre-birth, may determine predisposition and lead to immunologic sensitization. The underlying mechanisms possibly include changes in the epigenome, and they are hypothesized to be offset to a certain extent by early life endotoxin exposure, as stated in the hygiene hypothesis. Therapy is multitudinal and relies on allergen avoidance, symptomatic bronchus dilatation by pharmacological intervention, and anti-inflammation by administration of corticosteroids [9].

In 2009, Mattes and co-workers were the first to show the involvement of miRNAs in allergic airway disease. They made use of an established mouse model of house dust mite (HDM)-induced allergic asthma and revealed increased miR-16, -21, and -126 expression upon 24 h post-HDM challenge versus control [43]. Upregulation of these miRNAs could be further accentuated by repetitive HDM challenge. Increased expression of these miRNAs could be attributed to the TLR4/Myd88 signaling pathway, as hallmark features of allergic airway inflammation, such as airway hyper-responsiveness (AHR) to methacholine, could be observed in wild-type mice, but neither in TLR4$^{-/-}$ nor in Myd88$^{-/-}$ mice. Furthermore, airway recruitment of eosinophils, the number of mucus-producing cells, as well as IL-5 and IL-4 levels were reduced. This could be partly reproduced by treating wild-type mice with anti-miR-126, which yielded a strong reduction of IL-4, -5, and -13, which are commonly released from T$_H$2 cells during allergic airway inflammation. Screening for targets of miR-126 revealed several immunoglobulins and also Oct-binding factor 1 (OBF.1, a.k.a. B-cell Oct-binding protein 1, BOB.1). Furthermore, miR-126 also regulates PU.1 transcription factor, which is involved in repressing TLR4 and T$_H$2 responses by suppression of GATA3 [43]. Upregulation of miR-21 in murine airway walls upon HDM challenge could be shown to be accompanied by an increase of let-7b and miR-145 [44]. However, only inhibition of miR-145 attenuated mucus hypersecretion and eosinophilic inflammation, while inhibition of miR-21 or let-7b proved to be ineffective. Administration of anti-miR-145 to HDM-challenged mice also substantially reduced IL-5 and IL-13 secretion from T$_H$2 cells, while again antagonizing miR-21 or let-7b was futile. In another study of miRNA expression screening by Lu and colleagues, miR-21 induction was documented upon IL-13 induction in lung-specific IL-13 transgenic mice [45]. Among 21 differentially expressed miRNAs, miR-21 (up) and miR-1 (down) were most strongly regulated upon doxycycline-induced IL-13 expression. In follow-up studies using murine asthma models, these results were confirmed. By in situ hybridization, monocytes were found to be the main cell type responsible for miR-21 upregulation. Its main target appears to be IL-12p35, as suppression was observed through binding of miR-21 to cognate mRNA complementary sites [45]. Searching for circulating miRNAs specifically deregulated in the blood of asthmatic patients, Panganiban and colleagues performed high-throughput profiling comparing expression levels of 420 miRNAs in human plasma of asthmatic versus allergic but nonasthmatic patients and healthy volunteers [46]. In total, they identified 30 miRNAs being differentially expressed between the three groups of which miR-125b, miR-16, miR-299-5p, miR-126, miR-206, and miR-133b had the highest predictive capacity for the discrimination of asthmatic, allergic, and healthy subjects. In addition, cluster analyses of expression profiles revealed two main subgroups of asthmatic patients characterized by different blood eosinophil levels. Thus, analysis of circulating miRNAs may be a promising approach for noninvasive diagnosis of asthma as well as characterization of the disease subtypes [46]. Adding the effect of air pollution to the impact of allergen exposure in humans, Rider and colleagues performed miRNA and gene expression profiling in bronchial brushings from healthy volunteers with atopy to house dust mite upon allergen challenge, exposure to diesel exhaust, or coexposure [47]. Bioinformatics analyses revealed allergen exposure to predominantly affect both miRNA and mRNA expression in bronchial epithelial cells and showed a strong correlation between modulated miRNAs and target gene expression. Diesel exhaust, however, did not significantly affect miRNA or mRNA expression upon allergen induction [47]. Unexpectedly, miRNA expression appeared broadly unchanged in a study by Mark Lindsay's lab, which investigated global miRNA expression in airway biopsy samples taken from human patients with mild asthma versus healthy volunteers. Furthermore, corticoid treatment, which is effective to attenuate mild asthma, did not change the miRNA expression profiles of asthmatic patients [48]. As in general, 5–10 % of all asthmatic patients do not respond to steroid treatment, Kim and co-workers developed a mouse model of infection-induced severe, steroid-insensitive allergic airway disease (SSIAAD) and analyzed the roles of miR-21 and its downstream effectors in the pathogenesis of the disease [49]. In SSIAAD induced by *Chlamydia, Haemophilus influenzae*, influenza virus, or respiratory syncytial virus infection, respectively, expression analysis in lung tissue revealed persistently elevated miR-21 levels. Concurrently, expression of the miR-21 target phosphatase and tensin homolog (PTEN) and HDAC2, known to display dampened activity in severe, steroid-insensitive asthma [50], was reduced. As PTEN and HDAC2 are both associated with PI3K-dependent and steroid-mediated pathways, Kim and colleagues also investigated phosphoinositide 3-kinase (PI3K) activity, reflected by the phosphorylation of its target AKT. Immunoblot and densitometric analyses revealed increased nuclear levels of pAKT as well as reduced HDAC2 signals in the lungs of mice with SSIAAD. Accordingly, inhibiting miR-21 function by using a specific antagomir resulted in increased PTEN levels, reduced PI3K activity, and restored HDAC2 expression. Thus, Kim and co-workers identified miR-21 and the miR-21/PI3K/HDAC2 axis as a novel promising therapeutic target for treating SSIAAD [49]. Related to the model of virus-induced SSIAAD, Rupani and colleagues investigated the role of miRNAs in the impaired immune response of alveolar macrophages (AMs) from patients with severe, persistent asthma to viral infections [51]. In line with the underlying hypothesis that impaired IFN signaling to rhinovirus infection in BAL cells from patients with mild asthma [52] results from the disrupted TLR7-mediated response to single-stranded RNA viruses reported in peripheral blood mononuclear cells of asthma patients [53], expression analyses in alveolar macrophages (AMs) from BAL of patients with severe asthma revealed both significantly reduced IFN levels in response to rhinovirus infection and a significant decrease in TLR7 expression compared to healthy AMs. In contrast, expression levels of neither the TLR7 adaptor protein Myd88 nor other receptors relevant for rhinovirus recognition, such as TLR8, TLR3, RIG-1, or MDA5, were affected. Additionally, blocking TLR7 signaling in severe asthma AMs resulted in significantly reduced IFN expression. Investigating whether diminished TLR7 levels were caused by altered miRNA expression, microarray followed by bioinformatics analyses identified miR-150, miR-152, and miR-375 as being significantly elevated in AMs of severe asthma patients. In addition, reporter assays upon overexpression of miR-150, miR-152, or miR-375 revealed all three miRNAs to directly bind to the 3'UTR of TLR7. Moreover, concurrent overexpression of these three miRNAs led to an even more pronounced decrease in luciferase activity suggesting a cooperative effect on TLR7 expression. Consistently, in healthy AMs, cooperative depletion of miR-150, miR-152, and miR-375 significantly increased TLR7 expression, accompanied by significantly elevated IFN levels in response to subsequent rhinovirus infection [51]. Thus, blocking miR-150, miR-152, and miR-375 seems to restore the antiviral IFN response in AMs of severe asthma patients by elevating TLR7 levels, thereby representing a promising therapeutic approach in asthma therapy. Also focusing on the aberrant signal transduction in allergic asthma, Zhou and co-workers investigated the role of the phagocytic mannose receptor MRC1 expressed by macrophages and its intronic miRNA miR-511-3p in regulating the immune response to cockroach allergens [54]. In a former study in human fibrocytes, the same group demonstrated a functional interaction between MRC1 and cockroach allergens [55]. In lung macrophages from $Mrc1^{-/-}$ mice, the uptake of cockroach allergen was significantly reduced compared with wild-type mice. Additionally, in lung tissue of mice challenged with cockroach allergen, depletion of MRC1 resulted in exacerbated lung inflammation, accompanied by elevated serum levels of cockroach-allergen-specific IgE and increased T_H2/T_H17 cytokine levels (IL-4, IL-13, IL-17A, IL-6) in BAL fluid. Confirming a report of Squadrito and colleagues demonstrating *Mrc1* and miR-511-3p to be transcriptionally coregulated in tumor-associated macrophages [56], expression analysis in lung macrophages of allergen-treated wild-type mice or macrophages differentiated from bone marrow-derived macrophages (BMDMs), respectively, revealed a significant increase in miR-511-3p levels, which were dramatically reduced or virtually absent in macrophages from $Mrc1^{-/-}$ mice. Moreover, investigating a potential role of MRC1 in macrophage polarization through miR-511-3p, expression analysis in BMDMs from $Mrc1^{-/-}$ mice polarized to either M1 or M2 demonstrated a significant increase in M1 markers (IL-1β, IL-6, NOS2) accompanied by decreased expression levels of M2 markers (FIZZ1, Ym1, Arg-1) compared to wild-type BMDMs. In addition, overexpression of miR-511-3p resulted in a switch of wild-type BMDMs to M2 polarization and significantly reduced cockroach allergen-induced inflammation in mice. Moreover, global gene expression analysis in BMDMs overexpressing miR-511-3p identified 729 genes as being deregulated compared to wild-type BMDMs, of which the gene coding for the prostaglandin D_2 synthase (*Ptgds*) as well as its catalytic product PGD_2 were verified as being significantly downregulated by miR-511-3p. Based on the results in mice, Zhou and co-workers also analyzed miR-511-3p expression in plasma samples of human asthmatic patients. Compared to nonasthmatic allergic patients as well as healthy volunteers, expression levels of miR-511-3p were significantly reduced, suggesting miR-511-3p and MRC1 to play a crucial role in the protection against cockroach allergen-induced lung inflammation [54]. In a mouse model, focusing on the effect of chronic allergen-induced airway inflammation on the regulation of type 2 innate lymphoid cells (ILC2s), Johansson and co-workers identified miR-155 as the critical effector of IL-33-mediated ILC2 function [57]. Expression analysis in lung cells of chronically ovalbumin-treated wild-type mice revealed a significant upregulation of miR-155, which was in line with former results of the same group analyzing miR-155 expression in acute allergen-induced airway infection [58]. Moreover, allergen exposure also led to significantly increased IL-33 levels, being of crucial importance for ILC2 induction and activation [59]. In $miR-155^{-/-}$ mice, IL-33 expression in the lung was impaired in response to allergen challenge, and exposure to recombinant murine IL-33 did not affect ILC2 numbers, demonstrating miR-155 to be of vital importance for the induction of IL-33 and allergen-induced ILC2 expansion. In addition, IL-33 treatment of $miR-155^{-/-}$ mice dramatically reduced ILC2 proliferation, reflected by the number of GATA-3-expressing lung cells, and resulted in significantly decreased IL-13 expression [57]. Besides the miR-155-dependent induction of IL-33 in ILC2s, Yamazumi and colleagues identified the RNA-binding protein Mex-3B as the crucial inhibitor of the miR-487b-3p-mediated repression of IL-33 in murine lungs [60]. Based on the observation of Mex-3B being highly expressed in human as well as murine lung and thymus [60, 61], $Mex3b^{-/-}$ mice were generated and used in a model of ovalbumin-induced allergic airway inflammation. Expression analyses in lung tissues from wild-type mice revealed both increased Mex-3B and IL-33 levels in response to ovalbumin treatment. In contrast, in $Mex3b^{-/-}$ mice, ovalbumin challenge resulted in reduced IL-33 induction, demonstrating Mex-3B being crucial for the inflammation-induced expression of IL-33 in murine lungs. Moreover, depletion and RNA binding experiments in respiratory epithelial cell lines confirmed the Mex-3B-dependent expression of IL-33 and verified Mex-3B binding to the 3'-UTR of IL-33. Searching for miRNAs potentially competing with Mex-3B for IL-33 binding, Yamazumi and co-workers identified miR-487b-3p as directly targeting the 3'-UTR of IL-33, subsequently suppressing its expression. In addition, blocking miR-487b-3p in ovalbumin-treated $Mex3b^{-/-}$ mice by administering the specific LNA-anti-miR-487b-3p resulted in increased eosinophil infiltration and restored IL-33 levels, whereas in wild-type mice ovalbumin-induced airway inflammation was not affected. Thus, Yamazumi and co-workers identified miR-487b-3p as the effector posttranscriptionally regulating IL-33 expression and thereby suppressing allergic airway inflammation [60].

The advent of next-generation sequencing (NGS) enabled in-depth comparison of short transcripts in naïve versus allergen-challenged mouse lungs. This approach revealed a substantial increase in mature miRNA transcripts in the lungs upon allergen challenge. Irrespective of the allergic state, miRNAs of the let-7 family were most prominent in lung tissue [62]. Additionally, posttranscriptional miRNA modification occurred in naïve and in allergen-challenged mice. Computational target prediction analyses suggested *il-13* as the putative target of let-7 family miRNAs. Of note, IL-13 levels were substantially increased in the murine lung upon allergen challenge, but let-7 miRNA levels were unchanged. However, ex vivo analysis of T_H2 cells isolated from murine lungs did show high levels of *il-13* and, in line with the computational target prediction, low levels of let-7a (here exemplarily for the let-7 family). In an effort to shed more light on the issue, direct interaction of mmu-let-7a with the 3'UTR of *Il-13* was confirmed by luciferase reporter constructs. This finding could be extended using the human *IL-13* 3'UTR and hsa-let-7a. Nevertheless, repression of let-7 by administration of anti-let-7 locked nucleic acids to allergen-challenged mice attenuated the allergic manifestation, arguing for a proinflammatory role of the let-7 miRNA family in murine allergic lung disease [62]. Regulation of IL-13 by let-7 family miRNAs was confirmed by Kumar and colleagues in a later study [63]. Intranasal administration of let-7 mimics attenuated asthmatic symptoms in mice with allergic airway inflammation by the reduction of IL-13 levels, hinting at let-7 family-determined anti-inflammation in allergic airway disease. However, translation of these insights to human pathology requires further research efforts. In conclusion, we care to point out a study by Garbacki and colleagues, comprehensively analyzing miRNA and mRNA regulation in acute and chronic asthma models. Here, short-, intermediate-, and long-term allergen exposure was considered to compile a map of miRNA regulation, ranging from the onset of inflammation to late-stage chronically remodeled airways [64]. MiR-146b stood out as the only miRNA that was upregulated throughout all time points. MiRNAs found being regulated at two time points included miR-223, -690, -29c, -483, -574-5p, and -672. Computational analyses of putative miRNA-mRNA interaction pairs pointed to the regulation of various biological processes during the disease course. Examples are transcriptional regulation, cell cycle regulation, protein metabolism, apoptosis, immunity, inflammation, and cell signaling. The work helped to elucidate the molecular mechanisms underlying the development of asthma.

Besides investigating the role of miRNAs in asthma-associated inflammation, Sun and colleagues focused on analyzing the remodeling of airway tissue finally leading to airway obstruction [65]. In two former studies in rats and primary human fibroblasts, the same group demonstrated the protein arginine methyltransferase 1 (PRMT1) being upregulated in acute and chronic stages of inflammation, which was augmented through the ERK 1/2 MAPK signaling pathway [65, 66]. Unravelling the mechanism regulating PRMT1 activation, sequence analysis identified miR-19a as specifically binding to the 3'UTR of MAPK1 in airway smooth muscle cells (ASMCs) of asthmatic patients. Moreover, overexpression of miR-19a resulted in reduced protein levels of ERK 1/2 MAPK as well as PRMT1. Consistently, inhibiting miR-19a constitutively increased ERK 1/2 MAPK protein levels, leading to increased PRMT1 expression, thus, demonstrating miR-19a being of crucial importance for regulating PRMT1 activity, subsequently inducing ASMC proliferation, migration, and extracellular matrix deposition [65].

2.4 MiRNAs in Cystic Fibrosis

Cystic fibrosis (CF) is the most common genetic disease among Caucasians. The underlying genetic mutations (ΔF508 being the most common) are passed on in an autosomal recessive manner. While fatal at young age in the past, medical achievements have allowed an increasing number of patients to live to adulthood. The pathological manifestation of these mutations is causally rooted in the gene encoding the cystic fibrosis transmembrane conductance regulator (CFTR), a chloride channel expressed in the apical membrane glandular epithelium. Impaired function of this channel leads to increased viscosity of bronchial mucus ("mucoid impactions") and mucus in pancreas ducts, which is favorable for recurrent infections and chronic inflammation, leading to tissue destruction and organ failure [9].

In a miRNA expression profiling study in bronchial brushings from cystic fibrosis patients, Oglesby and colleagues found 93 significantly deregulated miRNAs when compared to non-CF brushings (56 miRNAs being downregulated, 36 miRNAs being upregulated in three out of five cases) [67]. Further research focused on the expression of miR-126, which seemed decreased to a significant extent in four out of five CF patients. Downregulation of miR-126 could furthermore be corroborated in a CF tracheal airway epithelial cell line (CFTE) and in a CF bronchial epithelial (CFBE) cell line versus their non-CF counterpart (HTE and HBE). The HBE cell line was the only system with robust expression of miR-126 compared to six other human non-lung cell lines. Induction of endoplasmic reticulum (ER) stress entailed downregulation of miR-126 in the normal bronchial epithelial cell line. TOM1 was established as a target of miR-126 by computational target prediction analyses. Furthermore, TOM1 expression strongly correlated with miR-126 expression in vitro and in vivo. Positive experimental evidence of the interaction between TOM1 and miR-126 was gathered from a luciferase reporter construct harboring the entire TOM1 3′ untranslated region (3′ UTR) [67]. This is given further relevance since TOM1 interacted with the adaptor protein Tollip in a two-hybrid screen [68]. Tollip has been documented to interact with IL-1-dependent signaling [69]. Consequently, overexpression of TOM1 was able to suppress IL-1ß- and TNFα-induced NFκB activation, hinting at TOM1 as a common repressor [70]. However, this is difficult to reconcile with the substantial inflammation in cystic fibrosis. Additional studies are required to shed light on the miR-126/TOM1 axis and their role in the pathology of cystic fibrosis.

The profound proinflammatory immune activation during CF is determined by high abundance of key proinflammatory mediators such as IL-8, IL-6, and TNFα in CF airways [71–73]. Bhattacharyya and colleagues established the miRNA profile of the IB3-1 CF lung epithelial cell line versus their wild-type CFTR-repaired daughter cell line, IB3-1/S9 [74]. Twenty-two out of 365 investigated miRNAs were significantly differentially expressed, eighteen of which showed increased levels, while four showed decreased expression in CF epithelial cells, miRNAs miR-155 and let-7c being most robustly induced with a relative fourfold upregulation. In an ex vivo CF bronchial brushing approach, upregulation of miR-155 was also observed when compared to healthy bronchial epithelial cells. The same was true in a comparison of CF neutrophils versus healthy neutrophils. Functional analysis pointed to a direct interaction of miR-155 with phosphatidylinositol-3,4,5-triphosphate

5-phosphatase 1 (SHIP 1, an actor in PI3K signaling to Akt) by showing that miR-155 upregulation confers downregulation to SHIP 1 transcript, causing reduction of SHIP 1 and thereby enhanced PI3K signaling to Akt, ultimately resulting in stabilization of IL-8 transcript and elevated IL-8 protein levels [74].

As a closing remark, it is noteworthy that the CFTR transcript is also subject to miRNA-mediated posttranscriptional regulation, as found by two groups [75, 76]. Both performed computational motive search to identify miRNA-responsive elements (MREs) in the CFTR 3' UTR. Despite some remaining controversy regarding the interaction between the CFTR 3' UTR and miR-101, both groups confirmed the yet putative miR-494 as a negative regulator of CFTR. Investigating miRNA responsiveness of the CFTR 3' UTR with a reporter plasmid system in various cell lines, Gillen and colleagues arrived at conclusively showing that miR-145, miR-331-3p, and miR-494 are substantially expressed in primary human airway epithelial cells [75]. While some partially inconsistent results remain, miR-494 has also been identified in a screen by Megiorni and co-workers [76].

Besides analyzing CFTR as the central cause of CF, Sonneville and co-workers focused on stimulating the expression and function of the calcium-activated chloride channel protein ANO1 [77]. Like for CFTR, expression, migration, and activity of ANO1 were reduced in CF patients and downregulation of ANO1 increased disease severity [78]. Bioinformatics analyses identified miR-9 as a promising effector targeting ANO1 and luciferase assays confirmed its binding to the 3' UTR of ANO1. Moreover, miR-9 expression was increased in a CF bronchial epithelial cell line as well as in primary and fully differentiated cells from CF patients, whereas expression levels of ANO1 were significantly reduced. Consistently, overexpression of miR-9 in non-CF cells resulted in reduced ANO1 expression, chloride activity, and cell migration [77]. Blocking miR-9 binding to the 3' UTR of ANO1 by a specific target site blocker significantly increased ANO1 expression in primary and fully differentiated CF cells. In CF mice, blockage of miR-9 binding to ANO1 resulted in improved chloride activity, migration rate, and mucus dynamics. Thus, Sonneville and colleagues identified a promising alternative therapeutic strategy for CF, which compensates for CFTR deficiency by increasing ANO1 activity through the specific blockage of miR-9 binding.

2.5 MiRNAs in Infection

Numerous studies have implicated miRNAs in host responses to both bacteria and viruses. One of the first studies monitoring global miRNA expression changes in response to live bacteria detected downregulation of miRNAs of the let-7 family as the common denominator of ncRNA regulation in *Salmonella typhimurium*-infected macrophages and epithelial cells [79]. Let-7 miRNAs were shown to repress IL-6 and IL-10 production through 3'

UTR-binding sites in the mRNAs of both cytokines. This suggests a feed-forward mechanism, wherein the downregulation of let-7 miRNAs upon bacterial stimulation removes a posttranscriptional brake on cytokine expression. Furthermore, well-established TLR feedback regulators such as miR-146 and miR-155 (*see* above) were confirmed to be activated upon live microbial stimulation of macrophages. Exemplarily, *Streptococcus pneumoniae* has been found to induce miR-146a in primary human macrophages, repressing members of the TLR-signaling pathway (e.g., IRAK-1) and inflammatory mediators (e.g., COX2, IL1β) [80]. Additionally, bacterial products, like outer membrane vesicles of Gram-negative bacteria, have been shown to induce this negative feedback loop [81].

Besides a role of miRNAs in the mere sensing of microbial presence by host cells, several studies have provided evidence for manipulation of miRNA expression by bacterial pathogens. *Salmonella typhimurium*, for instance, was shown to actively downregulate members of the miR-15 family in infected HeLa cells [82]. This promotes a derepression of cyclin D1 expression, which subsequently promotes cell accumulation in G1/S phase, a favorable state for intracellular *Salmonella* infection. Furthermore, *Mycobacterium tuberculosis* was found to downregulate miR-26a in invaded macrophages [83]. MiR-26a targets Krüppel-like factor 4 (KLF4), a transcription factor involved in M2 polarization of macrophages. Thus, miR-26a repression by *Mycobacterium tuberculosis* was suggested to promote anti-inflammatory M2 rather than bactericidal M1 polarization to the benefit of the bacterium. These studies suggest that controlling miRNA expression may pose a significant microbial virulence strategy to overcome host immune defense. Not surprisingly, viruses were found to follow similar strategies to promote infection. HIV-1, for example, was found to induce miR-34a expression in infected host cells, a miRNA targeting phosphatase 1 nuclear-targeting subunit (PNUTS). Viral suppression of PNUTS expression through miR-34a was suggested to promote HIV-1 replication in T cells [84]. Besides manipulation of miRNA-dependent defense pathways, viruses may also employ miRNAs as host factors. Hepatitis C virus (HCV) requires the liver-specific miRNA miR-122 for stabilization of its genome and establishment of long-term infection [85]. Thus, inhibition of miR-122 might pose a promising therapeutic option to interfere with HCV infection. Together, the examples presented here suggest that the immune response to pathogens is tightly controlled by miRNAs, which in turn can be hijacked by both bacterial and viral pathogens. Thus, therapeutic strategies to inhibit or promote the activity of specific miRNAs in vivo might open novel treatment options in microbial and viral infection settings.

3 Long Noncoding RNAs

Like miRNAs, long noncoding RNAs (lncRNAs) have been implicated both in innate and adaptive immune responses. Different from miRNAs, however, the functions of lncRNAs in the immune system are not limited to posttranscriptional control mechanisms. Rather, most lncRNAs seem to establish agonistic and antagonistic interactions with proteins to control a variety of cellular processes ranging from cytosolic receptor signaling cascades to the opening or closing of local chromatin structures in *cis* and in *trans*. Since comprehensive lncRNA gene annotations became available only around the year 2012, research on these biomolecules and their roles in the immune system still represents a developing field. Literature published so far however suggests a vital role of lncRNAs in inflammatory and infectious diseases. The specific involvements of lncRNAs in innate and adaptive immune responses reported so far are summarized in the following sections.

3.1 Long Noncoding RNAs in Innate Immunity

Several reports have suggested a vital role of lncRNAs in positive and negative control of innate immune gene expression in response to inflammatory signals such as cytokines or Toll-like receptor ligands. Within the Il1β locus for instance, several lncRNAs were shown to contribute to a balanced expression of the encoded cytokine. Anti-IL1β, a lncRNA expressed from the same promoter as IL1β but into the opposite direction, was shown to be induced in murine macrophages on treatment with TLR4 ligand lipopolysaccharide (LPS) with similar kinetics as IL1β mRNA. Ectopic expression of anti-IL1β had a negative effect on IL1β mRNA expression, likely through suppression of promoter H3K4 tri-methylation [86]. Another study identified additional noncoding RNAs transcribed in proximity to the IL1β mRNA. A transcript originating from a downstream enhancer region (IL1b-eRNA) and an upstream transcript (IL1β-RBT46) were further studied and shown to be induced in human monocytes by LPS [87]. Knockdown of both noncoding RNAs attenuated IL1β expression. Thus, induction of the major proinflammatory cytokine IL1β on TLR stimulation seems to be controlled both positively (anti-IL1β) and negatively (IL1β-eRNA and IL1β-RBT46) by several lncRNAs encoded in the same genomic locus.

Like the IL1β gene, the locus encoding the major immune gene Cox-2 contains several immune-regulatory lncRNAs. PACER (p50 associated Cox-2 extragenic RNA) binds the p50 subunit of NFκB to prevent inhibitory p50 dimer formation and to promote assembly of p50/p65 heterodimers that bind to NFκB target gene promoters to activate transcription. Since PACER itself was found to be induced by LPS it was suggested to function as a positive regulator promoting the expression of inflammatory genes such as

Cox2 [88]. Another lncRNA transcribed in proximity to the Cox2 locus and induced by TLR agonists is LincRNA-Cox2. Through interaction with hnRNPA/B and hnRNPA2/B1, this lincRNA was shown to repress transcription of key immune genes such as Ccl5 [89]. Different from PACER, lincRNA-Cox2 did not affect the expression of its neighboring gene, Cox2. In follow-up studies LincRNA-Cox2 was shown to interact with further protein complexes such as Mi-2/NuRD and SWI/SNF to promote epigenetic changes in the promoter regions of further immune genes [90, 91]. Together these reports suggest that similar to the IL1β locus-derived lncRNAs, noncoding RNAs transcribed in proximity to the Cox2 gene may function both as activators and suppressors of innate immune responses.

Besides transcripts originating in proximity to the IL1β and Cox2 loci, further lncRNAs have been described to function downstream of PRRs and cytokine receptors within the innate immune system. LncRNA FIRRE for instance was suggested to function as an LPS-induced positive regulator of innate immune gene expression by promoting mRNA stability through its interaction partner hnRNPU [92]. Similarly, lncRNA Carlr was found to be induced following NFκB activation in macrophages and to promote the expression of immune genes. Lethe is another lncRNA, activated following NFκB activation, and similar to PACER interacts with this transcription factor. Different from PACER however, Lethe is a negative regulator of NFκB-dependent gene activation, presumably by acting as a decoy binding to the p65 subunit of the transcription factor [93]. Another lncRNA proposed to act as a negative feedback regulator of the innate inflammatory response is THRIL, which interacts with hnRNPL to promote the expression of immune-response genes. TLR activation of macrophages evokes a downregulation of this lncRNA [94]. Similarly, lincRNA-EPS interacts with hnRNPL and is downregulated by TLR stimuli. Unlike THRIL however, LincRNA-EPS acts as a suppressor of inflammatory gene activation [95]. A regulatory mechanism similar to lincRNA-EPS has been proposed for lnc13, which interacts with hnRNPD to suppress the activation of inflammatory response genes and is downregulated by LPS stimulation of macrophages. Interestingly, lnc13 expression is reduced in the intestines of patients with celiac disease and a disease variant of lnc13 displays reduced binding to its natural interaction partner hnRNPD [96]. Together these reports suggest that the innate immune system employs a variety of lncRNAs up- or downregulated by pro-inflammatory stimuli to confer both positive and negative control of immune gene expression and thereby a balanced inflammatory response.

At the interface of innate and adaptive immunity, lincRNA lnc-DC plays a role in T-cell priming upon LPS activation of dendritic cells. Knockdown of lnc-DC in human dendritic cells reduced IL-12 expression and proliferation of co-incubated CD4+ T cells. In line with these results, co-stimulatory molecules such as CD40, CD80, CD86, and HLA-DR were downregulated upon lnc-DC knockdown. Thus, lncRNAs seem to play a major role as regulators of the initiation of immune responses triggered by molecular pathogen patterns and cytokines. Transcriptome profiling data suggest that innate immune cells such as macrophages regulate dozens of lncRNAs in response to PRR stimulation; thus further reports on lncRNA functions in the innate immune system are to be expected.

3.2 Long Noncoding RNAs in Adaptive Immunity

Similar to the innate immune system, lncRNAs have been described to contribute to proper differentiation and activation of cells of the adaptive immune system such as T- or B-lymphocytes. In an attempt to systematically identify lncRNAs regulated during T-cell development, lincR-Ccr2-5'AS was identified as a GATA-3-dependent, T_H2-specific lncRNA. LincR-Ccr2-5'AS knockdown experiments revealed a function in promoting expression of T_H2-related genes [97]. Similarly, the T_H2-LCR lncRNA cluster is highly expressed in T_H2 cells and promotes WDR5-dependent deposition of H3K4 trimethylation marks at promoter sites of the IL4 and IL13 genes. This ensures specific expression of important cytokines in the T_H2 lineage of T lymphocytes [98]. Linc-MAF4 on the other hand is a lncRNA expressed preferentially in T_H1 cells and functions to suppress MAF, a T_H2-specific transcription factor. Knockdown of Linc-MAF4 skewed T-cell differentiation toward T_H2. Most likely, suppression of MAF expression involves an association of Linc-MAF4 with suppressive chromatin modifiers such as LSD1 and EZH2 at the MAF promoter [99]. Besides the T_H1 and T_H2 lineages CD4$^+$ T cells may differentiate into T_H17 and regulatory T cells. The transcription factor RORγt is required for T_H17-cell differentiation. RORγt was found to promote expression of T_H17 genes through association with the RNA helicase enzyme DDX5. The interaction between RORγt and DDX5 in turn was found to be promoted by the conserved lncRNA Rmrp [100]. Interestingly, the Rmrp locus carries specific mutations in patients with cartilage-hair hypoplasia, which among others suffer from defective immunity. Finally, regulatory T-cell function was reported to be controlled by lncRNA Flicr. Treg differentiation and function depend on the transcription factor Foxp3. Expression of this lncRNA was found to be specific to a subset of regulatory T cells, where it functions in *cis* to counterregulate Foxp3 expression [101]. Negative control of Foxp3 by Flicr might serve to dampen Treg activity to overcome immune tolerance where required.

Besides regulating differentiation and activity of specific T-cell populations, lncRNAs may also function to control lymphocyte apoptosis. Fas-dependent apoptosis in lymphocytes is counterregulated by the sFas protein, generated through alternative splicing of

Fas mRNA. A lncRNA-expressed antisense to Fas was found to promote alternative splicing of Fas mRNA to promote sFas expression and protect against Fas-mediated cell death in B- and T-lymphocytes [102, 103].

In summary, development, proper function, and death of lymphocytes are controlled by a network of lncRNAs, many of which act in a cell-type-specific manner. Similar to cells of the innate immune system, lymphocytes were reported to express plenty of lncRNAs, most of which are currently of unknown function. Thus, adaptive immune cell development and function might be even more tightly controlled by regulatory noncoding RNA than currently known.

3.3 Long Noncoding RNAs in Infection

The first lncRNA to be implicated in immune defense against bacterial infections was NeST [104]. Through mouse crossbreeding and overexpression experiments, NeST lncRNA was found to act in *trans* to confer protection against *Salmonella* infection. The authors could ascribe this observation to increased production of interferon-γ (IFNγ) by CD8$^+$ T cells, a potent antimicrobial cytokine. Mechanistically, NeST was shown to interact with WDR5 to promote histone H3K4 tri-methylation at the IFNγ locus in CD8+ T cells. Upon *Mycobacterium bovis* infection of macrophages, lncRNA Meg3 was shown to be downregulated. Knockdown of Meg3 promoted autophagy and bacterial eradication [105]. Thus, Meg3 downregulation during infections serves to promote macrophage antimicrobial defense.

In the context of viral infection NEAT1 lncRNA was shown to play a role in host defense. During infection of T cells with human immunodeficiency virus 1 (HIV-1), NEAT1 lncRNA was reported to be induced to limit bacterial replication, probably by retaining a set of viral transcripts in the nucleus [106]. Other lncRNA regulators of antiviral immunity were found to be downregulated during viral infection. LncRNA NRAV constitutes a suppressor of antiviral immune responses, likely through association with RNA-binding protein ZONAB. During viral infection NRAV is downregulated, suggesting this lncRNA to act as a brake, preventing from premature activation of antiviral immunity [107]. LncRNA # 32 is another lncRNA downregulated during the antiviral immune responses in hepatocytes. LncRNA # 32 functions as a positive regulator of antiviral gene expression, likely through association with the chromatin factor ATF2 [108]. Thus, different from NRAV, lncRNA # 32 downregulation does not support but rather limits antiviral immunity. Taken together these reports suggest that lncRNAs vitally participate in antimicrobial and antiviral immune responses both as positive and negative regulators of host defense pathways.

Besides their above-delineated functions in the host immune response, lncRNAs may also be employed by pathogens to promote infection. LncRNA-CD244 was shown to be expressed specifically in CD244/CD8+ double-positive T cells, a subpopulation emerging during tuberculosis. LncRNA-DC244 inhibits the expression of IFNγ and defense against *Mycobacterium tuberculosis*. Consequently lncRNA-CD244 knockdown promoted bacterial eradication [109]. NRON is a lncRNA highly expressed in CD4+ T cells and its expression is regulated by HIV-1 to time viral transcription [110]. Mechanistically, NRON was shown to target HIV transactivator protein Tat to the proteasome to suppress viral transcription and promote latency [111]. LncRNA EGOT is increased in liver tissue of patients infected with hepatitis C virus (HCV) compared to healthy subjects. EGOT suppresses the expression of interferon-stimulated genes and favors viral replication. The mechanism through which HCV benefits from EGOT expression still needs to be determined [112]. LncRNA ACOD1 was found to be induced by several viruses independent of type I interferon. ACOD1 binds to and promotes the activity of the metabolic regulator GOT2 and thereby promotes viral infection [113]. Interestingly, Karposi's sarcoma-associated herpes virus (KSHV) was found to encode noncoding RNA PAN, an endogenous lncRNA regulator of virus replication. PAN functions as a decoy to inhibit latency-associated protein LANA and thereby promote virus reactivation [114].

Taken together, lncRNAs may function both as positive and negative regulators to promote a balanced immune response outcome conferring effective defense while protecting from excessive inflammation. Importantly, several reports suggest that lncRNAs may also be employed by pathogens to redirect host immune responses and promote infection. Thus, pharmacological targeting of this class of biomolecules might result in the development of novel antiviral and antimicrobial therapeutics.

4 Conclusion

The role of miRNAs and lncRNA as regulators of inflammation has been extensively elucidated during the last years. It remains to be seen if circular RNAs are similarly important for inflammatory or infectious diseases. An important challenge is the development of specific, easy-to-apply, and safe strategies to interfere with RNA networks as treatments for human diseases.

Acknowledgments

We thank many collaborators for fruitful discussion, especially Annalisa Marsico, Julio Vera Gonzales, Martin Vingron, and Xin Lai. Part of this work has been funded by BMBF (ERACoSysMed2 SysMed-COPD—FKZ 031L0140, JPIAMR Pneumo-AMR-Protect—FKZ 01KI1702, e:Med CAPSYS—FKZ 01X1304E/

01ZX1304F), DFG (SFB/TR-84), and LOEWE (LOEWE Medical RNomics—FKZ 519/03/00.001-(0003)) to B.S. and by von Behring-Röntgen-Stiftung (vBR project 63-0036), DFG (SFB/TR-84) and Forschungsförderfonds, Philipps Universität Marburg, to L.N.S. We would like to apologize to all colleagues whose excellent contributions to the field could not be included in this text due to space constraints.

References

1. Lee RC, Feinbaum RL, Ambros V (1993) The C. elegans heterochronic gene lin-4 encodes small RNAs with antisense complementarity to lin-14. Cell 75(5):843–854. https://doi.org/10.1016/0092-8674(93)90529-y

2. Wightman B, Ha I, Ruvkun G (1993) Post-transcriptional regulation of the heterochronic gene lin-14 by lin-4 mediates temporal pattern formation in C. elegans. Cell 75 (5):855–862

3. Derrien T, Johnson R, Bussotti G, Tanzer A, Djebali S, Tilgner H, Guernec G, Martin D, Merkel A, Knowles DG, Lagarde J, Veeravalli L, Ruan X, Ruan Y, Lassmann T, Carninci P, Brown JB, Lipovich L, Gonzalez JM, Thomas M, Davis CA, Shiekhattar R, Gingeras TR, Hubbard TJ, Notredame C, Harrow J, Guigo R (2012) The GENCODE v7 catalog of human long noncoding RNAs: analysis of their gene structure, evolution, and expression. Genome Res 22(9):1775–1789. https://doi.org/10.1101/gr.132159.111

4. Moss EG, Lee RC, Ambros V (1997) The cold shock domain protein LIN-28 controls developmental timing in C-elegans and is regulated by the lin-4 RNA. Cell 88 (5):637–646. https://doi.org/10.1016/S0092-8674(00)81906-6

5. Chen CZ, Li L, Lodish HF, Bartel DP (2004) MicroRNAs modulate hematopoietic lineage differentiation. Science 303(5654):83–86. https://doi.org/10.1126/science.1091903

6. Havelange V, Garzon R (2010) MicroRNAs: emerging key regulators of hematopoiesis. Am J Hematol 85(12):935–942. https://doi.org/10.1002/ajh.21863

7. Malumbres R, Lossos IS (2010) Expression of miRNAs in lymphocytes: a review. Methods Mol Biol 667:129–143. https://doi.org/10.1007/978-1-60761-811-9_9

8. Navarro F, Lieberman J (2010) Small RNAs guide hematopoietic cell differentiation and function. J Immunol 184(11):5939–5947. https://doi.org/10.4049/jimmunol.0902567

9. Sittka A, Schmeck B (2013) MicroRNAs in the lung. Adv Exp Med Biol 774:121–134. https://doi.org/10.1007/978-94-007-5590-1_7

10. Taganov KD, Boldin MP, Chang KJ, Baltimore D (2006) NF-kappaB-dependent induction of microRNA miR-146, an inhibitor targeted to signaling proteins of innate immune responses. Proc Natl Acad Sci U S A 103(33):12481–12486. https://doi.org/10.1073/pnas.0605298103

11. Magilnick N, Reyes EY, Wang WL, Vonderfecht SL, Gohda J, Inoue JI, Boldin MP (2017) miR-146a-Traf6 regulatory axis controls autoimmunity and myelopoiesis, but is dispensable for hematopoietic stem cell homeostasis and tumor suppression. Proc Natl Acad Sci U S A 114(34): E7140–E7149. https://doi.org/10.1073/pnas.1706833114

12. Perry MM, Moschos SA, Williams AE, Shepherd NJ, Larner-Svensson HM, Lindsay MA (2008) Rapid changes in microRNA-146a expression negatively regulate the IL-1beta-induced inflammatory response in human lung alveolar epithelial cells. J Immunol 180 (8):5689–5698

13. Perry MM, Williams AE, Tsitsiou E, Larner-Svensson HM, Lindsay MA (2009) Divergent intracellular pathways regulate interleukin-1beta-induced miR-146a and miR-146b expression and chemokine release in human alveolar epithelial cells. FEBS Lett 583 (20):3349–3355. https://doi.org/10.1016/j.febslet.2009.09.038

14. O'Connell RM, Taganov KD, Boldin MP, Cheng GH, Baltimore D (2007) MicroRNA-155 is induced during the macrophage inflammatory response. Proc Natl Acad Sci U S A 104(5):1604–1609. https://doi.org/10.1073/pnas.0610731104

15. Mantuano E, Brifault C, Lam MS, Azmoon P, Gilder AS, Gonias SL (2016) LDL receptor-related protein-1 regulates NFkappaB and microRNA-155 in macrophages to control the inflammatory response. Proc Natl Acad

16. Tili E, Michaille JJ, Cimino A, Costinean S, Dumitru CD, Adair B, Fabbri M, Alder H, Liu CG, Calin GA, Croce CM (2007) Modulation of miR-155 and miR-125b levels following lipopolysaccharide/TNF-alpha stimulation and their possible roles in regulating the response to endotoxin shock. J Immunol 179(8):5082–5089. https://doi.org/10.4049/jimmunol.179.8.5082

17. Schulte LN, Westermann AJ, Vogel J (2013) Differential activation and functional specialization of miR-146 and miR-155 in innate immune sensing. Nucleic Acids Res 41 (1):542–553. https://doi.org/10.1093/nar/gks1030

18. Janga H, Aznaourova M, Boldt F, Damm K, Grunweller A, Schulte LN (2018) Cas9-mediated excision of proximal DNaseI/H3K4me3 signatures confers robust silencing of microRNA and long non-coding RNA genes. PLoS One 13(2):e0193066. https://doi.org/10.1371/journal.pone.0193066

19. Moschos SA, Williams AE, Perry MM, Birrell MA, Belvisi MG, Lindsay MA (2007) Expression profiling in vivo demonstrates rapid changes in lung microRNA levels following lipopolysaccharide-induced inflammation but not in the anti-inflammatory action of glucocorticoids. BMC Genomics 8:240. https://doi.org/10.1186/1471-2164-8-240

20. Neudecker V, Brodsky KS, Clambey ET, Schmidt EP, Packard TA, Davenport B, Standiford TJ, Weng T, Fletcher AA, Barthel L, Masterson JC, Furuta GT, Cai C, Blackburn MR, Ginde AA, Graner MW, Janssen WJ, Zemans RL, Evans CM, Burnham EL, Homann D, Moss M, Kreth S, Zacharowski K, Henson PM, Eltzschig HK (2017) Neutrophil transfer of miR-223 to lung epithelial cells dampens acute lung injury in mice. Sci Transl Med 9(408). https://doi.org/10.1126/scitranslmed.aah5360

21. Kim JH, Suk MH, Yoon DW, Kim HY, Jung KH, Kang EH, Lee SY, Lee SY, Suh IB, Shin C, Shim JJ, In KH, Yoo SH, Kang KH (2008) Inflammatory and transcriptional roles of poly (ADP-ribose) polymerase in ventilator-induced lung injury. Crit Care 12 (4):R108. https://doi.org/10.1186/cc6995

22. Kramer NJ, Wang WL, Reyes EY, Kumar B, Chen CC, Ramakrishna C, Cantin EM, Vonderfecht SL, Taganov KD, Chau N, Boldin MP (2015) Altered lymphopoiesis and immunodeficiency in miR-142 null mice. Blood 125(24):3720–3730. https://doi.org/10.1182/blood-2014-10-603951

23. Boldin MP, Taganov KD, Rao DS, Yang L, Zhao JL, Kalwani M, Garcia-Flores Y, Luong M, Devrekanli A, Xu J, Sun G, Tay J, Linsley PS, Baltimore D (2011) miR-146a is a significant brake on autoimmunity, myeloproliferation, and cancer in mice. J Exp Med 208 (6):1189–1201. https://doi.org/10.1084/jem.20101823

24. Teng G, Hakimpour P, Landgraf P, Rice A, Tuschl T, Casellas R, Papavasiliou FN (2008) MicroRNA-155 is a negative regulator of activation-induced cytidine deaminase. Immunity 28(5):621–629. https://doi.org/10.1016/j.immuni.2008.03.015

25. Yang L, Boldin MP, Yu Y, Liu CS, Ea CK, Ramakrishnan P, Taganov KD, Zhao JL, Baltimore D (2012) miR-146a controls the resolution of T cell responses in mice. J Exp Med 209(9):1655–1670. https://doi.org/10.1084/jem.20112218

26. Rodriguez A, Vigorito E, Clare S, Warren MV, Couttet P, Soond DR, van Dongen S, Grocock RJ, Das PP, Miska EA, Vetrie D, Okkenhaug K, Enright AJ, Dougan G, Turner M, Bradley A (2007) Requirement of bic/microRNA-155 for normal immune function. Science 316(5824):608–611. https://doi.org/10.1126/science.1139253

27. Huffaker TB, Hu RZ, Runtsch MC, Bake E, Chen XJ, Zhao J, Round JL, Baltimore D, O'Connell RM (2012) Epistasis between microRNAs 155 and 146a during T cell-mediated antitumor immunity. Cell Rep 2 (6):1697–1709. https://doi.org/10.1016/j.celrep.2012.10.025

28. Mehta A, Baltimore D (2016) MicroRNAs as regulatory elements in immune system logic. Nat Rev Immunol 16(5):279–294. https://doi.org/10.1038/nri.2016.40

29. McDonough JE, Yuan R, Suzuki M, Seyednejad N, Elliott WM, Sanchez PG, Wright AC, Gefter WB, Litzky L, Coxson HO, Pare PD, Sin DD, Pierce RA, Woods JC, McWilliams AM, Mayo JR, Lam SC, Cooper JD, Hogg JC (2011) Small-airway obstruction and emphysema in chronic obstructive pulmonary disease. N Engl J Med 365(17):1567–1575. https://doi.org/10.1056/NEJMoa1106955

30. Sethi S, Murphy TF (2008) Infection in the pathogenesis and course of chronic obstructive pulmonary disease. N Engl J Med 359 (22):2355–2365. https://doi.org/10.1056/NEJMra0800353

31. Izzotti A, Calin GA, Arrigo P, Steele VE, Croce CM, De Flora S (2009) Downregulation of microRNA expression in the lungs of rats exposed to cigarette smoke. FASEB J 23

(3):806–812. https://doi.org/10.1096/fj. 08-121384
32. Izzotti A, Calin GA, Steele VE, Croce CM, De Flora S (2009) Relationships of microRNA expression in mouse lung with age and exposure to cigarette smoke and light. FASEB J 23(9):3243–3250. https://doi.org/10. 1096/fj.09-135251
33. Schembri F, Sridhar S, Perdomo C, Gustafson AM, Zhang XL, Ergun A, Lu JN, Liu G, Zhang XH, Bowers J, Vaziri C, Ott K, Sensinger K, Collins JJ, Brody JS, Getts R, Lenburg ME, Spira A (2009) MicroRNAs as modulators of smoking-induced gene expression changes in human airway epithelium. Proc Natl Acad Sci U S A 106 (7):2319–2324. https://doi.org/10.1073/pnas.0806383106
34. Conickx G, Mestdagh P, Avila Cobos F, Verhamme FM, Maes T, Vanaudenaerde BM, Seys LJ, Lahousse L, Kim RY, Hsu AC, Wark PA, Hansbro PM, Joos GF, Vandesompele J, Bracke KR, Brusselle GG (2017) MicroRNA profiling reveals a role for microRNA-218-5p in the pathogenesis of chronic obstructive pulmonary disease. Am J Respir Crit Care Med 195(1):43–56. https://doi.org/10. 1164/rccm.201506-1182OC
35. Davidson MR, Larsen JE, Yang IA, Hayward NK, Clarke BE, Duhig EE, Passmore LH, Bowman RV, Fong KM (2010) MicroRNA-218 is deleted and downregulated in lung squamous cell carcinoma. PLoS One 5(9): e12560. https://doi.org/10.1371/journal.pone.0012560
36. Sato T, Liu X, Nelson A, Nakanishi M, Kanaji N, Wang X, Kim M, Li Y, Sun J, Michalski J, Patil A, Basma H, Holz O, Magnussen H, Rennard SI (2010) Reduced miR-146a increases prostaglandin E(2)in chronic obstructive pulmonary disease fibroblasts. Am J Respir Crit Care Med 182 (8):1020–1029. https://doi.org/10.1164/rccm.201001-0055OC
37. Togo S, Holz O, Liu X, Sugiura H, Kamio K, Wang X, Kawasaki S, Ahn Y, Fredriksson K, Skold CM, Mueller KC, Branscheid D, Welker L, Watz H, Magnussen H, Rennard SI (2008) Lung fibroblast repair functions in patients with chronic obstructive pulmonary disease are altered by multiple mechanisms. Am J Respir Crit Care Med 178 (3):248–260. https://doi.org/10.1164/rccm.200706-929OC
38. Osei ET, Florez-Sampedro L, Tasena H, Faiz A, Noordhoek JA, Timens W, Postma DS, Hackett TL, Heijink IH, Brandsma CA (2017) miR-146a-5p plays an essential role in the aberrant epithelial-fibroblast cross-talk in COPD. Eur Respir J 49(5). https://doi.org/10.1183/13993003.02538-2016
39. Van Pottelberge GR, Mestdagh P, Bracke KR, Thas O, van Durme YM, Joos GF, Vandesompele J, Brusselle GG (2011) MicroRNA expression in induced sputum of smokers and patients with chronic obstructive pulmonary disease. Am J Respir Crit Care Med 183(7):898–906. https://doi.org/10.1164/rccm.201002-0304OC
40. Naz S, Kolmert J, Yang M, Reinke SN, Kamleh MA, Snowden S, Heyder T, Levanen B, Erle DJ, Skold CM, Wheelock AM, Wheelock CE (2017) Metabolomics analysis identifies sex-associated metabotypes of oxidative stress and the autotaxin-lysoPA axis in COPD. Eur Respir J 49(6). https://doi.org/10.1183/13993003.02322-2016
41. Lee JY, Donaldson AV, Lewis A, Natanek SA, Polkey MI, Kemp PR (2017) Circulating miRNAs from imprinted genomic regions are associated with peripheral muscle strength in COPD patients. Eur Respir J 49(4). https://doi.org/10.1183/13993003.01881-2016
42. Garros RF, Paul R, Connolly M, Lewis A, Garfield BE, Natanek SA, Bloch S, Mouly V, Griffiths MJ, Polkey MI, Kemp PR (2017) MicroRNA-542 promotes mitochondrial dysfunction and SMAD activity and is elevated in intensive care unit-acquired weakness. Am J Respir Crit Care Med 196(11):1422–1433. https://doi.org/10.1164/rccm.201701-0101OC
43. Mattes J, Collison A, Plank M, Phipps S, Foster PS (2009) Antagonism of microRNA-126 suppresses the effector function of TH2 cells and the development of allergic airways disease. Proc Natl Acad Sci U S A 106 (44):18704–18709. https://doi.org/10.1073/pnas.0905063106
44. Collison A, Mattes J, Plank M, Foster PS (2011) Inhibition of house dust mite-induced allergic airways disease by antagonism of microRNA-145 is comparable to glucocorticoid treatment. J Allergy Clin Immunol 128 (1):160–U251. https://doi.org/10.1016/j.jaci.2011.04.005
45. Lu TX, Munitz A, Rothenberg ME (2009) MicroRNA-21 is up-regulated in allergic airway inflammation and regulates IL-12p35 expression. J Immunol 182(8):4994–5002. https://doi.org/10.4049/jimmunol.0803560
46. Panganiban RP, Wang Y, Howrylak J, Chinchilli VM, Craig TJ, August A, Ishmael FT (2016) Circulating microRNAs as

biomarkers in patients with allergic rhinitis and asthma. J Allergy Clin Immunol 137 (5):1423–1432. https://doi.org/10.1016/j.jaci.2016.01.029

47. Rider CF, Yamamoto M, Gunther OP, Hirota JA, Singh A, Tebbutt SJ, Carlsten C (2016) Controlled diesel exhaust and allergen coexposure modulates microRNA and gene expression in humans: effects on inflammatory lung markers. J Allergy Clin Immunol 138(6):1690–1700. https://doi.org/10.1016/j.jaci.2016.02.038

48. Williams AE, Larner-Svensson H, Perry MM, Campbell GA, Herrick SE, Adcock IM, Erjefalt JS, Chung KF, Lindsay MA (2009) MicroRNA expression profiling in mild asthmatic human airways and effect of corticosteroid therapy. PLoS One 4(6):e5889. https://doi.org/10.1371/journal.pone.0005889

49. Kim RY, Horvat JC, Pinkerton JW, Starkey MR, Essilfie AT, Mayall JR, Nair PM, Hansbro NG, Jones B, Haw TJ, Sunkara KP, Nguyen TH, Jarnicki AG, Keely S, Mattes J, Adcock IM, Foster PS, Hansbro PM (2017) MicroRNA-21 drives severe, steroid-insensitive experimental asthma by amplifying phosphoinositide 3-kinase-mediated suppression of histone deacetylase 2. J Allergy Clin Immunol 139(2):519–532. https://doi.org/10.1016/j.jaci.2016.04.038

50. Hew M, Bhavsar P, Torrego A, Meah S, Khorasani N, Barnes PJ, Adcock I, Chung KF (2006) Relative corticosteroid insensitivity of peripheral blood mononuclear cells in severe asthma. Am J Respir Crit Care Med 174(2):134–141. https://doi.org/10.1164/rccm.200512-1930OC

51. Rupani H, Martinez-Nunez RT, Dennison P, Lau LC, Jayasekera N, Havelock T, Francisco-Garcia AS, Grainge C, Howarth PH, Sanchez-Elsner T (2016) Toll-like receptor 7 is reduced in severe asthma and linked to an altered microRNA profile. Am J Respir Crit Care Med 194(1):26–37. https://doi.org/10.1164/rccm.201502-0280OC

52. Sykes A, Edwards MR, Macintyre J, del Rosario A, Bakhsoliani E, Trujillo-Torralbo MB, Kon OM, Mallia P, McHale M, Johnston SL (2012) Rhinovirus 16-induced IFN-alpha and IFN-beta are deficient in bronchoalveolar lavage cells in asthmatic patients. J Allergy Clin Immunol 129(6):1506–1514.e6. https://doi.org/10.1016/j.jaci.2012.03.044

53. Roponen M, Yerkovich ST, Hollams E, Sly PD, Holt PG, Upham JW (2010) Toll-like receptor 7 function is reduced in adolescents with asthma. Eur Respir J 35(1):64–71. https://doi.org/10.1183/09031936.00172008

54. Zhou Y, Do DC, Ishmael FT, Squadrito ML, Tang HM, Tang HL, Hsu MH, Qiu L, Li C, Zhang Y, Becker KG, Wan M, Huang SK, Gao P (2018) Mannose receptor modulates macrophage polarization and allergic inflammation through miR-511-3p. J Allergy Clin Immunol 141(1):350–364.e8. https://doi.org/10.1016/j.jaci.2017.04.049

55. Tsai YM, Hsu SC, Zhang J, Zhou YF, Plunkett B, Huang SK, Gao PS (2013) Functional interaction of cockroach allergens and mannose receptor (CD206) in human circulating fibrocytes. PLoS One 8(5):e64105. https://doi.org/10.1371/journal.pone.0064105

56. Squadrito ML, Pucci F, Magri L, Moi D, Gilfillan GD, Ranghetti A, Casazza A, Mazzone M, Lyle R, Naldini L, De Palma M (2012) miR-511-3p modulates genetic programs of tumor-associated macrophages. Cell Rep 1(2):141–154. https://doi.org/10.1016/j.celrep.2011.12.005

57. Johansson K, Malmhall C, Ramos-Ramirez P, Radinger M (2017) MicroRNA-155 is a critical regulator of type 2 innate lymphoid cells and IL-33 signaling in experimental models of allergic airway inflammation. J Allergy Clin Immunol 139(3):1007–1016.e9. https://doi.org/10.1016/j.jaci.2016.06.035

58. Malmhall C, Alawieh S, Lu Y, Sjostrand M, Bossios A, Eldh M, Radinger M (2014) MicroRNA-155 is essential for T(H)2-mediated allergen-induced eosinophilic inflammation in the lung. J Allergy Clin Immunol 133(5):1429–1438., 1438.e1–7. https://doi.org/10.1016/j.jaci.2013.11.008

59. Bartemes KR, Iijima K, Kobayashi T, Kephart GM, McKenzie AN, Kita H (2012) IL-33-responsive lineage- CD25+ CD44 (hi) lymphoid cells mediate innate type 2 immunity and allergic inflammation in the lungs. J Immunol 188(3):1503–1513. https://doi.org/10.4049/jimmunol.1102832

60. Yamazumi Y, Sasaki O, Imamura M, Oda T, Ohno Y, Shiozaki-Sato Y, Nagai S, Suyama S, Kamoshida Y, Funato K, Yasui T, Kikutani H, Yamamoto K, Dohi M, Koyasu S, Akiyama T (2016) The RNA binding protein Mex-3B is required for IL-33 induction in the development of allergic airway inflammation. Cell Rep 16(9):2456–2471. https://doi.org/10.1016/j.celrep.2016.07.062

61. Buchet-Poyau K, Courchet J, Le Hir H, Seraphin B, Scoazec JY, Duret L, Domon-

Dell C, Freund JN, Billaud M (2007) Identification and characterization of human Mex-3 proteins, a novel family of evolutionarily conserved RNA-binding proteins differentially localized to processing bodies. Nucleic Acids Res 35(4):1289–1300. https://doi.org/10.1093/nar/gkm016

62. Polikepahad S, Knight JM, Naghavi AO, Oplt T, Creighton CJ, Shaw C, Benham AL, Kim J, Soibam B, Harris RA, Coarfa C, Zariff A, Milosavljevic A, Batts LM, Kheradmand F, Gunaratne PH, Corry DB (2010) Proinflammatory role for let-7 microRNAS in experimental asthma. J Biol Chem 285(39):30139–30149. https://doi.org/10.1074/jbc.M110.145698

63. Kumar M, Ahmad T, Sharma A, Mabalirajan U, Kulshreshtha A, Agrawal A, Ghosh B (2011) Let-7 microRNA-mediated regulation of IL-13 and allergic airway inflammation. J Allergy Clin Immunol 128 (5):1077–1085.e1–10. https://doi.org/10.1016/j.jaci.2011.04.034

64. Garbacki N, Di Valentin E, Huynh-Thu VA, Geurts P, Irrthum A, Crahay C, Arnould T, Deroanne C, Piette J, Cataldo D, Colige A (2011) MicroRNAs profiling in murine models of acute and chronic asthma: a relationship with mRNAs targets. PLoS One 6(1): e16509. https://doi.org/10.1371/journal.pone.0016509

65. Sun Q, Liu L, Mandal J, Molino A, Stolz D, Tamm M, Lu S, Roth M (2016) PDGF-BB induces PRMT1 expression through ERK1/2 dependent STAT1 activation and regulates remodeling in primary human lung fibroblasts. Cell Signal 28(4):307–315. https://doi.org/10.1016/j.cellsig.2016.01.004

66. Sun Q, Liu L, Roth M, Tian J, He Q, Zhong B, Bao R, Lan X, Jiang C, Sun J, Yang X, Lu S (2015) PRMT1 upregulated by epithelial proinflammatory cytokines participates in COX2 expression in fibroblasts and chronic antigen-induced pulmonary inflammation. J Immunol 195(1):298–306. https://doi.org/10.4049/jimmunol.1402465

67. Oglesby IK, Bray IM, Chotirmall SH, Stallings RL, O'Neill SJ, McElvaney NG, Greene CM (2010) miR-126 is downregulated in cystic fibrosis airway epithelial cells and regulates TOM1 expression. J Immunol 184 (4):1702–1709. https://doi.org/10.4049/jimmunol.0902669

68. Yamakami M, Yoshimori T, Yokosawa H (2003) Tom1, a VHS domain-containing protein, interacts with tollip, ubiquitin, and clathrin. J Biol Chem 278 (52):52865–52872. https://doi.org/10.1074/jbc.M306740200

69. Burns K, Clatworthy J, Martin L, Martinon F, Plumpton C, Maschera B, Lewis A, Ray K, Tschopp J, Volpe F (2000) Tollip, a new component of the IL-1RI pathway, links IRAK to the IL-1 receptor. Nat Cell Biol 2 (6):346–351. https://doi.org/10.1038/35014038

70. Yamakami M, Yokosawa H (2004) Tom1 (target of Myb1) is a novel negative regulator of interleukin-1-and tumor necrosis factor-induced signaling pathways. Biol Pharm Bull 27(4):564–566. https://doi.org/10.1248/bpb.27.564

71. Dean TP, Dai Y, Shute JK, Church MK, Warner JO (1993) Interleukin-8-concentrations are elevated in bronchoalveolar lavage, sputum, and sera of children with cystic fibrosis. Pediatr Res 34(2):159–161. https://doi.org/10.1203/00006450-199308000-00010

72. Richman-Eisenstat JB, Jorens PG, Hebert CA, Ueki I, Nadel JA (1993) Interleukin-8: an important chemoattractant in sputum of patients with chronic inflammatory airway diseases. Am J Phys 264(4 Pt 1): L413–L418. https://doi.org/10.1152/ajplung.1993.264.4.L413

73. Bonfield TL, Panuska JR, Konstan MW, Hilliard KA, Hilliard JB, Ghnaim H, Berger M (1995) Inflammatory cytokines in cystic fibrosis lungs. Am J Respir Crit Care Med 152(6 Pt 1):2111–2118. https://doi.org/10.1164/ajrccm.152.6.8520783

74. Bhattacharyya S, Balakathiresan NS, Dalgard C, Gutti U, Armistead D, Jozwik C, Srivastava M, Pollard HB, Biswas R (2011) Elevated miR-155 promotes inflammation in cystic fibrosis by driving hyperexpression of interleukin-8. J Biol Chem 286 (13):11604–11615. https://doi.org/10.1074/jbc.M110.198390

75. Gillen AE, Gosalia N, Leir SH, Harris A (2011) MicroRNA regulation of expression of the cystic fibrosis transmembrane conductance regulator gene. Biochem J 438 (1):25–32. https://doi.org/10.1042/BJ20110672

76. Megiorni F, Cialfi S, Dominici C, Quattrucci S, Pizzuti A (2011) Synergistic post-transcriptional regulation of the cystic fibrosis transmembrane conductance regulator (CFTR) by miR-101 and miR-494 specific binding. PLoS One 6(10):e26601. https://doi.org/10.1371/journal.pone.0026601

77. Sonneville F, Ruffin M, Coraux C, Rousselet N, Le Rouzic P, Blouquit-Laye S,

Corvol H, Tabary O (2017) MicroRNA-9 downregulates the ANO1 chloride channel and contributes to cystic fibrosis lung pathology. Nat Commun 8(1):710. https://doi.org/10.1038/s41467-017-00813-z

78. Ruffin M, Voland M, Marie S, Bonora M, Blanchard E, Blouquit-Laye S, Naline E, Puyo P, Le Rouzic P, Guillot L, Corvol H, Clement A, Tabary O (2013) Anoctamin 1 dysregulation alters bronchial epithelial repair in cystic fibrosis. Biochim Biophys Acta 1832(12):2340–2351. https://doi.org/10.1016/j.bbadis.2013.09.012

79. Schulte LN, Eulalio A, Mollenkopf HJ, Reinhardt R, Vogel J (2011) Analysis of the host microRNA response to Salmonella uncovers the control of major cytokines by the let-7 family. EMBO J 30(10):1977–1989. https://doi.org/10.1038/emboj.2011.94

80. Griss K, Bertrams W, Sittka-Stark A, Seidel K, Stielow C, Hippenstiel S, Suttorp N, Eberhardt M, Wilhelm J, Vera J, Schmeck B (2016) MicroRNAs constitute a negative feedback loop in Streptococcus pneumoniae-induced macrophage activation. J Infect Dis 214(2):288–299. https://doi.org/10.1093/infdis/jiw109

81. Jung AL, Stoiber C, Herkt CE, Schulz C, Bertrams W, Schmeck B (2016) Legionella pneumophila-derived outer membrane vesicles promote bacterial replication in macrophages. PLoS Pathog 12(4):e1005592. https://doi.org/10.1371/journal.ppat.1005592

82. Maudet C, Mano M, Sunkavalli U, Sharan M, Giacca M, Forstner KU, Eulalio A (2014) Functional high-throughput screening identifies the miR-15 microRNA family as cellular restriction factors for Salmonella infection. Nat Commun 5:4718. https://doi.org/10.1038/ncomms5718

83. Sahu SK, Kumar M, Chakraborty S, Banerjee SK, Kumar R, Gupta P, Jana K, Gupta UD, Ghosh Z, Kundu M, Basu J (2017) MicroRNA 26a (miR-26a)/KLF4 and CREB-C/EBPbeta regulate innate immune signaling, the polarization of macrophages and the trafficking of Mycobacterium tuberculosis to lysosomes during infection. PLoS Pathog 13(5):e1006410. https://doi.org/10.1371/journal.ppat.1006410

84. Kapoor R, Arora S, Ponia SS, Kumar B, Maddika S, Banerjea AC (2015) The miRNA miR-34a enhances HIV-1 replication by targeting PNUTS/PPP1R10, which negatively regulates HIV-1 transcriptional complex formation. Biochem J 470(3):293–302. https://doi.org/10.1042/BJ20150700

85. Jopling CL, Yi M, Lancaster AM, Lemon SM, Sarnow P (2005) Modulation of hepatitis C virus RNA abundance by a liver-specific MicroRNA. Science 309(5740):1577–1581. https://doi.org/10.1126/science.1113329

86. Lu J, Wu X, Hong M, Tobias P, Han J (2013) A potential suppressive effect of natural antisense IL-1beta RNA on lipopolysaccharide-induced IL-1beta expression. J Immunol 190(12):6570–6578. https://doi.org/10.4049/jimmunol.1102487

87. Ilott NE, Heward JA, Roux B, Tsitsiou E, Fenwick PS, Lenzi L, Goodhead I, Hertz-Fowler C, Heger A, Hall N, Donnelly LE, Sims D, Lindsay MA (2014) Long non-coding RNAs and enhancer RNAs regulate the lipopolysaccharide-induced inflammatory response in human monocytes. Nat Commun 5:3979. https://doi.org/10.1038/ncomms4979

88. Krawczyk M, Emerson BM (2014) p50-associated COX-2 extragenic RNA (PACER) activates COX-2 gene expression by occluding repressive NF-kappaB complexes. eLife 3:e01776. https://doi.org/10.7554/eLife.01776

89. Carpenter S, Aiello D, Atianand MK, Ricci EP, Gandhi P, Hall LL, Byron M, Monks B, Henry-Bezy M, Lawrence JB, O'Neill LA, Moore MJ, Caffrey DR, Fitzgerald KA (2013) A long noncoding RNA mediates both activation and repression of immune response genes. Science 341(6147):789–792. https://doi.org/10.1126/science.1240925

90. Tong Q, Gong AY, Zhang XT, Lin C, Ma S, Chen J, Hu G, Chen XM (2016) LincRNA-Cox2 modulates TNF-alpha-induced transcription of Il12b gene in intestinal epithelial cells through regulation of Mi-2/NuRD-mediated epigenetic histone modifications. FASEB J 30(3):1187–1197. https://doi.org/10.1096/fj.15-279166

91. Hu G, Gong AY, Wang Y, Ma S, Chen X, Chen J, Su CJ, Shibata A, Strauss-Soukup JK, Drescher KM, Chen XM (2016) LincRNA-Cox2 promotes late inflammatory gene transcription in macrophages through modulating SWI/SNF-mediated chromatin remodeling. J Immunol 196(6):2799–2808. https://doi.org/10.4049/jimmunol.1502146

92. Lu Y, Liu X, Xie M, Liu M, Ye M, Li M, Chen XM, Li X, Zhou R (2017) The NF-kappaB-

responsive long noncoding RNA FIRRE regulates posttranscriptional regulation of inflammatory gene expression through interacting with hnRNPU. J Immunol 199 (10):3571–3582. https://doi.org/10.4049/jimmunol.1700091

93. Rapicavoli NA, Qu K, Zhang J, Mikhail M, Laberge RM, Chang HY (2013) A mammalian pseudogene lncRNA at the interface of inflammation and anti-inflammatory therapeutics. elife 2:e00762. https://doi.org/10.7554/eLife.00762

94. Li Z, Chao TC, Chang KY, Lin N, Patil VS, Shimizu C, Head SR, Burns JC, Rana TM (2014) The long noncoding RNA THRIL regulates TNFalpha expression through its interaction with hnRNPL. Proc Natl Acad Sci U S A 111(3):1002–1007. https://doi.org/10.1073/pnas.1313768111

95. Atianand MK, Hu W, Satpathy AT, Shen Y, Ricci EP, Alvarez-Dominguez JR, Bhatta A, Schattgen SA, McGowan JD, Blin J, Braun JE, Gandhi P, Moore MJ, Chang HY, Lodish HF, Caffrey DR, Fitzgerald KA (2016) A long noncoding RNA lincRNA-EPS acts as a transcriptional brake to restrain inflammation. Cell 165(7):1672–1685. https://doi.org/10.1016/j.cell.2016.05.075

96. Castellanos-Rubio A, Fernandez-Jimenez N, Kratchmarov R, Luo X, Bhagat G, Green PH, Schneider R, Kiledjian M, Bilbao JR, Ghosh S (2016) A long noncoding RNA associated with susceptibility to celiac disease. Science 352(6281):91–95. https://doi.org/10.1126/science.aad0467

97. Hu G, Tang Q, Sharma S, Yu F, Escobar TM, Muljo SA, Zhu J, Zhao K (2013) Expression and regulation of intergenic long noncoding RNAs during T cell development and differentiation. Nat Immunol 14(11):1190–1198. https://doi.org/10.1038/ni.2712

98. Spurlock CF 3rd, Tossberg JT, Guo Y, Collier SP, Crooke PS 3rd, Aune TM (2015) Expression and functions of long noncoding RNAs during human T helper cell differentiation. Nat Commun 6:6932. https://doi.org/10.1038/ncomms7932

99. Ranzani V, Rossetti G, Panzeri I, Arrigoni A, Bonnal RJ, Curti S, Gruarin P, Provasi E, Sugliano E, Marconi M, De Francesco R, Geginat J, Bodega B, Abrignani S, Pagani M (2015) The long intergenic noncoding RNA landscape of human lymphocytes highlights the regulation of T cell differentiation by linc-MAF-4. Nat Immunol 16(3):318–325. https://doi.org/10.1038/ni.3093

100. Huang W, Thomas B, Flynn RA, Gavzy SJ, Wu L, Kim SV, Hall JA, Miraldi ER, Ng CP, Rigo F, Meadows S, Montoya NR, Herrera NG, Domingos AI, Rastinejad F, Myers RM, Fuller-Pace FV, Bonneau R, Chang HY, Acuto O, Littman DR (2015) DDX5 and its associated lncRNA Rmrp modulate TH17 cell effector functions. Nature 528 (7583):517–522. https://doi.org/10.1038/nature16193

101. Zemmour D, Pratama A, Loughhead SM, Mathis D, Benoist C (2017) Flicr, a long noncoding RNA, modulates Foxp3 expression and autoimmunity. Proc Natl Acad Sci U S A 114(17):E3472–E3480. https://doi.org/10.1073/pnas.1700946114

102. Yan MD, Hong CC, Lai GM, Cheng AL, Lin YW, Chuang SE (2005) Identification and characterization of a novel gene Saf transcribed from the opposite strand of Fas. Hum Mol Genet 14(11):1465–1474. https://doi.org/10.1093/hmg/ddi156

103. Sehgal L, Mathur R, Braun FK, Wise JF, Berkova Z, Neelapu S, Kwak LW, Samaniego F (2014) FAS-antisense 1 lncRNA and production of soluble versus membrane Fas in B-cell lymphoma. Leukemia 28 (12):2376–2387. https://doi.org/10.1038/leu.2014.126

104. Gomez JA, Wapinski OL, Yang YW, Bureau JF, Gopinath S, Monack DM, Chang HY, Brahic M, Kirkegaard K (2013) The NeST long ncRNA controls microbial susceptibility and epigenetic activation of the interferon-gamma locus. Cell 152(4):743–754. https://doi.org/10.1016/j.cell.2013.01.015

105. Pawar K, Hanisch C, Palma Vera SE, Einspanier R, Sharbati S (2016) Down regulated lncRNA MEG3 eliminates mycobacteria in macrophages via autophagy. Sci Rep 6:19416. https://doi.org/10.1038/srep19416

106. Zhang Q, Chen CY, Yedavalli VS, Jeang KT (2013) NEAT1 long noncoding RNA and paraspeckle bodies modulate HIV-1 posttranscriptional expression. MBio 4(1): e00596–e00512. https://doi.org/10.1128/mBio.00596-12

107. Ouyang J, Zhu X, Chen Y, Wei H, Chen Q, Chi X, Qi B, Zhang L, Zhao Y, Gao GF, Wang G, Chen JL (2014) NRAV, a long noncoding RNA, modulates antiviral responses through suppression of interferon-stimulated gene transcription. Cell Host Microbe 16 (5):616–626. https://doi.org/10.1016/j.chom.2014.10.001

108. Nishitsuji H, Ujino S, Yoshio S, Sugiyama M, Mizokami M, Kanto T, Shimotohno K (2016) Long noncoding RNA #32

contributes to antiviral responses by controlling interferon-stimulated gene expression. Proc Natl Acad Sci U S A 113 (37):10388–10393. https://doi.org/10.1073/pnas.1525022113

109. Wang Y, Zhong H, Xie X, Chen CY, Huang D, Shen L, Zhang H, Chen ZW, Zeng G (2015) Long noncoding RNA derived from CD244 signaling epigenetically controls CD8+ T-cell immune responses in tuberculosis infection. Proc Natl Acad Sci U S A 112(29):E3883–E3892. https://doi.org/10.1073/pnas.1501662112

110. Imam H, Bano AS, Patel P, Holla P, Jameel S (2015) The lncRNA NRON modulates HIV-1 replication in a NFAT-dependent manner and is differentially regulated by early and late viral proteins. Sci Rep 5:8639. https://doi.org/10.1038/srep08639

111. Li J, Chen C, Ma X, Geng G, Liu B, Zhang Y, Zhang S, Zhong F, Liu C, Yin Y, Cai W, Zhang H (2016) Long noncoding RNA NRON contributes to HIV-1 latency by specifically inducing tat protein degradation. Nat Commun 7:11730. https://doi.org/10.1038/ncomms11730

112. Carnero E, Barriocanal M, Prior C, Pablo Unfried J, Segura V, Guruceaga E, Enguita M, Smerdou C, Gastaminza P, Fortes P (2016) Long noncoding RNA EGOT negatively affects the antiviral response and favors HCV replication. EMBO Rep 17 (7):1013–1028. https://doi.org/10.15252/embr.201541763

113. Wang P, Xu J, Wang Y, Cao X (2017) An interferon-independent lncRNA promotes viral replication by modulating cellular metabolism. Science 358(6366):1051–1055. https://doi.org/10.1126/science.aao0409

114. Campbell M, Kim KY, Chang PC, Huerta S, Shevchenko B, Wang DH, Izumiya C, Kung HJ, Izumiya Y (2014) A lytic viral long noncoding RNA modulates the function of a latent protein. J Virol 88(3):1843–1848. https://doi.org/10.1128/JVI.03251-13

Chapter 2

p73-Governed miRNA Networks: Translating Bioinformatics Approaches to Therapeutic Solutions for Cancer Metastasis

Stella Logotheti, Stephan Marquardt, and Brigitte M. Pützer

Abstract

The transcription factor p73 synthesizes a large number of isoforms and presents high structural and functional homology with p53, a well-known tumor suppressor and a famous "Holy Grail" of anticancer targeting. *p73* has attracted increasing attention mainly because (a) unlike *p53*, *p73* is rarely mutated in cancer, (b) some p73 isoforms can inhibit all hallmarks of cancer, and (c) it has the ability to mimic oncosuppressive functions of p53, even in *p53*-mutated cells. These attributes render p73 and its downstream pathways appealing for therapeutic targeting, especially in mutant *p53*-driven cancers. p73 functions are, at least partly, mediated by microRNAs (miRNAs), which constitute nodal components of p73-governed networks. p73 not only regulates transcription of crucial miRNA genes, but is also predicted to affect miRNA populations in a transcription-independent manner by developing protein-protein interactions with components of the miRNA processing machinery. This combined effect of p73, both in miRNA transcription and maturation, appears to be isoform-dependent and can result in a systemic switch of cell miRNomes toward either an anti-oncogenic or oncogenic outcome. In this review, we combine literature search with bioinformatics approaches to reconstruct the p73-governed miRNA network and discuss how these crosstalks may be exploited to develop next-generation therapeutics.

Key words ncRNAs, p73 isoforms, miRNome, miRNA transcription, miRNA maturation, Computational analysis

1 Introduction

The majority of human tumors present dysfunction of the *p53* tumor suppressor gene. Over 50% of all tumors carry *p53* loss-of-function mutations, leading to synthesis of functionally impaired protein products which are unable to transactivate genes that induce cell cycle arrest and apoptosis in response to oncogenic stress. The remaining tumors that maintain wild-type *p53* exhibit other types of p53 inactivation, such as hyperactivation of its endogenous repressors MDM2 and MDM4. Besides loss of its tumor suppressive properties, *p53* can also acquire gain-of-function (GOF) mutations and switch from a tumor suppressor to an oncogene, which promotes cancer cell survival, proliferation, and invasion. Thus, a single point mutation acts in a "two-birds-with-one-stone" manner, because *p53*-mutated cancer cells not only evade one of the most potent tumor suppressive gene networks but also acquire, at the same time, novel malignant properties. Restoring normal p53 functionality in tumors by either disrupting the p53/MDM2 protein-protein interactions or targeting acquired genetic susceptibilities precipitated by GOF *p53*, a strategy termed as synthetic lethality with mutant *p53*, poses as the Holy Grail of cancer management, with a quest continuing for several decades [1].

However, *p53* mutations present intertumoral and intratumoral heterogeneity, which further complicates p53-targeting strategies. On the one hand, the existence of different *p53* mutations across individual cancer patients [2] can hinder the development of a universal p53-targeting drug, since a drug against a specific p53 mutation can be efficient only for the small percentage of a patient population carrying the particular mutation. On the other hand, the several types of *p53* mutations that may co-exist within the same tumor [3, 4] increase the possibilities for evolution of resistance against p53-targeting drugs: if the cells bearing one particular type of *p53* mutation are effectively eliminated by a specific drug, this treatment may lay the ground for selection and expansion of cell subpopulations carrying other types of drug-refractory *p53* mutations. In light of these obstacles, alternative molecules, which are as efficient as p53 but can be more resilient to acquiring GOF genetic mutations, pose as a promising alternative. In this respect, herein we review the potential of p73, a structural and functional homologue of p53, as an alternative therapeutic target, with particular emphasis on its noncoding RNA effectors.

2 p73-Networks: An Appealing Surrogate to the "Holy Grail" of Cancer Therapeutics

The transcription factor (TF) p73 is a p53 family member with high structural and functional similarity to the prototypical tumor suppressor p53. The *p73* gene synthesizes numerous isoforms by (a) use of its extrinsic (P1) or an alternative intrinsic (P2) promoter, generating TA and ΔN classes of isoforms, correspondingly, (b) alternative splicing in the 5′ end, resulting in amino-truncated ΔTA isoforms that partially or entirely lack the transactivation domain and, together with ΔN, constitute the DN isoforms, and (c) alternative splicing in the 3′ end, putting forth several C-terminal variants (α, β, γ, δ, ε, ζ, η, η*, η1, and θ) [5]. TAp73 isoforms are anti-oncogenic, while DNs exert oncogenic properties. Unlike TA and ΔN, which are also detected in normal tissues, the class of ΔTA is cancer-specific [6]. The mechanisms of function of p73 isoforms extend beyond those of a typical TF and are of paramount importance. It is well documented that TAp73 isoforms bind to their corresponding p73 responsive elements (RE) and directly transactivate target genes, while DNp73 isoforms antagonize this function either by competing for p73 binding sites or by forming inactive TAp73/DNp73 hetero-oligomers [7]. In addition to this typical mode of direct transactivation or transrepression, p73 isoforms also represent an emerging, yet understudied, protein-protein interaction (PPI)-based mode of gene regulation [6, 8].

From the therapeutics' point of view, p73 shows several advantageous features. First of all, unlike *p53*, *p73* is rarely mutated in cancer [5], a fact that renders its targeting uncomplicated from intertumoral and intratumoral heterogeneity of p73 mutations. Second, the TAp73 isoforms inhibit all hallmarks and enabling characteristics of cancer, and they also enhance responsiveness to standard radio- and chemotherapies [5]. Third, the fact that ΔTA are cancer-specific isoforms can be exploited, because these variants could be specifically inhibited in malignant tumors, leaving the surrounding non-pathological tissue unaffected. Another attractive characteristic of p73 is that it uses pathways that are either fully or partially overlapping with those of p53 to mediate the same antioncogenic effect. This essentially enables p73 to install a "back-up" network for p53, in analogy with a UPS circuit, which uses both common and alternative networking of the main power circuit, to ensure uninterrupted function. On the one hand, p73 faithfully replaces p53 functions by transactivating common genes and inducing the same axes. By regulating stress-response pathways fully overlapping with those of p53 (p53/p73-dependent pathways), p73 keeps them active even in the absence of functional *p53* gene [9]. On the other hand, it uses p53-independent pathways to recapitulate anti-oncogenic effects of p53. Upon DNA damage, for instance, p53 translocates to mitochondria and physically interacts with bcl-2 to induce mitochondrial death. Under the same conditions, p73 does not faithfully imitate this circuit, but instead, alternatively, transactivates GRADD4, a p73-specific target, which translocates to mitochondria and interacts with bcl-2 to induce mitochondrial decay. This alternative, p53-independent circuit enables cells to attain the ability for mitochondrial death-induced apoptosis even if p53 is defective [10]. Thus, by activating pathways only partially overlapping with those of p53 (p73-dependent/p53-independent pathways), p73 can circumvent blocks attributed to impairment of effectors downstream of p53.

Overall, p73 networks can offer functional redundancy to p53 circuits through several possible ways and are of particular interest as they provide alternative means of restoring defective p53-regulated pathways and/or networks in cancer cells. Noncoding RNAs, mainly microRNAs (miRNAs), emerge as key components of the p73 functional network modules. The identity and contribution of p73-associating miRNAs in cancer has just started to be unveiled. In the next sections, these components will be uncovered based on literature studies and computational predictions to reconstruct a map of the cancer p73/miRNA interactome.

3 MiRNAs and Factors Dictating Their Behavior in Networks

MiRNAs constitute a class of small, conserved RNA molecules of 19–27 nucleotide length, which are encoded by genes located within introns/exons of protein-coding genes or in intergenic areas. Biosynthesis of mature miRNAs is performed by the miRNA processing machinery. First, microRNA genes are transcribed into kilobase-long primary miRNAs (pri-miRNAs). These are next cleaved into 60–90 nucleotide-long precursor miRNAs (pre-miRNAs) by the nuclear microprocessor complex formed by the RNase III Drosha and DGCR8 (DiGeorge syndrome critical region gene 8), which cleaves the primary pri-miRNA to the characteristic hairpin stem-loop structure of pre-miRNA. Pre-miRNAs are transported into the cytoplasm, where they are cleaved by RNase III Dicer into a 22 nucleotide long miRNA duplex and unwound by helicase. The passenger strand is degraded and the selected guide strand together with Ago protein activates RISC (RNA-induced silencing-like complex), resulting in mRNA degradation or translational inhibition of mRNA targets. The nuclear processing efficacy of the Drosha/DGCR8/Dicer-complex is further fine-tuned by other accessory RNA-binding proteins, such as DDX5 and DDX17 (also known as p68 and p72, respectively), which both physically associate with Drosha/DGCR8 to form a larger nuclear processing complex (also termed super microprocessor) [11, 12].

Mature miRNAs hybridize with complementary regions in the 3′ UTRs of target mRNAs and induce either degradation of fully complementary targets or inhibition of translation of partially complementary targets, thus simultaneously modulating the expression of dozens of coding genes on the post-transcriptional level. Each miRNA has hundreds of mRNA targets and can potentially participate in numerous divergent functions. MiRNAs participate in networks and thereby fine-tune several biological processes, and their deregulation is common in cancer and other diseases. The majority of miRNAs does not lead to obviously abnormal phenotypes if individually mutated despite their potential involvement in numerous processes [13]. This is because the behavior of a miRNA within a network is influenced by several parameters, including the cellular milieu, topological network features, the genomic position, and the evolutionary age.

3.1 Cellular Milieu

It has now become evident that a miRNA co-operates with other miRNAs, and thus, the miRNome in which a particular miRNA is expressed is a major determinant of its functionality. A single mRNA can be targeted by several miRNAs and therefore pairs or combinations of miRNAs can repress its translation co-operatively leading to an enhanced effectiveness and specificity in target repression. Notably, two co-operating miRNAs may also bind to adjacent positions in the 3′ UTR and form a thermodynamically stable triplex with their common mRNA target to synergistically increase inhibition effectiveness [14, 15]. Vice versa, one miRNA can simultaneously inhibit several mRNA targets that participate in the same process, e.g., components of the same biological pathway. This way, it exerts a co-ordinated and systemic regulatory effect. In addition, a complex interplay among miRNA and its associating molecules, mainly its upstream TFs and downstream mRNA targets, takes place within the cell milieu to determine the actual function in which a miRNA will ultimately be engaged. These components are interconnected and form functional network modules [16], which may often comprise the general pattern "TF/miRNA/target genes" [17]. For instance, the effect of both p53 and p73 on EMT factors follows, at least in part, this network pattern [18, 19].

3.2 Network Topology

The arrangement of a miRNA in a network and the degree of its interconnection with other components has a significant impact on its function. The nodes of the networks can be occupied by miRNAs or mRNAs and are typically connected to many other nodes. The nodes with high numbers of connections are termed "hubs" and exert dominant regulatory roles in the network, reflecting sites of signaling convergence. On the one hand, mRNAs occupying hub positions tend to have longer 3′ UTRs with a higher density of target sites as compared to "housekeeping" and highly expressed tissue-specific genes. This indicates that they have been evolutionary selected as prominent points of direct miRNA-mediated suppression. On the other hand, only a small fraction of miRNAs occupies hub network positions, but these miRNAs play important roles in establishing and maintaining gene expression patterns. For instance, only 21 hub miRNAs were predicted to target 70% of the genes that are differentially expressed between grade II and grade III–IV gliomas [20].

3.3 Genomic Location and Gene Promoter Organization

In general, MIRNA genes that are co-localized, belong to large polycistronic clusters, or have promoters responsive to the same transcription factors are likely to exhibit co-functionality. Functionally associated miRNA genes tend to be co-transcribed and co-operate by targeting either the same gene or different ones in the same pathways [20]. The co-expressed MIRNA genes tend to be located in neighboring positions or be parts of the same MIRNA clusters. There are also MIRNA genes that although they are not residing in close proximity, they are still co-transcribed because they bear regulatory elements for the same TFs [13].

3.4 Evolutionary Age

Neither have all MIRNAs the same age, nor have they been recruited at the same time into a network. The time-point when each component has been recruited in the network and interconnected with the other components emerges as an essential consideration for the thorough characterization of cancer-related networks [21]. The behavior of MIRNAs largely depends on their evolutionary age [22]. It is plausible that MIRNAs that originated in more distant ancestors had adequate evolutionary time to develop robust interconnections within the network, eliminate stochastic effects on random mRNA targets, deterministically co-evolve with crucial targets, and occupy hub positions. Only a small proportion of MIRNAs are of prevertebrate origin, while an increased "birth rate" of novel miRNAs has occurred within the mammalian clade [23]. This MIRNA repertoire expansion is associated with major body plan innovations and phenotypic variations [24] and has also orchestrated regulation in mammalian nervous tissues [25]. Strikingly, 55% of all human MIRNA genes are primate-specific and have emerged after the rodent-primate split [22]. Compared to evolutionarily "old" MIRNAs, evolutionarily "young" specimens are lowly expressed, tissue-specific, present milder phenotypes upon knockdown, and are kept under tight transcriptional regulation [22]. There is a trend of miRNAs of different age to target the same or overlapping sets of genes [22], suggesting that the evolutionarily young miRNAs might have been recruited to further assist the function of evolutionarily old miRNAs [19]. Human-specific miRNAs and primate-specific miRNAs can decrease cancer susceptibility by co-operating with existing miRNAs or by enabling formation of novel regulatory circuits that modulate the function of cancer genes. This additional redundancy in miRNA regulation of common targets by evolutionarily old and young miRNAs also renders cells more resilient to loss of an individual miRNA [23].

4 The p73/MIRNA-Interactome in Cancer

Thus far, several miRNAs have been experimentally identified to associate with p73 in cancer networks. The p73/miRNA interactome includes miRNAs that can inhibit p73, MIRNAs that are directly transactivated by p73, or miRNAs targeting proteins that stabilize p73 (Fig. 1).

4.1 MiRNAs Targeting p73

A nodal relationship is established among p73 and MIR193A, whereby inhibition of MIR193A restores p73-mediated cisplatin sensitivity in particularly chemoresistant tumor types, such as squamous cell carcinomas, osteosarcoma, and Ewing sarcoma. In detail, MIR193A is a direct transcriptional target of p73. In turn, the miR-193a-5p mature product of MIR193A binds to the 3′ UTR

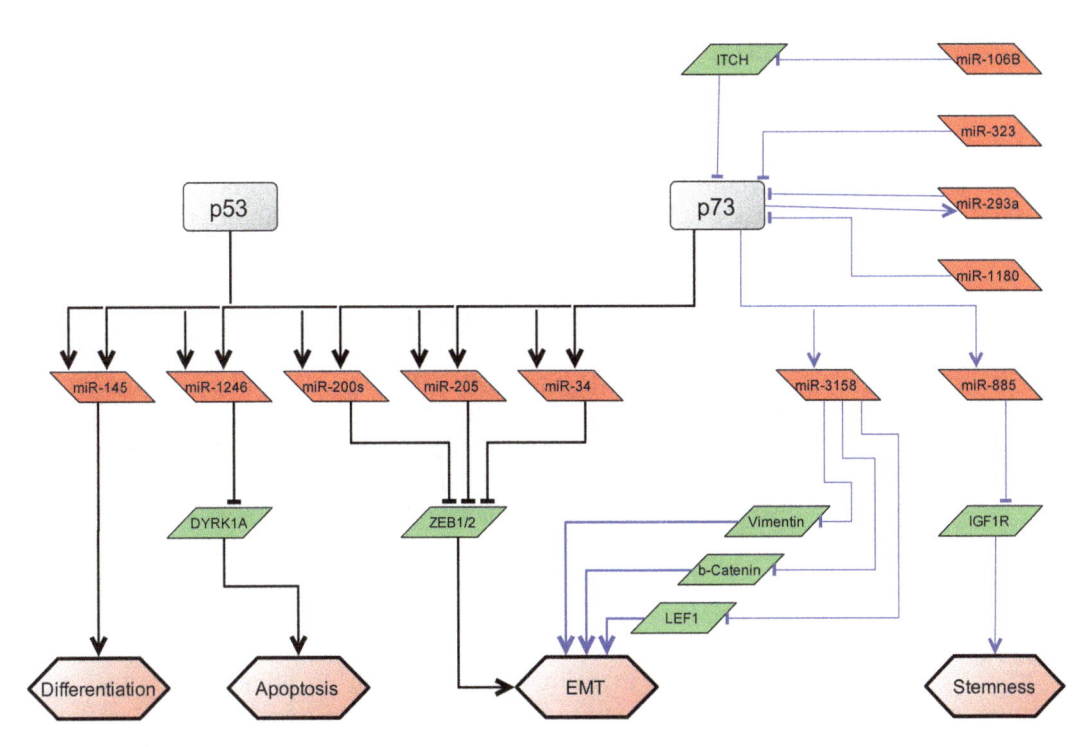

Fig. 1 p73-miRNA interactions build-up a network encompassing both p53-shared circuits (black lines) and p73-dedicated circuits (blue lines), which sustain differentiation and apoptosis and inhibit EMT and stemness. This way p73 generates a circuit that uninterruptibly sustains crucial anti-oncogenic and anti-metastatic pathways, in a redundant manner, even upon p53 failure. MiRNA components of this network appear in red, while mRNA/protein components appear in green

and downregulates p73 mRNA levels, thereby generating a negative feedback loop that increases resistance to cisplatin. Inhibition of MIR193A can perturb this feedback loop, thus suppressing tumor cell viability and enhancing chemosensitivity [26, 27]. A striking feature of miR-193a is that it is selectively sorted to exosomes [28]. Exosomes are secreted in large quantities into the circulation, are key for intercommunication between cancer cells and their surrounding and distant environments, and can act as miRNA carriers. Exosomally transferred miRNAs alter the behavior of recipient tumor and stromal cells by modulating immune responses, inhibiting apoptosis, and promoting angiogenesis [29]. Indeed, exosomally-transferred miR-193a increases the risk of liver metastasis in colon cancer patients, while its accumulation in the exosomal donor cells instead of the exosomes suffices to inhibit tumor progression [28]. This fact plausibly suggests that miR-193a may horizontally transfer the ability of p73 inhibition to ensure "spreading" of the chemoresistance trait from cancer cells in primary sites to cells in the surrounding tumor microenvironment or in distant sites. Another p73-inhibiting miRNA is miR-323, which can repress p73 and promote prostate cancer growth and docetaxel resistance in prostate cancer cells [30]. In a similar manner, miR-1180 targets p73 to inhibit apoptosis in Wilms' tumors [31].

4.2 MIRNAs Targeted by p73

p73 directly transactivates MIRNAs with a tumor suppressive role. First of all, p73 and p53 share common targets that are highly conserved master regulators of tumor suppression. More specifically, p53 is able to directly transactivate MIR200 family members [32] and MIR205 [33], which co-operatively regulate epithelial-to-mesenchymal transition (EMT) by targeting the E-cadherin inhibitors ZEB1 and SIP1 [34]. p73 can also transactivate MIR200s [35] and MIR205 [36]. For instance, it was recently shown that activation of MIR200A/B and suppression of ZEB1/2 in breast cancer cell lines are mediated by physical interaction between p73 and restin on MIR200 gene promoters to upregulate gene expression. Such a co-activator complex is specific for the p73-regulated networks, since restin does not physically associate with p53 to co-activate MIR200 [37]. This p73-dedicated co-activator could plausibly provide an alternative to the p53/MIR200 axis, perhaps to maintain miR-200 anti-oncogenic effects upon p53 deficiency. In addition, MIR205 is transactivated by p53, but most likely in a cell context-dependent manner, since it can be transactivated in triple negative cancer cell lines, but not in melanoma cells. However, in melanoma cell lines, MIR205 is instead transactivated by p73, implying that any blockade in the p53/MIR205 axis can be circumvented by a surrogate p73/MIR205 axis [36]. In a similar manner, p73 transactivates the tumor suppressor miR-34a, another direct p53 target which is evolutionarily old and exerts apoptotic and anti-metastatic behavior by directly and indirectly repressing several oncogenes [38]. MiR-3158, a newly identified primate-specific, p73-dependent/p53-independent target, is suggested to functionally associate with miR-34a in EMT inhibition by suppressing vimentin, lef1, and β-catenin [19]. Theoretically, even if a p53/miR-34a/ZEB1 circuit is disrupted, a p73/miR-3158/vimentin/lef1/β-catenin-circuit could, at least in part, surrogate for EMT inhibition and sustain prevention of metastatic transformation of cells. An additional p73-dependent/p53-independent miRNA gene is MIR885. It is upregulated by TAp73 to inhibit expression of IGF1R, a receptor that enhances downstream stemness cascades. The cancer-specific ΔTAp73 isoform enhances metastatic transformation by removing the TAp73/miR-885-5p-mediated blockade of IGF1R [39].

p73 reproduces other p53 pathways by transactivating the same MIRNA targets. MIR1246 and MIR145 are also p53/p73-dependent targets. In response to DNA damage, p53 transactivates MIR1246, which in turn reduces the level of DYRK1A, a Down-Syndrome-associated protein kinase. This leads to the nuclear retention of NFATc1, a protein substrate of DYRK1A, which ultimately triggers apoptotic response. p73 was shown to induce the same pathway via MIR1246 transactivation [40, 41]. The tumor suppressor miR-145, another p53 target, is also transactivated by p73 isoforms, and its inhibition has been associated with attenuation of differentiation in acute promyeloid leukemia [42].

4.3 MiRNAs Targeting Modulators of p73 Stability and Activity

p73 function largely depends on co-regulators and/or post-translational modifiers that determine its stability and apoptotic efficiency. These factors physically associate with unique motifs and interacting surfaces in the C-terminus of p73 isoforms and modify p73's ability to activate target genes and induce apoptosis [5]. To this end, miRNA-mediated inhibition of these factors directly reflects on the behavior of p73 in networks underlying cancer initiation and progression. For instance, the E3 ubiquitin ligase Itch binds to the PPPPY motif in the C-terminus of specific p73 isoforms, destabilizes them, and enhances their degradation [43]. Activation of MIR106B upon treatment with deacetylase inhibitors downregulates Itch and leads to reciprocal accumulation of TAp73. Consequently, p73 enhances PUMA transactivation and restores the apoptotic response of chronic lymphocytic leukemia cells [44]. To date, miRNAs targeting other post-translational modifiers and co-regulators of p73 remain largely elusive. According to a recently proposed concept, the function of TFs is determined by their interactions with co-activators or co-repressors, which physically associate with them on the target gene promoters to amplify or attenuate, respectively, the TFs' transactivation efficiency. Such co-regulators serve as molecular switches that link upstream signaling events to the downstream transcriptional programs [45]. In this respect, miRNAs that target p73 co-regulators can affect p73-controlled transcriptional programs by modifying the co-regulators' levels and availability. Hence, their identification would be of high relevance for the detailed characterization of cancer p73/miRNA network.

5 A Computational Approach Deciphers Isoform-Specific p73 Effects on Cancer Cell miRNomes

Though TAp73 isoforms are traditionally anti-oncogenic, they are not imitating each other in terms of function. Rather, these isoforms are functionally nonequivalent, with each presenting differences in transactivation efficiencies, protein interactions with co-regulators, and responses to stress stimuli. These differences ultimately lead to divergent cellular phenotypes and are attributable to the C-terminus, which is unique for each TAp73 variant [5]. Data on the most profoundly studied TAp73α and TAp73β isoforms clearly highlight that not only they activate the same targets to a different extent but also they have distinct gene targets

[46]. Having this in mind, we used bioinformatics approaches to meta-analyze and integrate publicly available transcriptomics data in order to explore whether their functional divergence is mediated by distinct isoform-responsive miRNAs. In detail, to characterize mRNA/miRNA networks responsive to each TAp73 isoform, we used miRNA [19] and RNA-Seq [46] data produced in a well-established study model system, which comprises doxycycline-inducible Saos2 clones stably transfected with TAp73α or TAp73β isoforms. Due to the p53-null background of Saos2 osteosarcoma cells, this study model is advantageous for identifying direct and indirect TAp73 targets in an isoform-specific manner and unbiased from any p53 interference [47]. The miRNomes of Saos-TAp73α and Saos-TAp73β clones are quite distinct, because the ≥2-fold-upregulated miRNAs they have in common are profoundly fewer than their non-common ≥2-fold-upregulated miRNAs. Overall we revealed that p73 isoform-specific functions can be achieved in two ways.

5.1 Co-operativity of Isoform-Specific miRNomes on Downstream mRNA Targets

The TAp73α and TAp73β isoforms upregulate different groups of miRNAs [19]. These TAp73α-specific and TAp73β-specific miRNomes were analyzed for potential co-operativity on downstream targets. First, the putative mRNA targets of either TAp73α- or TAp73β-specific miRNomes were collected using the DIANA microT-CDS platform. Of those, we assessed which targets are significantly downregulated in the corresponding RNA-Seq experiments. Of note, more than half of these mRNAs (53% and 57%, correspondingly) are targeted and potentially downregulated by TAp73α- and TAp73β-responsive miRNomes, underscoring that p73 effects on downstream mRNA targets are mediated by miRNAs. Co-operativity analysis was done by collecting the mRNA targets of all combinations of miRNAs of the TAp73α- or TAp73β-specific miRNomes. Then, the overlap of predicted co-operatively targeted mRNAs with the downregulated ones from the RNA-Seq experiments with the TAp73α- or TAp73β-expressing clones was calculated. This analysis yielded 252 predicted downregulated mRNA targets on which TAp73α-specific miRNomes can act co-operatively and 373 predicted downregulated mRNA targets on which TAp73β-specific miRNomes can act co-operatively (Fig. 2). "Hot-spot" mRNA targets of co-operativity of the miRNomes compared with each other were predicted as follows: 17 genes are simultaneously targeted by more than five miRNAs of TAp73α-specific miRNomes (ABCA1, ABTB1, ADD2, BMPR1B, C1R, C1S, CLSTN2, F13A1, HUNK, LHX4, LONRF2, NFIB, NPNT, RNF152, SOX6, SYTL2, TMEM132B), while 15 genes are simultaneously targeted by over five miRNAs of TAp73β-specific miRNomes (ADCY1, ALX4, CGNL1, HMCN1, KSR2, MSI1, NTRK3, PAPPA, PMEPA1, PTPRN, RADIL, RPL37, SCUBE3, TMEM132B,

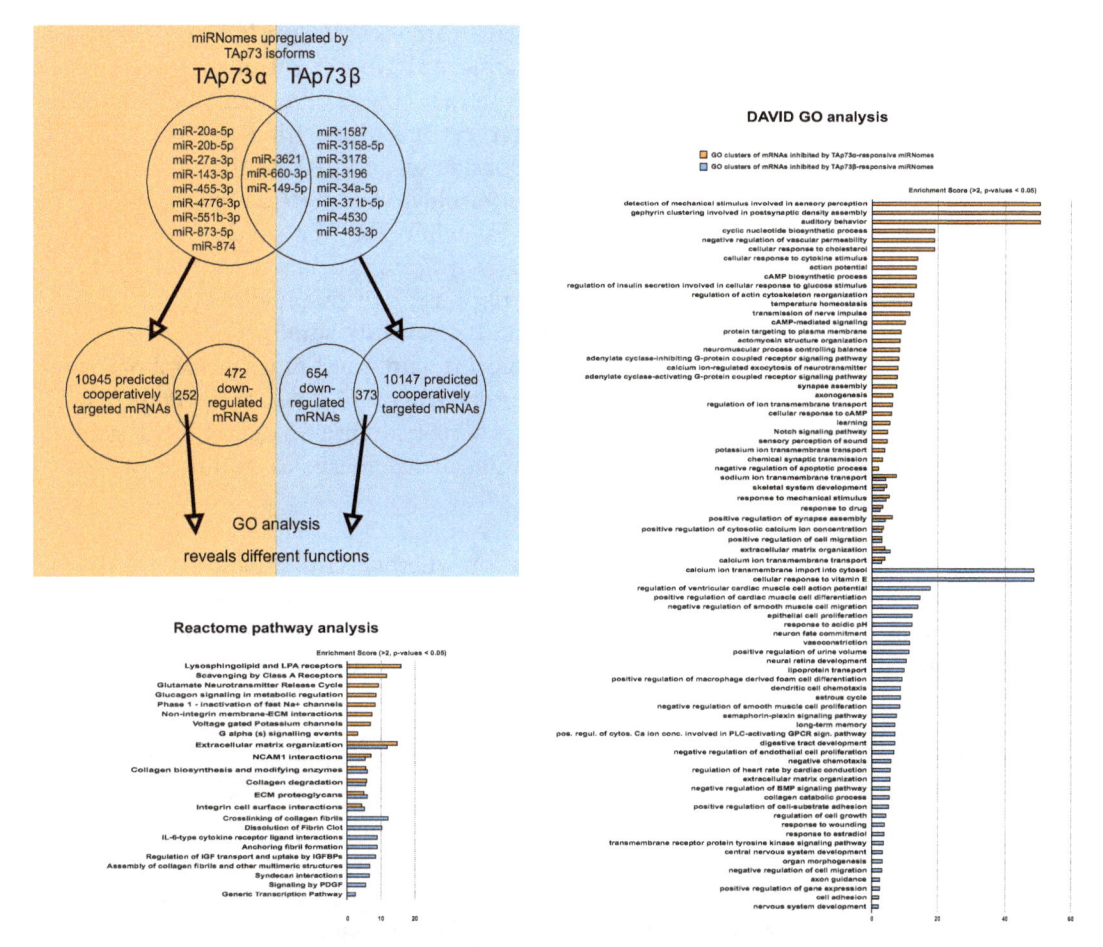

Fig. 2 The pipeline of computational analysis to predict mRNAs co-operatively inhibited by TAp73α- or TAp73β-responsive miRNA populations. The TAp73α- and TAp73β-specific miRNomes [19] were analyzed for their validated (TarBase v7.0 and v8.0) and putative (microT-CDS database) targets using the DIANA platform. Co-operativity analysis was done by collecting the mRNA targets of all combinations of miRNAs of the TAp73α- or TAp73β-specific miRNomes. The overlap of predicted co-operatively targeted mRNAs with the downregulated ones in RNAseq results of TAp73α- or TAp73β-expressing clones [46] was calculated. This analysis produced 252 mRNA targets of the TAp73α-specific miRNome and 373 mRNA targets of the TAp73β-specific miRNome. Pathway analysis (Reactome) and GO analysis (DAVID) of these co-operative targets reveals divergent functions for TAp73α- and TAp73β-specific miRNomes

TRDMT1). Additionally, a synergistic co-operativity of all combinations of miRNA pairs of the TAp73α- or TAp73β-specific miRNomes on these downregulated predicted mRNA targets was estimated using Triplex RNA prediction tool [14]. We found four isoform-specific synergistically targeted mRNAs in both sets of miRNome-specifically regulated mRNAs that are downregulated in the respective RNA-Seq analyses. The minimum overlap between both the co-operated mRNA targets and the "hot-spot" co-operated mRNA targets of the α- and β-miRNomes indicates that the functional divergence of the p73 isoforms is likely attributed to distinct downstream miRNA/mRNA cascades, where isoform-specific co-operating miRNomes ensure silencing of distinct downstream mRNA targets (Fig. 2). Further GO enrichment analysis of the mRNA targets of co-operating miRNAs revealed that TAp73α triggers a specific set of miRNAs that co-operatively inhibits genes involved in mechanotransduction, gephyrin clustering, and auditory behavior while TAp73β triggers a distinct set of miRNAs that co-operatively inhibits genes implicated in calcium ion import and cellular response to vitamin E (Fig. 2). Notably, our analysis also predicted previously unnoticed associations of TAp73α and TAp73β isoforms with novel functions and pathways. Further mechanistic investigations on how p73 triggers these pathways and if these pathways surrogate for p53 deficiency or gain-of-function mutations in tumor cells emerge as a subject of fruitful research.

5.2 Direct Transactivation of Isoform-Specific MIRNA Targets

Subsequent examinations on whether TAp73α and TAp73β could directly transactivate distinct MIRNA genes, which, in turn, will inhibit different groups of mRNA targets, ultimately lead to divergent phenotypes. Four MIRNAs encompassed in the TAp73-α-responsive and TAp73β-responsive miRNomes bear putative p73-responsive elements within their corresponding promoters [19]. Of those, MIR660, an established tumor suppressor, is upregulated in both TAp73α- and TAp73β-responsive miRNomes, thereby emerging as a putative common target. MIR874 is predicted as TAp73α-specific direct target, whereas MIR34A and MIR3158 are verified to be TAp73β-specific [19]. The mRNA targets of the putative TAp73α-transactivated MIR874 that are downregulated in the respective clone were defined by merging the predicted mRNA targets with the significantly downregulated mRNAs in the TAp73α-expressing clone and subjected to pathway analysis and GO enrichment analysis. In a similar manner, the mRNA targets of the TAp73β-transactivated MIR34A and MIR3158 that are downregulated in the TAp73β-expressing clone were defined and subjected to pathway analysis and GO enrichment analysis. While the TAp73α miRNA/mRNA networks were found to be associated with the Hedgehog pathway and potassium channels, the TAp73β miRNA/mRNA networks are associated with the organization and assembly of components of the tumor microenvironment. Overall, isoform-specific miRNA targets underlie different functionality of p73 isoforms (Fig. 3).

6 Transcription-Independent Regulation of p73 miRNomes: Predicted PPIs Between p73 and the miRNA Processing Machinery

Since miRNAs exhibit a high degree of co-operativity, synergism, and redundancy, a phenotypic change is relevant with massive

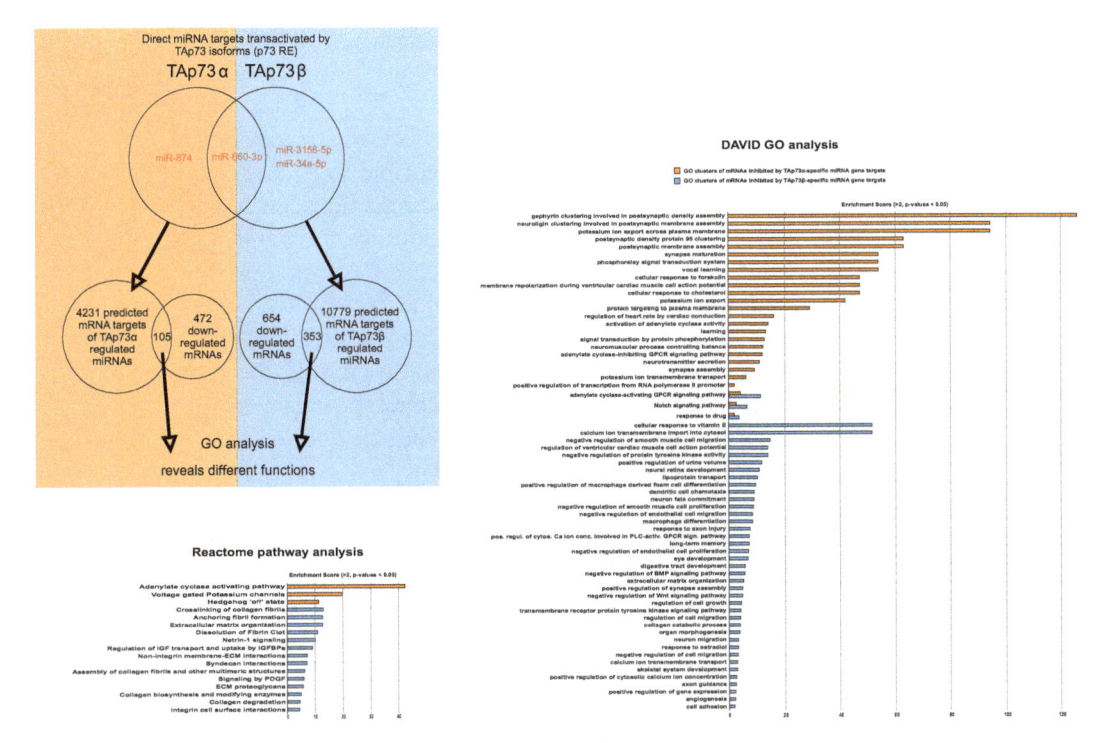

Fig. 3 The pipeline of computational analysis to predict potential direct MIRNA gene targets of TAp73α or TAp73β isoforms and their downstream effectors. Of the isoform-specific miRNomes, four MIRNAs were identified having putative p73-responsive elements within their corresponding promoters [19]. The validated and predicted (DIANA tools TarBase v7.0 & v8.0 and microT-CDS database) mRNA targets of TAp73α- or TAp73β-transactivated miRNAs that are downregulated after overexpression of the respective isoform were defined by overlapping the mRNA targets with the significantly downregulated mRNAs in TAp73α- or TAp73β-expressing clones. Pathway (Reactome) and GO analysis (DAVID) of the targets reveals divergent functions for TAp73α-specific and TAp73β-specific miRNomes

miRNA alterations rather than changes in individual miRNAs. Hence, factors that systemically alter the constitution of miRNA populations have a more profound impact on miRNA-mediated cellular outcomes. Such systemic intervention can be achieved through proteins that interact with the miRNA processing complex to dictate which pre-miRNAs will be selected to be ultimately matured to functional miRNAs. Importantly, the miRNA processing machinery is susceptible to p53. In detail, p53 selectively enhances the post-transcriptional maturation of tumor suppressive miRNAs, including miR-16-1, miR-143, and miR-145, in response to DNA damage. This is achieved through protein-protein interactions between p53 and the DDX5 accessory protein of the miRNA microprocessing complex, which increase the rate of maturation of pri-miRs to pre-miRs and augment response to DNA damaging agents. Transcriptionally inactive p53 mutants interfere with a functional assembly between the Drosha complex and DDX5, thereby attenuating miRNA processing activity [48]. In this way, p53 can specifically accelerate the maturation of selected combinations of miRNAs, which potentiate downstream anti-oncogenic cascades.

Intriguingly, a computational analysis revealed that the miRNA processing complex is also susceptible to p73. This is predicted to occur via protein-protein interactions between some p73 isoforms and DGCR8 [49]. More specifically, the DGCR8 subunit is a nucleus-localized, highly conserved component of the miRNA processing machinery, which binds pri-miRNA to stabilize it for processing by Drosha [50]. Importantly, DGCR8 is a miRNA processing dedicated protein, indispensable for miRNA maturation. Cells of *Dgcr8* knockout mice possess neither fully mature nor intermediate pre-miRNA products and, as a result, these mice cannot differentiate and arrest early in development [51]. DGCR8 possesses a WW domain that potentially interacts with the PPPPY motif of the C-terminal domains of p73α and p73β [49], a highly conserved motif that enhances the apoptotic response to DNA damaging agents [5]. Collectively, these data provide evidence that PPPPY-bearing p73 isoforms may crosstalk with the miRNA processing machinery to control the quality and quantity of the mature miRNA populations by physically associating with DGCR8. This association could provide an alternative pathway to sustain the effect of p53 on miRNA maturation. Even if p53 deficiency renders the p53/DDX5-mediated modification of miRNA processing impossible, maturation of tumor suppressive miRNAs can perhaps be sustained through an alternative p73/DGCR8 pathway.

7 Translational Opportunities, Challenges, and Future Perspectives for p73/MIRNA Cancer Networks

The rise of the "miRNA world" has reformed the landscape of cancer therapeutics in less than two decades. Favorable characteristics of miRNAs, such as the ability to function as master regulators of the genome and lack of reports on serious adverse events in preclinical studies, render them attractive as next-generation drugs. Various strategies have been developed to replenish tumor suppressive miRNAs using miRNA mimics or to suppress oncogenic miRNAs using antimiR inhibitors. MiRNAs follow the footsteps of siRNAs, which have meanwhile passed clinical trials to finally reach the bedside, with most representative success story being the anti-hypercholesterolaemia drug Mipomersen. The experience gained during the development of siRNA-based therapeutics is being exploited to the fullest, toward catalyzing establishment of miRNA-based therapeutics [52]. Indeed, several miRNAs have already entered clinical trials, whereas an increase has been observed in the number of the miRNA-related application filings for USA patents [53]. For instance, two miRNAs are currently in phase I cancer clinical studies: the miR-16 mimic against mesothelioma and non-small cell lung cancer and the antimiR-155 against cutaneous T-cell lymphoma. An earlier clinical study on the miR-34a mimic, which was however terminated in phase 2, highlights the significance of taking into account immune-related toxicities in the future miRNA-focused clinical trials [52].

One major challenge of miRNA-based therapeutics is the selection of a favorable drug target, which will achieve a combination of maximal efficacy with minimal off-target effects. MiRNAs occupying network hubs are more likely to induce a robust cell outcome; however due to their high degree of interconnections, several off-target effects are anticipated. There are also miRNAs that present a "janous" behavior in different cell contexts [54]. We suggest that the p73-governed miRNome could pinpoint toward miRNAs that are mediators of its consistent anti-oncogenic effects even in *p53*-mutated backgrounds. For instance, the newly identified p73 target MIR3158 can suppress EMT-mediated invasiveness in a functional analogy with its co-operating MIR34A. In addition, due to its tissue-specific nature, miR-3158 is expected to cause deleterious effects upon ectopic addition [23]. As a result, the tumor-specific delivery of a miR-3158-mimic might catch cancer cells off-guard inducing their death, because a cancer cell might not be preconditioned to deal with the ectopic expression of a miRNA, which was not supposed to be expressed in this tissue [19]. Another case of p73-responsive miRNA, MIR885 which can downregulate IGF1R and selectively disrupt a metastatic DNp73-IGF1R alliance [39], could be exploited in metastatic melanomas to overcome IGF1R-triggered acquired resistance to BRAF inhibitors [55]. The p73α-responsive MIR874 emerges as a consistent tumor suppressor, with no evidence of janous behavior so far and could, therefore, be considered as an anticancer target, especially if p73α-specific anti-oncogenic pathways need to be restored.

Other challenges for miRNA therapeutics include (a) increasing the half-life of therapeutic miRNAs in the bloodstream and (b) improving the efficacy and specificity of their delivery to the tumor site. Stability can be increased by chemical modifications such as the addition of a 2'-O-methyl group or locked nucleic acids (LNAs). Regarding delivery of the chemically modified mimics and antagomiRs, in vivo strategies encapsulating the stabilized miRNA modulators in a variety of delivery vehicles, mainly viral vectors, synthetic nanocells, or nanoparticles, have been developed [52]. Notably, nanocarrier-mediated delivery emerges as a safer approach, showing less toxicity than viral-based delivery. Both synthetic and natural polymers have been used for miRNA delivery. For achieving tumor-specific delivery and eliminate off-target effects, the miRNA-carrying nanoparticles are surface coated with antibodies or ligands against proteins specifically expressed in cancer cells [56]. Interestingly, an elegant nanoparticle-based miRNA delivery method has been developed recently, where the nanoparticle shell is exploited to induce anti-oncogenic cascades in addition to the delivery of a tumor suppressor miRNA mimic [57]. In detail, self-immolative nanoparticles have been prepared from a biodegradable cationic prodrug, named DSS-BEN, which derives from polymerization of N1,N11-bisethylnorspermine (BENSpm). Upon entering the cytoplasm, the nanoparticles are selectively disassembled to release the miRNA modulator while, in parallel, the DSS-BEN polymer sheath gets degraded to release BENSpm. The resulting intracellular BENSpm can regulate polyamine metabolism and increase the anti-oncogenic response of the targeted cells [57]. Overall, a continuously developing nanocarrier platform could be combined with efficient miRNA modulators to manipulate the p73/MIRNA interactome of cancer patients toward anti-invasive and anti-metastatic outcomes.

8 What's Around the Corner for p73-Regulated Networks?

The discovery of lncRNA genes, as a corollary of the decoding of the human genome, has substantiated the significance of noncoding RNA networks in the main cancer hallmarks. LncRNAs genes encode RNA transcripts longer than 200 bps which, however, are not translated to protein products. LncRNAs develop a repertoire of divergent mechanisms to exert their sophisticated functions and regulate nodal tumor suppressors or oncogenes. For instance, they modulate chromatin states by acting as recruiters, tethers, or scaffolds of epigenetic complexes; they control mRNA transcription by acting as decoys of transcription factors, transcriptional co-regulators, or polymerase II inhibitors; they are structural components of nuclear bodies; and they modulate alternative splicing, protein stability, and/or protein subcellular localization. Notably, lncRNAs affect miRNomes by acting as "sponges" that sequester specific miRNAs away from their targets, as hosts of MIRNA genes or as precursors of small mRNAs [58]. This remarkable functional plasticity has enabled them to participate in a wide range of biological processes, several of which are commonly deregulated in cancer. They exhibit deregulated expression in cancer and have been shown to be involved in cancer initiation and progression [59]. Several lncRNAs participate in p53 tumor suppressor regulatory networks either as upstream p53 regulators or as downstream p53 effectors [60]. In contrast, to date, no p73-associated lncRNA has been identified. However, given that p73 tends to share common patterns of functional network modules with the prototype family member p53, it is plausible that p73-responsive lncRNAs will be identified in the future as an additional level of complexity of p73-governed cancer networks. The characterization of p73-responsive lncRNAs could open novel and exciting roads for future development of diagnostics and therapeutics.

9 Conclusion

Activation of p73-regulated anti-oncogenic networks could be an alternative/substitute to the "Holy Grail" in cancer-targeted therapeutics with a potential to circumvent the obstacles of p53-based targeting. p73 exerts anti-oncogenic activities both via p53/p73-dependent and through p73-dedicated pathways. This functional redundancy is, at least in part, achieved via noncoding RNA components, mainly miRNAs, which fine-tune the p73 effects on their downstream targets, often by acting co-operatively. This strategy appears to be isoform-sensitive, since p73 variants with different C-termini have distinct miRNomes to mediate their effects on divergent pathways. This creates a reservoir of p73-governed miRNAs which, combined with the state-of-the-art technology of miRNA delivery, hold therapeutic promise for the management of p53-deficient tumors.

Acknowledgments

This work was supported by the German Cancer Aid, Dr. Mildred Scheel Stiftung [grant 70112353], the German Research Foundation (DFG) [grant PU188/17-1], Wilhelm Sander-Stiftung [grant 2015.036.1], German Federal Ministry of Education and Research (BMBF) grant 0316171 as part of the project eBio:SysMet, and Rostock University Medical Faculty for the project Systems Medicine of Cancer Invasion and Metastasis to B.M.P.

References

1. Abraham CG, Espinosa JM (2015) The crusade against mutant p53: does the COMPASS point to the holy grail? Cancer Cell 28(4):407–408. https://doi.org/10.1016/j.ccell.2015.09.019
2. Olivier M, Hollstein M, Hainaut P (2010) TP53 mutations in human cancers: origins, consequences, and clinical use. Cold Spring Harb Perspect Biol 2(1):a001008. https://doi.org/10.1101/cshperspect.a001008
3. Giaretti W, Rapallo A, Sciutto A, Macciocu B, Geido E, Hermsen MA, Postma C, Baak JP, Williams RA, Meijer GA (2000) Intratumor heterogeneity of k-ras and p53 mutations among human colorectal adenomas containing early cancer. Anal Cell Pathol 21(2):49–57
4. Ren ZP, Olofsson T, Qu M, Hesselager G, Soussi T, Kalimo H, Smits A, Nistér M (2007) Molecular genetic analysis of p53 intratumoral heterogeneity in human astrocytic brain tumors. J Neuropathol Exp Neurol 66(10):944–954. https://doi.org/10.1097/nen.0b013e318156bc05
5. Logotheti S, Pavlopoulou A, Galtsidis S, Vojtesek B, Zoumpourlis V (2013) Functions, divergence and clinical value of TAp73 isoforms in cancer. Cancer Metastasis Rev 32(3–4):511–534. https://doi.org/10.1007/s10555-013-9424-x
6. Engelmann D, Meier C, Alla V, Pützer BM (2015) A balancing act: orchestrating amino-truncated and full-length p73 variants as

7. Stiewe T, Zimmermann S, Frilling A, Esche H, Pützer BM (2002) Transactivation-deficient DeltaTA-p73 acts as an oncogene. Cancer Res 62(13):3598–3602
8. Ming L, Sakaida T, Yue W, Jha A, Zhang L, Yu J (2008) Sp1 and p73 activate PUMA following serum starvation. Carcinogenesis 29(10):1878–1884. https://doi.org/10.1093/carcin/bgn150
9. Chakraborty J, Banerjee S, Ray P, Hossain DM, Bhattacharyya S, Adhikary A, Chattopadhyay S, Das T, Sa G (2010) Gain of cellular adaptation due to prolonged p53 impairment leads to functional switchover from p53 to p73 during DNA damage in acute myeloid leukemia cells. J Biol Chem 285(43):33104–33112. https://doi.org/10.1074/jbc.M110.122705
10. John K, Alla V, Meier C, Pützer BM (2011) GRAMD4 mimics p53 and mediates the apoptotic function of p73 at mitochondria. Cell Death Differ 18(5):874–886. https://doi.org/10.1038/cdd.2010.153
11. Nelson P, Kiriakidou M, Sharma A, Maniataki E, Mourelatos Z (2003) The microRNA world: small is mighty. Trends Biochem Sci 28(10):534–540. https://doi.org/10.1016/j.tibs.2003.08.005
12. Shen J, Hung MC (2015) Signaling-mediated regulation of MicroRNA processing. Cancer Res 75(5):783–791. https://doi.org/10.1158/0008-5472.CAN-14-2568
13. Xiao Y, Xu C, Guan J, Ping Y, Fan H, Li Y, Zhao H, Li X (2012) Discovering dysfunction of multiple microRNAs cooperation in disease by a conserved microRNA co-expression network. PLoS One 7(2):e32201. https://doi.org/10.1371/journal.pone.0032201
14. Schmitz U, Lai X, Winter F, Wolkenhauer O, Vera J, Gupta SK (2014) Cooperative gene regulation by microRNA pairs and their identification using a computational workflow. Nucleic Acids Res 42(12):7539–7552. https://doi.org/10.1093/nar/gku465
15. Lai X, Gupta SK, Schmitz U, Marquardt S, Knoll S, Spitschak A, Wolkenhauer O, Pützer BM, Vera J (2018) MiR-205-5p and miR-342-3p cooperate in the repression of the E2F1 transcription factor in the context of anticancer chemotherapy resistance. Theranostics 8(4):1106–1120. https://doi.org/10.7150/thno.19904
16. Nazarov PV, Reinsbach SE, Muller A, Nicot N, Philippidou D, Vallar L, Kreis S (2013) Interplay of microRNAs, transcription factors and target genes: linking dynamic expression changes to function. Nucleic Acids Res 41(5):2817–2831. https://doi.org/10.1093/nar/gks1471
17. Sengupta D, Bandyopadhyay S (2013) Topological patterns in microRNA-gene regulatory network: studies in colorectal and breast cancer. Mol BioSyst 9(6):1360–1371. https://doi.org/10.1039/c3mb25518b
18. Hermeking H (2012) MicroRNAs in the p53 network: micromanagement of tumour suppression. Nat Rev Cancer 12(9):613–626. https://doi.org/10.1038/nrc3318
19. Galtsidis S, Logotheti S, Pavlopoulou A, Zampetidis CP, Papachristopoulou G, Scorilas A, Vojtesek B, Gorgoulis V, Zoumpourlis V (2017) Unravelling a p73-regulated network: the role of a novel p73-dependent target, MIR3158, in cancer cell migration and invasiveness. Cancer Lett 388:96–106. https://doi.org/10.1016/j.canlet.2016.11.036
20. Bracken CP, Scott HS, Goodall GJ (2016) A network-biology perspective of microRNA function and dysfunction in cancer. Nat Rev Genet 17(12):719–732. https://doi.org/10.1038/nrg.2016.134
21. Cheng F, Jia P, Wang Q, Lin CC, Li WH, Zhao Z (2014) Studying tumorigenesis through network evolution and somatic mutational perturbations in the cancer interactome. Mol Biol Evol 31(8):2156–2169. https://doi.org/10.1093/molbev/msu167
22. Zhu Y, Skogerbø G, Ning Q, Wang Z, Li B, Yang S, Sun H, Li Y (2012) Evolutionary relationships between miRNA genes and their activity. BMC Genomics 13:718. https://doi.org/10.1186/1471-2164-13-718
23. Koufaris C (2016) Human and primate-specific microRNAs in cancer: evolution, and significance in comparison with more distantly-related research models: the great potential of evolutionary young microRNA in cancer research. BioEssays 38(3):286–294. https://doi.org/10.1002/bies.201500135
24. Niwa R, Slack FJ (2007) The evolution of animal microRNA function. Curr Opin Genet Dev 17(2):145–150. https://doi.org/10.1016/j.gde.2007.02.004
25. Meunier J, Lemoine F, Soumillon M, Liechti A, Weier M, Guschanski K, Hu H, Khaitovich P, Kaessmann H (2013) Birth and expression evolution of mammalian microRNA

26. Ory B, Ramsey MR, Wilson C, Vadysirisack DD, Forster N, Rocco JW, Rothenberg SM, Ellisen LW (2011) A microRNA-dependent program controls p53-independent survival and chemosensitivity in human and murine squamous cell carcinoma. J Clin Invest 121 (2):809–820. https://doi.org/10.1172/JCI43897

27. Jacques C, Calleja LR, Baud'huin M, Quillard T, Heymann D, Lamoureux F, Ory B (2016) miRNA-193a-5p repression of p73 controls Cisplatin chemoresistance in primary bone tumors. Oncotarget 7 (34):54503–54514. https://doi.org/10.18632/oncotarget.10950

28. Teng Y, Ren Y, Hu X, Mu J, Samykutty A, Zhuang X, Deng Z, Kumar A, Zhang L, Merchant ML, Yan J, Miller DM, Zhang HG (2017) MVP-mediated exosomal sorting of miR-193a promotes colon cancer progression. Nat Commun 8:14448. https://doi.org/10.1038/ncomms14448

29. Tran N (2016) Cancer exosomes as miRNA factories. Trends Cancer 2(7):329–331. https://doi.org/10.1016/j.trecan.2016.05.008

30. Gao Q, Zheng J (2018) microRNA-323 upregulation promotes prostate cancer growth and docetaxel resistance by repressing p73. Biomed Pharmacother 97:528–534. https://doi.org/10.1016/j.biopha.2017.10.040

31. Jiang X, Li H (2018) MiR-1180-5p regulates apoptosis of Wilms' tumor by targeting. Onco Targets Ther 11:823–831. https://doi.org/10.2147/OTT.S148684

32. Kim T, Veronese A, Pichiorri F, Lee TJ, Jeon YJ, Volinia S, Pineau P, Marchio A, Palatini J, Suh SS, Alder H, Liu CG, Dejean A, Croce CM (2011) p53 regulates epithelial-mesenchymal transition through microRNAs targeting ZEB1 and ZEB2. J Exp Med 208 (5):875–883. https://doi.org/10.1084/jem.20110235

33. Piovan C, Palmieri D, Di Leva G, Braccioli L, Casalini P, Nuovo G, Tortoreto M, Sasso M, Plantamura I, Triulzi T, Taccioli C, Tagliabue E, Iorio MV, Croce CM (2012) Oncosuppressive role of p53-induced miR-205 in triple negative breast cancer. Mol Oncol 6(4):458–472. https://doi.org/10.1016/j.molonc.2012.03.003

34. Gregory PA, Bert AG, Paterson EL, Barry SC, Tsykin A, Farshid G, Vadas MA, Khew-Goodall Y, Goodall GJ (2008) The miR-200 family and miR-205 regulate epithelial to mesenchymal transition by targeting ZEB1 and SIP1. Nat Cell Biol 10(5):593–601. https://doi.org/10.1038/ncb1722

35. Knouf EC, Garg K, Arroyo JD, Correa Y, Sarkar D, Parkin RK, Wurz K, O'Briant KC, Godwin AK, Urban ND, Ruzzo WL, Gentleman R, Drescher CW, Swisher EM, Tewari M (2012) An integrative genomic approach identifies p73 and p63 as activators of miR-200 microRNA family transcription. Nucleic Acids Res 40(2):499–510. https://doi.org/10.1093/nar/gkr731

36. Alla V, Kowtharapu BS, Engelmann D, Emmrich S, Schmitz U, Steder M, Pützer BM (2012) E2F1 confers anticancer drug resistance by targeting ABC transporter family members and Bcl-2 via the p73/DNp73-miR-205 circuitry. Cell Cycle 11(16):3067–3078. https://doi.org/10.4161/cc.21476

37. Lu Z, Jiao D, Qiao J, Yang S, Yan M, Cui S, Liu Z (2015) Restin suppressed epithelial-mesenchymal transition and tumor metastasis in breast cancer cells through upregulating mir-200a/b expression via association with p73. Mol Cancer 14:102. https://doi.org/10.1186/s12943-015-0370-9

38. Agostini M, Knight RA (2014) miR-34: from bench to bedside. Oncotarget 5(4):872–881. https://doi.org/10.18632/oncotarget.1825

39. Meier C, Hardtstock P, Joost S, Alla V, Pützer BM (2016) p73 and IGF1R regulate emergence of aggressive cancer stem-like features via miR-885-5p control. Cancer Res 76 (2):197–205. https://doi.org/10.1158/0008-5472.CAN-15-1228

40. Zhang Y, Liao JM, Zeng SX, Lu H (2011) p53 downregulates Down syndrome-associated DYRK1A through miR-1246. EMBO Rep 12 (8):811–817. https://doi.org/10.1038/embor.2011.98

41. Liao JM, Zhou X, Zhang Y, Lu H (2012) MiR-1246: a new link of the p53 family with cancer and Down syndrome. Cell Cycle 11 (14):2624–2630. https://doi.org/10.4161/cc.20809

42. Batliner J, Buehrer E, Fey MF, Tschan MP (2012) Inhibition of the miR-143/145 cluster attenuated neutrophil differentiation of APL cells. Leuk Res 36(2):237–240. https://doi.org/10.1016/j.leukres.2011.10.006

43. Rossi M, De Laurenzi V, Munarriz E, Green DR, Liu YC, Vousden KH, Cesareni G, Melino G (2005) The ubiquitin-protein ligase Itch regulates p73 stability. EMBO J 24 (4):836–848. https://doi.org/10.1038/sj.emboj.7600444

44. Sampath D, Calin GA, Puduvalli VK, Gopisetty G, Taccioli C, Liu CG, Ewald B,

Liu C, Keating MJ, Plunkett W (2009) Specific activation of microRNA106b enables the p73 apoptotic response in chronic lymphocytic leukemia by targeting the ubiquitin ligase Itch for degradation. Blood 113(16):3744–3753. https://doi.org/10.1182/blood-2008-09-178707

45. O'Malley BW, Kumar R (2009) Nuclear receptor coregulators in cancer biology. Cancer Res 69(21):8217–8222. https://doi.org/10.1158/0008-5472.CAN-09-2223

46. Koeppel M, van Heeringen SJ, Kramer D, Smeenk L, Janssen-Megens E, Hartmann M, Stunnenberg HG, Lohrum M (2011) Crosstalk between c-Jun and TAp73alpha/beta contributes to the apoptosis-survival balance. Nucleic Acids Res 39(14):6069–6085. https://doi.org/10.1093/nar/gkr028

47. Nakano K, Bálint E, Ashcroft M, Vousden KH (2000) A ribonucleotide reductase gene is a transcriptional target of p53 and p73. Oncogene 19(37):4283–4289

48. Suzuki HI, Yamagata K, Sugimoto K, Iwamoto T, Kato S, Miyazono K (2009) Modulation of microRNA processing by p53. Nature 460(7254):529–533. https://doi.org/10.1038/nature08199

49. Boominathan L (2010) The tumor suppressors p53, p63, and p73 are regulators of microRNA processing complex. PLoS One 5(5):e10615. https://doi.org/10.1371/journal.pone.0010615

50. Yeom KH, Lee Y, Han J, Suh MR, Kim VN (2006) Characterization of DGCR8/Pasha, the essential cofactor for Drosha in primary miRNA processing. Nucleic Acids Res 34(16):4622–4629. https://doi.org/10.1093/nar/gkl458

51. Wang Y, Medvid R, Melton C, Jaenisch R, Blelloch R (2007) DGCR8 is essential for microRNA biogenesis and silencing of embryonic stem cell self-renewal. Nat Genet 39(3):380–385. https://doi.org/10.1038/ng1969

52. Rupaimoole R, Slack FJ (2017) MicroRNA therapeutics: towards a new era for the management of cancer and other diseases. Nat Rev Drug Discov 16(3):203–222. https://doi.org/10.1038/nrd.2016.246

53. van Rooij E, Purcell AL, Levin AA (2012) Developing microRNA therapeutics. Circ Res 110(3):496–507. https://doi.org/10.1161/CIRCRESAHA.111.247916

54. Skourti E, Logotheti S, Kontos CK, Pavlopoulou A, Dimoragka PT, Trougakos IP, Gorgoulis V, Scorilas A, Michalopoulos I, Zoumpourlis V (2016) Progression of mouse skin carcinogenesis is associated with the orchestrated deregulation of mir-200 family members, mir-205 and their common targets. Mol Carcinog 55(8):1229–1242. https://doi.org/10.1002/mc.22365

55. Villanueva J, Vultur A, Lee JT, Somasundaram R, Fukunaga-Kalabis M, Cipolla AK, Wubbenhorst B, Xu X, Gimotty PA, Kee D, Santiago-Walker AE, Letrero R, D'Andrea K, Pushparajan A, Hayden JE, Brown KD, Laquerre S, McArthur GA, Sosman JA, Nathanson KL, Herlyn M (2010) Acquired resistance to BRAF inhibitors mediated by a RAF kinase switch in melanoma can be overcome by cotargeting MEK and IGF-1R/PI3K. Cancer Cell 18(6):683–695. https://doi.org/10.1016/j.ccr.2010.11.023

56. Ganju A, Khan S, Hafeez BB, Behrman SW, Yallapu MM, Chauhan SC, Jaggi M (2017) miRNA nanotherapeutics for cancer. Drug Discov Today 22(2):424–432. https://doi.org/10.1016/j.drudis.2016.10.014

57. Xie Y, Murray-Stewart T, Wang Y, Yu F, Li J, Marton LJ, Casero RA, Oupický D (2017) Self-immolative nanoparticles for simultaneous delivery of microRNA and targeting of polyamine metabolism in combination cancer therapy. J Control Release 246:110–119. https://doi.org/10.1016/j.jconrel.2016.12.017

58. Wilusz JE, Sunwoo H, Spector DL (2009) Long noncoding RNAs: functional surprises from the RNA world. Genes Dev 23(13):1494–1504. https://doi.org/10.1101/gad.1800909

59. Huarte M (2015) The emerging role of lncRNAs in cancer. Nat Med 21(11):1253–1261. https://doi.org/10.1038/nm.3981

60. Zhang A, Xu M, Mo YY (2014) Role of the lncRNA-p53 regulatory network in cancer. J Mol Cell Biol 6(3):181–191. https://doi.org/10.1093/jmcb/mju013

Part II

Computational Methods and Workflows for ncRNA Discovery and Annotation Based on High-Throughput Sequencing Data

Chapter 3

Methods for Annotation and Validation of Circular RNAs from RNAseq Data

Disha Sharma, Paras Sehgal, Judith Hariprakash, Sridhar Sivasubbu, and Vinod Scaria

Abstract

Circular RNAs are an emerging class of transcript isoforms created by unique back splicing of exons to form a closed covalent circular structure. While initially considered as product of aberrant splicing, recent evidence suggests unique functions and conservation across evolution. While circular RNAs could be largely attributed to have little or no potential to encode for proteins, recent evidence points to at least a small subset of circular RNAs which encode for peptides. Circular RNAs are also increasingly shown to be biomarkers for a number of diseases including neurological disorders and cancer. The advent of deep sequencing has enabled large-scale identification of circular RNAs in human and other genomes. A number of computational approaches have come up in recent years to query circular RNAs on a genome-wide scale from RNA-seq data. In this chapter, we describe the application and methodology of identifying circular RNAs using three popular computational tools: FindCirc, Segemehl, and CIRI along with approaches for experimental validation of the unique splice junctions.

Key words Circular RNA, circRNAs, FindCirc, CIRI, Segemehl

1 Introduction

Circular RNAs (circRNAs) are recently discovered class of RNAs which do not follow the canonical model of linear splicing, but instead back splice from 5' donor end to 3' acceptor end forming a circular loop structure. Circular RNAs could originate from exons, introns, as well as UTRs [1]. Earlier considered as product of aberrant splicing [2] circRNAs are now contemplated to play a role in regulating gene expression. Circular RNAs have been recently shown to act as miRNA sponges [3], act as vehicle for RNA-binding proteins (RBPs) [4], and control the translation rate of respective linear mRNA transcripts [5]. While circular RNAs are largely classified as noncoding RNAs, recent evidence suggests that a small proportion of the candidate circRNAs could have a potential to encode peptides [6]. Circular RNAs are also increasingly reported as putative biomarkers in neurological diseases including Alzheimer's [7], Parkinson's, cancer [8], and cardiovascular diseases [9] reiterating their importance in biomedical research.

Earlier studies of circRNAs were based on amplification of cDNA using divergent primers and polymerase chain reaction (PCR) [2]. The recent advent of deep sequencing has enabled the genome-wide discovery of circular RNAs using computational methodologies for analysis of transcriptome datasets. Currently, there are over 20 different computational tools available for the identification and analysis of circRNAs. Table 1 summarizes the list of computational resources for identification of circRNA. The computational approaches that identify circRNAs leverage the presence of unique back-splice junctions, a characteristic feature of circRNAs [10]. Back-splice junctions could also arise out of tandem repeats within the genome, genome duplication, as well as trans-splicing and reverse transcription (RT) template switching [10]. It is therefore important to use appropriate methods as to avoid identification of false-positive candidates.

The published computational pipelines use different approaches including the mapping software and use of fusion reads/chimeric reads to identify circRNAs. These pipelines use the information on mapped/unmapped reads from sequence alignment mapped (SAM) file to identify back-splice junctions from reads which do not align end to end or are not mapped. A number of alignment tools including Bowtie [11], BWA [12], Star [13], Kallisto [14], and TopHat [15] (alternate-splicing aligner that uses bowtie as base aligner) are extensively used to align reads. Difference in sequences can also affect the alignment efficiency of reads by the aligners; for example TopHat-Fusion [16] and MapSplice [17] identify fusion reads or reads which are part of back-splice junctions. Sensitivity, specificity, and accuracy are major factors for the evaluation of the computational pipelines for circRNA identification. Hansen et al. have recently compared the prevalent computational pipelines for identification of circRNA [18].

In this chapter, we have highlighted three tools/pipelines based on the popularity, significance, and ease of description of underlying concept or algorithm. The basic concept of circRNAs is summarized in Fig. 1. The first pipeline FindCirc published by Memczak and colleagues [19] is one of the popular tools used to identify circular RNAs. This pipeline uses anchor sequences to identify back-splice junctions. This pipeline was used for the identification of circRNA in human, mouse, and *C. elegans*. The second pipeline is Segemehl [20] which identifies back-splice junctions based on enhanced suffix array (ESA) and follows a de novo-based approach while the third pipeline CIRI [21] is based on paired chiastic clipping algorithm and is also a de novo approach for identification of circular RNAs.

Table 1
List of different circular RNA software/tools available

Software/tool	Method used to identify back splice	Link to the software	Reference
FindCirc	De novo, back splice using anchor sequences	http://www.circbase.org/	[19]
Segemehl	Identifies fusion reads	http://www.bioinf.uni-leipzig.de/Software/segemehl/segemehl_0_2_0.tar.gz	[20]
MapSplice	Segment-based mapping	http://www.netlab.uky.edu/p/bioinfo/MapSpliceDownload	
CIRCExplorer	Annotating tool, identifies reads from back-splice exons and intron lariats	https://github.com/YangLab/CIRCexplorer	[22]
SUPeR-seq	Identifies polyA+ mRNA and polyA− RNAs from single-cell sequencing	https://codeload.github.com/huboqiang/TanglabCircularRNAPipeline/zip/master	[23]
PcircRNA_finder	Circular RNA in plants by chiastic clipping mapping of PE reads, uses known gene annotation	http://ibi.zju.edu.cn/bioinplant/tools/PcircRNA_finder.zip	[24]
DCC	circRNA from chimeric reads	https://github.com/dieterich-lab/DCC	[25]
CIRI	Differentiate back-splice junction from non-BSJ using maximum likelihood estimation (MLE)	https://sourceforge.net/projects/ciri/files/latest/download	[21]
UROBORUS	Can identify circRNA supported by exon-exon junctions and miss from intron/intergenic region	https://github.com/WGLab/uroborus/	[26]
NCLscan	Detecting intergenic and intergenic noncollinear transcript	https://github.com/TreesLab/NCLscan	[27]
CircPRO	Identifies circRNA protein-coding potential	http://bis.zju.edu.cn/CircPro/	[28]
Sailfish-cir	Accepts output from circRNA identification tools (CIRI, KNIFE, circRNA_Finder) and compare linear and circRNA transcripts	https://github.com/zerodel/Sailfish-cir	[29]
hppRNA	Identifies SNP, lncRNA and circRNAs	https://sourceforge.net/projects/hpprna/	[30]
KNIFE	Quantifying circRNA and linear RNA splicing events at both annotated and unannotated exon boundaries	https://github.com/lindaszabo/KNIFE	[31]
CircComPara	Correlate expression of linear and circRNA	https://github.com/egaffo/CirComPara	
Circmarker	Based on k-mers rather than read mapping	https://github.com/lxwgcool/CircMarker	

(continued)

Table 1
(continued)

Software/tool	Method used to identify back splice	Link to the software	Reference
CircTest	GIve gene and circRNA count (recommended to use DCC output)	https://github.com/dieterich-lab/CircTest	
PTESFinder	Identifies putative posttranscriptional exon shuffling (PTES) structure by mapping RNAseq reads to sequence models	https://sourceforge.net/projects/ptesfinder-v1/	[32]
Accurate circRNA finder suite	Uses maximum entropy model	https://github.com/arthuryxt/acfs	[33]
FUCHS	Identifies alternative exon usage within the same circle boundaries	https://github.com/dieterich-lab/FUCHS	[34]
circRNA_finder	Prediction of circRNAs with very proximal splice sites (below 100 bps)	https://github.com/orzechoj/circRNA_finder	[35]
circRNAFinder	Python and R scripts	https://github.com/bioxfu/circRNAFinder	

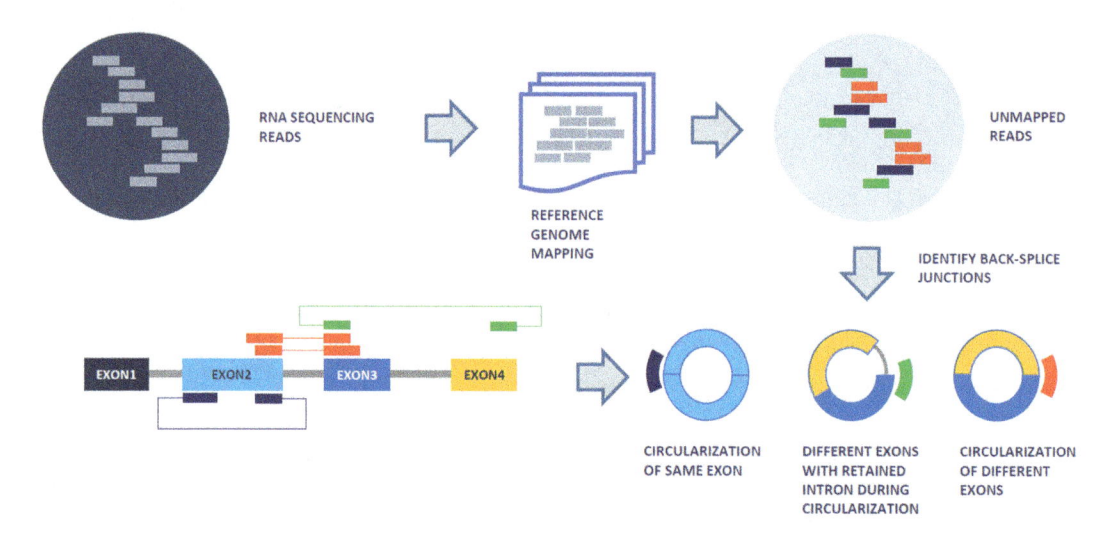

Fig. 1 The basic concept of identification of unique back-splice junctions to identify circRNAs

2 Materials

2.1 Pipeline1: FindCirc [19]

This pipeline is available on circBase Database (http://www.circbase.org/). The scripts are in Python and are easily downloadable. FindCirc identifies the reads overlapping the back-splice junctions. FindCirc uses bowtie2, a very fast short-read aligner. This tool uses very stringent criteria to identify circRNAs, therefore reducing the false positives. The pipeline uses a de novo approach and therefore is not dependent on any gene annotations. This pipeline scales efficiently with large datasets and the memory requirements are minimal (~8 GB). Studies suggest that this pipeline might be biased toward identification of highly expressed linear species, compared to other similar pipelines [9].

2.1.1 Hardware

A workstation with ~8 GB RAM and above should be sufficient to run this pipeline.

2.1.2 Software (see Note 1)

1. Bowtie2 (for alignment).
2. Samtools (processing aligned files).
3. Text editor (to open the scripts).
4. Python (with Numpy and pysam mdodules) (*see* **Note 2**).

2.1.3 Data Source

The tool uses FASTQ reads from RNA-sequencing data (*see* **Note 3**) for the respective samples. The reference genome for the organism can be downloaded in FASTA format from UCSC Genome Browser (http://genome.ucsc.edu/). It is advised to use the latest version of the reference genome so as to fetch updated information (*see* **Note 4**).

2.2 Pipeline2: Segemehl [20]

This pipeline uses FASTQ read sequences from RNA-sequencing data to identify splicing, trans-splicing, as well as fusion events. Segemehl has fewer false positives, especially those resulting from trans-splicing. This tool has very high sensitivity. But it needs very high RAM for large RNA-sequencing data. For large mammalian genomes, computer with less than 50 GB memory is not suggested. To reduce the computational load, one can run alignment for separate chromosomes.

2.2.1 Hardware

A workstation with 30–50 GB RAM should be sufficient to run this pipeline (*see* **Note 5**).

2.2.2 Software

1. Segemehl tool.

2.2.3 Data Source

The input for this tool is RNA-sequencing reads that should be non-polyadenylated/ribozero for better coverage of circRNAs. It is also advised to use RNase R-treated RNA for sequencing. The reference genome and annotation file can be downloaded from GENCODE/UCSC browser (*see* **Note 4**).

2.3 Pipeline3: CIRI [21]

This pipeline is based on de novo approach for identification of circular RNAs and uses BWA specifically for alignment of reads over reference genome. This pipeline uses unique chiastic clipping algorithm with multiple filters to remove false-positive circRNA candidates. No prior knowledge of exon-intron structure is required to run this pipeline. CircRNAs identified from CIRI could also be biased with respect to overexpressed linear species as in case of FindCirc. This increases the chances of false positives. This tool demands high-performance computing and may take a few days to analyze large datasets.

2.3.1 Hardware

A system of RAM >16 GB should be sufficient to complete this pipeline but any high-performance server would reduce the running time for large datasets.

2.3.2 Software

1. BWA aligner.
2. Perl ≥5.8.
3. CIRI.

2.3.3 Data Source

The reference genome and annotations can be downloaded from GENCODE or UCSC genome browser (*see* **Notes 6–8**). BWA aligner can be downloaded from http://bio-bwa.sourceforge.net/bwa.shtml and CIRI can be downloaded from https://sourceforge.net/projects/ciri/files/CIRI2/ (*see* **Note 4**).

2.4 Reagents for Experimental Validation

1. Trizol reagent (Invitrogen, USA, cat#15596018).
2. Chloroform (Sigma-Aldrich, USA, Cat#288306).
3. Isopropanol (Sigma-Aldrich, USA, Cat#190764).
4. Absolute ethanol (Merck Millipore, India, Part#107017).
5. Nuclease-free water (NFW) (Ambion, USA, Cat# AM9937).
6. Turbo DNase (Ambion, USA, Cat# AM2238).
7. 2× RNA loading dye (Thermo Fisher Scientific, USA, Cat# R0641).
8. Random hexamer primers (Fermentas, USA, Cat#S0142).
9. 5× First-strand buffer (Thermo Fisher Scientific, USA, Cat# 18080044).
10. 0.1 M DTT (Thermo Fisher Scientific, USA, Cat# 18080044).
11. 25 mM $MgCl_2$ (Thermo Fisher Scientific, USA, Cat#R0971).
12. 10 mM Equimolar dNTPs (Thermo Fisher Scientific, USA, Cat#R0192).
13. SuperScript II Reverse Transcriptase (SSIIRT) (Thermo Fisher Scientific, USA, Cat# 18064014).
14. RNase H (Thermo Fisher Scientific, USA, Cat# EN0202).
15. 10× PCR buffer (Thermo Fisher Scientific USA, Cat# 10342020).
16. RNase R (Epicenter Biotechnologies, USA, Cat# RNR07250).
17. 7.5 M LiCl (Ambion, USA, Cat# AM9480).
18. Spectrophotometer (Biospectrometer, Eppendorf, Germany, Cat# 6135).

3 Methods

3.1 Pipeline1: FindCirc [12]

3.1.1 Indexing the Reference Genome

1. Download the reference genome (http://genome.ucsc.edu/) in FASTA format.

   Reference genome can be downloaded using command line as well:

   ```
   wget -c <http://LINKtoUCSCfastaFile.tar.gz>
   ```

   where "http://LINKtoUCSCfastaFile.tar.gz" is the link to the FASTA file from UCSC browser.
   wget will fetch the fasta from the link to the system with -c for uninterrupted download.

2. Install Bowtie2 in your system. This can be installed using a simple command:

   ```
   sudo apt-get install bowtie2
   ```

   Or bowtie2 can be installed using binary packages on source code package available for Windows, Linux, or Macintosh.

3. Use command to extract and index the genome:

   ```
   gzip -dc <Reference_genome.zip> > Reference_genome.fa
   (if zip folder)
   ```

   Or

   ```
   tar -xvzf Reference_genome.tar.gz > Reference_genome.fa
   (if *.tar.gz)
   bowtie2-build <Reference_genome.fa> Reference_genome
   ```

   This command will create the index for reference genome with the index prefix `Reference_genome`.

3.1.2 Alignment Using Reference Genome

In this step, the reads that do not align contiguously and full length to the reference genome are discarded. To align the data over the reference genome, the command is mentioned below:

For single-end reads

```
bowtie2 -p 16 --very-sensitive --mm -M20 --score-min=C,-15,0
-q -x <Reference_genome_index> -U <reads.qfa> 2> <bowtie2.
log> | samtools view -hbuS - | samtools sort -
<sample_vs_genome>
```

For paired-end reads

```
bowtie2 -p 16 --very-sensitive --mm -D20 --score-min=C,-15,0
-q -x <Reference_genome_index> -1 <reads_R1.qfa> -2
```

```
<reads_R2.qfa> 2> <bowtie2.log> | samtools view -hbuS - |
samtools sort - <sample_vs_genome>
```

In this command, bowtie2 is a fast short-read aligner that aligns reads on the reference genome with options including

-p	Number of alignment threads to launch
--very-sensitive	-D 20 -R 3 -N 0 -L 20 -i S,1,0.50
-D	Give up extending after 20 failed extends in row
-R	For reads with repetitive seeds
-N	Maximum number of mismatches in the seed alignment
-L	Sets the length of reads substring to align
-i	Sets the function for seed interval (S is --sensitive mode)
--mm	Use memory-mapped I/O for index
-M20	Gives up extending after 20 nucleotide fails to extend in a row
--score-min	Minimum acceptable alignment score w.r.t. read length
-q	Reads are in FASTQ format
-x	Prefix for index filename
-U	Input file name
Bowtie2.log	Saves the output for alignment details in this file

Output for bowtie alignment is SAM file that can be manipulated using various utilities provided by samtools (*see* **Notes 9–11**). These include sort, merge, extracting mapped/unmapped/unique reads, alignment in per-position format, and many more.

Here in above case, output is piped to samtools where

view	Converts sam to bam or vice-versa
-h	Includes header in SAM output
-b	Output BAM
-u	Uncompressed BAM output
-S	Auto-detects input format

This output is then forwarded to samtools sort option where alignment file is sorted and saved in output file named "sample_vs_genome".

3.1.3 Extracting the Unmapped Reads and Anchor Sequences

This step is to discard the reads that are aligned full length and contiguously to the reference genome. Unmapped reads are taken forward and split into anchor sequences. Anchor sequences are

20mer from both ends of the reads extracted from unmapped reads that are aligned over reference genome and extended further in such a way that the complete read is aligned and breakpoint has GU/AG splice site.

1. To complete the above-mentioned step, run the following command:

```
samtools view -hf 4 <sample_vs_genome.bam> | samtools view -Sb - > <unmapped_sample.bam>
```

view	To convert sam to bam or vice-versa
-h	Includes header in the output file
-f 4	Only includes reads with flag integer, i.e., 4 in this case representing unmapped reads
Sample_vs_genome	Input aligned file from above step
-S	Auto-detect the format of file
-b	Output BAM
Unmapped_reads.bam	Contains the reads that are unmapped

2. Output file from above step contains unmapped reads that are split from both head and tail ends with 20mer sequences that are used as anchor sequences for further extension till full reads align with GU/AG splice site:

```
./unmapped2anchors.py <unmapped_sample.bam> | gzip > <sample_anchors.qfa.gz>
```

Unmapped2anchors.py is the customized script provided in FindCirc folder to extract anchor sequences from the unmapped reads.

3.1.4 Identifying Circular RNA Junctions from Unmapped Reads

Next step is to align and extend the anchor sequences over the reference genome and use customized script findcirc.py to identify reads aligning the circRNA junction sequences.

1. `mkdir <sample>` new directory is formed to save the output from next step.

2. Split the reference genome as separate chromosomes in the directory using the following command:

```
awk '{ if(NR==1 && $0 ~/>/){c=$0;gsub(">","",c);print $0 > c".fa";} else if($0~/>/ && NR>1){close(c".fa"); c=$0;gsub (">","",c);print $0 > c".fa"} else { print $0 > c".fa";}}' <Reference_genome.fa>
```

3. Realign the anchor sequences generated on reference genome to identify back-splice junctions.

```
bowtie2 -p 16 --reorder --mm -D20 --score-min=C,-15,0 -q -x
<reference_genome_index> -U <sample_anchors.qfa.gz> |
./find_circ.py -G genome -p <sample_> -s <sample/sites.
log> > <sample/sites.bed> 2> <sample/sites.reads>
```

Anchor sequences from both head and tail ends are aligned on the reference genome in paired ordering.

--reorder	SAM output to match with order of input reads
-U	sample_anchors.qfa.gz input FASTQ as 20mers anchor sequences
-G	Folder containing reference genome in FASTA format with separate FASTA for every chromosome
-p	Prefix for the label ID for each circRNA junction
-s	sites.log contains alignment log file saved in the directory
	sites.bed contains the read coordinates for reads fulfilling the criteria mentioned in the commands
	sites.reads contains the read sequences that are aligned over reference genome

This step gives the output for the reads aligned over the reference genome in opposite orientation or in back-splice junction format.

3.1.5 Filtering Putative circRNA Junctions

The next step is to filter out the putative circRNA junctions with cutoffs including

1. GU/AG splice sites
2. Unambiguous breakpoint junctions
3. Maximum of two mismatches in the extension step from anchors
4. No more than two nucleotide breakpoints in 20mer anchor alignment
5. Two reads cutoff
6. Unique anchor alignment with one anchor aligning with next best alignment
7. Distance between splice sites of no more than 100 kb:

```
grep circ <sample>/sites.bed | grep -v chrM | ./sum.py -2,3 | ./
scorethresh.py -16 1 | ./scorethresh.py -15 2 | ./scorethresh.py
-14 2 | ./scorethresh.py 7 2 | ./scorethresh.py 8,9 35 | ./
scorethresh.py -17 100000 > <sample>/circ_candidates.bed
```

This command is to filter out the above-mentioned criteria with customized script "scorethresh.py" that sets the threshold for each step and filters out the putative circRNA junctions.

3.2 Pipeline2: Segemehl [20]

3.2.1 Install Segemehl Tool

1. Download the zip file from the segemehl website (http://www.bioinf.uni-leipzig.de/Software/segemehl/).

2. Unzip the zip folder:
   ```
   tar -xvzf segemehl*
   ```

3. Go to the segemehl directory and type
   ```
   make
   ```
   This step should finish without any error.

3.2.2 Indexing the Reference Genome Using Segemehl

1. Download the reference genome using UCSC genome browser. The FASTA-formatted reference genome is indexed using segemehl.x in the parent directory:
   ```
   ./segemehl.x -x <Reference_genome.idx> -d </pathToReferenceGenome/Reference_genome.fa>
   ```
   where -x is the index name to be generated and -d is the FASTA for reference genome.

2. Align the FASTQ reads using segemehl.x as the command given below:
   ```
   ./segemehl.x -i <reference_genome_prefix.idx> -d <reference_genome.fa> -q <reads.fa> -S > <output.sam>
   ```
   (single-end reads)
   ```
   ./segemehl.x -i <reference_genome_prefix.idx> -d <reference_genome.fa> -q <reads_1.fa> -p <reads_2.fa> -S > <output_sampleName.sam>
   ```
   (paired-end reads).
   -q is for RNA-sequencing reads.
   -p is for query-mate sequence (applicable only in paired-end read).
   -s for splice reads.

3. Sort the sam file and circular RNA identification from SAM file can be done using testrealign.x:
   ```
   samtools sort -O sam -T sample.sort -o <sampleName.sort.sam> <output_sampleName.sam>
   ./testrealign.x -d <genome.fa> -q <output.sam> -n -U <splicesite_sampleName.bed> -T <Transalign_SampleName.bed> -o <output_sampleName.sam>
   ```

   -d is reference database.
   -q is aligned sam file.
   -n no-realign the sequences.

   This will generate two files: splice.bed and trans.bed.

3.3 Pipeline3: CIRI [21]

CIRI (circRNA identifier) uses novel chiastic clipping algorithm to identify circRNA using single as well as paired-end reads. CIRI uses multiple filters for prediction of circRNAs and supports multiple threads that makes it one of the most efficient tools to identify circRNAs.

3.3.1 Trimming Bad-Quality Reads

1. RNA-sequenced reads in FASTQ format can be uploaded in FASTQC to check if the quality of the reads is good (*see* **Notes 12–14**). If the sample has bad-quality reads, you can use trimmomatic software (http://www.usadellab.org/cms/?page=trimmomatic) (*see* **Notes 15** and **16**). Trimmomatic supports both single and paired-end reads. To use trimmomatic, first download the .jar file from the link http://www.usadellab.org/cms/uploads/supplementary/Trimmomatic/Trimmomatic-Src-0.36.zip.

2. Extract the zip file.

3. Access the jar file using java with the command given below:

```
java -jar trimmomatic-0.30.jar SE <input.fastq> <output.fastq> ILLUMINACLIP:TruSeq3-SE:2:30:10 LEADING:3 TRAILING:3 SLIDINGWINDOW:4:15 MINLEN:36
```

This command is for single-end reads where

SE	Single end
Input.fastq	Input FASTQ reads
Output.fastq	Output FASTQ file after trimming
ILLUMINACLIP	Cut adapter and Illumina-related sequences
TruSeq3-SE	Single-end Truseq adaptor sequence present in the "adaptors" in Trimmomatic directory

For paired-end reads

```
java -jar trimmomatic-0.30.jar PE <input_fwd.fastq> <input_rev.fastq> <output_fwd_paired.fastq> <output_fwd_unpaired.fastq> <output_rev_paired.fastq> <output_rev_unpaired.fastq> ILLUMINACLIP:TruSeq3-SE:2:30:10 LEADING:3 TRAILING:3 SLIDINGWINDOW:4:15 MINLEN:36
```

where output_fwd_paired and output_rev_paired.fastq are output fastq files for successfully trimmed reads that are paired between forward and reverse fastq reads and unpaired.fastq, the successfully trimmed reads whose mate pair was dropped by the tool for bad quality.

3.3.2 Aligning the Reads over Reference Genome (Split-Mapping Algorithm)

1. Install BWA. For Linux systems, you can use the command
   `sudo apt-get install bwa`.

2. Download the reference genome using UCSC genome browser in FASTA format (https://genome.ucsc.edu/cgi-bin/hgGateway?hgsid=609668381_BWFbDvyKkTHI4mIGa$32#Qur9o9rgKbl&redirect=manual&source=genome.ucsc.edu).

3. Index the reference genome:

   `bwa index -a bwtsw <Reference_genome.fa>`

4. Align the reads over the reference genome using BWA-MEM:
   `bwa mem -T 19 <ref.fa> <reads.fastq> > <aln-se.sam>` (for single end)
   `bwa mem -T 19 <ref.fa> <output_fwd_paired.fastq> <output_rev_paired.fastq> 1> <aln-pe.sam> 2> <aln-pe.log>` (for paired end).
   -T option is to give the minimum alignment score for the output (*see* **Note 10**).

3.3.3 Identification of CircRNAs Using CIRI

The next step is to run CIRI Perl script; before that we need the annotation file that can be downloaded from UCSC table browser or GENCODE annotations. To run CIRI, use

`perl CIRI2.pl -I <aln-se.sam> -O <outfile> -F <reference_genome.fa> -A <genome_annotation.gtf>`

where

-I	Input Sam file name generated by BWA-MEM
-O	Output circRNA list name
-F	FASTA file for reference sequences
-R	Directory containing the reference sequences
-G	Output log file
-A	Annotation file (GTF format)
-T	Number of threads to use

The manual for CIRI can be checked in the downloaded parent directory of the tool (*see* **Note 17**).

3.4 Merging and Comparing Annotations

To merge and compare the annotations among different tools, we need to put the prefix of the tool name with the circIDs/save the output in one specific folder and then first add the tool name to the output text files. Then use any text mining tool, for example Excel/tableau or customize one-line scripts to process the files.

1. Fetch the genome coordinates along with the tool name in the format given below:

   chromosomeName Start End Strand SoftwareName

   bed files for each sample compiled to one file as `<SoftwareName_circJunctions>` as above-mentioned format.
2. Run this script to compare the annotations by different tools:

```
gawk '{ key = $1 FS $2 FS $3 FS $4; names[$NF] = 1; has_name
[key][$5] = 1 }
    END {
    PROCINFO["sorted_in"] = "@ind_str_asc"
    printf "Name Value1 Value2 Strand"
    for (v in names) printf " %s", v
    print ""
    for (key in has_name) {
    printf "%s", key
    for (v in names) printf " %s", has_name[key][v] ? "Yes" : "No"
    print ""
    }
    }
' FindCirc_circJunctions CIRI_circJunctions Segemehl_circ-
Junctions > merged_CircRNA_file
```

3.5 Validation of Circular RNAs

Experimental validation is vital to estimate the specificities of different bioinformatics pipelines for identification of circular transcripts. There are different approaches which one could opt for the confirmation of the formation of circular transcripts from a given parent gene.

3.5.1 Experimental Validation of the Predicted Circular Transcripts Using Polymerase Chain Reaction (PCR)

Polymerase chain reaction is a simple but powerful technique to confirm the presence of back-spliced junctions in the circular transcripts. The circular transcripts can be differentiated from the linear transcripts by modifying the design of primers. The convergent primers are the canonical ones designed facing inwards to amplify the linear transcripts. The noncanonical divergent primers facing outwards can selectively amplify those types of transcripts which contain back-spliced junctions. Hence the divergent and convergent primers can be designed to selectively amplify the circular and linear transcripts, respectively, using coordinates predicted through the bioinformatics pipeline. The following methodology can be applied for PCR-based validation.

Primer Designing

Design and order the convergent primers facing outward near the junction which are predicted to get back spliced identified by the

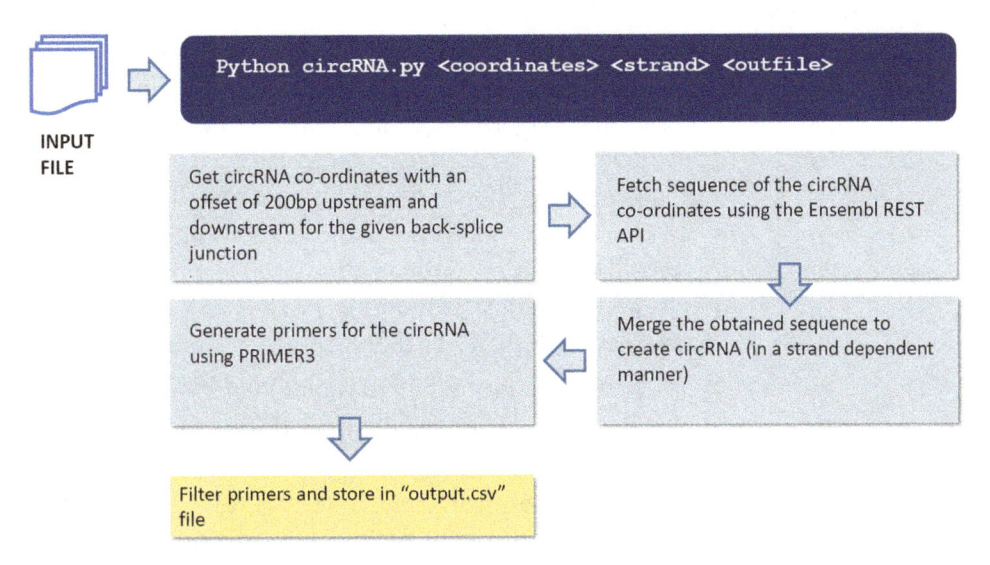

Fig. 2 Workflow to make primers using the Python script

bioinformatics analysis. The steps to be followed are shown in Fig. 2.

Designing PCR for circular RNAs.

Download link: https://github.com/judithhariprakash/circRNA_primer

circRNA primer is a Python program to generate primers for back-spliced circRNAjunctions. The only requirement for executing the program is Python 2.7, Python 3.x, or Python 2.6.

To execute the file:

```
python circRNA.py coordinates strand outfile
```

Upon user input of back-splice coordinates for circular RNA in GRCh38 along with strand information (exp: 10:25,000,000–25,001,000, +), the program generates inverse primers around the back-spliced coordinates with an offset of 200 bp upstream and downstream. The primer sequence, GC content, Tm, and sequence length will be stored to a .csv output file.

The user can change the species and/or the version of the genome in the script if necessary. The primers are designed to create a product size of 100–400 bp, with primer length between 18 and 23 bp and Tm ranging from 57 to 63°. Other specifications for the primer designing can be viewed in circrna_primers_settings.p3 file.

RNA Isolation from the Specific Tissue/Cells

1. Lyse the tissue/cells using 500 μL of Trizol reagent. For efficient lysis grind the tissue/cells using homogenizer and/or tincture using 26G needle.

2. After the lysis, separate the proteins using chloroform-based phase separation at room temperature. Add 200 μL of chloroform, mix vigorously, and incubate for 10 min at room temperature.
3. Centrifuge the above mix at 13,200 RPM (18,000 × g) for 15 min at 4 °C and transfer the aqueous phase to a fresh vial carefully.
4. Precipitate the RNA using 500 μL of isopropanol and pellet it down at 13,200 RPM (18,000 × g) for 15 min at 4 °C.
5. Wash the pellet using 500 μL of ice-cold 70% ethanol.
6. Finally dissolve the pellet using 20–100 μL of nuclease-free water (NFW).
7. For DNase treatment, treat with 0.1 U/μL of Turbo DNase for 15 min as per the manufacturer's instructions.
8. Repurify the RNA using **steps 2–5** again.
9. Finally dissolve the pellet using 20–100 μL of nuclease-free water.
10. The isolated RNA can be stored at −20 °C.

Quality Check of the Isolated RNA

RNA once isolated can be quality checked using the following methods.

First, we check RNA integrity by agarose gel electrophoresis.

1. Mix 1 μL of the isolated RNA with 8 μL of nuclease-free water and 1 μL of 2X RNA loading dye.
2. Incubate the above mix at 65 °C for 5 min followed by a snap chill of 10 min on ice.
3. Load the snap-chilled RNA on 1% agarose gel in 1× TAE buffer and run at 100 volts (V) for 25 min.
4. Intact RNA can be visualized on the agarose gel as three distinct bands corresponding to 28S, 18S, and 5S rRNA. The demonstration of gel image is shown in Fig. 3a.

Then, we performed quantitative analysis of RNA quality.

1. Prepare the RNA samples at a dilution of 1:100 using nuclease-free water and measure the absorbance, A260/280 and A260/230, in a spectrophotometer.

Complementary First-Strand Synthesis

1. Normalize the RNA to a final concentration of 500 ng/μL.
2. Mix 1 μL of RNA with 1 μL (500 ng/μL) of random hexamer primers, make up the volume to 12 μL with NFW, and incubate it at 70 °C for 5 min, followed by 1 min at 4 °C.
3. To this mixture, add 7 μL of the reverse transcriptase reaction mixture and incubate at 42 °C for 5 min. Reverse transcriptase

Fig. 3 The figure shows (**a**) gel image of RNA. (**b**) Demonstration of gel image for circular RNA validation PCR with *actb* as control

mixture comprises 2 μL of 10× first-strand buffer, 2 μL 0.1 M DTT, 2 μL of 25 mM MgCl$_2$, and 1 μL of 10 mM equimolar dNTPs.

4. Add 0.5 μL of SuperScript II Reverse Transcriptase (SSIIRT) enzyme to the above reaction mixture and incubate at 42 °C for 50 min, followed by 15 min at 70 °C.

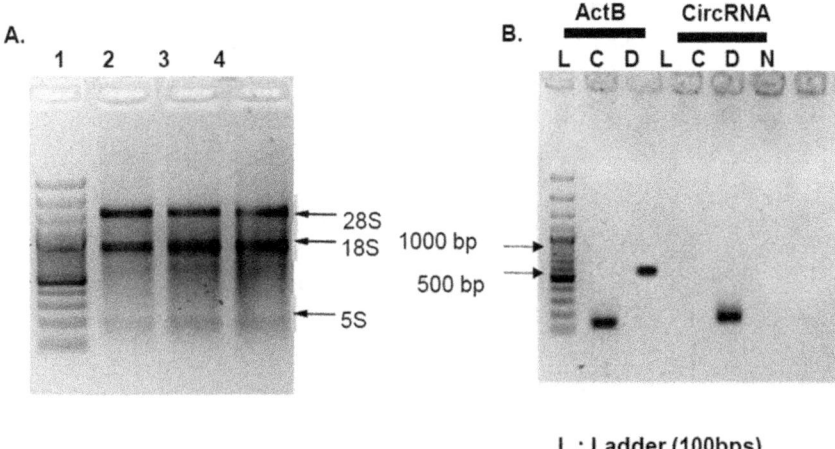

5. Add 0.5 μL of RNase H enzyme to the above reaction mixture and incubate at 37 °C for 20 min. The final cDNA can be stored at −20 °C until further use.

Validation by PCR Amplification

Using the convergent primers designed around the back-spliced junctions amplify the transcripts from the cDNA synthesized from RNA isolated from respective tissues.

1. For the PCR amplification set the reaction as shown in Table 2.
2. Genomic DNA can be used as a control in the reaction, to exclude false positives.
3. Run the reaction mixtures as shown in Table 3.
4. After the PCR reactions run the samples on 1.5% agarose gel in 1X TAE for 30 min at 100 volts (V).
5. If the prediction is correct, we should get an amplified product when cDNA is used as template. There should not be any amplification in genomic DNA template. PCR product with cDNA template indicates the joining of back-splice junctions at RNA level leading to the formation of circular RNAs.

Table 2
Demonstration of PCR amplification reaction to be used

10× PCR buffer	2.5 μL
25 mM MgCl$_2$	1.5 μL
10 mM dNTPs	1 μL
10 μM Forward primer	1 μL
10 μM Reverse primer	1 μL
Taq Pol (5 U/μL)	0.5 μL
cDNA/DNA	1 μL
NFW	16.5 μL
Total reaction volume	25 μL

Table 3
Conditions for PCR reaction steps with temperature and time

Steps	Temperature (°C)	Time	Cycle
Initial denaturation	95	3 min	1
Denaturation	95	30 s	35
Annealing	56	30 s	
Extension	72	30 s	
Final extension	72	7 min	1
Hold	4	∞	

6. Amplification of housekeeping gene such as actin B (actb) gene in cDNA and DNA sample can be used as a positive control for the PCR amplification. The presence of product in only cDNA sample validates the presence of circular transcript. A prototypic gel image is shown in Fig. 3b.

3.5.2 RNase R Enzyme Treatment and Validation of Circular Transcripts

RNase R is an exonuclease, which cleaves all the linear transcripts. After the digestion of RNA transcripts with the RNase R enzyme the circular transcripts remain safe. To validate the circular transcripts using RNase R enzyme the following methodology can be adopted.

RNA Isolation and Quality Check

For RNA isolation follow the same protocol as mentioned in Subheading 3.5.1, **steps 2** and **3**.

RNase R Treatment

For RNase R set the reaction as shown in Table 4.

Table 4
Reaction composition for RNase R treatment

10× Buffer	2 μL
RNase R enzyme[a]	0.5 μL
RNA (2.5 μg)	2.5 μg
NFW	Up to 20 μL
Total	20 μL

Incubate the above reaction mixture at 37 °C
[a]Set one control reaction also without enzyme

Purification of RNA

1. After the digestion of RNA with RNase R, add 10 μL of 7.5 M LiCl to the reaction mixture and incubate at −80 °C for at least 2 h.
2. After incubation centrifuge the mixture at 14,000 RPM (20,000 × g) for 15 min at 4 °C.
3. Discard the supernatant and add 500 μL of ice-cold 70% ethanol to the pellet.
4. Centrifuge at 14,000 RPM (20,000 × g) for 10 min at 4 °C.
5. Discard the supernatant, dissolve the pellet in 10 μL of NFW, and store in −20 °C.

Complementary First-Strand Synthesis

Synthesize the cDNA of RNase R-digested and control sample using the same method as mentioned in Subheading 3.5.1, **step 4**.

PCR Amplification for the Validation

Amplify the queried loci using divergent primers on cDNA synthesized above using the same methodology as mentioned in Subheading 3.5.1, **step 5**.

4 Notes

1. It is advised to use Linux operating system to run the scripts for ease of availability and installation of various modules.
2. While running the scripts, error might occur because of older or different versions in Perl/Python. Make sure that you have checked and installed the correct versions of modules.
3. RNA-sequencing reads should be polyA/ribo- and preferable RNase R treated. This will help in better coverage of circRNAs.
4. All the three pipelines have test datasets within the software package. Run the test package before running the sample of interest.
5. For large datasets, it is advised to use workstation with high RAM or high-performance computing facilities.

6. Annotation file should be in GTF or GFF format.
7. It should be taken under proper consideration that reference genome and annotation file must be of the same version.
8. It is difficult to identify circular RNAs from poorly annotated or de novo-assembled genomes, due to insufficient information.
9. Large datasets generates bigger SAM output file. So it is advised to have enough storage space beforehand.
10. Do not sort/modify/manipulate the SAM output generated from the alignment in any pipelines as the pipelines are designed to take the file in a specific format. If any changes are required, this must be already added in the script.
11. If, while running FindCirc, the following error shows: [bam_-header_read] EOF marker is absent. The input is probably truncated. This error is because of incompatible samtools version. Please check if the samtools version installed in your system fulfills the requirement of the pipeline you want to use.
12. Upload fastq reads on FastQC, quality control tool of fastq files (https://www.bioinformatics.babraham.ac.uk/projects/fastqc/). This will provide base quality as well as other related information about reads for example, read length, GC distribution, duplicate reads, adapters or any other overrepresented sequence, and presence of K-mers.
13. It is advised to remove K-mers, if present in the sequence. This affects the proper alignment of the reads and data can be lost. K-mers can be removed using Trimmomatic with options like HEADCROP with number of base pairs to crop.
14. Remove overrepresented sequences and adapters before aligning.
15. Make sure that you have good-quality RNA-sequencing reads. Bad-quality reads can affect the alignment and accuracy.
16. Trim the reads with phred score >20 and preferably >30.
17. Read the manuals for the scripts before using the tool.

References

1. Lasda E, Parker R (2014) Circular RNAs: diversity of form and function. RNA 20:1829–1842. https://doi.org/10.1261/rna.047126.114
2. Cocquerelle C, Mascrez B, Hetuin D, Bailleul B (1993) Mis-splicing yields circular RNA molecules. FASEB J 7:155–160
3. Kulcheski FR, Christoff AP, Margis R (2016) Circular RNAs are miRNA sponges and can be used as a new class of biomarker. J Biotechnol 238:42–51. https://doi.org/10.1016/j.jbiotec.2016.09.011
4. Qu S, Yang X, Li X et al (2015) Circular RNA: a new star of noncoding RNAs. Cancer Lett 365:141–148. https://doi.org/10.1016/j.canlet.2015.06.003
5. Vidal AF, Sandoval GTV, Magalhaes L et al (2016) Circular RNAs as a new field in gene regulation and their implications in translational research. Epigenomics 8:551–562. https://doi.org/10.2217/epi.16.3

6. Kos A, Dijkema R, Arnberg AC et al (1986) The hepatitis delta (delta) virus possesses a circular RNA. Nature 323:558–560. https://doi.org/10.1038/323558a0
7. Maoz R, Garfinkel BP, Soreq H (2017) Alzheimer's disease and ncRNAs. Adv Exp Med Biol 978:337–361. https://doi.org/10.1007/978-3-319-53889-1_18
8. Kumar L, Shamsuzzama, Jadiya P, et al (2018) Functional characterization of novel circular RNA molecule, circzip-2 and its synthesizing gene zip-2 in C. elegans model of Parkinson's disease. Mol Neurobiol 55 6914–6926. doi: https://doi.org/10.1007/s12035-018-0903-5
9. Wang H, Yang J, Yang J et al (2016) Circular RNAs: novel rising stars in cardiovascular disease research. Int J Cardiol 202:726–727
10. Jeck WR, Sharpless NE (2014) Detecting and characterizing circular RNAs. Nat Biotechnol 32:453–461. https://doi.org/10.1038/nbt.2890
11. Langmead B, Salzberg SL (2012) Fast gapped-read alignment with Bowtie 2. Nat Methods 9:357–359. https://doi.org/10.1038/nmeth.1923
12. Li H, Durbin R (2009) Fast and accurate short read alignment with Burrows-Wheeler transform. Bioinformatics 25:1754–1760. https://doi.org/10.1093/bioinformatics/btp324
13. Dobin A, Gingeras TR (2015) Mapping RNA-seq reads with STAR. Curr Protoc Bioinformatics 51:11.14.1–11.1419. https://doi.org/10.1002/0471250953.bi1114s51
14. Bray NL, Pimentel H, Melsted P, Pachter L (2016) Near-optimal probabilistic RNA-seq quantification. Nat Biotechnol 34:525–527. https://doi.org/10.1038/nbt.3519
15. Trapnell C, Pachter L, Salzberg SL (2009) TopHat: discovering splice junctions with RNA-Seq. Bioinformatics 25:1105–1111. https://doi.org/10.1093/bioinformatics/btp120
16. Kim D, Salzberg SL (2011) TopHat-Fusion: an algorithm for discovery of novel fusion transcripts. Genome Biol 12:R72. https://doi.org/10.1186/gb-2011-12-8-r72
17. Wang K, Singh D, Zeng Z et al (2010) MapSplice: accurate mapping of RNA-seq reads for splice junction discovery. Nucleic Acids Res 38:e178. https://doi.org/10.1093/nar/gkq622
18. Hansen TB, Veno MT, Damgaard CK, Kjems J (2016) Comparison of circular RNA prediction tools. Nucleic Acids Res 44:e58. https://doi.org/10.1093/nar/gkv1458
19. Memczak S, Jens M, Elefsinioti A et al (2013) Circular RNAs are a large class of animal RNAs with regulatory potency. Nature 495:333–338. https://doi.org/10.1038/nature11928
20. Hoffmann S, Otto C, Doose G et al (2014) A multi-split mapping algorithm for circular RNA, splicing, trans-splicing and fusion detection. Genome Biol 15:R34. https://doi.org/10.1186/gb-2014-15-2-r34
21. Gao Y, Wang J, Zhao F (2015) CIRI: an efficient and unbiased algorithm for de novo circular RNA identification. Genome Biol 16:4. https://doi.org/10.1186/s13059-014-0571-3
22. Zhang X-O, Wang H-B, Zhang Y et al (2014) Complementary sequence-mediated exon circularization. Cell 159:134–147. https://doi.org/10.1016/j.cell.2014.09.001
23. Dang Y, Yan L, Hu B et al (2016) Tracing the expression of circular RNAs in human pre-implantation embryos. Genome Biol 17:130. https://doi.org/10.1186/s13059-016-0991-3
24. Chen L, Yu Y, Zhang X et al (2016) PcircRNA_finder: a software for circRNA prediction in plants. Bioinformatics 32:3528–3529. https://doi.org/10.1093/bioinformatics/btw496
25. Cheng J, Metge F, Dieterich C (2016) Specific identification and quantification of circular RNAs from sequencing data. Bioinformatics 32:1094–1096. https://doi.org/10.1093/bioinformatics/btv656
26. Song X, Zhang N, Han P et al (2016) Circular RNA profile in gliomas revealed by identification tool UROBORUS. Nucleic Acids Res 44:e87. https://doi.org/10.1093/nar/gkw075
27. Chuang T-J, Wu C-S, Chen C-Y et al (2016) NCLscan: accurate identification of non-co-linear transcripts (fusion, trans-splicing and circular RNA) with a good balance between sensitivity and precision. Nucleic Acids Res 44:e29. https://doi.org/10.1093/nar/gkv1013
28. Meng X, Chen Q, Zhang P, Chen M (2017) CircPro: an integrated tool for the identification of circRNAs with protein-coding potential. Bioinformatics 33:3314–3316. https://doi.org/10.1093/bioinformatics/btx446
29. Li M, Xie X, Zhou J et al (2017) Quantifying circular RNA expression from RNA-seq data using model-based framework. Bioinformatics 33:2131–2139. https://doi.org/10.1093/bioinformatics/btx129
30. Wang D (2017) hppRNA-a Snakemake-based handy parameter-free pipeline for RNA-Seq analysis of numerous samples. Brief Bioinform 19:622–626. https://doi.org/10.1093/bib/bbw143

31. Szabo L, Morey R, Palpant NJ et al (2015) Statistically based splicing detection reveals neural enrichment and tissue-specific induction of circular RNA during human fetal development. Genome Biol 16:126. https://doi.org/10.1186/s13059-015-0690-5
32. Izuogu OG, Alhasan AA, Alafghani HM et al (2016) PTESFinder: a computational method to identify post-transcriptional exon shuffling (PTES) events. BMC Bioinformatics 17:31. https://doi.org/10.1186/s12859-016-0881-4
33. You X, Conrad TO (2016) Acfs: accurate circRNA identification and quantification from RNA-Seq data. Sci Rep 6:38820. https://doi.org/10.1038/srep38820
34. Metge F, Czaja-Hasse LF, Reinhardt R, Dieterich C (2017) FUCHS-towards full circular RNA characterization using RNAseq. PeerJ 5:e2934. https://doi.org/10.7717/peerj.2934
35. Westholm JO, Miura P, Olson S et al (2014) Genome-wide analysis of drosophila circular RNAs reveals their structural and sequence properties and age-dependent neural accumulation. Cell Rep 9:1966–1980. https://doi.org/10.1016/j.celrep.2014.10.062

Chapter 4

Methods to Study Long Noncoding RNA Expression and Dynamics in Zebrafish Using RNA Sequencing

Samatha Mathew, Ambily Sivadas, Paras Sehgal, Kriti Kaushik, Shamsudheen K. Vellarikkal, Vinod Scaria, and Sridhar Sivasubbu

Abstract

Long noncoding RNAs (lncRNAs) belong to a class of RNA transcripts that do not have the potential to code for proteins. LncRNAs were largely discovered in the transcriptomes of human and several model organisms, using next-generation sequencing (NGS) approaches, which have enabled a comprehensive genome scale annotation of transcripts. LncRNAs are known to have dynamic expression status and have the potential to orchestrate gene regulation at the epigenetic, transcriptional, and posttranscriptional levels. Here we describe the experimental methods involved in the discovery of lncRNAs from the transcriptome of a popular model organism zebrafish (*Danio rerio*). A structured and well-designed computational analysis pipeline subsequent to the RNA sequencing can be instrumental in revealing the diversity of the lncRNA transcripts. We describe one such computational pipeline used for the discovery of novel lncRNA transcripts in zebrafish. We also detail the validation of the putative novel lncRNA transcripts using qualitative and quantitative assays in zebrafish.

Key words RNA sequencing, Transcriptome, Noncoding RNA, Long noncoding RNA, Zebrafish

1 Introduction

Zebrafish (*Danio rerio*) has emerged as an excellent model organism for exploring the vertebrate non-protein-coding transcriptome. The employment of high-throughput deep sequencing approaches to characterize the transcriptome of zebrafish has revealed a previously unidentified subset of transcripts that do not have the potential to code for proteins. These transcripts are largely classified as noncoding RNAs (ncRNAs). The long noncoding RNAs (lncRNAs) constitute one of the largest subset of the ncRNAs in zebrafish (*Reviewed in ref.* 1). In recent years several studies have dissected the transcriptome of zebrafish using RNA sequencing approaches [2–4] with the intention to identify lncRNAs. Several candidate lncRNAs derived from these independent studies have been validated for their tissue-specific expression patterns and functionalities. LncRNAs have thus been implicated to be of fundamental importance in the development, differentiation, and maintenance of zebrafish tissues and organs.

We describe here the methodology to obtain high-quality RNA from zebrafish embryos and tissues for preparing cDNA libraries for next-generation sequencing application. The experimental protocols employed for zebrafish RNA sequencing is explained in detail, following which we also describe a tailored and structured bioinformatics workflow for processing the data obtained from the RNA sequencing studies for enabling meaningful insights. Further the methodologies for experimental validation of the putative zebrafish lncRNAs through quantitative RT-PCR and whole mount in situ hybridization are also detailed. In addition to confirming the existence of the lncRNAs in zebrafish, these techniques also provide insights into the dynamic and spatio-temporal expression profiles of lncRNAs.

Briefly, this chapter provides the methodologies for zebrafish RNA isolation, RNA sequencing, and the subsequent bioinformatics analyses for reliable and efficient discovery of annotated and un-annotated RNA transcripts including lncRNAs. The identities of these computationally predicted transcripts can be validated using molecular and biological assays in zebrafish. Note that the NGS library preparation kits and bioinformatic tools mentioned in this chapter may be updated time to time by developers, the details of which will be available at the primary source.

2 Materials and Equipment

2.1 RNA Isolation, cDNA Preparation, and Quality Check (QC) Reagents

1. TRIzol (Invitrogen, USA).
2. Tricaine (Sigma-Aldrich, USA).
3. 10× phosphate buffered saline (PBS): 1.37 M NaCl, 27 mM KCl, 100 mM Na_2HPO_4, 18 mM KH_2PO_4, pH 7.4. Dilute autoclaved 10× PBS to 1× with RO grade water.
4. Homogenizer (Thermo Scientific, USA).
5. Chloroform.
6. Absolute ethanol (Molecular grade).
7. RNeasy Mini kit (Qiagen, USA).
8. Nuclease-free water (NFW).
9. RNA loading dye.
10. Agarose.
11. 50× Tris-acetate-EDTA buffer (TAE): 2 M Tris acetate, 0.05 M EDTA, pH 8.4. Dilute to 1× with RO grade water.
12. 100 bp DNA ladder.
13. Spectrophotometer (Biospectrometer, Eppendorf, Germany).
14. Random hexamers (Invitrogen, USA).

15. SuperScript II Reverse Transcriptase (Invitrogen, USA).
16. RNase H (Invitrogen, USA).
17. DNA gel loading dye.

2.2 mRNA Library Preparation and RNA Sequencing Reagents

1. Thermal cycler.
2. Truseq RNA sample preparation kit v2 (Illumina Inc., USA).
3. Magnetic stand (1.5 mL magnetic stand, Invitrogen, USA).
4. SuperScript II Reverse Transcriptase (Invitrogen, USA).
5. Agencourt AMPure XP beads (Beckman Coulter Genomics, USA).
6. Absolute ethanol (Molecular grade).
7. Certified agarose gel (Affymetrix USB, USA).
8. 50× Tris-Acetate-EDTA buffer (TAE): 2 M Tris acetate, 0.05 M EDTA, pH 8.4. Dilute to 1× with RO grade water.
9. Ultra pure ethidium bromide (Invitrogen, USA).
10. DNA gel loading dye.
11. 100 bp DNA ladder.
12. Qubit apparatus (Qubit fluorometer, Invitrogen, USA).
13. High sensitivity DNA (HS-DNA) Qubit kit (Invitrogen, USA).
14. Sodium hydroxide.
15. TruSeq PE Cluster Kit v2-cBot-HS (Illumina, USA).
16. Elution buffer (Qiagen Inc., Netherlands).
17. Illumina NGS platforms: Novaseq 6000, X Ten, Hiseq 2500, Nextseq, Miseq, GAIIx, etc. (Illumina Inc. San Diego, USA).
18. Software: listed in Table 1.

2.3 Bioinformatic Analysis Pipeline Requisites for lncRNA Prediction from RNA Sequencing Data

1. Hardware:
   (a) Operating system: Linux or Macintosh OS X with Internet connection.
   (b) System configuration: at least four to eight cores with 8 GB RAM and at least 50 GB free space on hard drive.

Table 1
Software required for data generation from Illumina Hi-seq 2500 based RNA sequencing

Tool	Version	Download URL
Illumina Experimental Manager	IEM v1.8	http://support.illumina.com/sequencing/sequencing_software/experiment_manager/downloads.html
BCL2FASTQ	v1.8.4	http://support.illumina.com/downloads/bcl2fastq_conversion_software_184.html

Table 2
Software required for the bioinformatics analysis of the reads generated from Illumina Hi-seq 2500 based RNA sequencing for the prediction of lncRNAs

Tool	Version	Download URL
FastQC [5]	0.11.2	http://www.bioinformatics.babraham.ac.uk/projects/fastqc/
Trimmomatic [6]	0.32	http://www.usadellab.org/cms/?page=trimmomatic
Tophat [7]	2.0.13	http://ccb.jhu.edu/software/tophat/index.shtml
Cufflinks [8]	2.2.1	http://cole-trapnell-lab.github.io/cufflinks/
GetORF [EMBOSS] [9]	6.6.0	http://emboss.sourceforge.net/download/#Stable/
CPC [10]	0.9	http://cpc.cbi.pku.edu.cn/download/
PhyloCSF [11]	–	https://github.com/mlin/PhyloCSF/wiki
BedTools [12]	2.22.0	https://github.com/arq5x/bedtools2/releases/tag/v2.22.0
HMMER3 [13]	3.1b1	http://hmmer.janelia.org/
BLASTX [14]	2.2.30	ftp://ftp.ncbi.nlm.nih.gov/blast/executables/blast+/LATEST/

Table 3
Files required for the prediction and annotation of lncRNAs from Illumina Hi-seq 2500 based RNA sequencing data

File	Format	Source	Download URL
Reference DNA sequence (Zv9)	Fasta	UCSC Genome Browser	http://hgdownload.soe.ucsc.edu/goldenPath/danRer7/bigZips/
Reference transcriptome annotation (RefSeq)	GTF	UCSC Table Browser	http://genome.ucsc.edu/cgi-bin/hgTables?command=start (Select *clade:*Vertebrate *genome:*Zebrafish assembly: Jul.2010 (Zv9/danRer7) *group:* Genes and Gene Predictions *track:* RefSeq Genes *output format:* GTF)
PFAM protein database [15]	Fasta	PFAM	ftp://ftp.ebi.ac.uk/pub/databases/Pfam/current_release

2. Software: listed in Table 2.
3. Prerequisite files: listed in Table 3.

2.4 Quantitative Real-Time PCR and Whole Mount In Situ Hybridization Reagents

1. LightCycler 480 Real-Time PCR system (Roche Life Science, Germany).
2. LightCycler 480 SYBR Green I Master (Roche Life Science, Germany).
3. MiniElute PCR purification kit (Qiagen, USA).
4. TOPO-TA cloning kit (Life technologies, USA).

5. Luria Bertani Agar.
6. Luria Bertani Broth.
7. Carbenicillin.
8. QIAprep Spin Miniprep kit (Qiagen, USA).
9. Nuclease-free water (NFW).
10. *Eco*RI (NEB, UK).
11. *Not*I (NEB, UK).
12. *Stu*I (NEB, UK).
13. Agarose.
14. 50× Tris-Acetate-EDTA buffer (TAE): 2 M Tris acetate, 0.05 M EDTA, pH 8.4. Dilute autoclaved to 1× with RO grade water.
15. 1 kb DNA Ladder.
16. MEGAscript™ T7 Transcription Kit (Invitrogen, USA).
17. MEGAscript™ T3 Transcription Kit (Invitrogen, USA).
18. Absolute ethanol (molecular grade).
19. Ammonium acetate.
20. Chloroform.
21. Phenol:Chloroform:Isoamyl alcohol.
22. RNA loading dye.
23. Nanodrop spectrophotometer (Nanodrop1000 Spectrophotometer, Life Technologies, USA).
24. Pronase (Roche Life Science, Germany).
25. Paraformaldehyde (PFA).
26. 4% PFA: prepare by dissolving 100% PFA in 1× PBS. Aliquot into 10 mL fractions and store at −20 °C until use. Aliquots once thawed can be stored at 4 °C.
27. Phosphate buffered saline-Tween-20 (PBT) solution: 0.1% Tween-20 in 1× PBS, pH 7.4.
28. Methanol.
29. Proteinase K.
30. 20× Saline Sodium-citrate buffer (SSC): 3 M NaCl, 300 mM Sodium citrate, pH 7.0.
31. Heparin.
32. Yeast tRNA (Ambion, Life technologies, USA).
33. Formamide.
34. Hybridization mix (HM) with tRNA and Heparin: 50% Formamide (v/v), 5× SSC, 50 μg/μL Heparin, 500 μg/μL tRNA, 0.1% Tween 20 (v/v), 1% 1 M Citric Acid (v/v).

35. Hybridization mix without tRNA and Heparin: 50% Formamide (v/v), 5× SSC, 0.1% Tween 20 (v/v), 1% 1 M Citric Acid (v/v).
36. Bovine Serum Albumin (BSA).
37. Fetal Bovine Serum (FBS).
38. Blocking solution: 2% FBS and 2 mg/mL BSA in PBT.
39. Alkaline phosphatase-tagged anti-digoxigenin (DIG) antibody (Roche Life Science, Germany).
40. Alkaline Tris buffer: 100 mM Tris–HCl, pH 9.5, 50 mM $MgCl_2$, 100 mM NaCl, 0.1% Tween-20 (v/v).
41. Nitro Blue Tetrazolium (NBT) (Sigma-Aldrich, USA).
42. 5-Bromo-4-chloro-3-indolyl phosphate *p*-toluidine salt (BCIP) (Sigma-Aldrich, USA).
43. NBT/BCIP Labeling: 0.225 μg/mL NBT and 0.175 μmg/mL BCIP in Alkaline Tris buffer.
44. Methyl cellulose (Sigma-Aldrich, USA).
45. 1.5% methyl cellulose: 1.5% (w/v) methyl cellulose in RO grade water.

3 Methods

3.1 RNA Isolation

High-quality RNA is the prerequisite for good sequencing outputs. The trick to obtain intact and pure RNA starts with careful preservation of the source tissue, prevention of RNase contamination, and optimal storage of the isolated RNA. A thorough quality check of the RNA ensures derivation of conclusive and reliable data from RNA sequencing.

3.1.1 Isolation and Preservation of Zebrafish Embryos and Tissues

1. Transfer the early embryonic stages of zebrafish to 1.5 mL microfuge tubes (approximately 30 animals per tube), remove excess water, add 500 μL TRIzol reagent, and immediately freeze at −20 °C. For longer storage, transfer to −80 °C (*see* **Note 1**).

2. Anesthetize adult zebrafish by treatment with 0.4% Tricaine and dissect out individual tissues according to the requirement (minimum 3 mg of tissue can yield about 4 μg of RNA). Wash the tissues in 1× PBS several times to clean up any debris. Place the tissues in 1.5 mL microfuge tubes, remove excess PBS, add 500 μL TRIzol, and immediately freeze at −20 °C. For longer storage, transfer to −80 °C (*see* **Note 2**).

3.1.2 RNA Isolation Using RNeasy Mini Kit

1. Homogenize the fresh or frozen tissue samples in TRIzol using a homogenizer with pestle or by the manual method. In the manual method first homogenize using sterile pestle, followed by resuspension using sterile syringes with needles, till the tissue is entirely homogenized. Incubate at room temperature (RT) for 5–10 min.

2. Add 200 μL of chloroform and gently mix by flicking the tube; place at RT for 10 min.

3. Centrifuge at ≥8000 × *g* at 4 °C for 15 min and transfer the supernatant to a fresh tube. Add 500 μL of 70% ethanol and mix by gently inverting the tube several times.

4. Add 700 μL of the homogenate to the Qiagen column provided with the kit and centrifuge at ≥8000 × *g* for 1 min at RT. Discard the flow through. Repeat the step for the remaining volume of the homogenate.

5. Add 500 μL of wash buffer I (RWI) to the column and centrifuge at ≥8000 × *g* for 1 min. Discard the eluted fraction.

6. Add 500 μL of wash buffer II (RPE buffer added with absolute alcohol as per the manufacturer's instructions) to the column and centrifuge at ≥8000 × *g* for 1 min. Repeat the step one more time.

7. Centrifuge at ≥8000 × *g* for 1 min to remove the traces of wash buffer II.

8. Place the column in a 1.5 mL microfuge tube and add 30 μL of nuclease-free water (NFW) at the center of the column and incubate for 2 min at RT (*see* **Note 3**).

9. Centrifuge at ≥8000 × *g* for 1 min at RT.

3.1.3 Quality Check of the Isolated RNA

The quality of the isolated RNA can be checked using the following methods.

1. *Checking RNA integrity by agarose gel electrophoresis.*
   (a) Mix 1 μL of the isolated RNA with 8 μL of NFW and 1 μL of RNA gel loading dye.
   (b) Incubate the above mix at 65 °C for 5 min followed by a snap chill of 10 min on ice.
   (c) Load the snap chilled RNA on 1.5% agarose gel in 1× TAE buffer and run at 100 volts (V) for 25 min.
   (d) Intact RNA can be visualized on the agarose gel as three distinct bands corresponding to 28S, 18S, and 5S rRNA (refer Fig. 1) (*see* **Note 4**).

2. *Quantitative analysis of RNA quality*
   (a) Prepare the RNA samples at a dilution of 1:100 using NFW and measure the absorbance in a spectrophotometer (*see* **Note 5**).

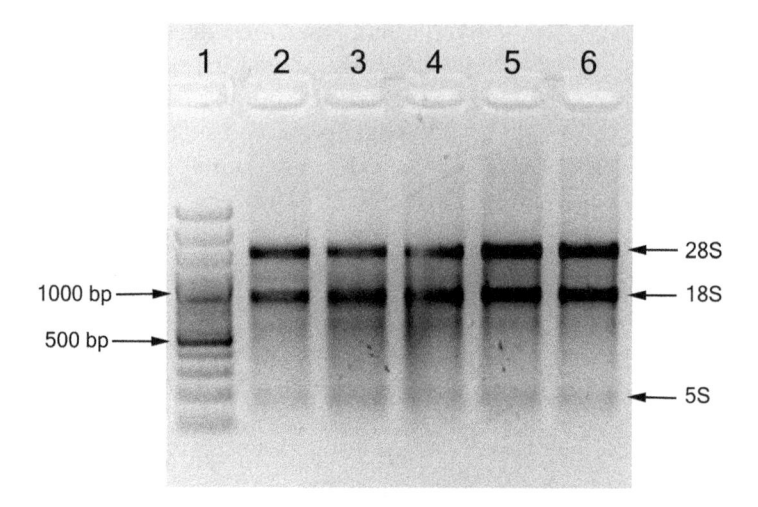

Fig. 1 Gel electrophoresis of isolated RNA samples on 1% agarose gel showing intact RNA. *1*—100 bp DNA ladder (Thermo Scientific, USA), *2–6*—isolated RNA samples. The intact bands corresponding to 28S, 18S, and 5S rRNA confirm the integrity of the RNA

    (b) On confirmation of acceptable RNA quality, proceed to validation and complementary DNA (cDNA) library preparation for Illumina-based RNA sequencing.

3. *Quality check of tissue samples through PCR*
   Prepare the cDNA from individual RNA samples using the cDNA preparation reagents as described below.

   (a) Normalize the RNA to a final concentration of 500 ng/μL.

   (b) Mix 1 μL of RNA with 1 μL of random hexamer primers and make up the volume to 12 μL with NFW and incubate it at 70 °C for 5 min, followed by 1 min at 4 °C.

   (c) To this mixture, add 7 μL of the reverse transcriptase reaction mixture and incubate at 42 °C for 5 min. Reverse transcriptase mixture comprises 2 μL of 10× first strand buffer, 2 μL 0.1 M Dithiothreitol (DTT), 2 μL of 25 mM MgCl$_2$, and 1 μL of 10 mM equimolar dNTPs.

   (d) Add 0.5 μL of SuperScript II Reverse Transcriptase (SSIIRT) enzyme to the above reaction mixture and incubate at 42 °C for 50 min, followed by 15 min at 70 °C.

   (e) Add 0.5 μL of RNase H enzyme to above reaction mixture and incubate at 37 °C for 20 min. The final cDNA can be stored at −20 °C until further use.

   (f) Check for the expression of key marker genes (such as *cmlc2* for heart, *tft* for liver, *murcb* for muscle, *mdka* for brain, and *tal1* for blood) in individual tissue cDNAs by PCR amplification using gene-specific primers. The

LncRNAs in Zebrafish 85

amplification of an expected product from only the corresponding tissue cDNA confirms the absence of cross-contamination from another tissue source.

3.2 RNA Sequencing and Data Generation

The preparation of libraries is one of the most crucial steps for RNA sequencing. The method of library preparation for sequencing is instrumental in determining the quality of data, abundance of transcripts, and uniform coverage. Broadly, there are two methods to prepare the total RNA library for RNA sequencing to identify lncRNAs: by ribosomal RNA depletion, which captures both poly-A and non-poly-A tailed transcripts, or by enrichment of poly-A tailed transcripts. Here we have described the poly-A enrichment based RNA library preparation for RNA sequencing using the Illumina platform (refer Fig. 2). Poly-A enriched RNA sequencing can be performed by using the oligo-dT-conjugated streptavidin magnetic beads. All the protocols are standardized for 1–2 μg of

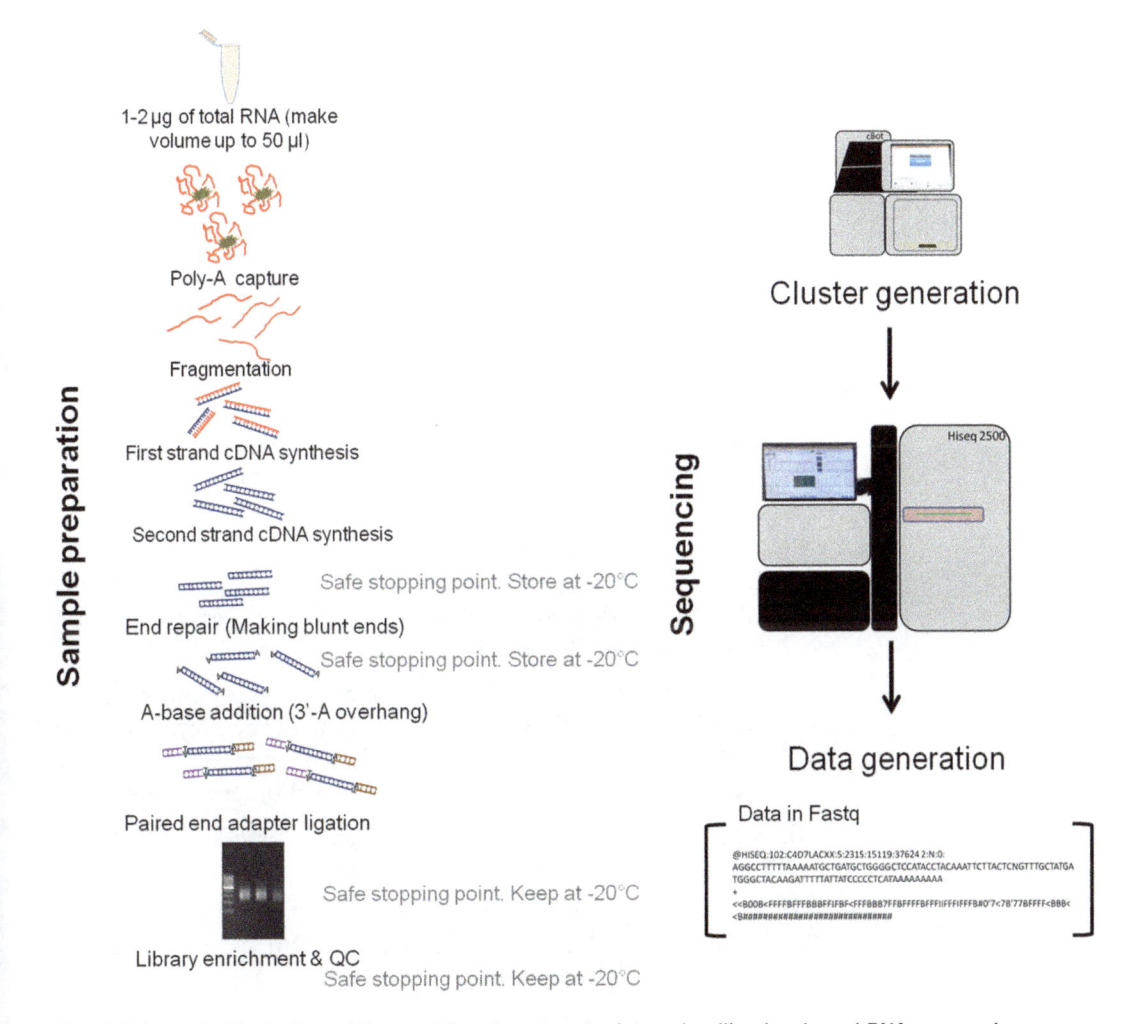

Fig. 2 Schematic illustration of the workflow to generate data using Illumina-based RNA sequencing

total RNA. (If the RNA concentrations are varying, the reagent utility should also be changed accordingly.) Sample preparation can be performed using Truseq RNA sample preparation kit v2, according to the manufacturer supplied protocol (Illumina Inc., USA, Part#15026495 Rev. F).

3.2.1 Poly-A Enriched RNA Library Preparation

*Poly-A capture and fragmentation using Illumina TruSeq RNA sample preparation kit v2 (see **Note 6**).*

1. Set the PCR thermal cycler temperature at 65 °C for 5 min followed by 4 °C on hold. Pause when the temperature reaches 65 °C.

2. Dilute 1–2 μg of high-quality, DNA-free total RNA to 50 μL by using ultrapure water (provided in the kit).

3. Add 50 μL of RNA purification beads (poly-dT-conjugated streptavidin beads supplied) into each sample and mix well by gently pipetting up and down at least five to six times (*see* **Note 7**).

4. Place the tubes in the thermal cycler and resume the run with the heated lid option.

5. Once the run is completed, remove the tubes from thermal cycler and place them on a magnetic stand for 5 min in order to separate the poly-A RNA fraction bound beads. Then remove the supernatant completely without disturbing the pellet.

6. Set the thermal cycler at 80 °C for 2 min followed by 25 °C on hold and start the program. Pause the run once the temperature reaches 80 °C.

7. Remove the tubes from the magnetic stand and add 200 μL of bead washing buffer, gently mix by pipetting five to six times to wash the beads, and place the tube again in the magnetic stand for 5 min at RT.

8. Remove the supernatant and add 50 μL of thawed elution buffer, gently mix by pipetting five to six times, place the tubes in the thermal cycler preset at 80 °C. Resume the run with the heated lid option.

9. At the end of the run, add 50 μL of bead binding buffer to the tubes in order to allow specific rebinding of the poly-A fraction RNA to the oligo dT magnetic beads. This step will remove the nonpolyadenylated RNA, including the abundant rRNA.

10. Mix the contents of the tubes by gently pipetting five to six times and place on the magnetic stand for 5 min at RT.

11. Meanwhile, set the thermal cycler at 94 °C for 8 min followed by 4 °C on hold. Start the program and pause at 94 °C and proceed to the next step.

12. After the 5 min incubation, carefully remove the supernatant without disturbing the pellet.

13. Remove the tubes from the magnetic stand and add 19.5 µL of thawed elute-fragment-prime-mix. Gently mix the solution by pipetting five to six times.

14. Place the tubes at 94 °C in the preset thermal cycler from **step 11** and start the program. In this step the poly-A RNA fractions will be enriched, fragmented, and hybridized to the random hexamers in the presence of cations at elevated temperature.

15. Immediately proceed to first strand cDNA preparation.
    (a) *First strand cDNA preparation*

    - Remove the first strand cDNA master mix from −20 °C, thaw, vortex thoroughly, and snap centrifuge.
    - Set a PCR thermal cycler program of 25 °C for 10 min, 42 °C for 50 min, 70 °C for 15 min followed by hold at 4 °C with the heating lid (100 °C) setting. Start the PCR program, pause the step at 25 °C, and proceed to the next step.
    - Add 50 µL of SSIIRT into the whole first strand cDNA synthesis mix (*see* **Note 8**).
    - Briefly place the centrifuged elute-fragment-prime mix samples in a magnetic stand for 5 min at RT.
    - Transfer 17.5 µL of clean supernatant to a fresh 200 µL tube or in a PCR plate with proper labeling or barcoding.
    - Add 8 µL of SSIIRT-first strand cDNA master mix to each tube, mix gently by pipetting five to six times, and centrifuge briefly. Place the tubes in preset PCR thermal cycler and start the program.
    - Once the thermal cycler reaches 4 °C, remove the tubes and keep on ice and proceed for second strand cDNA synthesis.

    (b) *Second strand cDNA preparation*

    - Set the thermal cycler at 16 °C for 1 h and pause the cycle once it reaches 16 °C.
    - Thaw the second strand master mix and briefly centrifuge after vortex mixing.
    - Add 25 µL of second strand master mix to product tube from first strand cDNA and mix gently by pipetting five to six times.
    - Briefly centrifuge the tubes and incubate the tubes at 16 °C in the prewarmed thermal cycler for 1 h.

- Transfer the Agencourt AMPure XP beads from 4 °C and keep at RT for at least half an hour before proceeding to the purification of double stranded cDNA (*see* **Note 9**).
- At the end of the incubation, add 90 µL of well-mixed AMPure XP beads to each tube and mix gently by pipetting the complete volume 10–12 times.
- Incubate the tubes at RT for 15 min.
- Place the tubes in the magnetic stand for 5 min at RT to allow complete adherence of AMPure beads on the magnet.
- Remove the supernatant carefully without disturbing the magnetic beads. Add 200 µL of freshly prepared 80% ethanol and incubate for 30 s without disturbing the pellet.
- Remove the supernatant completely without disturbing the beads. (Use 10 µL tips for complete removal of residual volumes.)
- Repeat **steps 9** and **10** for one or more times to remove all the residual impurities.
- Air-dry the beads by incubating at RT for 5 min (*see* **Note 10**).
- Add 52.5 µL of resuspension buffer (RSB) to each tube, remove the tubes from magnetic stands, and mix by gently pipetting for five to six times. Incubate the tubes for 2 min at RT.
- Place the tubes in the magnetic stand for 1 min and collect 50 µL of supernatant, without disturbing the pellet, into fresh labeled tubes (*see* **Note 11**).
- Store the labeled tubes at −20 °C until proceeding with end repair process.

(c) *End repair of cDNA fragments*
- Preheat the thermal cycler to 30 °C and set for 30 min.
- Add 10 µL of RSB to each tube.
- Thaw the end repair mix, briefly centrifuge, and add 40 µL of it to each tube. Gently mix by pipetting five to six times.
- Incubate the tubes in the preheated thermal cycler at 30 °C for 30 min. Transfer the AMPure XP beads from 4 °C to RT and keep for at least half an hour prior to proceeding to the next step.

- At the end of the incubation, add 160 μL of well-mixed AMPure XP beads to each tube and mix gently by pipetting complete volume 10–12 times.
- Perform **steps 7–12** from the section *Second strand cDNA preparation*.
- Add 17.5 μL of RSB to each tube, remove the tubes from the magnetic stand, and mix by gently pipetting for about five to six times. Incubate the tubes for 2 min at RT.
- Place the tubes back into the magnetic stand for 1 min and collect 15 μL of supernatant into fresh labeled tubes, without disturbing the pellet.
- Store the properly labeled tubes at −20 °C until proceeding to the process for A-base addition.

(d) *A-base addition and ligation of index adapters*
- Transfer the AMPure XP beads from 4 °C to RT and keep for at least half an hour prior to proceeding to purification of adapter ligated product.
- Set the PCR thermal cycler at 37 °C for 30 min followed by 70 °C for 5 min. Pause the thermal cycler once the machine attains 37 °C.
- Thaw the A-tailing mix, briefly centrifuge, and keep on ice.
- Add 2.5 μL of RSB and add 12.5 μL of A-tailing mix into each tube and mix by gently pipetting for five to six times.
- Incubate the tubes in the preset thermal cycler.
- Proceed immediately to adapter ligation after the run. Place the samples on ice.
- Set the PCR thermal cycler for 30 °C for 10 min and pause the cycle once it reaches 30 °C.
- Thaw the index adapters; spin the ligation buffer briefly and place on ice (*see* **Note 12**).
- Add 2.5 μL of RSB to each of the A-tailed tubes.
- Remove ligation mix from −20 °C and briefly centrifuge. Add 2.5 μL of ligation mix to each tube.
- Add 2.5 μL of respective adapter index mix to each sample mix by gently pipetting for five to six times. Care should be taken to avoid contamination of the index adapter.
- After brief centrifugation, incubate the tubes at 30 °C in the preheated thermal cycler for 10 min.

- Remove the tubes from the thermal cycler and immediately add 5 μL stop ligation buffer and mix thoroughly by pipetting up and down for five to six times in order to deactivate the ligase.
- Add 42 μL of well-mixed AMPure XP beads to each tube and mix gently by pipetting the entire volume 10–12 times.
- Perform **steps 7–12** from the section *Second strand cDNA preparation*.
- Add 52.5 μL of RSB to each tube, remove the tubes from the magnetic stand, and mix by gently pipetting for five to six times. Incubate the tubes for 2 min at RT.
- Place the tubes in the magnetic stand for 1 min and collect 50 μL of supernatant without disturbing the pellet into a fresh labeled tube.
- Add 50 μL of well-mixed AMPure XP beads to each tube and mix gently by pipetting the entire volume 10–12 times.
- Perform **steps 7–12** from the section *Second strand cDNA preparation*.
- Add 22.5 μL of RSB to each tube, remove the tubes from the magnetic stand, and mix by gently pipetting for five to six times. Incubate the tubes for 2 min at RT.
- Place the tubes in the magnetic stand for 1 min and collect 20 μL of supernatant without disturbing the pellet into a fresh labeled tube.
- Store the labeled tubes at −20 °C until proceeding to library enrichment.

(e) *Final enrichment of library*
- Thaw the PCR master mix and primer cocktail, briefly centrifuge, and place on ice.
- Program the PCR thermal cycler for 98 °C for 30 s followed by 15 cycles of 98 °C for 10 s, 60 °C for 30 s, and 72 °C for 30 s. The final extension step needs to be set at 72 °C for 5 min and the final hold at 10 °C. The lid temperature should be set at 100 °C.
- Add 5 μL of primer cocktail to each tube followed by 25 μL PCR master mix. Mix well by gently pipetting for five to six times and briefly centrifuge. Place the tubes in the preprogramed PCR thermal cycler and start the program.
- Transfer AMPure XP beads from 4 °C to RT and incubate for at least 30 min prior to proceeding for the next step.

- Add 50 μL of well-mixed AMPure XP beads to each tube and mix gently by pipetting the entire volume 10–12 times.
- Perform **steps 7–12** from the section *Second strand cDNA preparation.*
- Add 32.5 μL of RSB to each tube, remove the tubes from the magnetic stand and mix by gently pipetting for five to six times. Incubate the tubes for 2 min at RT.
- Place the tubes back in the magnetic stand for 1 min and collect 30 μL of supernatant into a fresh labeled tube, without disturbing the pellet.
- Store the properly labeled tubes at −20 °C until library QC protocol.

(f) *Library QC*
- *Agarose gel visualization*
  - Prepare a 2% low-melting agarose gel by using low-melting point agarose containing appropriate quantity of Ethidium bromide.
  - Load 3 μL of each library sample into wells along with 1 μL of gel loading dye. Load 1 μL of 100 bp ladder into the first and last wells for size reference.
  - Run at 80 V for 15 min and observe under an UV transilluminator.
- *Quantification using HS-Qubit platform*
  - According to the number of libraries, prepare the dye-buffer mix in a 1.5 mL microfuge tube covered with aluminum foil. (The dye is photosensitive.) For each sample add 199 μL of HS-Qubit buffer and 1 μL of dye and mix well by vortexing.
  - For each sample, take 198 μL of HS-dye-buffer mix in a Qubit tube. Add 2 μL of library and mix by gently pipetting five to six times, avoiding air bubble formation. (Flick the tubes to remove any air bubble formed.) Incubate at RT in the dark.
  - Measure the absorbance and calculate the concentration as per the HS menu in the Qubit apparatus.
  - A good library should have a library concentration between 5 ng/μL and 30 ng/μL and a clear product in between ~225 and 380 bp on a 2% low-melting point agarose gel (refer Fig. 3). Now the library can be used for sequencing in any of the Illumina NGS sequencing platforms such as Novaseq 6000, X Ten, Hiseq 2500, Nextseq, Miseq, GAIIx, etc.

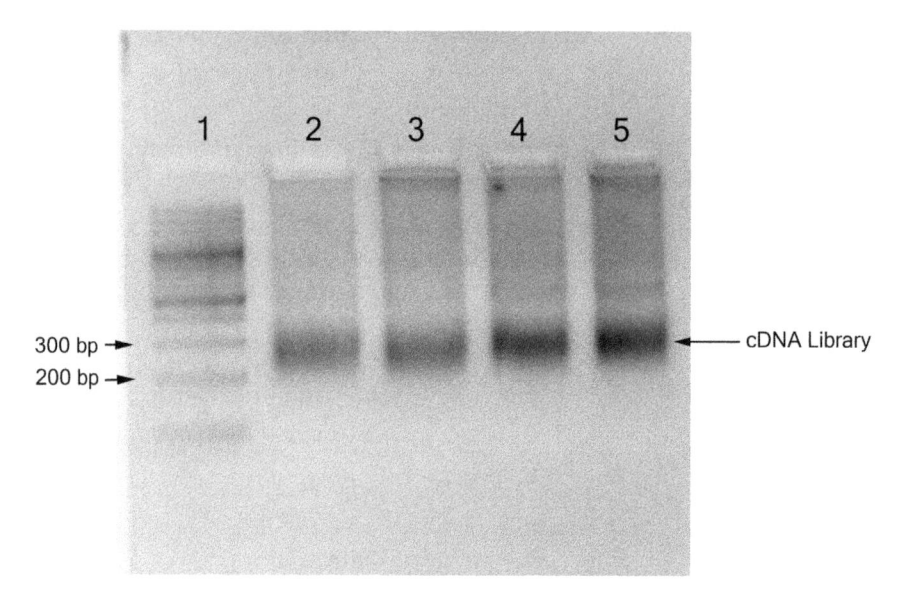

Fig. 3 Gel electrophoresis of cDNA library on 2% agarose gel. *1*—100 bp DNA ladder (Thermo Scientific, USA), *2–5*—cDNA library preparations. The uniform distribution of the fragments between ~225 and 380 bp (seen as the thick band) confirms the good quality of the library

3.2.2 Poly-A Enriched RNA Sequencing

1. NGS platforms require the appropriate concentration of library (generally in picomolar) which is achieved with the stepwise dilution. Convert the library concentration (in ng/μL) to X nanomolar (nM) by using the formula:

$$\frac{\text{Concentration (ng/μL)} \times 10^6}{\text{Fragment size} \times \text{molecular weight of a base pair}} = X \text{ nM}$$

2. Dilute the X nM to obtain 10 nM, and further to 2 nM by using elution buffer.
3. Pool 10 μL of each library with competent adapter index for sequencing (refer to the TruSeq Sample Preparation Pooling Guide Part #15042173, Illumina Inc., USA) and mix well by gently pipetting.
4. Take 10 μL of 0.1 N NaOH and 10 μL of pooled 2 nM library and mix well.
5. Incubate at RT for 5 min to denature the library.
6. Add 980 μL of hybridization buffer (HT1) (provided in the TruSeq PE cluster kit v2). This will make the final concentration of library to 20 pM.
7. Prepare appropriate dilutions of the library in HT1 buffer. For example to make a final concentration of 8 pM, take 400 μL of 20 pM library and 600 μL of HT1.
8. Aliquot 125 μL of the library dilutions into a strip tube set of eight tubes.

9. Load the samples into the c-Bot machine according to the manufacturer's instructions (Illumina Inc., USA, cBot user guide Part #15006165 Rev.k) for cluster generation on the flow cell (*see* **Note 13**).

10. Once the c-Bot cluster generation is completed, clean the flow cell and load to the respective sequencer platforms according to manufacturer supplied protocols (Illumina Inc., USA) for the sequence generation.

3.2.3 RNA Sequencing Data Output Template Format

1. Prepare a sample sheet for data generation by using Illumina Experimental Manager (IEM v1.8) software package (Illumina Inc., USA, IEM user guide Part #15031335 Rev. D) or manually prepare by creating a ".csv" file containing the following information demarcated by commas: FCID, Lane, SampleID, SampleRef, Index, Descriptor, Control, Recipe, Operator, SampleProject. (Care should be taken to avoid any space or other special characters other than numbers and alphabets.)

2. At the end of the sequencing, generate a fastq file using configureBclToFastq.pl module of BCL2FASTQ (v1.8.4, Illumina Inc., USA, bcl2fastq converter user guide Part#15038058 Rev. A) software with the sample sheet. Raw fastq file output will be saved to the respective sample project ID folder inside the output folder.

3.3 Bioinformatic Approaches for lncRNA Discovery and Annotation

A standard computational pipeline for genome-scale discovery and annotation of lncRNAs comprises three main modules: (a) alignment of RNA-seq reads to the reference genome, (b) reconstruction of full-length transcripts and alternative isoforms, and (c) annotation of non-protein-coding transcripts longer than 200 nucleotides (nt) that lack an open reading frame (ORF) of significant length. The bioinformatic pipeline is summarized in the given illustration (refer Fig. 4).

3.3.1 Alignment of RNA-Sequencing Reads

1. Obtain the raw single or paired-end short reads or RNA sequences generated by the next-generation sequencing machine in Fastq file format (*see* **Note 14**).

2. Assess the quality scores across the read sequences and estimate adapter contamination using FastQC (v0.11.2) tool [5] which generates a detailed QC report (*see* **Note 15**).

3. Clip the sequencing platform-specific adapter sequences and trim leading and/or trailing low-quality bases from the sequencing reads using Trimmomatic (v32) [6]. Use the parameter ILLUMINACLIP to specify the file containing sequencing machine and experiment-specific adapter sequences in fasta format. Use the parameter LEADING and TRAILING to specify the minimum quality score threshold of at least 20 for

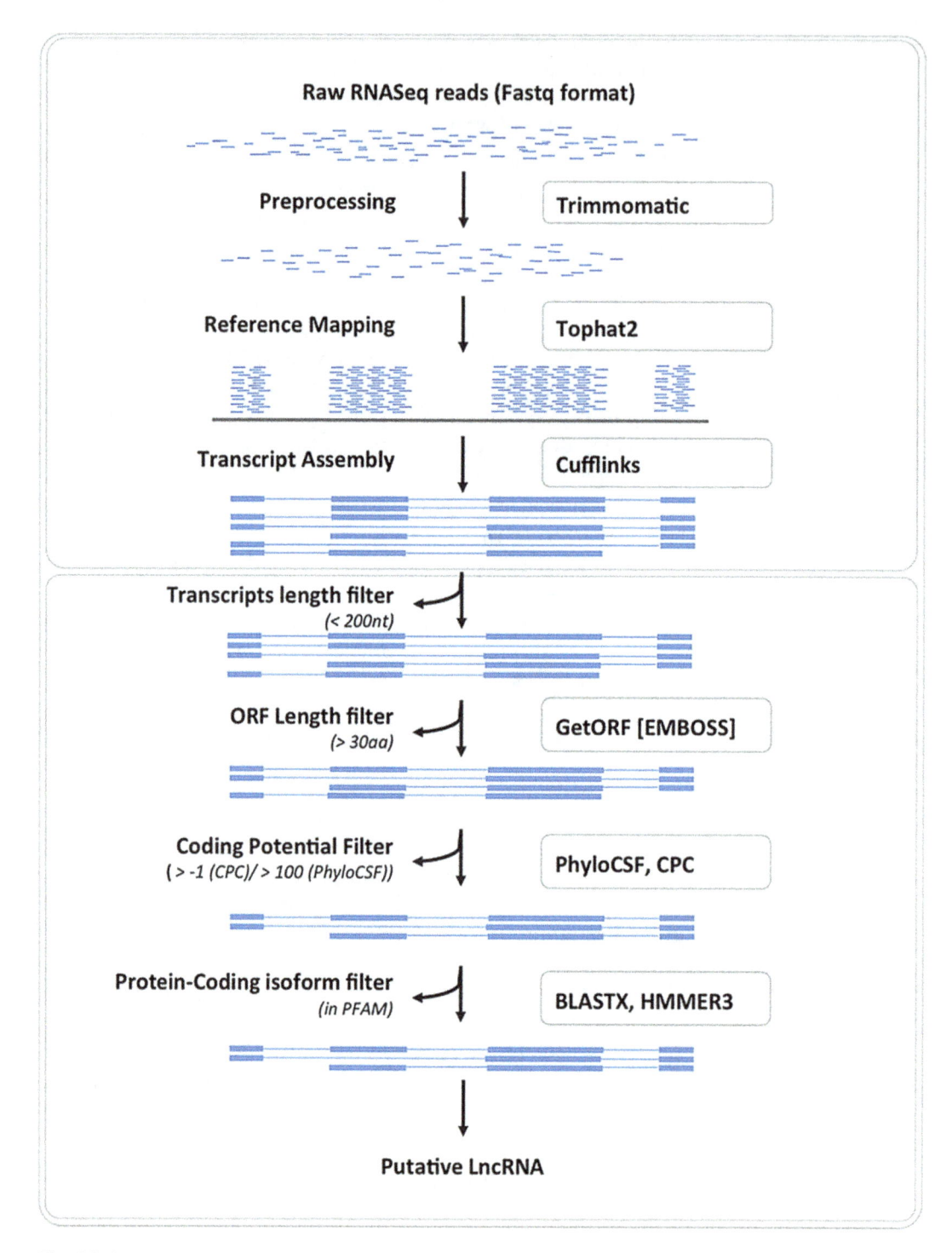

Fig. 4 Schematic illustration of the bioinformatic pipeline for the identification of Zebrafish lncRNAs from RNA sequencing data

read end trimming. Use the parameter MINLEN to set the minimum read length threshold after quality trimming.

4. Optionally, run FastQC analysis again on the trimmed read files to ensure that the quality thresholds have been met.

5. Align the trimmed reads to the reference genome using the spliced read aligner Tophat2 [7] using default parameters. In case of paired-end reads, use the parameter -r (--mate-inner-dist) to specify the inner mate distance and --mate-std-dev to set an appropriate standard deviation for the measure. Study the mapping statistics to ensure good mapping (*see* **Note 16**). The read alignments are generated in BAM file format (accepted_hits.bam).

3.3.2 Reconstruction of Full-Length Transcripts and Alternative Isoforms

1. Use Cufflinks (v2.2.1) [8] to perform a reference-based de novo assembly of the aligned reads (accepted_hits.bam) into full-length transcripts and alternate isoforms with default parameters. In case of strand-specific sequencing, use the parameter --library-type to specify the appropriate protocol used.

2. Use Cuffmerge [8] to merge this de novo transcript model (transcripts.gtf) with the reference transcript annotation from RefSeq or Ensembl (*see* **Note 17**).

3. Upload the transcript coordinates (merged.gtf) on the UCSC Table browser and download the transcript sequences including 5′ UTR, exons, and 3′ UTR. Run in-house scripts to select transcripts longer than 200 bases for downstream analysis.

3.3.3 Annotation of Non-protein-Coding Transcripts

1. Run "Getorf" program from the EMBOSS [9] suite to predict open reading frames (ORFs) within the transcript sequences. Remove transcripts with predicted ORFs longer than 30 amino acids as those could represent short functional peptides (*see* **Note 18**).

2. For the remaining transcripts, run tools such as Coding Potential Calculator (CPC) [10] and/or PhyloCSF [11] to assess the coding potential. Remove transcripts with CPC score <-1 and PhyloCSF score >100 as those could represent potential protein coding loci.

3. Use BEDTools [12] to find exonic overlap of the remaining noncoding transcript coordinates with RefSeq protein coding gene catalog to remove all protein-coding isoforms. In order to exclude transcripts with known protein domains, run HMMER3 [13] to query the predicted ORFs against Pfam protein database [15] with default parameters and remove transcripts with a significant Pfam hit (E-value <0.001). Also, perform a BLAST using the repeat-masked transcript sequences against the RefSeq protein database using Blastx [14] and filter out those which map with an E-value

<0.0001. The retained transcripts represent the set of putative lncRNAs in the sample.

4. Use BEDTools to intersect the predicted lncRNAs with known lncRNAs compiled from previously published studies and generate the final novel lncRNome.

3.4 Experimental Validation of the Bioinformatically Predicted lncRNA Transcripts

The bioinformatically predicted lncRNA transcripts are evaluated for their spatio-temporal and tissue-restricted expression using two basic molecular biology techniques: quantitative real-time PCR (qPCR) and whole mount in situ hybridization (*WISH*). qPCR validates the relative enrichment of a predicted lncRNA transcript in the tissue from which it was identified or the expression levels across the developmental stages. A visual interpretation of the localization of the lncRNA transcript can be attained by *WISH*, which provides an insight into the functional role of the lncRNA. The *WISH* is a technique that marks the expression of a particular transcript by using a complementary, labeled RNA probe which can be visualized using an antibody-based assay [16]. The protocol used by our group was primarily adopted from Thisse et al. [16], implementing the modifications included in the 2010 update from https://wiki.zfin.org. Improvisations of the protocol were derived wherever necessary as detailed in the following sections.

3.4.1 Quantitative Real-Time PCR for the Validation of lncRNA Expression

1. Prepare the cDNA from individual RNA samples as described in Subheading 3.1.3 (Under *Quality check of tissue samples through PCR*).

2. Set the qPCR reactions using primers designed against the lncRNA gene locus so as to obtain an amplicon with an approximate length of 200 bp. Each reaction should be composed of 1 μL of cDNA, 5 μL SYBR Green, 1 μL of 2.5 mM equimolar primers, and 3 μL of NFW. Perform each reaction in triplicates for assessing the significance of the signal and technical integrity of the readings.

3. Set an additional set of reactions concomitantly using primers against a ubiquitous house-keeping gene, such as beta-actin, in order to normalize the expression levels of the lncRNA transcript in individual cDNA samples.

4. Perform the qPCR reaction at set conditions according to the requirements of the primer pairs in the LightCycler 480 Real-Time PCR system.

5. At the end of the reaction, set a threshold level for background signals and retrieve the data as C_T values for individual reactions.

6. Calculate the relative expressions of the lncRNA transcripts in terms of fold change using the delta-delta C_T method [17]. Tissue-restricted lncRNA transcripts should show higher fold

change in the tissue type from which it was predicted to be derived. A similar trend is expected in the case of embryonic stage-specific lncRNA expression.

3.4.2 Whole Mount In Situ Hybridization for Visualization of lncRNA Expression

1. *Cloning of the gene loci of putative lncRNAs for probe synthesis*
   (a) From the lncRNA gene locus, amplify 500 bp to 1 kb region by standard PCR (*see* **Note 19**).
   (b) Purify the PCR product using MiniElute PCR purification kit (*see* **Note 20**).
   (c) To clone the purified amplicon, add 4 µL of amplicon to 1 µL of TOPO-TA vector and 1 µL of salt solution.
   (d) Incubate the mixture at RT for 30 min without disturbing.
   (e) For transformation, add above mix to 100 µL of TOP10 competent cells and incubate in ice for 30 min, followed by heat shock at 42 °C for 90 s and then in ice for 5 min.
   (f) Add 200 µL of SOC media (provided in the kit) to the above mix and incubate at 37 °C for 1–2 h.
   (g) Plate the cells on Luria Bertani agar with Carbenicillin (50 µg/mL) and incubate at 37 °C overnight (*see* **Note 21**).
   (h) Pick single isolated colonies from the plate and inoculate into 3–4 mL Luria-Bertani liquid media with Carbenicillin (50 µg/mL) and incubate at 37 °C overnight.
   (i) Isolate the plasmids using QIAprep Spin miniprep kit and elute in NFW and store at 4 °C.
   (j) Confirm successful transformation and the presence of the amplicon in the vector using *Eco*RI restriction digestion. Set up a restriction digestion reaction in a final volume of 25 µL with 2 µL of the good quality plasmid of approximate concentration 100 ng/µL, 10 units of the *Eco*RI enzyme, and 2.5 µL of appropriate manufacturer's buffer for 2–3 h at 37 °C (refer Fig. 5) (*see* **Note 22**).
   (k) Once confirmed, linearize the plasmid using *Not*I or *Stu*I for in vitro transcription (IVT) using T3 or T7 promoter, respectively. Essentially the plasmid is digested in two batches, for *Not*I and *Stu*I separately, each containing three sets of individual reactions. Each reaction is made up to a final volume of 25 µL with 6 µL of the good quality plasmid of approximate concentration 100 ng/µL, 10 units of the required enzyme, and 2.5 µL of appropriate manufacturer's buffer at 37 °C overnight (*see* **Note 23**).

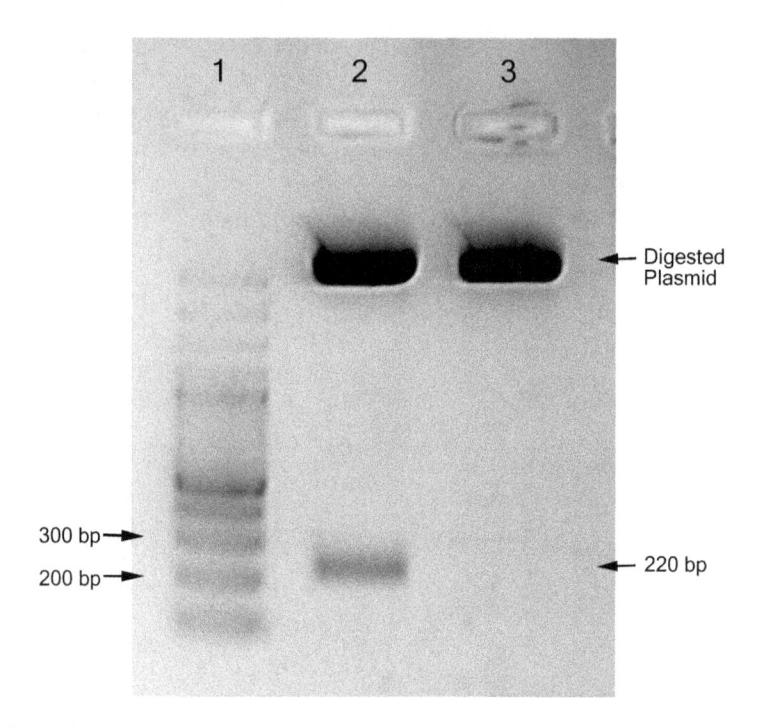

Fig. 5 Gel electrophoresis of *Eco*RI-digested plasmid on 1% agarose gel. *1*—100 bp DNA ladder (Thermo Scientific, USA), *2* and *3*—*Eco*RI-digested plasmids. An excision product of expected size (220 bp) in lane 2 confirms that the plasmid contains the insert (amplicon from lncRNA gene locus)

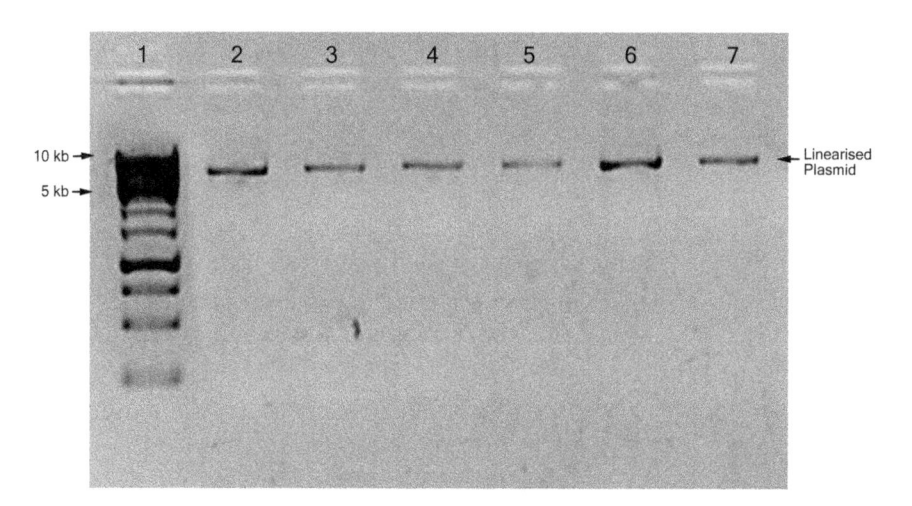

Fig. 6 Gel electrophoresis of linearized plasmid for IVT on 1% agarose gel. *1*—1 kb DNA ladder (Thermo Scientific, USA), *2–4*—linearized plasmid digested with *Not*I enzyme. *5–7*—linearized plasmid digested with *Stu*I enzyme

(1) To confirm complete digestion, load 1 μL of the linearized plasmid from each reaction on 1% agarose gel along with 1 kb ladder and run at 100 V for 30 min (refer Fig. 6) (*see* **Note 24**).

(m) For precipitation of linearized plasmid, pool the three digested sets and add double volume of absolute ethanol (approximately 144 µL for the three pooled reactions) and 1/10th volume of 7.5 M Ammonium acetate (approximately 7.2 µL), mix, and incubate at −80 °C for 1–2 h.

(n) Centrifuge the mix at ≥8000 × *g* for 15 min at 4 °C.

(o) Discard the supernatant and dissolve the pellet in 10 µL of NFW after air-drying. Check the quality of the linearized product by running on a 1% agarose gel along with 1 kb ladder and run at 100 V for 30 min.

2. *IVT reaction for RNA probe synthesis for putative lncRNAs of Zebrafish*

   (a) The IVT reaction was set up in a final reaction volume of 20 µL with 8 µL of the linearized plasmid template, 2 µL each of 5× transcription buffer, 10 mM dATP, dCTP, and dGTP, along with DIG-labeled dUTP and T3 or T7 polymerase, using MEGAscript in vitro transcription kit components (*see* **Note 25**).

   (b) Incubate the above reaction at 37 °C for 2 h (can vary according to length of probe).

   (c) Add 1 µL of TURBO DNase and further incubate at 37 °C for 30 min (*see* **Note 26**).

3. *Purification of RNA probes for the WISH protocol*

   (a) To the DNase-treated IVT reaction mix, add 15 µL of 7.5 M Ammonium acetate, 115 µL of NFW, and 150 µL of Phenol:Chloroform:Isoamyl alcohol, mix, and centrifuge at ≥8000 × *g* for 15 min at 4 °C.

   (b) Transfer the top aqueous layer carefully to a fresh tube and add equal amount of Chloroform; centrifuge it at ≥8000 × *g* for 15 min at 4 °C (*see* **Note 27**).

   (c) Transfer top aqueous layer to a fresh tube and add equal amount of Isopropanol; incubate it at −80 °C for 1–2 h.

   (d) Centrifuge the mix at ≥8000 × *g* for 15 min at 4 °C.

   (e) Discard the supernatant, air-dry the pellet and dissolve it in 15–20 µL of NFW.

4. *Quality check for the IVT products for preparing RNA probes*

   (a) *Agarose gel visualization*
   - Mix 1 µL of IVT product with 8 µL of NFW and 1 µL of RNA loading dye.
   - Incubate the above mix at 65 °C for 5 min followed by snap chill on ice for 10 min.

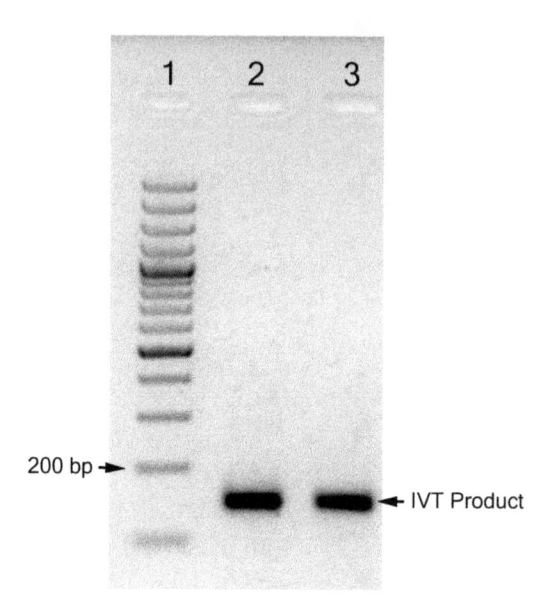

Fig. 7 Gel electrophoresis of IVT product on 1% agarose gel. *1*—100 bp DNA ladder (Thermo Scientific, USA), *2* and *3*—IVT product. Intact band at expected size (250 bases, RNA) confirms integrity of the IVT product

- Run the above snap chilled IVT product in an agarose gel (gel composition depends on IVT product size) along with 1 kb ladder and run at 100 V for 30 min. An intact band of the expected size confirms the integrity of the IVT product (refer Fig. 7).

(b) *Qualitative analysis of IVT products for RNA probes by spectrophotometry*
- Measure absorbance of the samples using nanodrop spectrophotometer.
- If IVT product displays high concentration and purity, proceed with the *WISH* protocol for visualization of lncRNA transcripts.

5. *Whole mount in situ hybridization protocol for embryonic stages of Zebrafish*
   (a) *Fixing of embryos at specific developmental stage*
   - Dechorionate Zebrafish embryos at desired developmental stage using 50 mg/mL of pronase (*see* **Note 28**).
   - Fix the dechorionated embryos in 4% PFA in PBS for 4 h at RT or overnight at 4 °C with continuous and gentle agitation (*see* **Note 29**).
   - Dehydrate the embryos with gradually increasing methanol gradation solutions (25%, 50%, 75%, and 100% in PBT). Each time the existing solution is carefully removed without disturbing the embryos, the

next solution is added and placed on a shaker for continuous mixing for 5 min at RT.

- Rehydrate the embryos with gradually decreasing methanol gradation solutions (100%, 75%, 50%, and 25% in PBT). Each time the existing solution is carefully removed without disturbing the embryos, the next solution is added and the tubes are placed on a shaker for continuous mixing for 5 min at RT.

(b) *Permeabilization of the Zebrafish embryos*

- Treat the embryos with 10 μg/mL Proteinase K in PBT at RT. The incubation period is determined by the stage of the embryos; embryos of the stages 12–24 h postfertilization (hpf), 24 hpf, 36 hpf, and 48 hpf should be treated for 5, 10, 15, and 20 min, respectively.
- Transfer the embryos to 4% PFA in PBS and incubate for 20 min with continuous gentle agitation (*see* **Note 30**).
- Wash the embryos five times 5 min each in PBT at RT.

(c) *Pre-hybridization treatment of Zebrafish embryos*

- Incubate the embryos in 0.6–1 mL of hybridization mix (HM) containing Heparin and yeast tRNA.
- Remove the HM and add fresh HM solution containing 500 μg of the DIG-labeled probe.
- Incubate at 65° C overnight.

(d) *Excess probe removal from the post-hybridization Zebrafish embryos*

- Perform a quick wash with HM (without tRNA and Heparin) at 65 °C.
- Remove excess probe by sequential washes with gradations of HM and 2× Saline Sodium-citrate buffer (SSC) at 65 °C, each for 10 min, with gentle agitation. Perform sequential washes using 100% HM, 75% HM in 2× SSC, 50% HM in 2× SSC, 25% HM in 2× SSC and 2× SSC.
- Wash the embryos twice for 30 min each in 0.2× SSC at 65 °C with gentle agitation.
- Perform following sequential washes with gradations of 0.2× SSC and PBT at RT, each for 1 min, with gentle agitation: 100% 0.2× SSC, 75% 0.2× SSC in PBT, 50% 0.2× SSC in PBT, 25% 0.2× SSC in PBT and PBT.

(e) *Blocking and anti-DIG antibody treatment for tagging of bound probes*
- Incubate the embryos in blocking solution for 2–5 h with continuous agitation at RT (*see* **Note 31**).
- Remove the solution and add fresh blocking solution containing 1:2000 (v/v) alkaline phosphatase-conjugated anti-DIG antibody.
- Incubate at RT for 4 h with continuous gentle agitation.
- Replace the solution with PBT and incubate at 4° C overnight.

(f) *Addition of chromogenic substrate of alkaline phosphatase NBT/BCIP and development of signal*
- Give a quick wash to the embryos in PBT at RT.
- Wash thrice for 15 min each in PBT at RT with gentle agitation.
- Wash thrice for 5 min each with alkaline Tris buffer at RT.
- Transfer the embryos to a multi-well plate (16 or 24 well) and replace the alkaline Tris buffer with the NBT/BCIP labeling solution.
- Place the plate in the dark with gentle agitation at RT (*see* **Note 32**).
- Observe the plate after every 30 min under a bright-field upright microscope till the signal develops (*see* **Note 33**).
- At the optimal contrast between the signal and the background, stop the reaction by replacing the labeling solution with PBT.

(g) *Imaging the Zebrafish embryos with probe signals of lncRNA expression patterns*
- Observe the embryos under a bright-field microscope and transfer the representative ones to 1.5% methyl cellulose.
- Position the embryos for convenient imaging and visualization and capture at different resolutions for a legibled view of the signal. Refer Fig. 8a for representative images of whole mount in situ hybridization showing brain-specific lncRNA transcript expression pattern in the 24 hpf embryonic of Zebrafish *(adapted with permission from ref. 4)*.

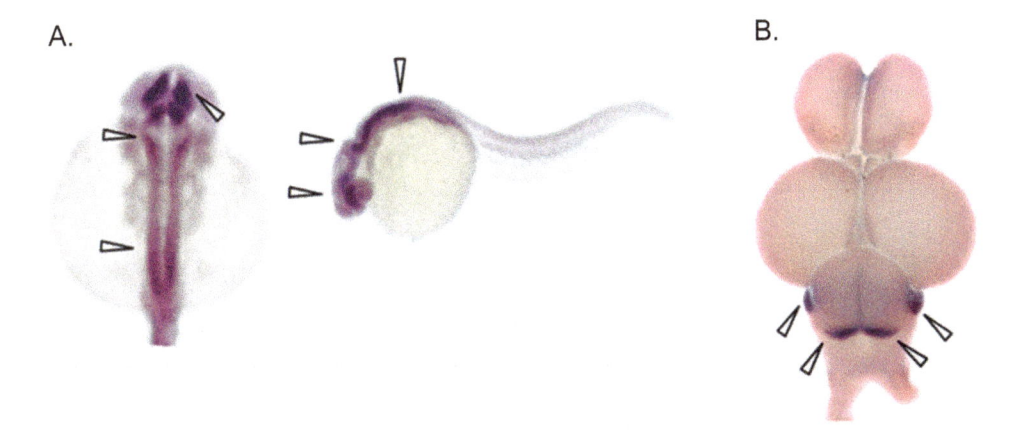

Fig. 8 Whole mount in situ hybridization of brain-specific Zebrafish lncRNAs. (**a**) Dorsal view (anterior up) and lateral view (anterior to the left) showing expression of lncRNA in brain tissue of 24 hpf zebrafish embryos. (**b**) Dorsal view (anterior up) of the adult zebrafish brain showing expression of lncRNA in specific regions. Arrowheads indicate expression patterns (Image adapted with permission from ref. 4)

(h) *Fixing and storage of embryos with probe signals of lncRNA expression patterns*
- Fix the signal by transferring the embryos to 4% PFA in PBS.
- Store the fixed embryos at 4 °C for further use.

(i) *Whole mount in situ hybridization protocol for adult tissues of Zebrafish*
The isolation of adult tissues from Zebrafish can be performed as described in **step 1** of Subheading 3.1.2. The WISH protocol for the adult tissues follows the same schema as for the embryonic stages. Note that the Proteinase K digestion step has to be extended to 20 min in the case of the adult tissues of Zebrafish. Refer Fig. 8b for a representative image of whole mount in situ hybridization showing brain-specific lncRNA transcript expression pattern in adult brain tissue of Zebrafish (*adapted with permission from ref. 4*).

4 Notes

1. For high-quality RNA, it is advisable to use freshly fixed embryonic stages and adult tissues of Zebrafish for the isolation protocol. However, samples stored at −80° C up to 1 month can also be used for RNA isolation.

2. Utmost care should be taken to ward off contamination to obtain pure homogenous samples for each tissue type. The major contaminating element in all tissues is the blood. In

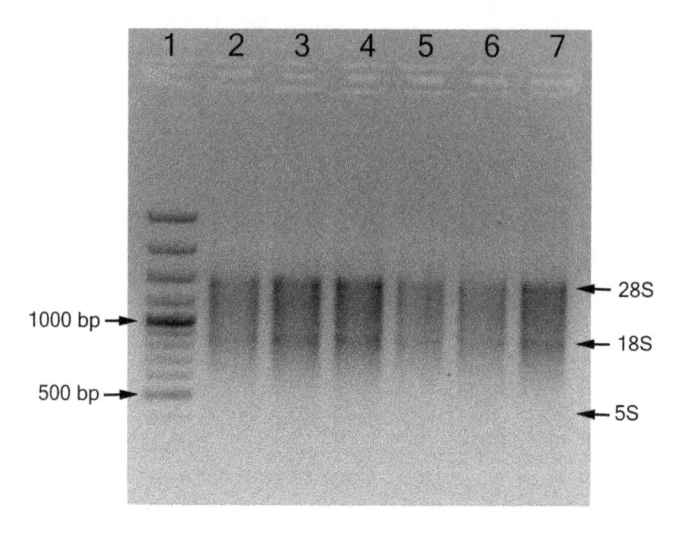

Fig. 9 Gel electrophoresis of isolated RNA samples on 1% agarose gel showing low-quality RNA. *1*—100 bp DNA ladder (Thermo Scientific, USA), *2–6*—isolated RNA samples. The shearing of the 28S, 18S, and 5S rRNA subunits indicates the degradation of the RNA

    case of tissues like heart, liver, and brain, a cross-contamination of some muscle tissue can also occur. Repeated washes of 1× PBS can ensure removal of most of the debris from muscle tissue and blood tissue.

3. All the centrifugation steps of the RNeasy Mini kit columns should be carried out at RT. For better yield of RNA, heat the NFW at 37 °C before elution. The minimum volume for elution should be 30 μL. A longer incubation period at RT after adding NFW can reduce the elute volume while optimal incubation leads to efficient high yield elution of the RNA from the column.

4. Careless handling during isolation of the RNA may lead to degradation and thus a compromised quality of the final isolated RNA. The degradation can be easily visualized on a gel (refer Fig. 9). The degraded RNA fails to show intact bands of 28S, 18S, and 5S rRNAs. Another problem with RNA quality is salt contamination. This arises due to improper washing of the column during RNA isolation. Salt contamination can be visualized as an even smear across the gel. Genomic DNA contamination is a serious issue that affects the quality of RNA. DNA contamination in RNA samples can be visualized on agarose gel near the wells. In such a situation, DNase treatment of the isolated RNA sample is recommended.

5. The quality of RNA can be determined from the A260/280 and A260/230 ratios. Any deviations from the acceptable values of these ratios (1.8–2.0 for A260/280 and 2.0 and above for A260/230) indicate contamination of RNA.

6. Illumina TrueSeq RNA sample preparation kit v2 supplies reagents for a total of 48 reactions. Aliquot the reagents into small quantities according to user's requirement in order to avoid repeated freeze-thaw. For multiple samples (high-throughput sample preparation) please refer to the manufacturer's HT protocol (Illumina Inc., San Diego, USA, Part # 15026495 Rev. F). Store the reagents at −20 °C or 4 °C, as specified. Place all the reagents and samples in the ice until specifically mentioned for other temperatures. After each step, the reagents should be transferred back immediately to their respective storage conditions.

7. Care should be taken to prevent inter-sample mixing and the sample handling should be gentle in order to prevent the library degradation.

8. For fewer numbers of samples, prepare the mix by adding 1 μL of SSIIRT to 9 μL of first strand master mix. Aliquot the mix into different tubes (8 μL per sample) and store at −20 °C in order to avoid repeated freeze and thaw.

9. Agencourt AMPure XP beads should be stored in 4 °C. Aliquot the well-mixed AMPure beads to small quantities to avoid repeated thawing. Incubate the AMPure beads at least 30 min at RT before using for any experiment.

10. For the drying of the bead pellet, leave the tube lids open and cover the magnetic stand with a large tray as to avoid contamination and sample degradation. Over-drying of the pellet is not advisable as it may give a reduced sample yield.

11. During collection of the supernatant from the bead pellet, pipette out a volume 5–10 μL less than the original volume followed by pipetting the remaining volume with a 10 μL pipette to avoid disturbing the pellet.

12. Extreme care should be exercised while handling adapter indexes in order to avoid cross-contamination.

13. Care should be taken to avoid over-clustering or under-clustering. Optimum cluster number can be obtained by calibrating different concentrations of the library by qPCR. Refer to the manufacturer instructions (Illumina Inc., USA, Sequencing Library qPCR Quantification user guide, Cat# SY-930-1010).

14. Sequencing reads are typically represented by a four-line entry in the fastq format as shown in Fig. 10. The first line starts with "@" character followed by the read identifier, which describes the machine, flow cell, cluster, grid coordinate, end and barcode for the read. In case of paired-end reads, the read identifiers will be identical except for the barcode. The second line is the raw nucleotide sequence. The third line begins with a "+"

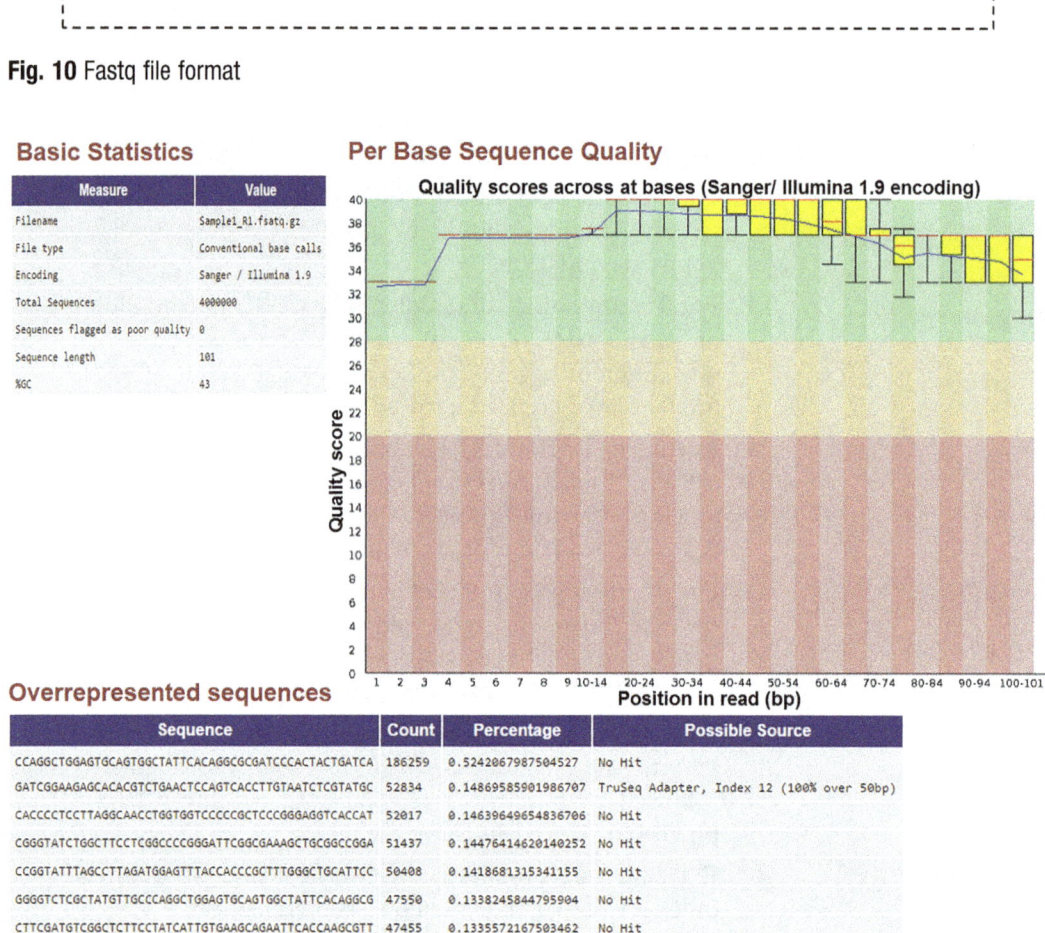

```
@HISEQ:78:C3L3MACXX:1:1101:14305:41771:N:0:CCGTCC
CACACACTTAGAACTACACTTTTCCTGGACTACAACTTCATACATGCTCTTCGCCCACACACCTGCTTTCTC
ATTTAGCTTG
+
BBBFFFFFFFFFFFFIIIIIIIIIIIIIIIIIIIIIIIIIIIIIIIIIIIIIIIIIIIIIIIIIIIIIFFFFF
FFFFFFFFFF
```

Fig. 10 Fastq file format

Fig. 11 Quality assessment reports generated by FastQC tool for data generation downstream to RNA sequencing

character and optionally followed by a sequence description. The fourth line is a sequence of ASCII-encoded quality values corresponding to every base in the sequence.

15. The FastQC tool generates multiple quality assessment reports including "Basic Statistics" report which records the total number of read sequences, read length, GC content, etcetera. "Per Base Sequencing Quality" report helps one decide quality-based read trimming parameters, and "Over-represented Sequences" report helps evaluate adapter contamination (refer Fig. 11).

16. A standard quality RNA-Seq experiment yields an average mapping percentage of at least 80%. Low mapping percentage could be explained by sample contamination or any other technical/sequencing artifacts. After a successful run, Tophat outputs the alignments in BAM file format, which is a binary version of the SAM format. The SAM file format is used to store information about alignment of reads to the genome. It starts with a header, describing the format version, sorting order (SO) of the reads, genomic sequences to which the reads were mapped. The file can contain both mapped and unmapped reads. Each line gives detailed information about the mapped read, its mapping position, alignment score, etc.

17. Merging the de novo transcript model with the reference transcript annotation helps one identify transcripts that have already been annotated and distinguish between known and novel transcripts. The novel transcripts can be identified by their gene id that will always start with XLOC_. Known genes will be marked by the gene id specified in the reference annotation.

18. ORF length cutoffs ranging from stringent 30–100 amino acids are used to remove putative protein-coding transcripts. However, lncRNAs may contain ORFs longer than 100 codons. Reciprocally, ORFs shorter than 100 amino acids may also be translated, given that functional peptides as small as 11 amino acids have been discovered in *Drosophila* species.

19. Longer the probe size, higher will be the specificity and lower the chances of background. Hence it is desirable to amplify a region of 500 bp or more as the template for probe synthesis. Care must be taken that the region amplified is unique in the Zebrafish genome. The PCR conditions may be set according to the amplicon size and the primer properties. The reaction volume is generally set to 50 µL to obtain higher amount of PCR product.

20. In order to obtain high concentration yield of amplicon required for cloning, elute the PCR product in a low volume of NFW (around 10 µL).

21. Plate the whole 200 µL of inoculated cells on LBA plate with 50 µg/mL of Carbenicillin. Incubate the plate for at least 16 h as TOP10 cells are a slow growing *E. coli* bacterial strain.

22. If plasmid digested with *Eco*RI does not yield an excision product of the expected size of the lncRNA gene locus, either screen few more colonies or repeat the cloning of the amplicon into the TOPO-TA vector.

23. In order to obtain a high concentration plasmid material for proceeding with the IVT, three different reactions for linearization of plasmid should be set.
24. Incubate the digestion reaction till the plasmids get completely digested. If the plasmid is not digested, then a longer transcript will be formed which will mostly utilize the transcriptional machinery as well as could also lead to nonspecific hybridization of the probe.
25. Use of filter tips for setting up the IVT reaction and purification of the IVT product is mandatory to keep the RNA product safe from degradation.
26. Over-incubation with the TURBO DNase can lead to nonspecific degradation of RNA.
27. Carefully remove the aqueous layer by avoiding contact with organic layer to minimize the chances of protein contamination.
28. Overexposure of pronase is harmful for the embryos. After pronase treatment, make sure to wash the embryos two to three times in embryo water.
29. Handle the embryos carefully using the Pasteur pipette. Always retain 10% of the solution in the tube while changing solutions without disturbing the embryos. For every treatment, place the tubes containing the embryos on a rocker with gentle agitation. Any harsh treatment can affect the morphology of embryos.
30. Embryos should be incubated in Proteinase K for permeabilization for a specific time period based on the developmental stage. Overexposure of the embryos can cause damage to the embryos while suboptimal incubation of embryos can lead to low permeability of the embryos to the RNA probe.
31. Prepare the blocking solution in PBT which should comprise 2% Sheep serum and 2 mg/mL BSA. Treatment of the embryos with the blocking reagent reduces the nonspecific binding of DIG antibody.
32. The NBT/BCIP labeling solution is light sensitive. Therefore the NBT/BCIP treatment of the embryos should be performed in the dark for rapid development of the signal.
33. Detection time for the labeling depends on the specificity of the probe and expression level of the lncRNA transcript. Over-incubation will give rise to high background signal. In order to avoid high background staining, continuous observation is mandatory such that the chromogenic reaction is stopped at the optimal contrast between the signal and the background. If overall reaction seems to be slow, replace the labeling solution and incubate at 4 °C overnight.

5 Conclusion

Several studies in zebrafish have uncovered a repertoire of long noncoding RNAs using RNA sequencing approaches [2–4]. We described a set of methodologies to congregate the differentially expressed lncRNA transcripts in zebrafish with high sensitivity. A custom-made pipeline for the identification of lncRNAs was designed with stringent parameters to discover the lncRNA transcripts from the total RNA library from various embryonic stages and adult tissues of zebrafish. The protocols for experimental validation of the lncRNAs have also been discussed. The protocols detailed here are thorough and systematic and can be used for the identification of lncRNA transcripts in cell and tissue systems, irrespective of the source. Our methods underline the significance of the instrumental role played by the twin toolset of RNA sequencing and computational biology toward the discovery and annotation of novel zebrafish transcripts including lncRNAs.

Acknowledgments

This work was funded by the Council of Scientific and Industrial Research (CSIR), India. S.M., A.S., P.S., and K.K. acknowledge Senior Research Fellowships from CSIR, India.

References

1. Haque S, Kaushik K, Leonard VE, Kapoor S, Sivadas A, Joshi A, Scaria V, Sivasubbu S (2014) Short stories on zebrafish long noncoding RNAs. Zebrafish 11(6):499–508. https://doi.org/10.1089/zeb.2014.0994
2. Pauli A, Valen E, Lin MF, Garber M, Vastenhouw NL, Levin JZ, Schier AF (2012) Systematic identification of long noncoding RNAs expressed during zebrafish embryogenesis. Genome Res 22(3):577–591. https://doi.org/10.1101/gr.133009.111
3. Ulitsky I, Shkumatava A, Jan CH, Sive H, Bartel DP (2011) Conserved function of lincRNAs in vertebrate embryonic development despite rapid sequence evolution. Cell 147(7):1537–1550. https://doi.org/10.1016/j.cell.2011.11.055
4. Kaushik K, Leonard VE, Kv S, Lalwani MK, Jalali S, Patowary A et al (2013) Dynamic expression of long non-coding RNAs (lncRNAs) in adult zebrafish. PLoS One 8(12):e83616. https://doi.org/10.1371/journal.pone.0083616
5. Andrews, S. (2010) FastQC: a quality control tool for high throughput sequence data. http://www.bioinformatics.babraham.ac.uk/projects/fastqc
6. Bolger AM, Lohse M, Usadel B (2014) Trimmomatic: a flexible trimmer for Illumina sequence data. *Bioinformatics* 30:2114–2120
7. Kim D, Pertea G, Trapnell C, Pimentel H, Kelley R, Salzberg SL (2013) TopHat2: accurate alignment of transcriptomes in the presence of insertions, deletions and gene fusions. Genome Biol 14(4):R36
8. Trapnell C, Roberts A, Goff L, Pertea G, Kim D, Kelley DR, Pachter L (2012) Differential gene and transcript expression analysis of RNA-seq experiments with TopHat and Cufflinks. Nat Protoc 7(3):562–578
9. Rice P, Longden I, Bleasby A (2000) EMBOSS: the European molecular biology open software suite. Trends Genet 16(6):276–277
10. Kong L, Zhang Y, Ye ZQ, Liu XQ, Zhao SQ, Wei L, Gao G (2007) CPC: assess the protein-

coding potential of transcripts using sequence features and support vector machine. Nucleic Acids Res 35(suppl 2):W345–W349

11. Lin MF, Jungreis I, Kellis M (2011) PhyloCSF: a comparative genomics method to distinguish protein coding and non-coding regions. Bioinformatics 27(13):i275–i282

12. Quinlan AR (2014) BEDTools: the Swiss-Army tool for genome feature analysis. Curr Protoc Bioinformatics 47:11.12.1–34

13. Eddy SR (2009) A new generation of homology search tools based on probabilistic inference. Genome Inform 23(1):205–211

14. Gish W, States DJ (1993) Identification of protein coding regions by database similarity search. Nat Genet 3(3):266–272

15. Punta M, Coggill PC, Eberhardt RY, Mistry J, Tate J, Boursnell C, Finn RD (2011) The Pfam protein families database. Nucleic Acids Res 40:D290–D301

16. Thisse C, Thisse B (2008) High-resolution *in situ* hybridization to whole-mount zebrafish embryos. Nat Protoc 3:59–69

17. Schmittgen TD, Livak KJ (2008) Analyzing real-time PCR data by the comparative C_T method. Nat Protoc 3:1101–1108

Chapter 5

Workflow Development for the Functional Characterization of ncRNAs

Markus Wolfien, David Leon Brauer, Andrea Bagnacani, and Olaf Wolkenhauer

Abstract

During the last decade, ncRNAs have been investigated intensively and revealed their regulatory role in various biological processes. Worldwide research efforts have identified numerous ncRNAs and multiple RNA subtypes, which are attributed to diverse functionalities known to interact with different functional layers, from DNA and RNA to proteins. This makes the prediction of functions for newly identified ncRNAs challenging. Current bioinformatics and systems biology approaches show promising results to facilitate an identification of these diverse ncRNA functionalities. Here, we review (a) current experimental protocols, i.e., for Next Generation Sequencing, for a successful identification of ncRNAs; (b) sequencing data analysis workflows as well as available computational environments; and (c) state-of-the-art approaches to functionally characterize ncRNAs, e.g., by means of transcriptome-wide association studies, molecular network analyses, or artificial intelligence guided prediction. In addition, we present a strategy to cover the identification and functional characterization of unknown transcripts by using connective workflows.

Key words Workflow, ncRNA, Transcript identification, Experimental RNA discovery, Data analysis, Next Generation Sequencing, Network analysis, Co-expression analysis, Machine learning

1 The Missing Link to Functionally Characterize ncRNAs

Before the 1980s, RNAs were seen as macromolecules that primarily support the protein synthesis and have been considered to be dormant—their regulative potential was unheard of. Our current knowledge about the human organism assumes 2% of the genome to encode for functional protein-coding RNAs, messenger RNAs (mRNAs), and more than 60% of the transcriptional output can be attributed to ncRNAs, which shows the actual importance of this formerly unrecognized molecule class [1]. Nowadays, we also know that the regulation of gene expression by noncoding RNAs (ncRNAs) via mRNA/ncRNA interactions is an essential and widespread phenomenon that occurs in almost all biological domains and, therefore, has become a basic principle in biology [2]. As of

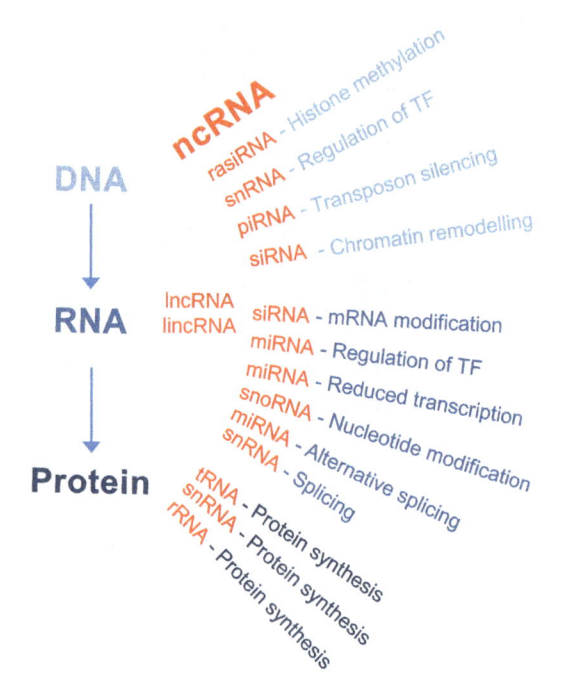

Fig. 1 Illustration to show the complexity and versatile role of ncRNA subtypes

early 2018, further ncRNA subtypes, e.g., circular RNAs (circRNAs), microRNAs (miRNAs), piwi-interacting RNAs (piRNAs), long noncoding RNAs (lncRNAs), and many more, have been recently characterized [3]. The currently known ncRNA subtypes and their respective functionalities are summarized in Fig. 1.

Next Generation Sequencing (NGS) technologies provide an attractive platform for the identification and quantification of genomes as compared to other high-throughput technologies and, furthermore, have been widely implemented for various applications such as DNA sequencing, de novo genome sequencing, epigenomics and transcriptomics profiling, and chromatin immunoprecipitation sequencing [4]. The application of NGS in clinical areas includes the identification of genetic variants, somatic or inherited mutations, as well as epigenetic changes to analyze an individual's disease-specific genome or tissue-specific transcriptome, where a comprehensive match of variants, like single-nucleotide polymorphisms (SNPs), can be easily detected [5]. Only very few of the ncRNAs under investigation can be interpreted and are known to be actionable, which means that a notable amount will be of either unknown or novel clinical importance and, therefore, holds the biological need and multifarious capability in a functional characterization [6].

Although state-of-the-art experimental technologies have yielded promising results in finding and characterizing novel ncRNAs, they are still subject to certain limitations, because the expression of most ncRNAs is lower than mRNA expression and, moreover, ncRNAs show tissue/stage-specific expression patterns [7, 8]. High-throughput sequencing generates an enormous amount of data and, thus, requires substantial computational power [9]. Algorithms for the identification are complementing experimental methods, allowing for a more focused approach for specific organisms and cell types. The following threefold difficulties of computational prediction that arise through the biological circumstances will be discussed throughout this chapter.

1. Variety of ncRNA subtypes: Advances in sequencing technologies have led to the discovery of a multitude of ncRNA subtypes, in which some are highly conserved (e.g., miRNAs, circRNAs) and others are generally lacking conservation across species, such as lncRNAs [10, 11].

2. Amount of uncharacterized ncRNAs: Only considering two of the most common ncRNA subtypes, many thousands of miRNAs have been discovered in many organisms. According to miRBase (http://www.mirbase.org/), there are currently 2,694 mature miRNAs in the human genome that are individually predicted to target hundreds of genes across multiple pathways [12]. In addition, the recently discovered lncRNAs are comprehensively summarized within the LNCipedia database (https://lncipedia.org/), which currently consists of 120,353 human lncRNA transcripts that are obtained from different sources, e.g., RefSeq, Ensembl, and Noncode [13]. As of March 2018, the LncRNA Database (http://www.lncrnadb.org/), a repository of lncRNAs curated from evidence and supported by the literature, lists 184 biologically validated lncRNAs in humans [14].

3. Versatility of functionality: The diverse biological impact of (1) ncRNAs toward multiple layers, such as chromatin remodeling (signal and/or scaffold), chromatin interactions, competing endogenous mRNAs, and natural antisense transcripts, shows that interactions at genomic, transcriptomic, and protein levels are possible and, thus, are very difficult to encounter by a single computational algorithm [15].

2 Experimental and Computational Identification of Novel ncRNAs

The first step in the characterization of ncRNAs—the discovery itself—is the most crucial one in the process, but is often based on inadequate sequencing experiments. The determination of the scientific problem rather than the affordability of the technology should drive the investigation [16]. For this reason, we are going to highlight common experimental and computational practices for

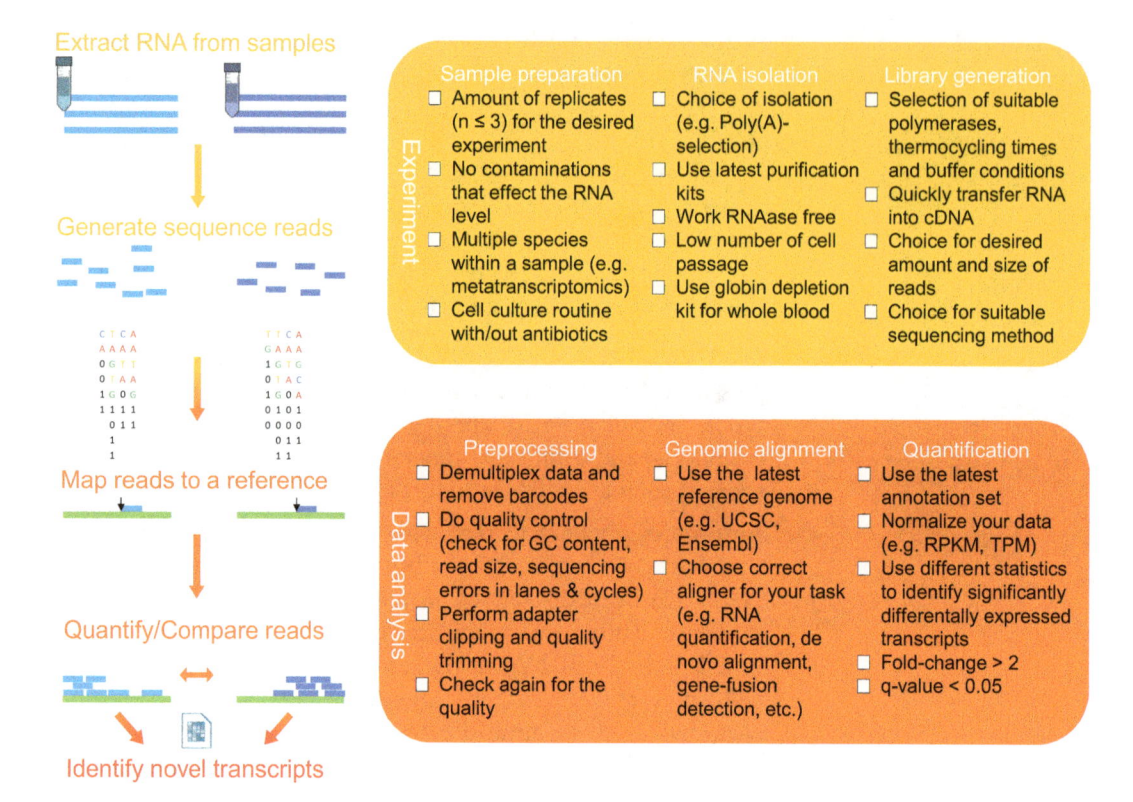

Fig. 2 Integrated experimental and computational workflow with specific checkboxes for the identification of ncRNAs from RNA-Seq datasets

ncRNA discovery. As an example, a workflow starting from sample preparation to data analysis and processing of already known and novel transcripts can be seen in Fig. 2.

2.1 Experimental Procedures for Proper ncRNA Identification

The initial step toward a successful experiment starts with the process of sample preparation, which involves already numerous decisions that are specifically dependent on the RNA subtype and species of interest. An overview of the extensive sequencing methodologies for the different platforms has been recently published by Tripathi et al. [17]. As an example to ensure clinical grade quality of the NGS technology, the Korean Society of Pathologists developed laboratory guidelines for NGS cancer panel testing procedures and requirements for clinical implementation of NGS [18]. The suggested laboratory part addresses important issues across multistep NGS cancer panel tests including the choice of the gene panel and platform, sample handling, nucleic acid management, sample identity tracking, library preparation, sequencing, data analysis, and reporting to the patient.

In research-oriented sequencing experiments, at least three biological replicates per condition are necessary to enable meaningful downstream statistical comparisons for the genome-wide detection of significant differences. In a recent review, Ouzain et al. [19]

found that for exploratory analyses of homogeneous samples (e.g., in cell lines or tissues from genetic mouse mutants and controls) in a highly controlled experimental setup, three biological replicates at high read depth can provide on the one hand sufficient reads to detect novel lncRNAs and on the other sufficient power to detect statistical differences between the conditions. However, the number of samples required will depend strongly on the biological and technical variability of an experiment and, of course, in a real-world setting involving the use of clinical patient samples many more samples would be required, but are rarely obtained, to achieve a similar statistical power [19]. Some other experimental aspects that have to be considered with caution are the isolation of the sample via commercially available purification kits itself, influence of the cell culture passage and tissue origin, poly(A)-tail selection of the RNA transcripts, globin depletion while working with native blood samples, or effect of antibiotics in cell culture in general [20].

After the RNA isolation step, the cDNA library preparation has to be done, which evokes the risk of biases resulting from genomes with high or low GC content. This should be in fact avoided by optimizing the preparation step through careful selection of polymerases for PCR amplification, thermocycling, condition, and buffer optimization [17]. The total amount of reads is yet another important parameter in determining the genomic coverage in RNA sequencing (RNA-Seq), because during the sequencing experiment different reads are generated from different RNA libraries and, thus, the overall coverage is defined by the number of times a genomic region, at the single base pair level, is covered by a read [16]. The combination of the desired read length and the amount of reads defines the throughput of an instrument in number of bases per run. NGS technologies are still, unless very low, prone to sequencing errors, but these most randomly occurring incorrect base calls (probability approximately 0.0001) can be compensated by sequencing the same region multiple times, which would result in an increased coverage. Increasing the read coverage likewise increases the confidence of existing variations (e.g., SNPs) in the genome/transcriptome under investigation. However, it was shown that a too high coverage might be problematic as well, because the absolute number of sequencing errors will jointly increase with the coverage and will impact the quality of the genome assembly [21]. The fractionation and sizing of the reads (based on the application of interest, e.g., de novo genome alignments will need longer reads), especially the impact of the sequencing depth and read length on single-cell RNA sequencing data, have to be considered [22].

2.2 A User-Friendly Environment for Computational Data Analysis

The application of computational analyses in the life science plays an increasingly important role. A comprehensive database of such analysis tools for different omics datasets can be obtained from the OMICtools community (https://omictools.com/), which aims to accelerate the selection of the most appropriate tools for specific use cases. An additional service of guided data analysis is provided by the German Network for Bioinformatics Infrastructure—de.NBI (https://www.denbi.de/)—which is a national infrastructure providing comprehensive, high-quality bioinformatics guidance to users in life science research and biomedicine. The European-wide bioinformatics support is coordinated and integrated by ELIXIR (https://www.elixir-europe.org/), which sustains bioinformatics resources across its member states and enables users in academia and industry to access computational services.

Reduced costs and increased accuracy of sequencing experiments enable the investigation of biological phenomena at a high resolution [23]. Despite the low technological entrance barriers of the already presented specialized experimental protocols, the challenge of proper, transparent, and reproducible data analyses is still a bottleneck [24]. With respect to the number of data analysis steps, including preprocessing, genomic alignment, and quantification, the complexity in tool selection, implementation, and benchmarking is increasing likewise, hence calling for more systematic approaches such as tool management frameworks [23, 25]. This means that many NGS tools being installed on desktop computers are inadequate for interdisciplinary collaborations and many researchers are eagerly looking for easily accessible cloud-computing solutions to provide scalable processing environments for sequencing data [26].

One of these cloud-computing solutions for RNA-centric research is the RNA workbench [27]. This platform is unique in combining available tools, workflows, and training material as well as providing easy access for experimentalists. The RNA workbench is built upon the Galaxy project, which is a framework that makes advanced computational tools accessible without the need of prior extensive training [28]. Galaxy seeks to make data-intensive research more accessible, transparent, and reproducible by providing a Web-based environment in which users can perform computational analyses and have all of the details automatically tracked for later inspection, publication, or reuse. In order to achieve long-term sustainability, it provides the essential resources on sustainable platforms such as BioConda (https://bioconda.github.io) and BioContainers [29], which are emerging solutions to deploy complete data analysis workflows, including all necessary tools and dependencies. Running the containerized RNA workbench simply requires installing Docker (https://www.docker.com/) and starting the Galaxy RNA workbench image [27]. For example, Wolfien et al. [30] and Schulz et al. [31] demonstrated successful implementations of a Galaxy/Docker-based workflow with discrete software applications for the analysis of NGS data. The provided layer of virtualization also allows the handling of user-defined input data in a secure and compartmentalized way, which is a key requirement for researchers working on sensitive data (e.g., patient data in clinics) [27].

2.3 Best Practices for Sequencing Data Analysis

Most NGS technologies currently available are based on sequencing a large number of fragments (thousands to millions) in parallel within a single-flow cell that pools multiple samples per run. Assuming demultiplexed RNA-Seq data, which means that the individual samples have been separated by specific barcodes, the processing usually starts with the quality control of the raw reads provided in the fastq format. In addition to the sequence of nucleotides, the fastq format also provides a quality value, i.e., Phred score, for each of the sequenced bases. In the first step, an evaluation of these quality values as well as the calculation of the GC content, read duplication levels and contaminations are crucial for any further analysis. The quality control tools FastQC [32], NGS QC Toolkit [33], or Qualimap2 [34] calculate multiple quality statistics and create visualizations for sequencing data, which can be used to fine-tune parameters and further downstream processing steps.

Adapter sequences that are added during the experimental steps do not provide any additional information and can therefore be removed by using various tools such as Cutadapt [35], Skewer [36], or Reaper from the Kraken package [37]. Often adapter clippers are already integrated into quality score trimming software like Trimmomatic [38] or TrimGalore! [39]. After removing the adapters, a quality trimming step that is removing low-quality parts of a read (usually quality score <20 is removed) is essential and improves the reliability of the subsequent analyses [23].

Numerous algorithms have been developed to align the individual reads onto a reference genome. The most popular tools are aligners like TopHat2 [40], HiSat2 [41], STAR [42], or Segemehl [43]. A comparison of different alignment tools regarding accuracy/mismatch frequency, splice site detection, and performance was already done [44]. Alignment-free quantification methods for the quantification of RNA-Seq such as Kallisto [45] and Salmon [46] are faster and additional resource-sparing analyses, because they efficiently use the structure of a reference sequence without performing full base-to-base alignments, which is most time consuming. Nevertheless, these alignment-free quantification algorithms perform only with similar accuracy compared to traditional approaches when applied to ordinary tasks like transcript quantifications, clustering, and isoform prediction [47]. With respect to the identification of noncoding transcripts, the reads should preferably not accumulate mismatches in seed regions, but can be truncated. Furthermore, most mapping tools allow for a multiple or unique mapping strategy, which means that reads may be aligned to multiple regions, pseudogenes, and regions of low complexity in genomic references or numerous isoforms in transcriptomic references [23, 48].

After aligning the reads onto the reference genome (e.g., from UCSC or Ensembl), they have to be quantified and compared across different samples to finally obtain the differentially expressed genes. To be able to compare two or more samples from the same or different sequencing runs, it is inevitable to normalize for the varying library/read size and length. For this reason, different statistical methodologies like *transcripts per million* (TPM) or *reads/fragments per kilobase per million mapped reads* (R/FPKM) have been developed [48]. Commonly used tools for the detection of differentially expressed genes are Cuffdiff [49], DESeq2 [50], and Sleuth [51] that are likewise used subsequently after applying TopHat2, STAR, or Kallisto. A recent study about the investigation of host-pathogen interactions, with a special focus on lncRNAs, showed the incorporation of an additional filter for enhancing the accuracy of differentially expressed genes [52].

Being able to get further self-paced training within the complex field of sequencing data analysis and the usage of evolving complex workflows, the Galaxy training network, which is a community-driven framework, enables interested users modern, interactive training for data analytics in life sciences and, therefore, facilitates the general use of NGS [53]. The Galaxy training network community combines online tutorials with a Web-based analysis framework to empower biomedical researchers to perform computational analyses themselves through a Web browser without the need to install software or search for tutorial datasets (http://galaxyproject.github.io/training-material/).

3 Linking Novel ncRNAs to Already Known Features

Ultimately, the functionality of (l)ncRNAs should be determined and/or tested by using experimental approaches, however, such experimental approaches like gene knockdown, overexpression, or CRISPR-Cas editing are typically too time and cost intense for an application toward an extensive pool of identified candidates [54]. Fortunately, numerous promising, cost-efficient in silico methods have been developed to overcome this experimental bottleneck by mathematical algorithms, structure and topological based features, network methodologies, or machine learning approaches. These individual approaches often do not require any novel generated experimental data, but are reusing publicly available data obtained from databases or repositories, e.g., Gene Ontology (http://www.geneontology.org/), STRING (https://string-db.org/), or Reactome (https://reactome.org/). A specific overview about computational life science databases was summarized from Hall et al. [55]. A comprehensive database for lncRNAs in human and mouse is LncRBase (http://bicresources.jcbose.ac.in/zhumur/lncrbase/), which hosts information on basic lncRNA transcript features, with additional details on genomic location, overlapping small noncoding RNAs, associated repeat elements,

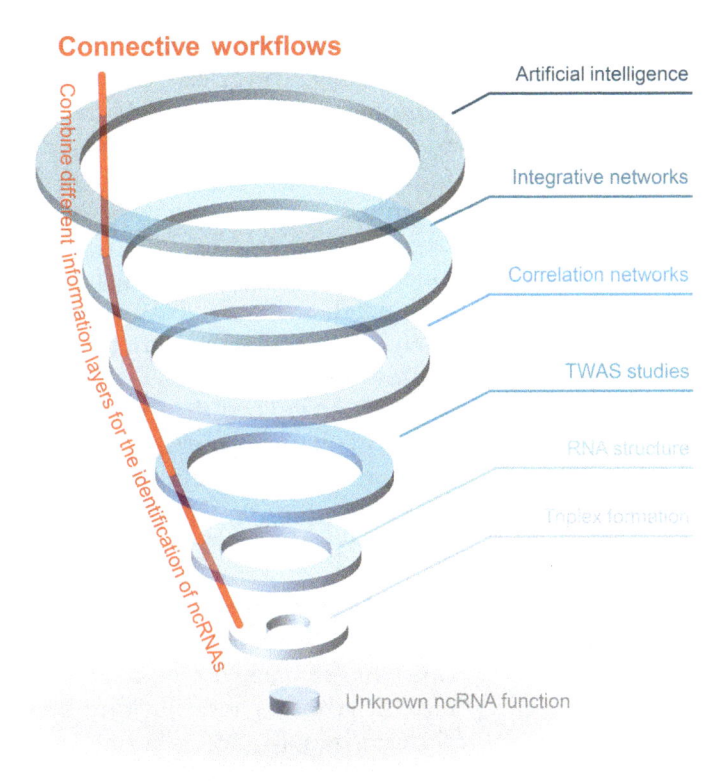

Fig. 3 Using connective workflows for the functional characterization of ncRNAs

imprinted genes, and lncRNA promoter information [56]. Users can also search for microarray probes mapped to specific lncRNAs and associated disease information as well as search for lncRNA expression in a wide range of tissues. In the following, we will show current approaches to integrate commonly used mathematical algorithms, topological features, and network modeling to characterize newly identified or uncharacterized ncRNAs (Fig. 3).

3.1 Combining Genomics Knowledge with Transcript Information: SNPs and TWAS

Genome-wide association studies (GWAS) are already applied to large numbers of individuals in a population or across multiple populations and result in associating their individual genome-wide genotypic variations to the personal respective phenotype. In a complementary manner, it is possible to perform a transcriptome-wide association study (TWAS) to identify significant expression-trait associations [57]. The obtained results from such a study showcase the power of integrating genotype and gene expression (mRNA and ncRNA) information together with phenotype data to gain insights into the genetic basis of complex traits.

Recently, Lopez-Meastre et al. [58] proposed a method that identifies, quantifies, and annotates SNPs without any reference genome from RNA-Seq data. The basis of their study is to identify the variants related to a phenotype, whereas SNPs are called de novo from the reads, without separating the steps of assembly and SNP calling. The clear advantage is that it can be applied to non-model species, without a reference genome being available. It was likewise shown that the SNP calling methods could be tailored to have a good precision, meaning that most of the reported SNPs are true SNPs. Clearly, only SNPs from transcribed regions can be targeted, but they arguably correspond to those with a more direct functional impact. RNA-Seq experiments may also provide very high depth at specific loci and, therefore, allow discovering infrequent alleles in highly expressed genes. Finally, pooling samples is already extensively used in DNA-Seq (sometimes termed Pool-Seq) [59]. They outlined that, even though the case studies presented included only two replicates, the method can be applied to any number of replicates and the key contribution is that they are able to produce a list of SNPs stratified by their impact on the protein sequence [58].

3.2 Using RNA-Triplexes to Refine Cooperative miRNA Target Prediction Possibilities

The biological phenomenon of cooperating miRNAs, which is a pair of miRNAs that is able to synergistically regulate mutual targets (mRNAs) to compel a more effective target repression, has been recently reported and shown to be meaningful for the functional characterization of formerly uncharacterized miRNAs [60, 61]. Their workflow can be used to identify so-called RNA triplexes and determine the respective functionality of cooperative target regulation by two miRNAs for each triplex [61]. The algorithm and the underlying data have been implemented in the triplexRNA database (https://triplexrna.org/), which contains predicted RNA triplexes composed of two cooperatively acting microRNAs (miRNAs) and their mutual target mRNAs for humans and mice. They derived experimentally confirmed miRNA-gene interactions from miRTarBase, a manually curated, literature-based database of validated miRNA-target interactions based on different experimental methods (release 6.0) [62] and complemented the set of validated miRNAs by predictions from the highly sensitive miRNA-target prediction algorithm miRanda [63]. An exemplary use case scenario for this application can be obtained from Lai et al. [64] who show the integration of bioinformatics, structural and kinetic modeling, as well as experimental validation to study the cooperative regulation of E2F1 by miRNA pairs in the context of anticancer chemotherapy resistance.

3.3 From RNA Structure to Biological Functionality

Structural versatility is another explanation for ncRNAs to have various different functions and, therefore, specific algorithms attributed to the primary, secondary, or tertiary structure provide new insights into the respective functionalities. Veneziano et al. [65] reviewed the computational methods for RNA structure prediction that have been adopted to analyze the structure of circRNAs, small ncRNA, as well as lncRNAs and have been already shown to provide indispensable information for further in-depth investigation. Another current review by Yan et al. [66] focuses on mainstream RNA structure prediction methods at the secondary and tertiary levels. They conclude that ongoing improvements in the accuracy of ncRNA structural prediction contribute to reliable predictions for the tertiary structure of small RNA molecules, but are lacking in the accurate prediction for the structure of large RNA molecules or those with complex topological structures [66]. A third review by Guo et al. [67] likewise concludes that the structures predicted by the computational methods still retain a high false-positive rate and the distinct structure–function relationships for many lncRNAs are still unknown, but with respect to the structure level of lncRNAs components discovered in the lncRNA secondary structures are of great value for further analysis, especially based on high-throughput sequencing technologies [67].

Two collective approaches to accumulate structure-based knowledge about RNAs are the Rfam database (http://rfam.sanger.ac.uk), which categorizes ncRNAs and their conserved primary sequence and RNA secondary structure through the use of multiple sequence alignments, consensus secondary structure annotation, as well as covariance models. In addition, with respect to the structure of lncRNAs, there is LNCipedia (http://www.lncipedia.org), a novel database for human lncRNA transcripts and genes [68]. LNCipedia offers 21,488 annotated human lncRNA transcripts obtained from different sources and includes basic transcript information, the gene structure, secondary structure information, protein coding potential, and miRNA-binding sites.

3.4 Integrative Network Approaches to Investigate the Regulative Potential of ncRNAs

Recent advances in systems biology shed light into the regulation of different pathways that investigate the interaction of genes, resulting in biological networks depicted as graphs. Especially, ncRNAs are highly connected within gene interaction networks and can therefore influence numerous targets to drive a specific biological response and the fate of cells or tissues. It has been shown that ncRNAs are particularly relevant in various research fields (e.g., cancer), in which ncRNAs act as main drivers or suppressors and can be seen as key regulators of physiological programs in developmental and disease-specific contexts [1]. It has become increasingly difficult to investigate ncRNA functionality in an isolated manner, because, e.g., miRNAs usually target mRNAs of various genes and, likewise, the mRNA of each gene can be targeted by multiple miRNAs that means these ncRNAs naturally link associated genes into regulatory networks [69, 70].

By the means of the previously discussed transcriptomics technologies, such as RNA-Seq, the activity of genes can be measured and integrated into a molecular network of choice to gain deeper insights into the gene expression in general. The recently published

KeyPathwayMiner enables the extraction and visualization of interesting subnetworks from a larger network based only on a series of gene expression datasets [71]. After applying the tool KeyPathwayMiner, one is able to identify the important subnetworks within a constructed large-scale molecular protein-protein interaction (PPI) network (e.g., based on all interactome information) to demonstrate known molecular interactions between significantly upregulated genes [72]. Once such a network is developed, even more sophisticated mathematical models can be applied to the network. Based on the available information, one can employ ordinary differential equations (ODE), discrete modeling, or hybrid modeling (composed of ODE and logic sub-modules) to dynamically analyze the networks for stimulus-response behavior and in silico perturbations [73]. It has been shown that these kinds of network approaches are able to identify disease-specific regulatory cores within large gene networks and, moreover, to predict receptor signatures associated with certain diseases [74].

In general, network enrichment methods combine experimental transcriptomic and proteomic data to be able to extract subnetworks from data-derived setups. Nevertheless, the integration of time series expression data with such network approaches is still challenging, thus limiting the identification of time-dependent responses. To overcome this limitation, Wiwie et al. combined human-augmented clustering with a novel approach for network enrichment to find temporal expression prototypes that are mapped to a network [75]. Their developed Time Course Network Enrichment (TiCoNE) methodology investigates enriched prototype pairs that interact more often than expected by chance.

So far, network-based computational studies investigate the hypothetic functions of lncRNAs through identifying molecules interacting with them, but since there are only a few molecular interactions known for multitudes of lncRNAs the application of these methods is rather difficult.

3.5 Correlation Networks to Link Uncharacterized ncRNAs to Known Annotations

It is well accepted that co-expressed genes are more likely to be co-regulated and, therefore, functionally related [76]. Thus identifying co-expressed protein-coding genes can help to assign the functions of uncharacterized ncRNAs [54], which has been successfully applied to study protein-coding genes, like the mammalian γ2 AMPK, that regulate the intrinsic heart rate [77].

For this reason, correlation networks are considered to be increasingly important bioinformatics applications, especially the weighted gene co-expression network analysis, which describes the correlation patterns among genes across multiple samples. This weighted correlation network analysis (WGCNA) is a guilt-by-association (GBA) approach for constructing co-expression networks based on gene expression data that is subsequently used for finding clusters (modules) of highly correlated genes [78]. This analysis supports not only the identification of co-regulated ncRNAs within specific modules, but also the investigation of intramodular hub genes (e.g., transcription factors) that may connect different pathways for relating different clustered modules together or even associate them to external sample traits (e.g., environmental factors) [78].

Based on public RNA-Seq datasets of four solid cancer types, Li et al. utilized WGCNA and proposed a strategy for exploring the functions of lncRNAs altered in more than two cancer types [79]. WGCNA in combination with DAVID (the database for annotation, visualization, and integrated discovery https://david.ncifcrf.gov/home.jsp) [80] can be used to explore the underlying associated functions of the identified modules. Their results indicate that cancer-expressed lncRNAs show high tissue specificity and likely play key roles in the multistep development of human cancers, covering a wide range of functions in genome stability maintenance, signaling, cell adhesion and motility, morphogenesis, cell cycle, immune, and inflammatory response, whereas the lncRNAs are lower expressed than protein-coding genes.

In addition to gene set enrichment analyses performed by the well-known DAVID tool, one can also use Enrichr (http://amp.pharm.mssm.edu/Enrichr), which currently contains a large collection of diverse gene set libraries available for analysis and download [81]. In total, Enrichr currently contains 234,849 annotated gene sets from 128 gene set libraries. New features have been added to Enrichr including the ability to submit fuzzy sets and upload BED files, improved application programming interface, and visualization of the results as clustergrams [82]. Overall, Enrichr is a resource for curated gene sets and a search engine that accumulates biological knowledge for further in-depth discoveries.

3.6 Artificial Intelligence-Guided Identification of ncRNA Functionality

Identifying meaningful information from huge data in bioinformatics, machine learning (ML) or artificial intelligence (AI) has evolved to one of the most dominant approaches in this area. Supervised and unsupervised ML algorithms are likewise using training data to look for characteristic patterns, build a mathematical model around the training data, and, finally, predict new data or data that has been left out for training (e.g., normalized by ten-fold cross-validation). The data is usually preprocessed by removing features with low variance and high correlation for initial dimension reduction and, therefore, following best practice recommendations [83]. Frequently used algorithms are support vector machines (SVMs), Bayesian networks (BNs), random forest (RF), boosting, or hidden Markov models (HMM) that have been applied across various omics fields [84, 85]. Small clinical datasets are often prone to overfitting, which is the reason why it is important to choose classifiers that are suitable for training on small datasets for a comparison of features given little training and choose the most appropriate algorithm according to accuracy and robustness toward overfitting [86]. Classical ML algorithms have limitations in processing the extensive amount of raw data that made researchers to predefine sets of suitable high-abstraction-level features, which can be used by the algorithms [87]. This time-consuming step could only be managed with considerable good domain expertise. In contrast to supervised ML algorithms, unsupervised approaches need no specific ground truth to train the actual model, but based on their nonlinear dimensional reduction they are less effective to identify a specific set of important features. These unsupervised statistical learning approaches, such as t-distributed stochastic neighbor embedding (t-SNE), assume that there are naturally occurring subclasses within sample sets that behave differently yet reproducibly across a number of populations and varying scenarios (e.g., treatment/control case, environments) [88]. In the following, we highlight specific ML-based tools and algorithms that have been recently applied to characterize ncRNAs.

iSeeRNA [89] is an SVM-based classifier, which can accurately and quickly identify lncRNAs from expression datasets (e.g., RNA-Seq) by using conservation open reading frame- and nucleotide sequence-based features in order to appropriately classify lncRNAs from protein-coding genes. Due to the lack of the aforementioned annotation for novel transcripts, they did not include homology search-based features, because this would enlarge the false-positive rate for predictions. In addition, customized SVMs for other species of interest can be trained and built upon on own datasets.

Xiao et al. predicted functions of lncRNAs through the construction of a regulatory PPI network between lncRNAs and protein-coding genes [90]. By integrating RNA-Seq data, they have been able to construct transcript profiles for the lncRNAs and protein-coding genes. After applying their Bayesian network approach, which implies dependency relations between lncRNAs and protein-coding genes, toward the initial regulatory network, a refined network was built. The integration of the highly connected coding genes and a single given lncRNA was subsequently used to predict functions of the lncRNA through functional enrichment within the PPI network [90].

The identification of ncRNAs within genomic regions can be done by a classification tool that was developed based on a hybrid RF and logistic regression model to classify short ncRNA sequences as well as long complex ncRNA sequences [91]. This classifier was trained and tested on a dataset with an achieved accuracy of 92.11%, sensitivity of 90.7%, and specificity of 93.5%. The authors also introduced a so-called SCORE feature, which is generated based on a logistic regression function that combines five significant features (structure, sequence, modularity, structural robustness, and coding potential) to enable an improved characterization of lncRNAs. They showed that the use of SCORE improved the performance of the formerly used RF-based classifier in the identification of Rfam lncRNA families [91].

Deep learning (DL), a subtype of machine learning algorithms, has emerged recently on the basis of big data, the power of parallel and distributed computing, and even more sophisticated mathematical algorithms. DL algorithms have overcome the former limitations of manually, handcrafted feature selection and are making major advances in diverse fields such as image recognition, speech recognition, and natural language processing [92]. Recently, there have been studies published that focused in particular on deep learning algorithms for the prediction of ncRNAs [93]. The main advantage of using DL approaches is that they do not require pre- and post-processing classification steps to handle the raw big data formats.

The developers of *deepTarget*, an end-to-end machine learning framework for miRNA target prediction, showed that even without any known features there are substantial performance boosts over existing miRNA target detectors [93]. DeepTarget uses deep recurrent neural networks and does not depend on any sequence alignment processing, which is being considered as indispensable in many bioinformatics workflows to identify meaningful differences between samples. As highlighted in the previous section, the numerous alignment algorithms involve parameter optimization strategies and the obtained results are often not reproducible, because the different alignment methods are not intercomparable in terms of allowance for mismatches and base-to-base comparison.

MiRTDL is another novel miRNA target prediction algorithm and based on convolutional neural networks (CNN) [94]. The authors showed that their CNN-based approach automatically extracts the most important information from the formerly created balanced training datasets and is then applied to 1606 experimentally validated miRNA target pairs. Finally, their results indicate that miRTDL performs better in comparison to existing target prediction algorithms and achieves significantly higher sensitivities.

4 Connective Workflow Development

Due to the advances in omics technologies, life science research is coming toward a detailed molecular level at single-nucleotide resolution. Each on its own, these technologies have contributed numerous advances in the field of genomics, transcriptomics, proteomics, and metabolomics. However, each technology individually cannot capture the entire biological complexity across all the given layers of most human diseases and, therefore, needs further integration of multiple levels as a combined approach to provide a more comprehensive view of the underlying biological phenomenon [95].

We already presented workflow management frameworks and cloud-computing services that are responsible for bridging the gap between tool developers and end users and showed that workflows facilitate the use of state-of-the-art computational tools, which would be difficult to access for nonexperts without graphical user interface frameworks [23]. However, the use of single workflows for specific tasks (e.g., the presented RNA-Seq analysis workflow) can be even more facilitated by the assembly of multiple workflows into a single connective workflow. Such universal connective workflows could be used to apply multilayered approaches and can incorporate several independent algorithms to test/benchmark different workflows against each other (Fig. 3). This would facilitate the certainty of the obtained knowledge, because independent algorithms, such as WGCNA and ML/DL, can be combined for the functional characterization of novel transcripts. The anticipated strategy for interoperable standards of workflows, namely the common workflow language, joins command-line tools across multiple platforms to workflows and, likewise, offers a modular concept for functional workflows that are built around containerized software solutions (documentation available at https://www.commonwl.org/). In order to adapt tools and workflows over time and ensure reusability and sustainability, we recommend keeping up track with the changes in the tools by a registration in platforms such as OMICtools (https://omictools.com/) or bio.tools (https://bio.tools/) [96], where tools are described by means of the meta-descriptive EDAM Ontology [97].

5 Conclusion

Each and every in silico predicted functionality of an ncRNA should be experimentally validated. Computational methods are used to accelerate the generation of new hypotheses or to narrow down specific molecular candidates. The interplay of model-driven experimentation and data-driven modeling could be seen as a guiding principle for the integration of the different interdisciplinary needs and expertise to achieve a task like characterization of a novel ncRNA [98].

The ideal validation of a new transcript would likewise involve the same layer of identification (e.g., ncRNAs validated via real-time PCR or Northern blot) as well as other layers of identification (e.g., ncRNA was co-immunoprecipitated with a protein) and, ultimately, a functional clarification by means of a knockout or overexpression experiment. If the elimination of an ncRNA affects a biological process, which is required for the proper development or homeostasis of the organism, can be checked very feasible with the advent of CRISPR/Cas technology [99]. In contrast to the golden path of experimental characterization, Palazzo and Lee [100] identified a scenario where experimental validation cannot give meaningful results, because a given ncRNA may only have a small impact on a biological process and, thus, results only in a small reduction of a relevant feature (e.g., reducing the number of offspring by 0.1%). Such small effects would be difficult to detect in a laboratory setting, but would be strongly selected against in the natural environment and would indicate that the specific ncRNA has a function [100].

The life sciences are mainly driven by technological and algorithmic developments that will both have steady impact toward the field. The best and most recent examples belong to the development of the NGS methodology and the still accelerating pace of AI algorithms within all research fields. The joint potential of both fields was demonstrated by the study of Scarano et al. [101], who showed the applicability of strand-specific RNA-Seq data in gene prediction. They also implied that libraries covering different organs, tissues, developmental stages, and a range of stress conditions are necessary to get meaningful annotation-specific genes. However, there is still a growing need in individual developments of both fields such as nanopore direct RNA-Seq, which allows for single-molecule sequencing that circumvents reverse transcription or amplification steps and, therefore, enables a real-time RNA-Seq technology [102]. Equally important are new DL algorithms that can be trained on genomic or transcriptomic data. Those algorithms can currently build predictive models of RNA-processing events such as splicing, transcription, and polyadenylation. When applied to clinical data, the algorithms were able to identify mutations and flag them as pathogenic, even though they have never seen clinical data for training [103].

To address the continuous growing number of identified ncRNAs by means of sequencing experiments, it is inevitable to look for computational guidance to rank, predict, or score the most likely functionally important transcripts. From experimental design to computational data analysis, it still needs specific domain expertise that can be satisfied only by interdisciplinary research teams or large community efforts.

Acknowledgments

We acknowledge the partners and management of the German Network for Bioinformatics Infrastructure (de.NBI) for continuous support and guidance. Financial support for this work by the German Federal Ministry for Education and Research (BMBF) and European Social Fund (ESF) is greatly acknowledged (Grant 031L0106C, 02NUK043C, ESF/14-BM-A55-0027/18).

References

1. Anastasiadou E, Jacob LS, Slack FJ (2017) Non-coding RNA networks in cancer. Nat Rev Cancer 18:5–18. https://doi.org/10.1038/nrc.2017.99
2. Delihas N (2015) Discovery and characterization of the first non-coding RNA that regulates gene expression, micF RNA: a historical perspective. World J Biol Chem 6:272. https://doi.org/10.4331/WJBC.V6.I4.272
3. Schmitz U, Naderi-Meshkin H, Gupta SK et al (2016) The RNA world in the 21st century—a systems approach to finding non-coding keys to clinical questions. Brief Bioinform 17:380–392. https://doi.org/10.1093/bib/bbv061
4. Tripathi R, Chakraborty P, Varadwaj PK (2017) Unraveling long non-coding RNAs through analysis of high-throughput RNA-sequencing data. Non-coding RNA Res 2:111–118. https://doi.org/10.1016/J.NCRNA.2017.06.003
5. Xuan J, Yu Y, Qing T et al (2013) Next-generation sequencing in the clinic: promises and challenges. Cancer Lett 340:284–295. https://doi.org/10.1016/j.canlet.2012.11.025
6. Metzker ML (2010) Sequencing technologies—the next generation. Nat Rev Genet 11:31–46. https://doi.org/10.1038/nrg2626
7. Bernhart SH, Hofacker IL (2009) From consensus structure prediction to RNA gene finding. Briefings Funct Genomics Proteomics 8:461–471. https://doi.org/10.1093/bfgp/elp043
8. Derrien T, Johnson R, Bussotti G et al (2012) The GENCODE v7 catalog of human long noncoding RNAs: analysis of their gene structure, evolution, and expression. Genome Res 22:1775–1789. https://doi.org/10.1101/gr.132159.111
9. Moran VA, Perera RJ, Khalil AM (2012) Emerging functional and mechanistic paradigms of mammalian long non-coding RNAs. Nucleic Acids Res 40:6391–6400. https://doi.org/10.1093/nar/gks296
10. Bejerano G, Pheasant M, Makunin I et al (2004) Ultraconserved elements in the human genome. Science 304:1321–1325. https://doi.org/10.1126/science.1098119
11. Johnsson P, Lipovich L, Grandér D, Morris KV (2014) Evolutionary conservation of long non-coding RNAs; sequence, structure, function. Biochim Biophys Acta 1840:1063–1071. https://doi.org/10.1016/J.BBAGEN.2013.10.035
12. Hammond SM (2015) An overview of microRNAs. Adv Drug Deliv Rev 87:3–14. https://doi.org/10.1016/j.addr.2015.05.001
13. Volders P-J, Verheggen K, Menschaert G et al (2015) An update on LNCipedia: a database for annotated human lncRNA sequences. Nucleic Acids Res 43:D174–D180. https://doi.org/10.1093/nar/gku1060
14. Quek XC, Thomson DW, Maag JLV et al (2015) lncRNAdb v2.0: expanding the reference database for functional long noncoding RNAs. Nucleic Acids Res 43:D168–D173. https://doi.org/10.1093/nar/gku988
15. Fang Y, Fullwood MJ (2016) Roles, functions, and mechanisms of long non-coding RNAs in cancer. Genomics Proteomics Bioinformatics 14:42–54. https://doi.org/10.1016/j.gpb.2015.09.006
16. Vincent AT, Derome N, Boyle B et al (2017) Next-generation sequencing (NGS) in the microbiological world: how to make the most of your money. J Microbiol Methods 138:60–71. https://doi.org/10.1016/J.MIMET.2016.02.016
17. Tripathi R, Sharma P, Chakraborty P, Varadwaj PK (2016) Next-generation sequencing revolution through big data analytics. Front Life Sci 9:119–149. https://doi.org/10.1080/21553769.2016.1178180
18. Kim J, Park W-Y, Kim NKD et al (2017) Good laboratory standards for clinical next-generation sequencing cancer panel tests. J Pathol Transl Med 51:191–204. https://doi.org/10.4132/jptm.2017.03.14
19. Ounzain S, Micheletti R, Beckmann T et al (2015) Genome-wide profiling of the cardiac transcriptome after myocardial infarction identifies novel heart-specific long non-coding RNAs. Eur Heart J 36:353–68a. https://doi.org/10.1093/eurheartj/ehu180
20. Ryu AH, Eckalbar WL, Kreimer A et al (2017) Use antibiotics in cell culture with caution: genome-wide identification of antibiotic-induced changes in gene expression and regulation. Sci Rep 7:7533. https://doi.org/10.1038/s41598-017-07757-w
21. Ekblom R, Wolf JBW (2014) A field guide to whole-genome sequencing, assembly and annotation. Evol Appl 7:1026–1042. https://doi.org/10.1111/eva.12178
22. Rizzetto S, Eltahla AA, Lin P et al (2017) Impact of sequencing depth and read length on single cell RNA sequencing data of T cells.

23. Lott SC, Wolfien M, Riege K et al (2017) Customized workflow development and data modularization concepts for RNA-sequencing and metatranscriptome experiments. J Biotechnol 261:85–96. https://doi.org/10.1016/j.jbiotec.2017.06.1203
24. Spjuth O, Bongcam-Rudloff E, Dahlberg J et al (2016) Recommendations on e-infrastructures for next-generation sequencing. Gigascience 5:26. https://doi.org/10.1186/s13742-016-0132-7
25. Lampa S, Dahlö M, Olason PI et al (2013) Lessons learned from implementing a national infrastructure in Sweden for storage and analysis of next-generation sequencing data. Gigascience 2:9. https://doi.org/10.1186/2047-217X-2-9
26. Celesti A, Celesti F, Fazio M et al (2017) Are next-generation sequencing tools ready for the cloud? Trends Biotechnol 35:486–489. https://doi.org/10.1016/J.TIBTECH.2017.03.005
27. Grüning BA, Fallmann J, Yusuf D et al (2017) The RNA workbench: best practices for RNA and high-throughput sequencing bioinformatics in Galaxy. Nucleic Acids Res 45:D626–D634. https://doi.org/10.1093/nar/gkx409
28. Afgan E, Baker D, van den Beek M et al (2016) The Galaxy platform for accessible, reproducible and collaborative biomedical analyses: 2016 update. Nucleic Acids Res 44:W3–W10. https://doi.org/10.1093/nar/gkw343
29. da Veiga Leprevost F, Grüning BA, Alves Aflitos S et al (2017) BioContainers: an open-source and community-driven framework for software standardization. Bioinformatics 33:2580–2582. https://doi.org/10.1093/bioinformatics/btx192
30. Wolfien M, Rimmbach C, Schmitz U et al (2016) TRAPLINE: a standardized and automated pipeline for RNA sequencing data analysis, evaluation and annotation. BMC Bioinformatics 17:21. https://doi.org/10.1186/s12859-015-0873-9
31. Schulz W, Durant T, Siddon A, Torres R (2016) Use of application containers and workflows for genomic data analysis. J Pathol Inform 7:53. https://doi.org/10.4103/2153-3539.197197
32. FASTQC (2010) Babraham Institute. https://www.bioinformatics.babraham.ac.uk/projects/fastqc/. Accessed 20 Jun 2018
33. Patel RK, Jain M (2012) NGS QC Toolkit: a toolkit for quality control of next generation sequencing data. PLoS One 7:e30619. https://doi.org/10.1371/journal.pone.0030619
34. Okonechnikov K, Conesa A, García-Alcalde F (2016) Qualimap 2: advanced multi-sample quality control for high-throughput sequencing data. Bioinformatics 32:292–294. https://doi.org/10.1093/bioinformatics/btv566
35. Martin M (2011) Cutadapt removes adapter sequences from high-throughput sequencing reads. EMBnet J 17:10. https://doi.org/10.14806/ej.17.1.200
36. Jiang H, Lei R, Ding S-W, Zhu S (2014) Skewer: a fast and accurate adapter trimmer for next-generation sequencing paired-end reads. BMC Bioinformatics 15:182. https://doi.org/10.1186/1471-2105-15-182
37. Wood DE, Salzberg SL (2014) Kraken: ultrafast metagenomic sequence classification using exact alignments. Genome Biol 15:R46. https://doi.org/10.1186/gb-2014-15-3-r46
38. Bolger AM, Lohse M, Usadel B (2014) Trimmomatic: a flexible trimmer for Illumina sequence data. Bioinformatics 30:2114–2120. https://doi.org/10.1093/bioinformatics/btu170
39. TrimGalore! (2012) Babraham Institute. https://www.bioinformatics.babraham.ac.uk/projects/trim_galore/. Accessed 20 Jun 2018
40. Kim D, Pertea G, Trapnell C et al (2013) TopHat2: accurate alignment of transcriptomes in the presence of insertions, deletions and gene fusions. Genome Biol 14:R36. https://doi.org/10.1186/gb-2013-14-4-r36
41. Kim D, Langmead B, Salzberg SL (2015) HISAT: a fast spliced aligner with low memory requirements. Nat Methods 12:357–360. https://doi.org/10.1038/nmeth.3317
42. Dobin A, Davis CA, Schlesinger F et al (2013) STAR: ultrafast universal RNA-seq aligner. Bioinformatics 29:15–21. https://doi.org/10.1093/bioinformatics/bts635
43. Hoffmann S, Otto C, Doose G et al (2014) A multi-split mapping algorithm for circular RNA, splicing, trans-splicing and fusion detection. Genome Biol 15:R34. https://doi.org/10.1186/gb-2014-15-2-r34
44. Engström PG, Steijger T, Sipos B et al (2013) Systematic evaluation of spliced alignment programs for RNA-seq data. Nat Methods

45. Bray NL, Pimentel H, Melsted P, Pachter L (2016) Near-optimal probabilistic RNA-seq quantification. Nat Biotechnol 34:525–527. https://doi.org/10.1038/nbt.3519
46. Patro R, Duggal G, Love MI et al (2017) Salmon provides fast and bias-aware quantification of transcript expression. Nat Methods 14:417–419. https://doi.org/10.1038/nmeth.4197
47. Robert C, Watson M (2015) Errors in RNA-Seq quantification affect genes of relevance to human disease. Genome Biol 16:177. https://doi.org/10.1186/s13059-015-0734-x
48. Conesa A, Madrigal P, Tarazona S et al (2016) A survey of best practices for RNA-seq data analysis. Genome Biol 17:13. https://doi.org/10.1186/s13059-016-0881-8
49. Trapnell C, Hendrickson DG, Sauvageau M et al (2013) Differential analysis of gene regulation at transcript resolution with RNA-seq. Nat Biotechnol 31:46–53. https://doi.org/10.1038/nbt.2450
50. Love MI, Huber W, Anders S (2014) Moderated estimation of fold change and dispersion for RNA-seq data with DESeq2. Genome Biol 15:550. https://doi.org/10.1186/s13059-014-0550-8
51. Pimentel H, Bray NL, Puente S et al (2017) Differential analysis of RNA-seq incorporating quantification uncertainty. Nat Methods 14:687–690. https://doi.org/10.1038/nmeth.4324
52. Riege K, Hölzer M, Klassert TE et al (2017) Massive effect on LncRNAs in human monocytes during fungal and bacterial infections and in response to vitamins A and D. Sci Rep 7:40598. https://doi.org/10.1038/srep40598
53. Batut B, Hiltemann S, Bagnacani A et al (2017) Community-driven data analysis training for biology. bioRxiv: 225680. doi: https://doi.org/10.1101/225680
54. Signal B, Gloss BS, Dinger ME (2016) Computational approaches for functional prediction and characterisation of long noncoding RNAs. Trends Genet 32:620–637. https://doi.org/10.1016/J.TIG.2016.08.004
55. Smalter Hall A, Shan Y, Lushington G, Visvanathan M (2013) An overview of computational life science databases & exchange formats of relevance to chemical biology research. Comb Chem High Throughput Screen 16:189–198
56. Chakraborty S, Deb A, Maji RK et al (2014) LncRBase: an enriched resource for lncRNA information. PLoS One 9:e108010. https://doi.org/10.1371/journal.pone.0108010
57. Gusev A, Ko A, Shi H et al (2016) Integrative approaches for large-scale transcriptome-wide association studies. Nat Genet 48:245–252. https://doi.org/10.1038/ng.3506
58. Lopez-Maestre H, Brinza L, Marchet C et al (2016) SNP calling from RNA-seq data without a reference genome: identification, quantification, differential analysis and impact on the protein sequence. Nucleic Acids Res 44:e148. https://doi.org/10.1093/nar/gkw655
59. Schlötterer C, Tobler R, Kofler R, Nolte V (2014) Sequencing pools of individuals — mining genome-wide polymorphism data without big funding. Nat Rev Genet 15:749–763. https://doi.org/10.1038/nrg3803
60. Lai X, Bhattacharya A, Schmitz U et al (2013) A systems' biology approach to study microRNA-mediated gene regulatory networks. Biomed Res Int 2013:703849. https://doi.org/10.1155/2013/703849
61. Schmitz U, Lai X, Winter F et al (2014) Cooperative gene regulation by microRNA pairs and their identification using a computational workflow. Nucleic Acids Res 42:7539–7552. https://doi.org/10.1093/nar/gku465
62. Chou C-H, Chang N-W, Shrestha S et al (2016) miRTarBase 2016: updates to the experimentally validated miRNA-target interactions database. Nucleic Acids Res 44:D239–D247. https://doi.org/10.1093/nar/gkv1258
63. Betel D, Koppal A, Agius P et al (2010) Comprehensive modeling of microRNA targets predicts functional non-conserved and non-canonical sites. Genome Biol 11:R90. https://doi.org/10.1186/gb-2010-11-8-r90
64. Lai X, Gupta SK, Schmitz U et al (2018) MiR-205-5p and miR-342-3p cooperate in the repression of the E2F1 transcription factor in the context of anticancer chemotherapy resistance. Theranostics 8:1106–1120. https://doi.org/10.7150/thno.19904
65. Veneziano D, Nigita G, Ferro A (2015) Computational approaches for the analysis of ncRNA through deep sequencing techniques. Front Bioeng Biotechnol 3:77. https://doi.org/10.3389/fbioe.2015.00077
66. Yan K, Arfat Y, Li D et al (2016) Structure prediction: new insights into decrypting long

noncoding RNAs. Int J Mol Sci 17:132. https://doi.org/10.3390/IJMS17010132
67. Guo X, Gao L, Wang Y et al (2016) Advances in long noncoding RNAs: identification, structure prediction and function annotation. Brief Funct Genomics 15:38–46. https://doi.org/10.1093/bfgp/elv022
68. Volders P-J, Helsens K, Wang X et al (2013) LNCipedia: a database for annotated human lncRNA transcript sequences and structures. Nucleic Acids Res 41:D246–D251. https://doi.org/10.1093/nar/gks915
69. Ebert MS, Sharp PA (2012) Roles for MicroRNAs in conferring robustness to biological processes. Cell 149:515–524. https://doi.org/10.1016/J.CELL.2012.04.005
70. Yamamura S, Imai-Sumida M, Tanaka Y, Dahiya R (2018) Interaction and cross-talk between non-coding RNAs. Cell Mol Life Sci 75:467–484. https://doi.org/10.1007/s00018-017-2626-6
71. Alcaraz N, Küçük H, Weile J et al (2011) KeyPathwayMiner: detecting case-specific biological pathways using expression data. Internet Math 7:299–313. https://doi.org/10.1080/15427951.2011.604548
72. Hausburg F, Jung JJ, Hoch M et al (2017) (Re-)programming of subtype specific cardiomyocytes. Adv Drug Deliv Rev 120:142–167. https://doi.org/10.1016/j.addr.2017.09.005
73. Khan FM, Schmitz U, Nikolov S et al (2014) Hybrid modeling of the crosstalk between signaling and transcriptional networks using ordinary differential equations and multi-valued logic. Biochim Biophys Acta 1844:289–298. https://doi.org/10.1016/J.BBAPAP.2013.05.007
74. Khan FM, Marquardt S, Gupta SK et al (2017) Unraveling a tumor type-specific regulatory core underlying E2F1-mediated epithelial-mesenchymal transition to predict receptor protein signatures. Nat Commun 8:198. https://doi.org/10.1038/s41467-017-00268-2
75. Wiwie C, Rauch A, Haakonsson A, et al (2017) Elucidation of time-dependent systems biology cell response patterns with time course network enrichment. arXiv.org arXiv:1710.10262
76. Stuart JM, Segal E, Koller D, Kim SK (2003) A gene-coexpression network for global discovery of conserved genetic modules. Science 302:249–255. https://doi.org/10.1126/science.1087447
77. Yavari A, Bellahcene M, Bucchi A et al (2017) Mammalian γ2 AMPK regulates intrinsic heart rate. Nat Commun 8:1258. https://doi.org/10.1038/s41467-017-01342-5
78. Langfelder P, Horvath S (2008) WGCNA: an R package for weighted correlation network analysis. BMC Bioinformatics 9:559. https://doi.org/10.1186/1471-2105-9-559
79. Li S, Li B, Zheng Y et al (2017) Exploring functions of long noncoding RNAs across multiple cancers through co-expression network. Sci Rep 7:754. https://doi.org/10.1038/s41598-017-00856-8
80. Huang DW, Sherman BT, Lempicki RA (2009) Systematic and integrative analysis of large gene lists using DAVID bioinformatics resources. Nat Protoc 4:44–57. https://doi.org/10.1038/nprot.2008.211
81. Chen EY, Tan CM, Kou Y et al (2013) Enrichr: interactive and collaborative HTML5 gene list enrichment analysis tool. BMC Bioinformatics 14:128. https://doi.org/10.1186/1471-2105-14-128
82. Kuleshov MV, Jones MR, Rouillard AD et al (2016) Enrichr: a comprehensive gene set enrichment analysis web server 2016 update. Nucleic Acids Res 44:W90–W97. https://doi.org/10.1093/nar/gkw377
83. Caicedo JC, Cooper S, Heigwer F et al (2017) Data-analysis strategies for image-based cell profiling. Nat Methods 14:849–863. https://doi.org/10.1038/nmeth.4397
84. Kuhn M (2008) Building predictive models in R using the caret package. J Stat Softw 28:1–26. https://doi.org/10.18637/jss.v028.i05
85. Ray SS, Maiti S (2015) Noncoding RNAs and their annotation using metagenomics algorithms. Wiley Interdiscip Rev Data Min Knowl Discov 5:1–20. https://doi.org/10.1002/widm.1142
86. Saeb S, Lonini L, Jayaraman A, et al (2016) Voodoo machine learning for clinical predictions. bioRxiv: 059774. https://doi.org/10.1101/059774
87. Yu N, Cho KH, Cheng Q, Tesorero RA (2009) A hybrid computational approach for the prediction of small non-coding RNAs from genome sequences. In: 2009 International Conference on Computational Science and Engineering. IEEE, pp 1071–1076
88. van der ML, Hinton G (2008) Visualizing Data using t-SNE. J Mach Learn Res 9:2579–2605
89. Sun K, Chen X, Jiang P et al (2013) iSeeRNA: identification of long intergenic non-coding RNA transcripts from transcriptome sequencing data. BMC Genomics 14(Suppl 2):S7.

https://doi.org/10.1186/1471-2164-14-S2-S7

90. Xiao Y, Lv Y, Zhao H et al (2015) Predicting the functions of long noncoding RNAs using RNA-Seq based on Bayesian network. Biomed Res Int 2015:1–14. https://doi.org/10.1155/2015/839590

91. Lertampaiporn S, Thammarongtham C, Nukoolkit C et al (2014) Identification of non-coding RNAs with a new composite feature in the Hybrid Random Forest Ensemble algorithm. Nucleic Acids Res 42:e93. https://doi.org/10.1093/nar/gku325

92. Abbas Q, Raza SM, Biyabani AA, Jaffar MA (2016) A review of computational methods for finding non-coding RNA genes. Genes (Basel) 7:113. https://doi.org/10.3390/genes7120113

93. Lee B, Baek J, Park S, Yoon S (2016) deepTarget: end-to-end learning framework for microRNA target prediction using deep recurrent neural networks. arXiv.org arXiv:1603.09123

94. Cheng S, Guo M, Wang C et al (2016) MiRTDL: a deep learning approach for miRNA target prediction. IEEE/ACM Trans Comput Biol Bioinform 13:1161–1169. https://doi.org/10.1109/TCBB.2015.2510002

95. Karczewski KJ, Snyder MP (2018) Integrative omics for health and disease. Nat Rev Genet 19:299–310. https://doi.org/10.1038/nrg.2018.4

96. Ison J, Rapacki K, Ménager H et al (2016) Tools and data services registry: a community effort to document bioinformatics resources. Nucleic Acids Res 44:D38–D47. https://doi.org/10.1093/nar/gkv1116

97. Ison J, Kalas M, Jonassen I et al (2013) EDAM: an ontology of bioinformatics operations, types of data and identifiers, topics and formats. Bioinformatics 29:1325–1332. https://doi.org/10.1093/bioinformatics/btt113

98. Wolkenhauer O (2014) Why model? Front Physiol 5:21. https://doi.org/10.3389/fphys.2014.00021

99. Doudna JA, Charpentier E (2014) Genome editing. The new frontier of genome engineering with CRISPR-Cas9. Science 346:1258096. https://doi.org/10.1126/science.1258096

100. Palazzo AF, Lee ES (2015) Non-coding RNA: what is functional and what is junk? Front Genet 6:2. https://doi.org/10.3389/fgene.2015.00002

101. Scarano D, Rao R, Corrado G (2017) In silico identification and annotation of non-coding RNAs by RNA-seq and de novo assembly of the transcriptome of Tomato Fruits. PLoS One 12:e0171504. https://doi.org/10.1371/journal.pone.0171504

102. Garalde DR, Snell EA, Jachimowicz D et al (2018) Highly parallel direct RNA sequencing on an array of nanopores. Nat Methods 15:201–206. https://doi.org/10.1038/nmeth.4577

103. Webb S (2018) Deep learning for biology. Nature 554:555–557. https://doi.org/10.1038/d41586-018-02174-z

Chapter 6

ncRNA Editing: Functional Characterization and Computational Resources

Giovanni Nigita, Gioacchino P. Marceca, Luisa Tomasello, Rosario Distefano, Federica Calore, Dario Veneziano, Giulia Romano, Serge Patrick Nana-Sinkam, Mario Acunzo, and Carlo M. Croce

Abstract

Noncoding RNAs (ncRNAs) have received much attention due to their central role in gene expression and translational regulation as well as due to their involvement in several biological processes and disease development. Small noncoding RNAs (sncRNAs), such as microRNAs and piwiRNAs, have been thoroughly investigated and functionally characterized. Long noncoding RNAs (lncRNAs), known to play an important role in chromatin-interacting transcription regulation, posttranscriptional regulation, cell-to-cell signaling, and protein regulation, are also being investigated to further elucidate their functional roles.

Next-generation sequencing (NGS) technologies have greatly aided in characterizing the ncRNAome. Moreover, the coupling of NGS technology together with bioinformatics tools has been essential to the genome-wide detection of RNA modifications in ncRNAs. RNA editing, a common human co-transcriptional and posttranscriptional modification, is a dynamic biological phenomenon able to alter the sequence and the structure of primary transcripts (both coding and noncoding RNAs) during the maturation process, consequently influencing the biogenesis, as well as the function, of ncRNAs. In particular, the dysregulation of the RNA editing machineries have been associated with the onset of human diseases.

In this chapter we discuss the potential functions of ncRNA editing and describe the knowledge base and bioinformatics resources available to investigate such phenomenon.

Key words A-to-I RNA editing, ncRNA editing, 3′UTR, Introns, miRNA, lncRNA, Bioinformatics, NGS

1 Regulation of Gene Expression in Humans: From Static to Dynamic

As a result of many attempts and thanks to the concurrent development and improvement of more innovative technologies, unprecedented advances have been made in the field of molecular biology since the determination of the double-helix structure of the DNA by J. Watson and F. Crick in 1953 [1]. Following those early discoveries, it was soon suggested that DNA was the real carrier of genetic information and that sequence of DNA bases (adenine (A) always pairing with thymine (T), and cytosine (C) always pairing with guanine (G)) on a DNA molecule contained specific genetic instructions. However, very little was known about the RNA at that time.

During the early 1940s, the biochemist J. Brachet and the cytologist T. Caspersson independently showed that the cytoplasmic environment was rich in RNA, and pointed out for the first time a strong correlation between ribonucleic acid levels and protein synthesis [2, 3]. About 10 years later, several works [4–6] shed light on the two most abundant noncoding RNA (ncRNA) classes, directly involved in protein synthesis, later renamed as ribosomal RNA (rRNA) and transfer RNA (tRNA), respectively. However, the concept of coding RNA still remained obscure. Altogether, these findings led F. Crick to set out the well-known "central dogma" of molecular biology in 1958 [7]; albeit this model is to be considered incomplete and "static" in light of the current knowledge, it still represents the cornerstone of the modern genomics and transcriptomics (see Fig. 1).

Consequent to the divulgation of these observations, several research groups shifted their focus on RNA, in the attempt to further define its role in protein synthesis and, more generally, in cell biology. This led to hard evidences about the functional role of coding RNA, or messenger RNA (mRNA) on the midway between 1960s and 1970s [8–10]. The discovery of this third class of RNA led to get a much clearer picture of how the genetic information contained into nuclear DNA is turned into proteins. In 1977, a few scientific works from the labs of R. Roberts and P. Sharp gave proof of mind-blowing evidences. Whereas bacterial mRNAs resulted to be faithful copies of their DNA counterpart, eukaryotic ones were not, since in these organisms genes are split by many interrupting sequences, not present in the "mature" version of coding transcripts [11–13]. Aware of this fact, scientists changed the way they thought about genome structure and defined the key concepts of exon and intron. However, it still remained totally unclear what caused the exclusion of introns from mature mRNAs and the splicing of exons.

Isolation of the non-rRNA, non-tRNA, and non-mRNA fraction consented the identification of yet unknown ncRNA classes. One of these included small, U-rich ncRNAs, termed small nuclear RNAs (snRNAs), found to be essential components of ribonucleoprotein (RNP) complexes, located into the cellular nucleus [14]. Soon after, the RNPs were identified as the basic components of that previously obscure molecular machinery responsible for processing precursors of mRNA (pre-mRNA) into mature mRNAs: the spliceosome [15, 16]. Similar experiments led to the discovery of small nucleolar RNAs (snoRNAs), found to have

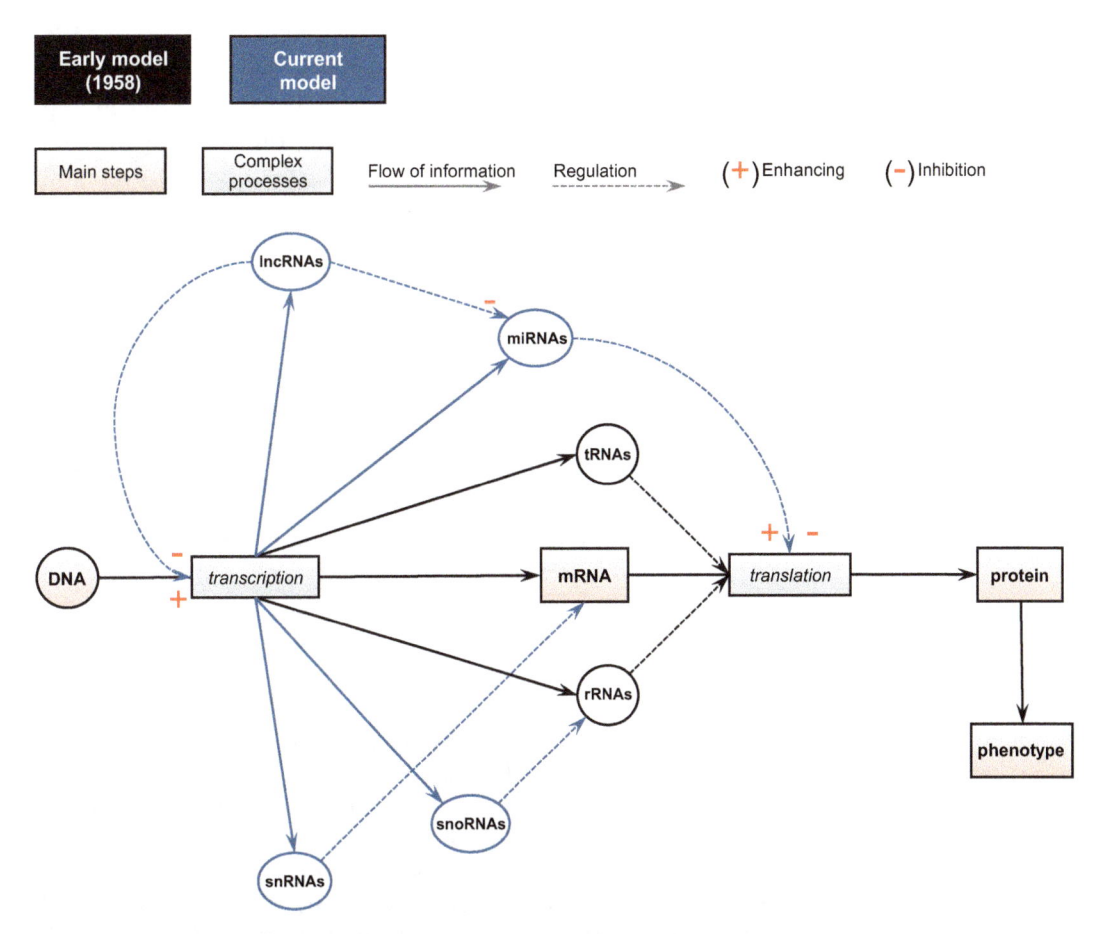

Fig. 1 Schematic chart showing the simplistic view of the early-enunciated "central dogma of molecular biology" model and the complexity of the current dynamic model. The chart intentionally highlights impact and roles of various ncRNA classes in modulation of gene expression

essential functional roles related to maturation, modification, and stabilization of rRNAs [17].

Other important classes of ncRNAs were identified on the midway between late 1980s and early 1990s, though their roles in cell biology had remained enigmatic for several years after their discovery. For instance, long noncoding RNAs (lncRNAs) were discovered during experiments of genomic imprinting with mice. At the beginning of their discovery, lncRNAs were considered as mRNA molecules. However, they soon showed noncoding features, but were rather involved in processes leading to regulation of gene expression [18–21]. Meanwhile, experiments carried out on the model organism *Caenorhabditis elegans* allowed the isolation of two previously unknown sncRNAs, later termed microRNAs (miRNAs), suggested to exert a basilar role in mechanisms leading to posttranscriptional gene silencing. In particular, these sncRNAs were shown to interact with elements contained, for the most part, in the 3′ untranslated region (UTR) of certain mRNAs by complementary base pairing, causing the blockage of the translational stage and/or degradation of the targeted RNA transcript [22, 23]. These new perspectives shed light on the many different roles that RNA plays, not only in every step of protein synthesis, but also in several mechanisms underlying modulation of gene expression.

Since then, knowledge on ncRNAs and their importance in cell biology has impressively increased. Much attention has been posed on ncRNAs having a role in regulatory processes like modulation of protein expression and induction of chromatin remodeling and new classes of ncRNA molecules have been discovered, such as that of piRNAs [24] and tRNA-derived fragments (tRFs) [25–28]. In support of research in this field, several dedicated computational sources have been created, allowing accessing of ncRNA-related information as well as to analyze data from experiments and predicting outcomes, as extensively reviewed below.

2 RNA Modification

In their basic form, the four nitrogenous bases adenosine (A), cytosine (C), guanosine (G), and uracil (U) usually referred to as "bases" represent the "four letters of the RNA alphabet." Each one of them, together with a ribose molecule bonded to a phosphate group through its 5′ carbon atom, constitutes a nucleotide. The resulting four different nucleotides are the building blocks of all RNA molecules and their ordered sequence itself determines the functional role of the coding or noncoding RNA (see Fig. 2).

A and G are classified as purines, consisting of a pyrimidine ring (i.e., a six-membered ring composed of carbon and nitrogen atoms) fused to an imidazole ring. C and U are classified as pyrimidines and are characterized by a unique pyrimidine ring. Unlike DNA, RNA is typically found as a single-stranded nucleic acid. However, RNA can easily interact with other nucleic acid strands since its four major bases dispose of base-specific patterns of hydrogen bond acceptors and donors (see Fig. 2).

This enables the mutual recognition of complementary regions, principally according to the canonical Watson–Crick base pairing. However, non-Watson and Crick thermodynamically stable base-pairing patterns also exist as is the case of the Hoogsteen [29] and the G-U wobble [30, 31] base pairing. Such acceptor-donor interactions are relatively strong and dictate transient RNA-RNA/RNA-DNA/RNA-protein interactions as well as single-stranded RNA (ssRNA) folding events, leading to the formation of double-stranded RNA (dsRNA) regions. This latter phenomenon induces higher structural levels in ssRNA molecules, i.e., secondary and tertiary structures, having great importance from the functional point of view. In particular, secondary

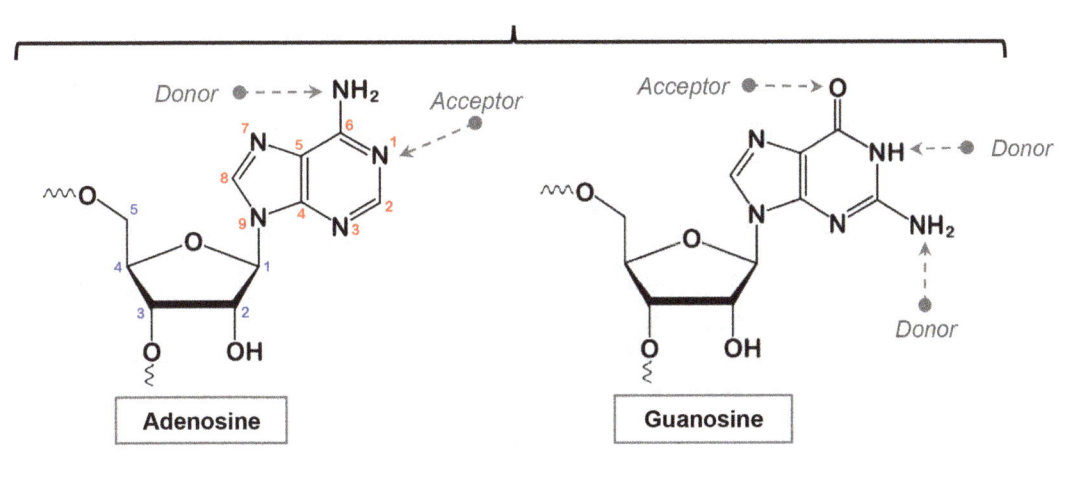

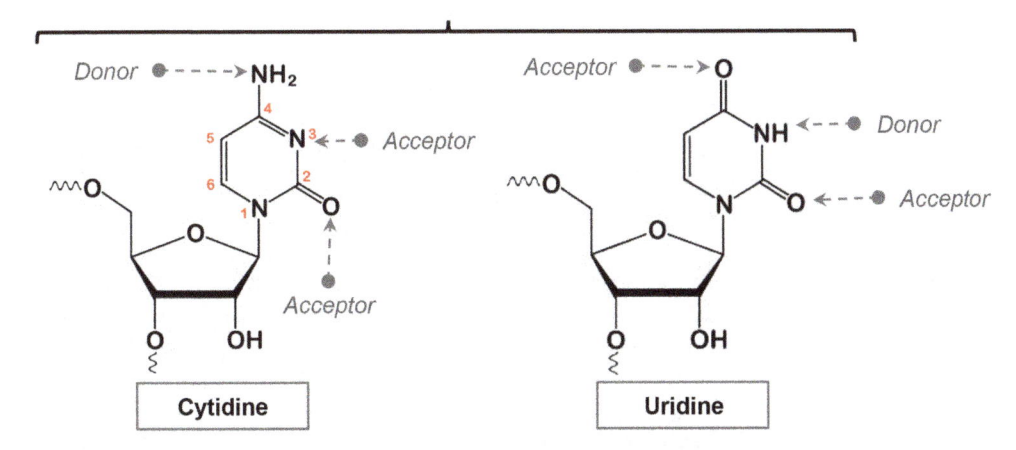

Fig. 2 The four basic nucleotides constituting RNA molecules. Their nitrogenous bases are divided into pyrimidines and purines. Red numbers show the IUPAC numbering for nucleosides of both categories. Blue numbers indicate the IUPAC numbering for the ribose. Chemical functional groups suitable for hydrogen bounds are indicated as acceptors and donors

structures are basically due to canonical base pairing of complementary regions of the same ssRNA molecule and are quite predictable. Differently, tertiary structures involve mainly noncanonical hydrogen-bonding interactions, including hydrogen bonds between specific bases' acceptors and the 2′-hydroxyl group of ribose, resulting in hardly predictable three-dimensional motifs [32].

Aside from the four basic RNA alphabet letters, several "minor letters" also exist, notably expanding the RNA vocabulary. These noncanonical bases are de facto obtained from canonical nucleotides through secondary modifications, as a consequence of RNA-processing events. The concept of RNA processing originated with concurrent findings on tRNA and rRNA indicates not only that these RNAs derive from longer precursors, but also that posttranscriptional modifications, such as methylation, thiolation, deamination, and isomerization, occur at the single-base level [33, 34]. A certain number of modification patterns in these ncRNAs have been found highly conserved across divergent taxa, highlighting a critical role in cell biology [35, 36], while others are taxon specific, potentially evidencing taxon-related molecular diversity (e.g., [37]). More importantly, several studies have revealed tissue specificity as well as condition-dependent modification patterns for tRNAs [38], suggesting that dynamic changes of tRNA modifications are influenced by, and exert influence on, different biological processes within the same organism. Following the same line, similar observations have been reported for other RNA classes, though many functional details still remain overshadowed.

Given the unquestionable importance of such issue, the research community has recently expressed an urgent need to gain more hints allowing the deciphering of this sophisticated RNA language in terms of biological functions. In this scenario, the blossoming research field of epitranscriptomics (i.e., the study of functionally relevant biochemical RNA modifications not involving the ribonucleotide sequence) is assuming an ever-growing importance.

2.1 Epitranscriptomics

In nature, RNA modifications are the outcome of either single-step or multistep biochemical transformations of canonical nucleosides (i.e., the unit comprising one of the four nitrogenous bases bonded to the 1′ of the ribose, but not the 5′ phosphate group), catalyzed by a series of enzymes acting in a well-defined order. Since the end of the 1980s, RNA modifications had increasingly been receiving attention from scientific community and this was attested by the birth and development of the first dedicated databases. In line with this, in 1999 McCloskey and coworkers illustrated the early online version of RNA Modification Database (RNAMDB) [39], reporting slightly more than 95 different nucleoside modifications in RNA. However, 93 out of those RNA modifications had been already described by the same research group back in 1994 [40]. For each modified nucleoside shown, the database reported basic information such as chemical structure, occurrence among RNA classes, occurrence among phylogenetic groups, and corresponding literature references and citations. Twelve years later, the 2011 release of RNAMDB comprised 109 different modified nucleosides, still reporting the same basic information for each noncanonical base [41]. In more recent times, 163 RNA modifications have been described on the MODOMICS database [42], with the addition of advanced information related to each modified nucleoside, such as location into the RNA sequence, biosynthetic pathway, and information on RNA-modifying enzymes.

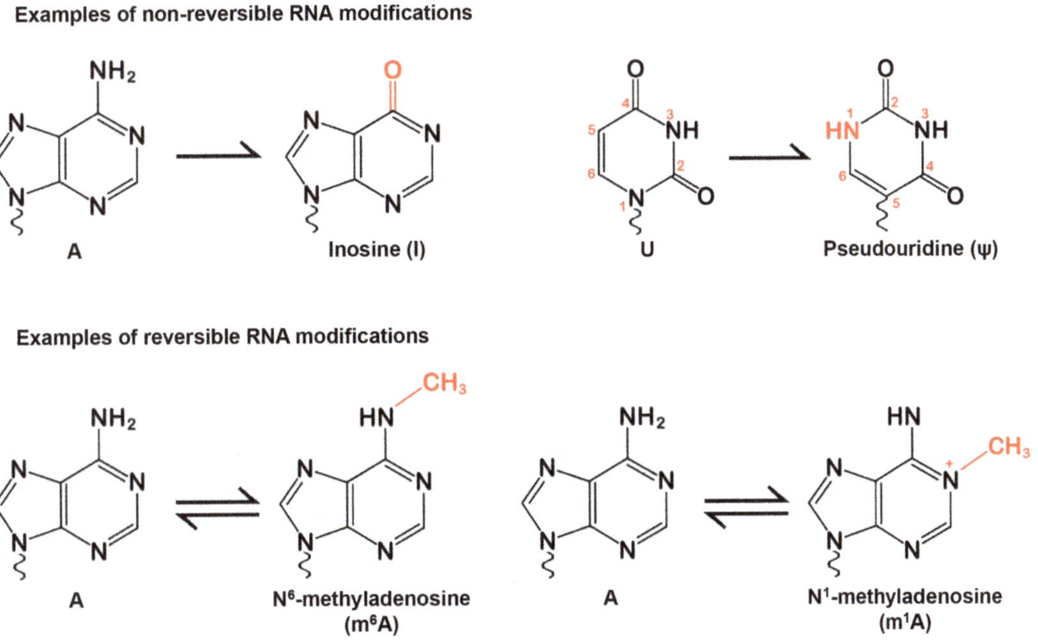

Fig. 3 Structural variations of nucleosides in some examples of reversible (m⁶A and m¹A RNA methylations) and nonreversible (A-to-I RNA editing and pseudouridylation) base modifications

As the three aforementioned editions of RNAMDB witnessed a somewhat linear increase in the amount of information collected over the years, the 2018 release of MODOMICS has evidenced a decisively exponential boost both in quality and quantity of RNA modification-related data. The explanation of that obviously lies in recent rapid improvements of high-throughput sequencing (HTS) technologies and also in the setting up of independent detection methods relying on principles like differential chemical reactivity, physicochemical properties, and differential enzymatic turnover [43, 44].

Nucleoside modifications could be classified as "nonreversible" or "reversible" (*see* Fig. 3).

As the very term indicates, nonreversible modifications lead to permanent changes of RNA properties, since no enzymes exist capable to revert the biochemical reaction exist. Well-known examples are deamination of adenosine, leading to formation of inosine (I) (*see* next sections), and isomerization of uridine to pseudouridine (ψ) [45] (*see* Fig. 3).

Differently, reversible modifications are associated with enzymes capable to bring back the modified bases into their original state once the modification had occurred. In this case, enzymes are conventionally divided into three different categories: (a) "writers," i.e., enzymes capable to transform a canonical base into a noncanonical one; (b) "erasers," i.e., enzymes allowing the reversion of a noncanonical base to its corresponding canonical base; (c) "readers," i.e., enzymes capable to bind the noncanonical nucleoside(s), transducing the chemical modification(s) into a biological function [46]. Notorious examples of reversible modified nucleosides are of N^6-methyladenosine (m^6A) [47] and N^1-methyladenosine (m^1A) [48] (*see* Fig. 3).

In line with DNA modification, RNA modification is a master regulator influencing several aspects of protein expression, molecular functions, cell growth, and homeostasis. Thus, any imbalance in RNA modification can lead to disruption of the physiological status. Nonetheless, epitranscriptomic modifications are significantly distinct when compared with epigenomic modifications [49].

Certainly, an important topic to be discussed is how RNA modifications can induce or modulate certain molecular functions. A key point in molecular biology is that the three-dimensional (3D) structure of any macromolecule determines its biological function, whereby the latter is in turn determined by an ordered series of atomic interactions. Thus, even a single modification at the atomic level can considerably affect the conformational status and/or the flexibility of the entire molecule. Applying this principle to epitranscriptomics, it has been possible to uncover several aspects regarding the close relationship between ribonucleotide sequence and RNA 3D structure, both at local and global scale [50, 51]. Nucleoside modifications can significantly influence the stereoelectronic effects between the base itself and the ribose, consequently determining the conformational preference of the latter. At a larger scale, such stereoelectronic changes can also affect base-base interactions. One of the most important examples regards the π-π stacking effects, showing that base modifications can increase or decrease such interactions between nitrogenous bases' aromatic rings and thus the RNA stability, at least locally [52]. Noteworthy, a considerable portion of the current knowledge on RNA modifications and their effect on RNA structure/functionality comes from a large amount of studies carried out on tRNAs, considered by far as the most important source of modified nucleosides [52–54].

2.2 RNA Editing

In general, RNA editing describes alterations of RNA sequence either by non-templated nucleotide insertion/deletion or through base conversion by deamination/amination [55]. More recently, however, this terminology has been principally used in literature to indicate the latter aspect. From this viewpoint, adenosine to inosine (A-to-I) and cytidine to uridine (C-to-U) RNA editing phenomena are the best characterized in humans, though detection of C-to-U editing in ncRNAs is much less frequent than A-to-I editing. In this regard, no extensive functional studies exist on C-to-U editing to this day and some groups has interpreted C-to-U editing of regulatory ncRNAs as false positives or as probable random events (e.g., [56, 57]). However, C-to-U editing will not be discussed in what follows.

Conversion of adenosine to inosine is due to hydrolytic deamination occurring on carbon (C)6 of adenosine. Such biochemical reaction is catalyzed by adenosine deaminase acting on RNA (ADAR) enzymes, able to bind dsRNA molecules (*see* below). ADAR family proteins are found in almost all metazoans, with vertebrata possessing three of these proteins, i.e., ADAR1-3, although not all of them are involved in catalyzing deamination of adenosine [58].

2.2.1 ADAR Family Members' General Features

In humans, ADAR1 and ADAR2 are the chief functional adenosine deaminases acting on RNA and the best characterized among the ADAR family members. The *ADAR1* gene is located on chromosome 1. Its transcription can start from three different promoters, with two of them being constitutive while the other being interferon inducible, leading to the expression of two different main isoforms [59]. The constitutively expressed isoform is $\approx$110 kDa in size (ADAR1p110) and lacks the N-terminal portion of the protein containing the Zα domain (*see* below), while the interferon (INF)-inducible isoform is a full-length protein having a molecular weight of $\approx$150 kDa (ADAR1p150). The former is primary located into the nucleolus, while the latter is mainly located into the cytoplasm, although both these isoforms can shuttle between the nucleus and the cytoplasm [60–62]. In addition to that, alternative splicing events contribute to the generation of further ADAR1 isoforms [63] which differ in catalysis of adenosine deamination due to differential distance between ADAR1's protein domains (e.g., [64, 65]).

The *ADAR2* gene is located on chromosome 21 and is primarily expressed in the brain [66], whereby its expression is regulated by the transcription factor CREB [67]. However, *ADAR2* can be expressed in other tissues as well [66]. For instance, in pancreatic β-cells ADAR2 expression is regulated by the activated c-Jun transcription factor [68]. However, it is likely that other transcriptional factors are involved in such event [68]. The ADAR2 enzyme is found in only one abundant form and has a predominant nucleolar localization, though it can shuttle between nucleolus and nucleoplasm [62, 69]. As in the case of ADAR1, alternative spliced variants of ADAR2 also exist, having altered enzymatic activity compared to the main isoform (e.g., [65, 70]).

Unlike the two previous adenosine deaminases, the ADAR3 protein, whose gene is located on chromosome 10, is catalytically inactive, although arrangement and sequence of its deaminase domain (*see* below) are very similar to the one of ADAR2

[71]. *ADAR3* expression is restricted to the human brain, where certain regions, such as amygdala and thalamus, show significantly higher expression levels of this gene with respect to other brain's regions [71]. Importantly, ADAR3 negatively correlates with editing levels in brain and causes variations in ADAR1- and ADAR2-related editing levels [72]. In particular, ADAR3 behaves as an editing inhibitor by directly competing for binding dsRNA substrates with the enzymatically active ADARs, thus playing an essential role in the editing regulation in the brain [73]. Consistent with this, alterations in *ADAR3* expression are very likely to impact processes related to memory formation and behavior, and may contribute to tumor development [73, 74].

2.2.2 Structures of the ADAR proteins

Human ADARs are modular proteins with multiple folded domains, some of which are shared by all the ADAR members. In fact, each ADAR has a C-terminal deaminase domain and up to three dsRBDs located on the N-terminal side. The deaminase domain is the ADARs' active site where the hydrolytic deamination takes place, probably through a base-flipping mechanism [75]. X-ray crystallography of the ADAR2 deaminase domain has revealed a protein structure comprising 10 α-helices and 11 β-strands interconnected by several loops. Such structure is ~400 amino acids (aa) in length and is disposed in a virtually spherical 40 Å diameter structure. Here, four amino acidic residues, i.e., His394, Glu396, Cys451, and Cys516, are involved in the formation of the catalytic core, all coordinated by a zinc ion [75]. Furthermore, an inositol hexakisphosphate molecule buried within the enzyme core function as a structural element contributing to the correct protein fold [75]. In terms of primary structure and length, ADAR1 deaminase domain is very similar to that of ADAR2 being formed by ~400 aa and scoring 39% identity and 59% similarity in relation to the latter, though the surface loop involved in RNA recognition significantly differs between these two ADARs [76].

ADARs' dsRNA-binding domains (dsRBDs) make direct contact with dsRNA molecules [71, 77]. The dsRBD is ~65 aa in length and has an αβββα secondary conformation, where the two alpha helices lie on one face of a three-stranded antiparallel beta sheet [78]. In general, ADARs require well-structured dsRBDs in order to carry out their catalytic function. However, not all the ADARs' dsRBDs influence A-to-I editing in the same way. Specifically, of the three dsRBDs of ADAR1, dsRBD1 and dsRBD3 play a relevant role in RNA editing while dsRBD2 seems to not influence the efficiency of ADAR1-mediated catalysis [79, 80]. In the case of ADAR2, dsRBD1 is only required for editing a subset of ADAR2 substrates, while dsRBD2 is always required for RNA editing [81]. dsRBDs are also required for ADAR homodimerization, an essential condition for the A-to-I RNA editing activity. Defects in dsRBDs of ADAR monomers prevent the cooperative interactions between ADAR homodimers and thus do not allow homodimer formation [82]. Differently from the enzymatically active ADARs, ADAR3 remains as a monomeric form [82]. Importantly, the dsDRB3 of ADAR1 contains motives interacting with transporter proteins, determining the intracellular localization of this enzyme [83, 84].

Despite the aforementioned protein motifs common to all ADARs, there are also ADAR-specific exclusive structural features. ADAR1p150 presents two additional domains, termed Zα and Zβ, located at its N-terminal region. The former, closer to the N-terminus region of the enzyme, has a common helix-turn-helix architecture with α/β topology and is capable to bind both DNA and RNA in left-handed Z conformation [85, 86]. Intriguingly, the binding of left-handed Z-RNA by ADAR1p150 increases the enzyme's catalytic activity and concomitantly alters the pattern of base modification along the RNA duplex as reaction times decrease [86]. Furthermore, the binding of ADAR1p150 Zα domain to certain DNA promoter regions is capable to regulate gene expression, while its binding to actively transcribing DNA regions may allow co-transcriptional editing activity on RNA [87, 88]. Such functional mechanisms are peculiar of the ADAR1p150 isoform and may underlie its role in interferon-induced antiviral defense pathway. Worthy of note, ADAR1p150 Zα domain plays an important role in intracellular location of ADAR1p150 as it includes a nuclear export signal (NES) [61]. Differently from Zα domain, Zβ domain is not exclusive of the ADAR1p150 isoform, but it is also present in the ADAR1p110 isoform. Similarly to the Zα domain, the ADAR1p150 Zβ domain also has a helix-turn-helix architecture with α/β topology, but with a different arrangement and with the addition of a fourth C-terminal helix (α1-β1-α2-α3-β2-β3-α4) [89]. Despite both ADAR1p150 Zα and Zβ domains being similar in fold, the latter cannot bind to nucleic acids as it lacks crucial residues [89].

ADAR3 also contains an exclusive domain at its amino-terminal region, rich in arginine residues and allowing the binding of ssRNA molecules, termed R domain [71]. To date no extensive information is available for this protein domain. However, it is thought that such R domain serves as a nuclear localization signal since it mediates interactions between ADAR3 and a transporter protein complex enabling ADAR3 to locate to the nucleus, where the majority of ADAR-mediated RNA editing is believed to occur [90].

2.2.3 The Product of ADAR Activity: A-to-I Editing

Conversion of adenosine into inosine on coding and noncoding transcripts is highly relevant in cell biology, since it alters structural and functional properties of RNA molecules. In particular, in the context of base pairing, inosine stably pairs with cytosine due to Watson–Crick base pairing, though also less stable pairings can occur with the other three bases [91]. For such reason, inosine is commonly recognized as a guanosine by both the translational machinery and the spliceosome, leading to remarkable effects at the protein level [92]. However, no studies have described the thermodynamic properties of inosine to date. For such reason, predictive studies on RNA structure following conversion of A-to-I have used G instead of I (e.g., [93]). A considerable body of evidence collected over the past years have brought out that ADAR-mediated A-to-I editing is not just a random event, but it shows well-defined patterns, as discussed hereafter.

ADAR-dependent editing has classically been considered as a posttranscriptional event exerting crucial regulation on coding and noncoding transcripts [92, 94]. Notwithstanding, strong evidence shows that A-to-I editing can occur at the co-transcriptional level, at least in the case of ADAR2 [95–97]. In such cases, interactions between ADAR2 and the C-terminal domain of the large subunit of RNA-polymerase II are required in order to coordinate efficient editing events [95, 96].

Distinct inverted repeats belonging to the class of "short interspersed nuclear element" (SINE) transposable elements, termed *Alu*, represent the main editing target for ADAR enzymes [98–100]. When located close enough to reversely oriented *Alu* sequences, *Alu* repeats induce the folding of single-stranded RNA (ssRNA) regions into double-stranded stem loops. Such RNA structures represent important editing inducer motifs [56, 101]. Distance between reversely oriented *Alu* and editing levels is inversely correlated by an exponential function, according to which the closer the neighboring *Alu* is, the stronger the stem-loop structure is, with the decay distance being ~800 bp in length. Nonetheless, the editing level starts decreasing if reversely oriented *Alu* is very close to each other, due to the RNA flexibility [101]. In this context, however, only specific adenosine sites within the *Alu* repeats are edited at high levels, while several other adenosine sites in the same region are edited at very low levels (<1%) [101, 102]. *Alus* are estimated to cover slightly more than 10% of the human genome and are primary harbored in noncoding regions flanking exons [103]. Because of that, almost the totality of A-to-I transitions occur in noncoding RNA regions, with introns, 3′UTRs, and ncRNA molecules, such as miRNA and lncRNAs, possessing the higher number of editing sites, while 5′UTRs and protein-coding sequences being edited at very low level [104–107].

Aside from the predominant *Alu* editing phenomenon, RNA editing can also occur in non-*Alu* repeats—mainly represented by non-*Alu* SINEs, long interspersed elements (LINEs), and long terminal repeats—as well as in non-repetitive sequences [104, 105, 108, 109]. Such other transcriptome regions are edited at much lower levels than *Alu* repeats, whereby editing of the latter represents ~90% of edited sites in the human transcriptome [104, 108, 109]. In this scenario, ADAR1 is primarily involved in the editing of both *Alu* and non-$32#*Alu* repetitive sites, whereas ADAR2 is the main editor of non-repetitive sites [72, 110].

At the sequence level, ADAR-binding motifs mainly consist of a triplet of nucleotides comprising both the -1 upstream ($5'$) and $+1$ downstream ($3'$) neighbor nucleotides in respect to the edited adenosine. ADAR-binding motifs differ between ADARs, although they show partial overlapping. ADAR1 has been shown to have $5'$ neighbor preference consisting of $U = A > C > G$, but no $3'$ neighbor preference has been observed [111]. $5'$ Neighboring preference identified for ADAR2 ($U \approx A > C = G$) is similar to that reported for ADAR1. However, differently from ADAR1, ADAR2 also shows a $3'$ neighboring preference, consisting of $U = G > C = A$ [112]. This causes ADAR1 and ADAR2 to target their own specific editing sites, with only a relatively small subset of shared substrates [72, 110, 112]. In general, the UAG triplet has been found to be the most favored, followed by some non-UAG triplets edited with lower efficacy [110, 113]. Within the dsRNA structure, C has been found to oppose the edited A about three times higher than expected by the natural occurrence of A::C mismatches, contributing to stabilization of the dsRNA structure once the conversion A-to-I has occurred [107, 110, 113, 114]. Such occurrence could lead to the idea that the A::C mismatch causes an increase in editing frequency. Despite the A::C punctual mismatch, both the -1 and the $+1$ neighbors of the UAG triplet are usually found to base pair according to the Watson–Crick rule, evidencing the importance of a local stable primary structure surrounding the adenosine residue [113].

All the aforementioned properties, i.e., RNA primary and secondary structures at the local scale, underlie the specificity of the editing events [115, 116]. Briefly, when long, completely base-paired dsRNAs undergo A-to-I editing, multiple adenosines are subjected to ADAR enzymatic activity, giving rise to promiscuous hyper-editing. When A pairs with U, deamination changes A::U base pairs to I::U mismatches that are thermodynamically less stable. Thus, once 50–60% of adenosines in long stretched dsRNA have been edited, several regions of the RNA substrate become single stranded, consequently interrupting ADAR activity. Differently, shorter RNA substrates, or those interrupted by single-stranded bulges and loops, tend to be selectively deaminated in one or few adenosines, since such events cause a thermodynamic destabilization of the double-stranded molecules, causing ADAR release [115]. However, a recent comprehensive study regarding the global effects of ADAR1-mediated editing on RNA secondary structure [114] brought out unexpected important hints. In summary, it seems that ADAR1 editing indeed causes an increase in dsRNA/ssRNA ratio of human cytoplasmic RNAs. However, the ratio is not generalizable, but is rather regulated in a tissue-specific manner. Such apparent contradiction would be explained by the fact that in long perfectly paired dsRNA structures the only possible effect of A-to-I editing is to make the structure less stable due to I::U mismatches. To the contrary, in dsRNA structures formed by imperfect *Alu* repeats, A-to-I editing of the overrepresented A::C mismatches leads to I::C pairs, which contributes to stabilize the double-strand structure [114]. Intriguingly, dysregulation of mechanisms determining the correct RNA folding of secondary and 3D structures can cause the onset of pathological status [114, 117, 118].

In addition to this, a recent model based on the glutamate receptor subunit A2 (GluA2) transcript has been proposed to show how the presence of "editing inducer elements" (EIEs) on RNA substrate increases editing efficiency at specific sites [97]. In this context, an important peculiarity of GluA2 transcripts is that ~100% are edited by ADAR2 at the so-called Q/R site (from glutamine to arginine) in the adult mammalian brain [119]. According to such model, a first long, stable dsRNA stem region (the EIE) is required in order to recruit ADAR, increasing its local concentration. A second short imperfect stem is located in proximity of the EIE, whereby these two dsRNA regions are separated by an internal loop. Once the EIE has recruited high concentration of ADARs, this dsRNA region undergoes promiscuous editing events. On the other hand, however, a highly efficient editing occurs on the selected adenosine posed on the shorter dsRNA. Although the proposed model would undoubtedly give a significant contribution to the knowledge regarding specific substrate recognition by ADAR enzymes, the precise mechanism through which EIEs lead to selective and efficient editing still remains to be further clarified.

As opposed to what is observed in other mammals, in humans no direct correlation is found between ADAR expression and editing levels [56, 72, 107]. One meaningful explanation to that is thought to lie in the existence of several protein-mediated regulatory mechanisms influencing ADARs' enzymatic activity. Overall, protein-mediated regulation of A-to-I editing is carried out either by preventing/facilitating ADAR accessibility to RNA or by directly acting on ADARs. The best known regulatory mechanism impacting the nuclear ADAR enzymatic activity relates on the previously mentioned ADAR shuttling between nucleus and cytoplasm. In this context, dsRBD3's flanking regions of ADAR1p110 and ADAR1p150 constituting the NLS motif and the Zα domain-related NES motif of ADAR1p150 allow interactions between ADAR1 enzymes and specific transporters [61, 83, 84, 120, 121]. Differently, intracellular localization of ADAR2 is determined by posttranslational modifications of such enzyme at several amino acidic residues. Specifically, two members of the importin alpha family, karyopherin subunit α1 (KPNA1) and α3 (KPNA3), are responsible for the nuclear import of the main ADAR2 isoform [90]. Once ADAR2 is translocated into the nuclear compartment, it undergoes phosphorylation of several N-terminal residues, among which Ser26, Ser31, and Thr32, probably due to the activity of a Pro-directed kinase [122]. Phosphorylation of these three amino acidic residues, and primary the one occurring at Thr32, is of particular importance as it forms the base for the peptidyl-prolyl *cis/trans* isomerase PIN1 to efficiently interact with ADAR2 [123]. As a consequence of such interaction, PIN1 catalyzes the cis/trans isomerization of the pThr32-Pro33 peptide bonds [123, 124]. Isomerization of P33 stabilizes ADAR2 causing its detention within the nucleolus, where it exerts its editing activity [123].

A few other molecular factors have been evidenced capable to enhance, reduce, or inhibit ADARs' enzymatic activity, though many details still remain evanescent. For instance, in certain circumstances ADAR1 has been found to undergo sumoylation by SUMO-1, resulting in a significant decrease of its deamination activity [125]. Another example regards the DSS1 protein, which has been found to enhance ADAR2-mediated editing up to 30% [126]. Finally, a plethora of dsRNA-binding proteins (dsRBPs) coexist in human cells, and each one of these could potentially compete or cooperate with ADARs for binding to dsRNAs. An example is ADAR3, which negatively regulates ADAR1-ADAR2-mediated editing by directly competing for binding dsRNA substrates, as explained in the previous section.

3 Functional Characterization of ncRNA Editing

For every single cell in the human body, continuous adjustments in protein expression are necessary in order to maintain homeostasis and properly regulate the intracellular/extracellular environment. Thus, under physiological conditions, cells must be able to rapidly respond to both external and internal stimuli. ncRNA-mediated regulation of protein expression represents a widespread, prompt, effective, and sophisticated regulatory mechanism allowing to achieve such goal [127, 128]. Alternative splicing of coding transcripts represents another powerful and dynamic means through which cells can regulate their physiological status [129]. Efficiency of both these mechanisms strictly relies on the primary sequence of specific RNA regions, derived from noncoding DNA regions. Thus, even a single-nucleotide variant (SNV) on their primary sequence can affect their functional role(s) in cell biology [130]. In this respect, SNVs occurring in noncoding DNA regions corresponding to intronic regions or regulatory ncRNAs are known to be the cause of several pathological conditions, including cancer [130]. A-to-I editing is an example of biochemical process that creates SNVs directly on RNA molecules, somehow influencing their functionality (*see* Fig. 4).

In what follows, validated and potential functions of RNA editing in both the noncoding regions of mRNAs and in ncRNA molecules are discussed.

3.1 A-to-I Editing in Noncoding Regions of Coding Transcripts

3.1.1 A-to-I Editing of Intronic Regions

Introns are noncoding regions included within genes. At a first, introns are transcribed into the early transcript, but are rapidly removed through co-transcriptional splicing events. This classically refers to coding transcripts, where the pre-mRNA has to be processed and its exons have to be spliced after the removal of introns. This leads to generation of mature mRNA, which is subsequently transported into the cytoplasm, where the translation takes place [131]. Introns are also present in some noncoding transcripts like rRNAs, tRNAs [132], and a fraction of lncRNAs [131]. Noteworthy, in the latter class of ncRNA as well as in coding transcripts undergoing alternative splicing, the removal of introns occurs post-transcriptionally instead of co-transcriptionally [131]. Splicing is performed through two consecutive transesterification reactions occurring between RNA nucleotides. For splicing to happen, the presence of two specific sequence motifs is necessary located, respectively, at the 5′ (donor site, usually GU) and at the 3′ (acceptor site, usually AG) of each intron. Moreover, the presence of a branching point, i.e., a nucleotide internal to the intron and exploited for transesterification with the first nucleotide of the donor site, is also required [133–135]. However, these basic splicing elements are estimated to contain at most half the information used for directing splice site recognition, while other informative sequence motifs lie outside these splices. Such motifs function as binding sites for specific RNA-binding proteins termed serine- and arginine-rich (SR) proteins, which in turn lead to enhanced or silenced splicing of well-defined RNA regions [134, 136]. In particular, exonic splicing silencers (ESSs) and exonic splicing enhancers (ESEs) are located in exonic regions flanking introns, whereas intronic splicing silencers (ISSs) and intronic splicing enhancers (ISEs) are located within introns [136]. Aside from being the main protagonists of splicing events, introns have been found to

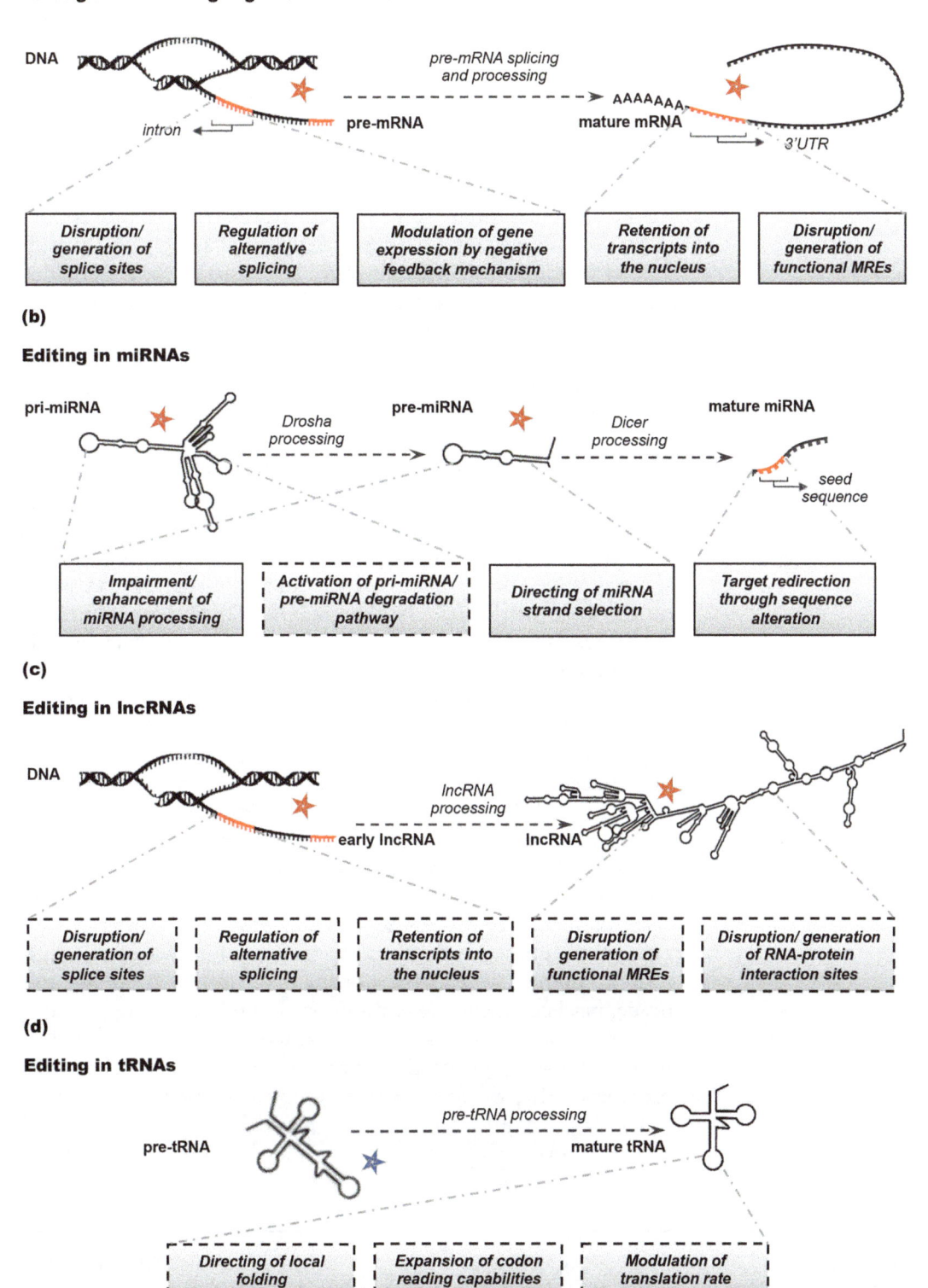

Fig. 4 Functional aspects of A-to-I RNA editing on both noncoding regions of mRNAs (**a**) and ncRNA molecules (**b–d**). For each RNA class is figured the maturation steps, respectively. Stars indicate at what level (co-transcriptional, posttranscriptional, pre-processing, or mature stage) does the A-to-I RNA editing occur.

play several other regulatory roles in cell biology [137], mainly through the posttranscriptional phenomenon of intronic retention (IR) in mature mRNAs [138]. IR events can occur at different positions in the mRNA and influence gene expression by inducing different molecular mechanisms. Briefly, introns included in 5′UTRs contain information for regulation of nuclear mRNA export, determining cytoplasmic mRNA levels [139]. Introns within 3′UTR often contain premature termination codons (PTCs), which trigger the nonsense-mediated decay (NMD) pathway, repressing protein expression [139, 140]. Introns retained within the coding sequence of mRNAs assume diverse biological meanings, since they can be used for (a) production of novel isoforms with specific functions, (b) regulation of protein expression through activation of the NMD pathway, and (c) determination of mRNA intracellular localization [138].

As previously mentioned, introns, along with 3′UTRs, have a high content in repetitive elements and thus are particularly prone to be edited [56, 98–100, 102]. A substantial number of editing events map to intronic *Alu* repeated elements within human pre-mRNAs and can potentially cause disruption or generation of splice sites by A-to-I conversion [98, 141]. In fact, I is read as a G during base pairing and so specific A-to-I editing of intronic sequences can generate noncanonical splice donors—when an AU duplet is converted into IU, read as GU—and acceptors—when an AA duplet is converted into AI, read as AG [92, 141]. This potentially leads to alternative splicing events and could be used by cells as regulatory mechanism that modulates protein expression. An important example consists of self-editing of ADAR2 [142, 143]. ADAR2 has the ability to edit its own pre-mRNA in response to yet undefined physiological stimuli. Here, ADAR2 targets a specific 3′ intronic site within intron 4, creating an alternative 3′ splice acceptor site. This eventually causes the abortion of translation and pre-mRNA degradation, resulting in an important negative feedback mechanism allowing modulation of ADAR2-related editing levels in the human brain [143]. Although this model has been demonstrated for rats and mice, the high (>90%) sequence conservation of inverted repeats in the predicted region as well as conservation of the alternative 3′ splice site potentially presuppose that the same splicing event may occur in humans as well [143]. According to the murine model, mice lacking ADAR2 autoregulation ability show altered editing levels of well-established

Fig. 4 (continued) Red stars represent ADAR-mediated editing (**a–c**), while blue stars represent ADAT-mediated editing (**d**). Repercussions at the functional level of ncRNAs editing are linked to each maturation step through gray dot lines. Solid-lined boxes represent evidence-based effects. Dot-lined boxes represent potential functional repercussions of editing for which no evidences are available to date

editing sites in coding ADAR2 targets, i.e., the R/G site of *GluA2* (different from the Q/R site), the D site of serotonin receptor *5-HT2CR*, and the I/V site of *K_v1.1* [143]. As editing of these transcripts is fundamental in the fine regulation of neurotransmission, altered ADAR2-related editing levels in the brain can lead to the development of neuropsychiatric disorders [144]. Another instance regards the effects of ADAR2-mediated editing in human 5-HT2CR transcripts [145]. Here, editing can occur in six different editing sites (termed A–F, ordered accordingly as 5′→3′). Five of them (A–E) are located on the exonic region (exon 5) of a stem loop, whereas the F site is located on the intronic region (intron 5) of the same stem loop. Different combinations of A-to-I editing of these six sites, including the F site, were shown to profoundly influence alternative splice site selection [145], though details are yet to be uncovered. A-to-I editing has been recently found to occur also in an intronic region (intron 12) of the heterogeneous nuclear ribonucleoprotein L-like (*HNRLL*) transcript, inducing alternative splicing of such transcript [146]. HNRLL is an RNA-binding protein that functions as regulator of alternative splicing for multiple target mRNAs. The "E12A" isoform of the HNRLL transcript, ingenerated through such alternative splicing event, presents an exonic splicing enhancer (ESE) that functions as binding site for the SR splicing factor 1 (SRSF1). SRSF1 is in turn involved in the modulation of both constitutive and alternative splicing events, and leads to the inclusion of "exon 12A" in the mature *HNRLL-E12A* transcripts edited at the specific intronic editing site. Differently from canonical HNRLL, a known tumor-suppressor protein, HNRLL-E12A prevalently targets a subset of cell growth-related genes and is thought to have an important role in tumor progression [146]. Moreover, it seems that the editing event generating the *HNRLL-E12A* variant would prioritize splicing of the latter at the expense of the splicing of the canonical form [146].

3.1.2 A-to-I Editing of 3′UTRs

The 3′ untranslated region is located downstream of protein-coding transcripts. This region contains several conserved motifs involved in numerous regulatory processes at the posttranscriptional level [147]. Such motifs can function as binding sites for a number of RNA-binding proteins, as is the case of AU- and GU-rich elements (AREs, GREs), and influence RNA folding, interactions, and stability [148, 149]. Length and structure of 3′UTRs have deep implications on translation efficiency, as both these features determine the 5′–3′-UTR interaction in mRNA, required for protein expression [150–152]. Also, 3′UTRs contain from few to tens of miRNA response elements (MREs), i.e., RNA sequences allowing partial (rarely perfect) base pairing with a well-delimited miRNA region, termed "seed sequence" (normally nt 2–8). miRNA-mRNA interactions are well known to cause negative regulation of protein expression at the posttranscriptional level by impairing the translational process or by enhancing mRNA degradation [153]. Nonetheless, during the past years a few rare cases have been described in which certain miRNAs induced translational upregulation [154–156]. Established mechanisms explaining the latter cases are (a) miRNA-mediated recruitment of protein complexes associated with translation upregulation [154, 155] and (b) facilitation of 5′–3′-UTR interaction in mRNA by simultaneously pairing 5′miRNA::3′UTR-related MREs and 3′miRNA::5′UTR-related MREs [156].

In some cases, A-to-I editing occurring in inverted repeated *Alus* within 3′UTRs can result in nuclear retention of the transcripts, probably due to editing-induced interactions between the edited 3′UTR and the NONO RNA-binding protein complex [157, 158]. However, the best known consequence of A-to-I conversion within 3′UTRs consists of either potential disruption (as reported for the *ARHGAP26* [159] and *MDM2* [160] genes) or generation (as reported for the *DFFA* gene [161]) of functional MREs. The basic thought is that miRNA-mRNA interactions require a thermodynamically stable base pairing between the MRE and specific miRNA(s). Here, Watson–Crick base pairs between the two interacting RNA regions would be the ideal condition. Nonetheless, in several instances, non-Watso–Crick base pairing, like the G::U wobbles, can sufficiently contribute to the thermodynamic stability of the RNA-RNA interaction. Furthermore, some mismatches are tolerated, primarily depending on the specific localization along the RNA molecule [162, 163]. In line with this, conversion of adenosine(s) to inosine(s) can subsequently cause thermodynamically important mismatches, preventing miRNA-MRE pairing, or give rise to a previously inexistent miRNA-specific MRE. This regulatory mechanism has important implications from the clinical viewpoint. For example, high ADAR1-mediated editing levels of the 3′UTR of *ARHGAP26* mRNA permit the expression of this protein at consistent levels [159]. ARHGAP26 is a Rho GTPase-activating protein, is thought to be a tumor suppressor, and functions as an essential modulator of the CLIC/GEEC endocytic pathway and of cell spreading as well. When ADAR1 expression is inhibited, miR-30b-3p or miR-573 pairs with complementary sequences included in the non-edited 3′UTR of *ARHGAP26* mRNA, exerting an inhibitory role. This leads to upregulation of RhoA and Cdc42, which have growth-promoting effects and a role in some human cancers. To the contrary, editing of the 3′UTR of this mRNA disrupts MREs for both miR-30b-3p and miR-573 [159]. A diametrically opposed effect was described for the E3 ubiquitin-protein ligase MDM2, an important negative regulator of the tumor suppressor p53 [160]. In this case, the non-edited 3′UTR of *MDM2* mRNA would be targeted by two members of the miR-200 family (i.e., miR-200b/c) allowing silencing of MDM2. Differently, editing of the *MDM2* 3′UTR would prevent base pairing with these two miRNAs, leading to a potential tumorigenic effect [160].

On the other hand, additional studies have elucidated the importance of target-site accessibility in miRNA target recognition as a result of RNA secondary structure stabilization/modification [164]. Aside from introducing changes into the primary structure of MREs, A-to-I editing modulates RNA secondary structures. The latter phenomenon can cause reduced accessibility of the miRNA-containing RNA-induced silencing complex (miR-RISC, next section) [107], i.e., the protein complex that incorporates miRNAs and exploits them to either inhibit the translation of coding transcripts or induce their degradation [165]. From this viewpoint, A-to-I editing in 3′UTRs seems to induce a generalized stabilization of cytosolic mRNAs' secondary structure at a local scale [100, 107, 166]. Although a considerable fraction of editing sites have been found at 10–100 nt of distance from 3′UTRs' MREs, they still caused inhibition of miRNA targeting if the 3′UTR due to high stabilization of the RNA duplex [107]. These facts likely evidence that substitutions of As with Gs at editing sites provide limited accessibility to the miR-RISC in the proximity of edited As. This effect has been observed either when A is opposed by U (leading to a I::U $\approx$ G::U wobble base pairing) or when A is opposed by C (leading to a I::C $\approx$ G::C Watson-Crick base pairing) [107].

3.2 A-to-I Editing of Noncoding RNA Molecules

3.2.1 A-to-I Editing of miRNAs

miRNAs currently represent the best characterized regulatory ncRNA class. At a first, miRNAs were identified in the model organism *Caenorhabditis elegans*, with the discovery of lin-4 [22] and let-7 [23]. However, it was soon demonstrated that the human genome also contains a large number of such regulatory elements [167–169]. To date, more than 2550 human miRNAs have been identified and described in literature (miRBase release 22) [170]. According to the canonical biogenesis pathway, RNA polymerase II (Pol II) is responsible for transcription of miRNAs. Intriguingly, a significant number of miRNAs are located in intronic regions of predicted or annotated protein-coding genes [171]. Primary miRNAs (pri-miRNAs), i.e., miRNA transcripts at their early stage of maturations, are much longer than mature miRNAs and are retained into the nucleus. Here, pri-miRNAs undergo a first maturation process carried out by the ribonuclease type III Drosha. In complex with its cofactor DGCR8, Drosha cleaves pri-miRNAs in ~60–70 nt long RNA molecules, termed precursor miRNAs (pre-miRNAs), which typically fold into an RNA hairpin structure. At this stage, pre-miRNAs are exported to the cytoplasm by the XPO5-RAN•GTP complex and are further processed by the ribonuclease type III Dicer, included into the RISC-loading complex (RLC) together with the Argonaute loading protein (AGO). After being cleaved by Dicer, pre-miRNAs are turned into mature miRNAs, consisting of 19–25 nt long ssRNAs. These are then incorporated into the miR-RISC complex, where they can exert their negative inhibitory role toward mRNA translation [165]. However, besides the canonical biogenesis pathway, a number of noncanonical pathways of miRNA biogenesis have also been reconstructed [165]. One of the most prominent leads to the generation of pre-miRNAs in the form of short hairpins during the process of intron removal from mRNAs. Thus, the very first two steps of the canonical biogenesis process, comprised of the transcription by Pol II and by the cleavage of the pri-miRNA by Drosha are bypassed for these miRNAs. Such intron-derived pre-miRNAs are termed mirtrons and are directly processed by Dicer [172]. Classically, miRNA::mRNA interactions leading to miRNA-mediated gene silencing are thought to occur between the miRNA's seed sequence and MREs located on the 3'UTR of mRNA transcripts. Indeed, it has been shown that miRNA-mediated interference can also occur through base pairing between the seed sequence and either complementary regions within 5'UTRs or coding sequences (e.g., [173, 174]). Moreover, noncanonical seed regions have been reported, where nucleotides 2–7, 3–8, or centered nucleotides (approximately 4–15) are involved in gene silencing [175, 176]. Nonetheless, pairing between the miRNA seed sequence and MREs on 3'UTRs is still considered the main regulatory event in miRNA-mediated gene silencing [176].

As previously mentioned, some miRNAs can be involved in upregulation of protein expression, contrary to what is commonly thought. For instance, some miRNAs are capable to interact with 5'UTRs through complementary sequences at their 3' end and, simultaneously, they interact with 3'UTRs through complementary sequences at their 5' end. miRNAs having this characteristic are termed miBridges [156]. Yet, under certain conditions, some miRNAs can induce upregulation of protein expression by binding particular motifs of the 3'UTR and recruiting specific proteins that enhance the translational process. This is the case of hsa-miR-369-3p, which under growth arrest conditions can bind the ARE located in the 3'UTR of tumor necrosis factor-α (TNF-α) transcripts and recruit translation-associated proteins [154].

During the last decade, a growing number of scientific studies have gathered much relevant information on miRNA editing. Nonetheless, many details still remain to be defined. In general, it is estimated that 10–20% of miRNAs undergo A-to-I editing at the pri-miRNA level [113, 177], whereby this process may occur co-transcriptionally other than posttranscriptionally, at least for a fraction of miRNAs [97]. The vast majority of detected sites are edited to a very low levels (<5%) and are thought to have not meaningful biological implication. Only a minor fraction is edited at significant (>5%) levels [56, 57, 178–180]. However, editing levels for each miRNA are not generalizable throughout all the human tissues, since certain miRNAs are highly edited in certain tissues while showing basal editing levels in other tissues [56, 178]. Moreover, high levels of miRNA editing in humans are not limited to neuronal tissues only, as in the case of the murine model, but they are rather observable in different types of tissues [56]. Pri-miRNAs harboring *Alu* repeats in their dsRNA stem's secondary structure show the highest editing levels (~17%) if compared with other pri-miRNAs, while editing levels gradually decrease going from pri-miRNAs harboring non-*Alu* repeats to pri-miRNAs with non-repetitive sequences. In these stem regions, promiscuous editing is carried out, often with an editing level <2% [56]. Differently, editing in mirtrons (pre-miRNAs) is highly site selective and occurs at much higher levels in comparison with non-mirtron miRNAs [56].

A-to-I editing at the pri- and pre-miRNA level can profoundly affect the fate of the corresponding mature miRNA, primarily in terms of miRNA expression [113, 181]. Considering highly edited stem regions of non-mature miRNAs, a large fraction of editing events occur in correspondence of the mature miRNA region. Nonetheless, expression of their mature edited counterpart appears to be relatively rare [56, 113, 181]. Intriguingly, it seems that miRNAs showing high difference in editing level between the non-mature transcript and its mature counterpart have a higher number of editing sites (3–4 on average) if compared with those showing lower difference (2 on average) [56]. An explanation of this phenomenon consists in the fact that editing of pri-miRNAs within the corresponding mature miRNA or in its relative flanking regions can prevent Drosha/Dicer ability to process the noncoding transcript [113, 182]. Using pre-miR-151 (wild-type vs. edited forms) as model molecule, a recent structural study demonstrated how local structural variations due to single A-to-I conversions can affect the pre-miRNA processing by Dicer [182]. Briefly, the human Dicer is a multi-domain enzyme, whose N-terminal region contains three protein domains: a DExH/D ATPase-helicase domain, a DUF283 domain, and a PAZ domain. DExH/D domain is involved in remodeling of RNA and is able to unwind RNA duplexes [183], while PAZ domain binds a single-strand, 5-nucleotide RNA, 5′→3′ oriented [184]. On the other hand, pre-miRNAs are typically structured in a hairpin conformation containing a terminal loop and a stem duplex, where the 5′ phosphate is unbound and 2 nt overhang at the 3′ end. During the pre-miRNA processing, the PAZ domain of Dicer binds the 5′-unbound phosphate and together with the two overhang nucleotide at the 3′ of its premature substrate [185]. Once this interaction has occurred, the DExH/D domain interacts with the terminal loop of the pre-miRNA and aligns such pre-miRNA's region to the RNase III catalytic site, setting a precise cleavage [186]. Pre-miR-151 can be edited at three different editing sites, i.e., pre-miR-151A1I (at −1 site), pre-miR-151A3I (at 3 site), and pre-miR-151A13I (at −1 and 3 sites) [187]. Compared with wild-type (WT) pre-miR-151 dicing, the cleavage of pre-miR-151A1I and pre-miR-151A13I is processed with strongly reduced rate, while dicing of pre-miR-151A3I does not show significant changes. A single-particle electron microscopy reconstruction has shown that such dicing rate changes depend on editing-induced structural changes in the pre-miRNA's terminal loop, which in turn induces a conformational change in the DExH/D domain [182]. More precisely, when Dicer binds WT pre-miR-151 or pre-miR-151A3I, its DExH/D domain is found in an open state, allowing the unwinding of the pre-miRNA's stem duplex and, consequently, the excision at the dicing site. To the contrary, when Dicer binds pre-miR-151A1I or pre-miR-151A13I, its DExH/D domain is found in a close state, impairing the remodeling of the stem duplex and, subsequently, the RNase III catalytic activity [182]. However, in spite of this, different processes, like editing-induced mature miRNA degradation, could also be hypothesized in order to explain such a difference [56]. Yet, cases in which pri- or pre-miRNA editing increases the Dicer cleavage rate of the ncRNA are also known [113]. The difference in editing level between non-mature miRNAs and their corresponding mature form is not kept constant among human tissues, but once again it is possible to observe tissue-dependent patterns [56]. miRNA editing seems to be involved also in strand selection during Dicer cleavage, although hard evidences are still missing [56, 113, 180]. In conclusion, A-to-I editing of miRNAs is very likely to be considered as a rapid and effective mechanism allowing modulation of miRNA expression.

If considering editing events falling into the mature miRNA sequence, these are particularly enriched within the seed region, which is primary involved in gene expression silencing according to the canonical model. From the functional viewpoint, this event can lead to a phenomenon known as target redirection [56, 180]. As discussed in the previous section, even a single-nucleotide variation in the seed sequence of a certain miRNA will change the thermodynamics of base pairing, at least to some degree, potentially causing the edited miRNA to shift toward a new set of targets [57, 179, 180, 188]. Such event can have a profound impact in cell biology and has important clinical significance, as demonstrated by some functional studies. For instance, in the human brain, the miR-376 cluster normally undergoes a widespread (from 50 to 100%) programmed seed sequence modification by ADAR2-mediated editing. Edited miR-376a-5p, which is abundant under physiological conditions, modulates expression of the autocrine motility factor receptor (AMFR), member of the E3 ubiquitin ligase family of proteins and membrane receptor capable to bind the tumor motility-stimulating AMF protein, secreted by tumor cells. In glioblastoma (GBM) cells, however, editing levels of miR-376a-5p tragically decrease, with subsequent upregulation of AMFR expression. Concurrently, the unedited miR-376a-5p has been found to target transcripts encoding for RAP2A, member of RAS oncogene family. This contributed to increased motility of GBM cells and has been correlated with decreased patient survival [189]. Very similar situation has been reported for miR-589–3p. This miRNA is almost fully edited by ADAR2 in the human brain under physiological conditions, while editing on the same miRNA is abolished in GBM. Edited miR-589–3p targets mRNAs encoding for the metalloproteinase ADAM12, involved in promotion of GBM cell invasion, while unedited miR-589–3p downregulates expression of the tumor suppressor PCDH9 [190]. However, there exist also circumstances in which editing of certain miRNAs can cause the onset of pathological status. For example, it has been reported that editing of miR-200b correlates with worsen prognosis in patients affected by several tumor types when edited. Wild-type miR-200b is normally unedited and functions as tumor suppressor of metastatic processes, since it silences the expression of a number of oncogenes. In particular, ZEB1 and ZEB2, both involved in transcription of master regulators of epithelial mesenchymal transition, are repressed by unedited miR-200b [191, 192]. Once miR-200b is edited, it shifts its functional role from tumor suppressor to oncogene, losing the ability to target several important oncogenes (among which ZEB1/ZEB2) and gaining the ability to target a new subset of coding transcripts, among which the leukemia inhibitory factor receptor LIFR, a well-known tumor suppressor [57]. Aside from these case reports, few other studies have demonstrated target redirection due to switch between unedited and edited forms of certain miRNAs [193, 194]. In general, however, comparison between predicted target sets shows that only about 10–35% of target mRNAs are shared between each unedited and edited miRNA, with widespread effects on pathway activation/deactivation [56, 57, 180]. Furthermore, a recent comprehensive study showed that miRNA editing is globally downregulated and, concomitantly, editing of 3'UTRs of mRNA is upregulated in cancer. In line with this, the same study reported a significant correlation between higher miRNA editing levels and longer survival of patients [180]. A complementary study has also showed that it may be possible to define specific miRNA editing patterns distinguishing cancer from non-cancer conditions, whereby certain cancer-associated patterns may be predictive of the specific cancer type [57].

3.2.2 A-to-I Editing of lncRNAs

Human cells are capable to produce several hundreds of lncRNAs, for which a plethora of roles in diverse biological functions have emerged during the last decade [195]. H19 was the first lncRNA to be isolated. At first considered to be an mRNA, H19 soon showed peculiar features in comparison with coding transcripts. In fact, though it contained small open reading frames, translation events were not observed for this RNA [18]. Also, gene expression studies showed that *H19* is normally expressed at the blastocyst stage, being accumulated in tissues of endodermal and mesodermal origin, while its expression during prenatal stages caused death in mouse embryos, suggesting an important role in developmental control [19]. Concomitantly, a second pseudo mRNA was identified, *Xists*, transcribed from the X-inactivation center locus of the X chromosome of mouse and involved in dosage compensation [20, 21]. To date, ~14.400 lncRNAs have been annotated in the human genome to this day [196]. Nonetheless, only ~180 of them have been biologically investigated [197], evidencing the need of further functional investigations for this ncRNA class. Conventionally, 5′-capped transcripts, longer than 200 nucleotides and lacking significant protein-coding capacity, are classified as lncRNAs. Most of these ncRNAs are polyadenylated at the 3′ end and spliced [198]. A considerable fraction of lncRNAs is transcribed from pseudogenes, and thus exhibits high sequence homology to their parent genes. A number of lncRNAs also include open reading frames (ORFs) in their sequences and can undergo translational process, challenging the view that they are noncoding RNAs. Initially, it was thought that lncRNA-derived peptides had no significant roles in cell biology and that they had a short half-life, being easily degradable [199, 200], but few contrasting evidences have been recently reported [201]. However, the remaining lncRNA fraction does not encode for any peptide. Depending on the adopted criterion of classification, lncRNAs can be categorized in several subclasses. For instance, established classification criteria are transcript length, genomic location with respect to protein-coding genes, genomic location with respect to specific DNA regulatory elements, intracellular localization, and biogenesis pathway [198]. Importantly, classification criteria for lncRNA also include association with specific biological processes and biological function [198]. Following the latter criterion, an important regulatory subclass of lncRNAs is represented by competing endogenous RNAs (ceRNAs), which include pseudogenes and circular RNAs (circRNAs). Due to their antisense RNA sequence content (including MREs), these lncRNAs act as sponges toward other RNA molecules and are capable to sequester miRNAs, potentially leading to a fine-tuning of protein expression [202]. Differently, decoy lncRNAs essentially regulate gene expression by acting on protein activities, exerting a role of ribo-repressors or ribo-enhancers. Such role is performed by interfering with proteins' binding sites and/or inducing allosteric modifications that in turn cause inhibition of catalytic activity [203–205]. A further mechanism by which lncRNAs can exert a regulatory role toward gene expression is through formation of RNA–DNA triplex structures, the latter having the potential to induce local modifications of the DNA structure, inducing the recruitment of specific transcription factors [206, 207]. This lncRNA subclass has been defined as guide lncRNAs.

To date, no strong evidences regarding the functional aspects of edited lncRNAs are available in literature, but only few generalized speculations have been made [104, 208] as this research field is still at an embryonal stage. Few recent studies have reported that a considerable amount of A-to-I editing events occur in lncRNAs [104, 208, 209]. Approximately 192.000 editing events falling into lncRNAs have been annotated in humans, according to information reported on the LNCediting database [208]. As in the case of mRNAs, here the majority of editing events occur in intronic regions and 3'UTRs of both poly(A)+ and poly(A)− lncRNAs, where *Alu* repeats are particularly enriched, while only a minority occur in exons [209, 210]. Interestingly, intronic editing sites of poly(A)− lncRNAs show distinct location characteristics in respect to poly(A)+ lncRNAs, and the portion of exonic editing sites in the former group is decreased when compared with that in the latter group [210]. Given that the functionality of lncRNAs is strongly dependent on their global and local secondary and tertiary structures (e.g., [211]), it is very likely that even a single exonic editing event will somehow influence their biological role at a certain extent through stabilization or destabilization of lncRNA-related structures [104, 208]. Finally, a preliminar study showed that, if analyzed at a global level, lncRNA editing patterns may be useful to distinguish between healthy and various carcinogenic status in patients with tumor growth [210].

3.2.3 A-to-I Editing of tRNAs

Transfer RNA, previously introduced as the "adaptor" or "soluble" RNA, is the most abundant RNA among all cellular RNAs together with ribosomal RNA. Fifty-one different tRNA types are present in humans, with ~510 estimated nuclear genes encoding for cytoplasmic tRNAs, of which ~430 are unique. In human cells, tRNAs are transcribed at millions of RNA copies by RNA polymerase III (Pol III), whereby tRNA gene expression is fine-regulated in a tissue-specific and cell-specific manner in order to modulate, in turn, the mRNA translation rate [212]. In protein biosynthesis, tRNA determines what amino acid is going to be inserted into the peptide sequence depending on the RNA sequence. This is made possible by the characteristic 3D structure of this ncRNA. The most essential event during peptide elongation regards the correct base pairing between a specific triplet of the tRNA, termed anticodon and located into the anticodon loop of the tRNA, and a specific codon in the mRNA transcript being processed. The specified amino acid is covalently bonded to the 3'-hydroxyl group of the terminal adenosine located in the 5'-CCA-3' tail at the 3' end of all tRNA molecules [213].

Among the various ncRNAs classes, tRNA is the most extensively modified one at the epitranscriptomic level, carrying an average of 13 modifications per molecule [38]. Such posttranscriptional modifications are of great importance for the maintenance of the correct folding of this molecule, and thus are fundamental for the role tRNAs play not only in protein biosynthesis, but also in other minor, less understood molecular mechanisms. Pseudouridylation and methylations on several nucleoside residues are the most frequently reported modifications [38]. However, A-to-I editing of specific human tRNAs has been reported as well (e.g., [214, 215]). Differently from the other ncRNA classes, tRNAs does not undergo A-to-I editing by ADARs. Here, deamination of adenosine is rather the result of the enzymatic activity of three distinct enzymes, termed adenosine deaminase acting on tRNAs (ADATs) [216]. ADAT1 is structurally related to ADARs and displays a very likely familiarity with ADAR2, since both enzymes require the presence of IP6 to carry out efficient catalysis, whereby several residues coordinating IP6 in ADAR2 are conserved in ADAT1 [75, 217]. This enzyme is specific for deamination of adenosine at position 37 (A37) in tRNA carrying alanine (tRNA^{Ala}). Such modification is then followed by a second epitranscriptomic modification, i.e., methylation of I37 by the tRNA methyltransferase 5 (Trm5), resulting in 1-methylinosine (m^{1}I) [218, 219]. m^{1}I has a peculiar property compared to non-modified inosines, being read as a T instead of G (e.g., [75]). ADAT2 and ADAT3 are related to the clade of apolipoprotein B mRNA editing enzymes (APOBECs) and form a heterodimer indicated as hetADAT [217]. While ADAT2 is thought to be the catalytic subunit of hetADAT, ADAT3 seems to function as a recognizer of the tRNA substrate, coordinating ADAT2 enzymatic activity [220]. hetADAT specifically catalyzes the conversion of A34, the first nucleotide (5′) of the anticodon of a number of tRNAs, to I34. This leads to a noncanonical wobble base pairing of the anticodon with complementary triplets on the mRNA, allowing tRNAs containing I34 to recognize more than one mRNA codon [30]. Importantly, conversion of A34 to I34 is performed at the precursor tRNA level of human tRNAs, during the maturation process [221].

Very little is known about the biological meanings of A-to-I editing in tRNAs. It has been speculated a role in expansion of codon reading capabilities and modulation of translation rate [216]. Furthermore, some reported data suggest a possible role for hetADAT in intellectual and visual abilities [222]. All considered, editing of tRNAs is a largely unexplored field, especially from the functional perspective, and further investigations are required.

4 Computational Resources to Investigate ncRNA Editing

4.1 Computational Pipelines for ncRNA Editing Detection

In the early 2000s, only a handful of RNA editing events were identified and studied [223]. By using comparative genomics, Dr. Reenan's group [224] detected and experimentally validated 16 novel RNA editing events in fruit fly and just one in human. Remarkably, these editing events are located in highly conserved exonic regions that form a dsRNA structure.

One year later, three independent groups [98–100] designed computational methods to systematically discover clustered A-to-I RNA editing sites in repetitive regions (*Alu* repeats) of the human transcriptome. Basically, their methods consist of the alignment of expressed sequence tags (EST) [225] against a reference genome, and collection of all those A-to-G mismatches representing putative A-to-I editing events. Moreover, in order to reach a good accuracy, they took into consideration all those mismatches that were not known single-nucleotide polymorphisms (SNPs) and located in long, stable dsRNAs. These efforts led to the uncovering of tens of thousands of RNA editing events.

With the progress of HTS technology, we have witnessed a drastic improvement in terms of detection of novel genes, as well as co-transcriptional and/or posttranscriptional modification events, such as RNA editing phenomena [226]. Starting from the first HTS-based application for RNA editing profiling employed by Li et al. [227], in 2009, until recently we went from tens of thousands to millions of editing events, whereby many of them have been seen to occur in noncoding RNA repetitive regions [228]. In 2010, de Hoon et al. designed one of the first strategies to systematically detect human miRNA editing from small RNA sequencing library, concluding that miRNA editing events are rare in animals [229]. In 2012, Alon et al. developed another HTS-based strategy through which they discovered known and novel editing sites in mature miRNAs [178, 230]. Successively, in 2015, Alon et al. developed a Web server [231], named DREAM (http://www.cs.tau.ac.il/~mirnaed/), implementing the pipeline to identify and quantify canonical and noncanonical editing events in mature miRNAs from small RNA-sequencing data.

Despite the substantial results achieved by the bioinformatics methods for RNA editing detection, none are able to distinguish a guanosine originating from an I-to-G replacement, from a guanosine as a product of sequencing errors, or SNP. In 2010 Sakurai et al. designed a biochemical method to overcome this limit, named inosine chemical erasing (ICE) [232]. Through this strategy they were able to successfully identify inosines on RNA molecules employing inosine-specific cyanoethylation with reverse transcription, PCR amplification, and direct sequencing. Lately, Sakurai et al. combined the ICE method with NGS technology (ICE-seq) to profile A-to-I editing sites [233].

4.2 Knowledge Base and Functional Characterization Resources for ncRNA Editing

After the previous mentioned birth, in 2004, of the computational methods for the RNA editing detection [98–100], mentioned above, the thousands of editing events discovered up to that point were collected in the first RNA editing database called DARNED (Database of RNa Editing) [234]. This database contains more than 40,000 putative human RNA editing events. According to DARNED (first release), Laganà et al. developed another database [235], named miR-EdiTar (http://microrna.osumc.edu/mireditar), of predicted miRNA-binding sites that could potentially be affected by A-to-I editing sites occurring in 3′UTRs.

With the advent of HTS technology [228], Kiran and Baranov generated a second release of DARNED, recording more than 300,000 editing events in human [236]. This has led to the rise of computational tools for both the visualization and annotation of RNA-Seq data with published editing events [237, 238].

Despite the fact that DARNED includes important information on published editing events, only a small fraction has been subsequently manually annotated [239, 240]. An improved resource was later developed by Ramaswami and Li [241], who provided the scientific community with the database RADAR, including rigorously annotated A-to-I editing events. Specifically, RADAR provides detailed manually curated information on each editing site, i.e., genomic coordinates, type of genomic region (intergenic, 3′- or 5′UTR, intron, or coding regions), type of repetitive element (i.e., *Alu* or others), conservation (chimpanzee, rhesus, mouse), and tissue-specific editing level, when known. Totally, RADAR reports more than one million editing sites in *Homo sapiens*, with only a small fraction associated to human ncRNA molecules (i.e., microRNAs). Recently, Gong et al. have developed LNCediting, a knowledge base on the functional effects of RNA editing events on lncRNAs [208] in human, rhesus, mouse, and fly, observing a significant impact of such events on secondary structures of lncRNAs, as well as on lncRNA-miRNA interactions.

5 Conclusions

In this chapter, we have provided an overview of the panorama of the validated and putative functions of RNA editing in both noncoding regions of mRNAs as well as in ncRNA molecules. Also, we have described the knowledge base and bioinformatics resources available to investigate such phenomenon. Progress in HTS-based technologies, together with improvements of several bioinformatics tools, has allowed to characterize a wide fraction of the noncoding RNAome, thus revealing its extremely large complexity from the functional standpoint. Such advances also led to the identification of a wide plethora of epitranscriptomic modifications and outlined at the same time their importance in cell biology. To the contrary of what was thought, RNA modifications seem to occur frequently during the cell life span and has been shown to be notably dynamic and highly regulated events. A-to-I editing of RNA is among the most common and well-known RNA modification. Editing of coding RNA sequences has been deeply studied during the past years and the consequences of such phenomenon are now well known. Differently, editing occurring in ncRNAs has been only recently taken into consideration, whereby several genome-wide studies have shown that A-to-I conversions in ncRNA represent the vast majority of RNA editing events, though a large part of their related functional spectrum remains still to be uncovered. Albeit current technologies allow single-nucleotide resolution and precise mapping of editing sites, there is an urgent need in developing of novel and more effective HTS-based methodologies and bioinformatics tools aiming to better identify and analyze editing events in ncRNAs. Specifically, robust prediction tools capable of assessing the potential functional impact of edited ncRNAs are still missing and more efforts should be made in the coming years, as they would provide the opportunity to better orient experiments on functional characterization.

References

1. Watson JD, Crick FH (1953) Molecular structure of nucleic acids; a structure for deoxyribose nucleic acid. Nature 171:737–738
2. Brachet J (1942) La localization des acides pentosenucleiques dans les tissues animaux et les oufs d'Amphibiens en voie de developpement. Archs Biol 53:207–257
3. Caspersson T (1947) The relations between nucleic acid and protein synthesis. Symp Soc Exp Biol 1:127–151
4. Palade GE, Siekevitz P (1956) Liver microsomes; an integrated morphological and biochemical study. J Biophys Biochem Cytol 2:171–200
5. Zamecnik PC, Keller EB, Littlefield JW et al (1956) Mechanism of incorporation of labeled amino acids into protein. J Cell Physiol Suppl 47:81–101
6. Hoagland MB, Stephenson ML, Scott JF et al (1958) A soluble ribonucleic acid intermediate in protein synthesis. J Biol Chem 231:241–257
7. Crick FH (1958) On protein synthesis. Symp Soc Exp Biol 12:138–163
8. Brenner S, Jacob F, Meselson M (1961) An unstable intermediate carrying information from genes to ribosomes for protein synthesis. Nature 190:576–581
9. Gros F, Hiatt H, Gilbert W et al (1961) Unstable ribonucleic acid revealed by pulse labelling of Escherichia Coli. Nature 190:581–585. https://doi.org/10.1038/190581a0
10. Nirenberg MW, Matthaei JH (1961) The dependence of cell-free protein synthesis in E. coli upon naturally occurring or synthetic polyribonucleotides. Proc Natl Acad Sci U S A 47:1588–1602
11. Chow LT, Roberts JM, Lewis JB, Broker TR (1977) A map of cytoplasmic RNA transcripts from lytic adenovirus type 2, determined by electron microscopy of RNA:DNA hybrids. Cell 11:819–836. https://doi.org/10.1016/0092-8674(77)90294-X
12. Berk AJ, Sharp PA (1977) Sizing and mapping of early adenovirus mRNAs by gel electrophoresis of S1 endonuclease-digested hybrids. Cell 12:721–732

13. Berget SM, Sharp PA (1977) A spliced sequence at the 5′-terminus of adenovirus late mRNA. Brookhaven Symp Biol 29:332–344
14. Busch H, Reddy R, Rothblum L, Choi YC (1982) SnRNAs, SnRNPs, and RNA processing. Annu Rev Biochem 51:617–654. https://doi.org/10.1146/annurev.bi.51.070182.003153
15. Krainer AR, Maniatis T (1985) Multiple factors including the small nuclear ribonucleoproteins U1 and U2 are necessary for pre-mRNA splicing in vitro. Cell 42:725–736
16. Konarska MM, Sharp PA (1986) Electrophoretic separation of complexes involved in the splicing of precursors to mRNAs. Cell 46:845–855
17. Eliceiri GL (1999) Small nucleolar RNAs. Cell Mol Life Sci 56:22–31
18. Brannan CI, Dees EC, Ingram RS, Tilghman SM (1990) The product of the H19 gene may function as an RNA. Mol Cell Biol 10:28–36
19. Bartolomei MS, Zemel S, Tilghman SM (1991) Parental imprinting of the mouse H19 gene. Nature 351:153–155. https://doi.org/10.1038/351153a0
20. Borsani G, Tonlorenzi R, Simmler MC et al (1991) Characterization of a murine gene expressed from the inactive X chromosome. Nature 351:325–329. https://doi.org/10.1038/351325a0
21. Kelley RL, Kuroda MI (2000) Noncoding RNA genes in dosage compensation and imprinting. Cell 103:9–12
22. Lee RC, Feinbaum RL, Ambros V (1993) The C. elegans heterochronic gene lin-4 encodes small RNAs with antisense complementarity to lin-14. Cell 75:843–854
23. Reinhart BJ, Slack FJ, Basson M et al (2000) The 21-nucleotide let-7 RNA regulates developmental timing in Caenorhabditis elegans. Nature 403:901–906. https://doi.org/10.1038/35002607
24. Siomi H, Siomi MC (2009) On the road to reading the RNA-interference code. Nature 457:396–404. https://doi.org/10.1038/nature07754
25. Thompson DM, Parker R (2009) The RNase Rny1p cleaves tRNAs and promotes cell death during oxidative stress in Saccharomyces cerevisiae. J Cell Biol 185:43–50. https://doi.org/10.1083/jcb.200811119
26. Garcia-Silva MR, Cabrera-Cabrera F, Güida MC, Cayota A (2012) Hints of tRNA-derived small RNAs role in RNA silencing mechanisms. Genes (Basel) 3:603–614. https://doi.org/10.3390/genes3040603
27. Martens-Uzunova ES, Olvedy M, Jenster G (2013) Beyond microRNA—novel RNAs derived from small non-coding RNA and their implication in cancer. Cancer Lett 340:201–211. https://doi.org/10.1016/j.canlet.2012.11.058
28. Kumar P, Anaya J, Mudunuri SB, Dutta A (2014) Meta-analysis of tRNA derived RNA fragments reveals that they are evolutionarily conserved and associate with AGO proteins to recognize specific RNA targets. BMC Biol 12:78. https://doi.org/10.1186/PREACCEPT-5867533061403216
29. Hoogsteen K (1963) The crystal and molecular structure of a hydrogen-bonded complex between 1-methylthymine and 9-methyladenine. Acta Crystallogr 16:907–916. https://doi.org/10.1107/S0365110X63002437
30. Crick FH (1966) Codon--anticodon pairing: the wobble hypothesis. J Mol Biol 19:548–555
31. Varani G, McClain WH (2000) The G x U wobble base pair. A fundamental building block of RNA structure crucial to RNA function in diverse biological systems. EMBO Rep 1:18–23. https://doi.org/10.1093/embo-reports/kvd001
32. Leontis NB, Westhof E (2001) Geometric nomenclature and classification of RNA base pairs. RNA 7:499–512
33. Bjork GR, Ericson JU, Gustafsson CED et al (1987) Transfer RNA modification. Annu Rev Biochem 56:263–285. https://doi.org/10.1146/annurev.bi.56.070187.001403
34. Maden BE (1990) The numerous modified nucleotides in eukaryotic ribosomal RNA. Prog Nucleic Acid Res Mol Biol 39:241–303
35. Phizicky EM, Hopper AK (2010) tRNA biology charges to the front. Genes Dev 24:1832–1860. https://doi.org/10.1101/gad.1956510
36. Sharma S, Lafontaine DLJ (2015) "View from a bridge": a new perspective on eukaryotic rRNA base modification. Trends Biochem Sci 40:560–575. https://doi.org/10.1016/j.tibs.2015.07.008
37. Liu M, Douthwaite S (2002) Methylation at nucleotide G745 or G748 in 23S rRNA distinguishes Gram-negative from Gram-positive bacteria. Mol Microbiol 44:195–204
38. Pan T (2018) Modifications and functional genomics of human transfer RNA. Cell Res 28:395–404. https://doi.org/10.1038/s41422-018-0013-y

39. Rozenski J, Crain PF, McCloskey JA (1999) The RNA modification database: 1999 update. Nucleic Acids Res 27:196–197
40. Limbach PA, Crain PF, McCloskey JA (1994) Summary: the modified nucleosides of RNA. Nucleic Acids Res 22:2183–2196. https://doi.org/10.1093/nar/22.12.2183
41. Cantara WA, Crain PF, Rozenski J et al (2011) The RNA modification database, RNAMDB: 2011 update. Nucleic Acids Res 39:D195–D201. https://doi.org/10.1093/nar/gkq1028
42. Boccaletto P, Machnicka MA, Purta E et al (2018) MODOMICS: a database of RNA modification pathways. 2017 update. Nucleic Acids Res 46:D303–D307. https://doi.org/10.1093/nar/gkx1030
43. Kellner S, Burhenne J, Helm M (2010) Detection of RNA modifications. RNA Biol 7:237–247. https://doi.org/10.4161/rna.7.2.11468
44. Helm M, Motorin Y (2017) Detecting RNA modifications in the epitranscriptome: predict and validate. Nat Rev Genet 18:275–291. https://doi.org/10.1038/nrg.2016.169
45. Davis FF, Allen FW (1957) Ribonucleic acids from yeast which contain a fifth nucleotide. J Biol Chem 227:907–915
46. Fu Y, Dominissini D, Rechavi G, He C (2014) Gene expression regulation mediated through reversible m6A RNA methylation. Nat Rev Genet 15:293–306. https://doi.org/10.1038/nrg3724
47. Jia G, Fu Y, He C (2013) Reversible RNA adenosine methylation in biological regulation. Trends Genet 29:108–115. https://doi.org/10.1016/j.tig.2012.11.003
48. Liu F, Clark W, Luo G et al (2016) ALKBH1-mediated tRNA demethylation regulates translation. Cell 167:816–828.e16. https://doi.org/10.1016/j.cell.2016.09.038
49. Frye M, Nishikura K, Jaffrey SR et al (2016) A-to-I editing of coding and non-coding RNAs by ADARs. Nat Rev Mol Cell Biol 17:83–96. https://doi.org/10.1038/nrm.2015.4
50. Gait MJ, Moore MJ, Zimmermann RA (1998) Incorporation of modified nucleotides into RNA for studies on RNA structure, function and intermolecular interactions. In: Modification and editing of RNA. American Society of Microbiology, Washington, DC, pp 59–84
51. Vermeulen A, McCallum SA, Pardi A (2005) Comparison of the global structure and dynamics of native and unmodified tRNAval. Biochemistry 44:6024–6033. https://doi.org/10.1021/bi0473399
52. Davis DR (1998) Biophysical and conformational properties of modified nucleosides in RNA (nuclear magnetic resonance studies). In: Modification and editing of RNA. American Society of Microbiology, Washington, DC, pp 85–102
53. Helm M (2006) Post-transcriptional nucleotide modification and alternative folding of RNA. Nucleic Acids Res 34:721–733. https://doi.org/10.1093/nar/gkj471
54. Motorin Y, Helm M (2010) tRNA stabilization by modified nucleotides. Biochemistry 49:4934–4944. https://doi.org/10.1021/bi100408z
55. Price DH, Gray MW (1998) Editing of tRNA. In: Modification and editing of RNA. American Society of Microbiology, Washington, DC, pp 289–305
56. Li L, Song Y, Shi X et al (2018) The landscape of miRNA editing in animals and its impact on miRNA biogenesis and targeting. Genome Res 28:132–143. https://doi.org/10.1101/gr.224386.117
57. Wang Y, Xu X, Yu S et al (2017) Systematic characterization of A-to-I RNA editing hotspots in microRNAs across human cancers. Genome Res 27:1112–1125. https://doi.org/10.1101/gr.219741.116
58. Jin Y-F, Zhang W, Li Q (2009) Origins and evolution of ADAR-mediated RNA editing. IUBMB Life 61:572–578. https://doi.org/10.1002/iub.207
59. George CX, Samuel CE (1999) Human RNA-specific adenosine deaminase ADAR1 transcripts possess alternative exon 1 structures that initiate from different promoters, one constitutively active and the other interferon inducible. Proc Natl Acad Sci U S A 96:4621–4626. https://doi.org/10.1073/pnas.96.8.4621
60. Eckmann CR, Neunteufl A, Pfaffstetter L, Jantsch MF (2001) The human but not the Xenopus RNA-editing enzyme ADAR1 has an atypical nuclear localization signal and displays the characteristics of a shuttling protein. Mol Biol Cell 12:1911–1924. https://doi.org/10.1091/mbc.12.7.1911
61. Poulsen H, Nilsson J, Damgaard CK et al (2001) CRM1 mediates the export of ADAR1 through a nuclear export signal within the Z-DNA binding domain. Mol Cell Biol 21:7862–7871. https://doi.org/10.1128/MCB.21.22.7862-7871.2001
62. Desterro JMP, Keegan LP, Lafarga M et al (2003) Dynamic association of RNA-editing

enzymes with the nucleolus. J Cell Sci 116:1805–1818. https://doi.org/10.1242/jcs.00371

63. Lykke-Andersen S, Piñol-Roma S, Kjems J (2007) Alternative splicing of the ADAR1 transcript in a region that functions either as a 5′-UTR or an ORF. RNA 13:1732–1744. https://doi.org/10.1261/rna.567807

64. Liu Y, George CX, Patterson JB, Samuel CE (1997) Functionally distinct double-stranded RNA-binding domains associated with alternative splice site variants of the interferon-inducible double-stranded RNA-specific adenosine deaminase. J Biol Chem 272:4419–4428. https://doi.org/10.1074/jbc.272.7.4419

65. Schmauss C, Zimnisky R, Mehta M, Shapiro LP (2010) The roles of phospholipase C activation and alternative ADAR1 and ADAR2 pre-mRNA splicing in modulating serotonin 2C-receptor editing in vivo. RNA 16:1779–1785. https://doi.org/10.1261/rna.2188110

66. Melcher T, Maas S, Herb A et al (1996) A mammalian RNA editing enzyme. Nature 379:460–464. https://doi.org/10.1038/379460a0

67. Peng PL, Zhong X, Tu W et al (2006) ADAR2-dependent RNA editing of AMPA receptor subunit GluR2 determines vulnerability of neurons in forebrain ischemia. Neuron 49:719–733. https://doi.org/10.1016/j.neuron.2006.01.025

68. Yang L, Huang P, Li F et al (2012) c-Jun amino-terminal kinase-1 mediates glucose-responsive upregulation of the RNA editing enzyme ADAR2 in pancreatic Beta-cells. PLoS One 7:e48611. https://doi.org/10.1371/journal.pone.0048611

69. Sansam CL, Wells KS, Emeson RB (2003) Modulation of RNA editing by functional nucleolar sequestration of ADAR2. Proc Natl Acad Sci U S A 100:14018–14023. https://doi.org/10.1073/pnas.2336131100

70. Filippini A, Bonini D, Giacopuzzi E et al (2018) Differential enzymatic activity of Rat ADAR2 splicing variants is due to altered capability to interact with RNA in the deaminase domain. Genes (Basel) 9:79. https://doi.org/10.3390/genes9020079

71. Chen C-X, Cho D-SC, Wang Q et al (2000) A third member of the RNA-specific adenosine deaminase gene family, ADAR3, contains both single- and double-stranded RNA binding domains. RNA 6:755–767

72. Tan MH, Li Q, Shanmugam R et al (2017) Dynamic landscape and regulation of RNA editing in mammals. Nature 550:249–254. https://doi.org/10.1038/nature24041

73. Oakes E, Anderson A, Cohen-Gadol A, Hundley HA (2017) Adenosine deaminase that acts on RNA 3 (ADAR3) binding to glutamate receptor subunit B pre-mRNA inhibits RNA editing in glioblastoma. J Biol Chem 292:4326–4335. https://doi.org/10.1074/jbc.M117.779868

74. Mladenova D, Barry G, Konen LM et al (2018) Adar3 is involved in learning and memory in mice. Front Neurosci 12:e118. https://doi.org/10.3389/fnins.2018.00243

75. Macbeth MR, Schubert HL, VanDemark AP et al (2005) Inositol hexakisphosphate is bound in the ADAR2 core and required for RNA editing. Science 309:1534–1539. https://doi.org/10.1126/science.1113150

76. Matthews MM, Thomas JM, Zheng Y et al (2016) Structures of human ADAR2 bound to dsRNA reveal base-flipping mechanism and basis for site selectivity. Nat Struct Mol Biol 23:426–433. https://doi.org/10.1038/nsmb.3203

77. Stefl R, Xu M, Skrisovska L et al (2006) Structure and specific RNA binding of ADAR2 double-stranded RNA binding motifs. Structure 14:345–355. https://doi.org/10.1016/j.str.2005.11.013

78. Chang KY, Ramos A (2005) The double-stranded RNA-binding motif, a versatile macromolecular docking platform. FEBS J 272:2109–2117. https://doi.org/10.1111/j.1742-4658.2005.04652.x

79. Lai F, Drakas R, Nishikura K (1995) Mutagenic analysis of double-stranded RNA adenosine deaminase, a candidate enzyme for RNA editing of glutamate-gated ion channel transcripts. J Biol Chem 270:17098–17105. https://doi.org/10.1074/jbc.270.29.17098

80. Liu Y, Lei M, Samuel CE (2000) Chimeric double-stranded RNA-specific adenosine deaminase ADAR1 proteins reveal functional selectivity of double-stranded RNA-binding domains from ADAR1 and protein kinase PKR. Proc Natl Acad Sci U S A 97:12541–12546. https://doi.org/10.1073/pnas.97.23.12541

81. Xu M, Wells KS, Emeson RB (2006) Substrate-dependent contribution of double-stranded RNA-binding motifs to ADAR2 function. Mol Biol Cell 17:3211–3220. https://doi.org/10.1091/mbc.E06-02-0162

82. Valente L, Nishikura K (2007) RNA binding-independent dimerization of adenosine deaminases acting on RNA and dominant negative effects of nonfunctional subunits on dimer functions. J Biol Chem 282:16054–16061. https://doi.org/10.1074/jbc.M611392200
83. Strehblow A, Hallegger M, Jantsch MF (2002) Nucleocytoplasmic distribution of human RNA-editing enzyme ADAR1 is modulated by double-stranded RNA-binding domains, a leucine-rich export signal, and a putative dimerization domain. Mol Biol Cell 13:3822–3835. https://doi.org/10.1091/mbc.E02-03-0161
84. Barraud P, Banerjee S, Mohamed WI et al (2014) A bimodular nuclear localization signal assembled via an extended double-stranded RNA-binding domain acts as an RNA-sensing signal for transportin 1. Proc Natl Acad Sci U S A 111:E1852–E1861. https://doi.org/10.1073/pnas.1323698111
85. Schwartz T, Rould MA, Lowenhaupt K et al (1999) Crystal structure of the Zalpha domain of the human editing enzyme ADAR1 bound to left-handed Z-DNA. Science 284:1841–1845
86. Koeris M, Funke L, Shrestha J et al (2005) Modulation of ADAR1 editing activity by Z-RNA in vitro. Nucleic Acids Res 33:5362–5370. https://doi.org/10.1093/nar/gki849
87. Oh D-B, Kim Y-G, Rich A (2002) Z-DNA-binding proteins can act as potent effectors of gene expression in vivo. Proc Natl Acad Sci U S A 99:16666–16671. https://doi.org/10.1073/pnas.262672699
88. Kang H-J, Le TVT, Kim K et al (2014) Novel interaction of the Z-DNA binding domain of human ADAR1 with the oncogenic c-Myc promoter G-quadruplex. J Mol Biol 426:2594–2604. https://doi.org/10.1016/j.jmb.2014.05.001
89. Athanasiadis A, Placido D, Maas S et al (2005) The crystal structure of the Zβ domain of the RNA-editing enzyme ADAR1 reveals distinct conserved surfaces among Z-domains. J Mol Biol 351:496–507. https://doi.org/10.1016/j.jmb.2005.06.028
90. Maas S, Gommans WM (2009) Identification of a selective nuclear import signal in adenosine deaminases acting on RNA. Nucleic Acids Res 37:5822–5829. https://doi.org/10.1093/nar/gkp599
91. Martin FH, Castro MM, Aboul-ela F, Tinoco I (1985) Base pairing involving deoxyinosine: implications for probe design. Nucleic Acids Res 13:8927–8938. https://doi.org/10.1093/nar/13.24.8927
92. Valente L, Nishikura K (2005) ADAR gene family and A-to-I RNA editing: diverse roles in posttranscriptional gene regulation. Prog Nucleic Acid Res Mol Biol 79:299–338. https://doi.org/10.1016/S0079-6603(04)79006-6
93. Liddicoat BJ, Piskol R, Chalk AM et al (2015) RNA editing by ADAR1 prevents MDA5 sensing of endogenous dsRNA as nonself. Science 349:1115–1120. https://doi.org/10.1126/science.aac7049
94. Siomi H, Siomi MC (2010) Posttranscriptional regulation of microRNA biogenesis in animals. Mol Cell 38:323–332. https://doi.org/10.1016/j.molcel.2010.03.013
95. Laurencikiene J, Källman AM, Fong N et al (2006) RNA editing and alternative splicing: the importance of co-transcriptional coordination. EMBO Rep 7:303–307. https://doi.org/10.1038/sj.embor.7400621
96. Ryman K, Fong N, Bratt E et al (2007) The C-terminal domain of RNA Pol II helps ensure that editing precedes splicing of the GluR-B transcript. RNA 13:1071–1078. https://doi.org/10.1261/rna.404407
97. Daniel C, Widmark A, Rigardt D, Öhman M (2017) Editing inducer elements increases A-to-I editing efficiency in the mammalian transcriptome. Genome Biol 18:195. https://doi.org/10.1186/s13059-017-1324-x
98. Athanasiadis A, Rich A, Maas S (2004) Widespread A-to-I RNA editing of Alu-containing mRNAs in the human transcriptome. PLoS Biol 2:e391. https://doi.org/10.1371/journal.pbio.0020391.st001
99. Kim DDY, Kim TTY, Walsh T et al (2004) Widespread RNA editing of embedded alu elements in the human transcriptome. Genome Res 14:1719–1725. https://doi.org/10.1101/gr.2855504
100. Levanon EY, Eisenberg E, Yelin R et al (2004) Systematic identification of abundant A-to-I editing sites in the human transcriptome. Nat Biotechnol 22:1001–1005. https://doi.org/10.1038/nbt996
101. Bazak L, Levanon EY, Eisenberg E (2014) Genome-wide analysis of Alu editability. Nucleic Acids Res 42:6876–6884. https://doi.org/10.1093/nar/gku414
102. Bazak L, Haviv A, Barak M et al (2014) A-to-I RNA editing occurs at over a hundred million genomic sites, located in a majority of human genes. Genome Res 24:365–376. https://doi.org/10.1101/gr.164749.113

103. Consortium IHGS (2001) Initial sequencing and analysis of the human genome. Nature 409:860–921. https://doi.org/10.1038/35057062
104. Picardi E, Manzari C, Mastropasqua F et al (2015) Profiling RNA editing in human tissues: towards the inosinome Atlas. Sci Rep 5:14941. https://doi.org/10.1038/srep14941
105. Zhang Q, Xiao X (2015) Genome sequence-independent identification of RNA editing sites. Nat Methods 12:347–350. https://doi.org/10.1038/nmeth.3314
106. Soundararajan R, Stearns TM, Griswold AJ et al (2015) Detection of canonical A-to-G editing events at 3′ UTRs and microRNA target sites in human lungs using next-generation sequencing. Oncotarget 6:35726–35736. https://doi.org/10.18632/oncotarget.6132
107. Brümmer A, Yang Y, Chan TW, Xiao X (2017) Structure-mediated modulation of mRNA abundance by A-to-I editing. Nat Commun 8:1255. https://doi.org/10.1038/s41467-017-01459-7
108. Ramaswami G, Lin W, Piskol R et al (2012) Accurate identification of human Alu and non-Alu RNA editing sites. Nat Methods 9:579–581. https://doi.org/10.1038/nmeth.1982
109. Ramaswami G, Zhang R, Piskol R et al (2013) Identifying RNA editing sites using RNA sequencing data alone. Nat Methods 10:128–132. https://doi.org/10.1038/nmeth.2330
110. Riedmann EM, Schopoff S, Hartner JC, Jantsch MF (2008) Specificity of ADAR-mediated RNA editing in newly identified targets. RNA 14:1110–1118. https://doi.org/10.1261/rna.923308
111. Polson AG, Bass BL (1994) Preferential selection of adenosines for modification by double-stranded RNA adenosine deaminase. EMBO J 13:5701–5711
112. Lehmann KA, Bass BL (2000) Double-stranded RNA adenosine deaminases ADAR1 and ADAR2 have overlapping specificities. Biochemistry 39:12875–12884. https://doi.org/10.1021/bi001383g
113. Kawahara Y, Megraw M, Kreider E et al (2008) Frequency and fate of microRNA editing in human brain. Nucleic Acids Res 36:5270–5280. https://doi.org/10.1093/nar/gkn479
114. Solomon O, Di Segni A, Cesarkas K et al (2017) RNA editing by ADAR1 leads to context-dependent transcriptome-wide changes in RNA secondary structure. Nat Commun 8:1440. https://doi.org/10.1038/s41467-017-01458-8
115. Bass BL (2002) RNA editing by adenosine deaminases that act on RNA. Annu Rev Biochem 71:817–846. https://doi.org/10.1146/annurev.biochem.71.110601.135501
116. Wahlstedt H, Öhman M (2011) Site-selective versus promiscuous A-to-I editing. Wiley Interdiscip Rev RNA 2:761–771. https://doi.org/10.1002/wrna.89
117. Halvorsen M, Martin JS, Broadaway S, Laederach A (2010) Disease-associated mutations that alter the RNA structural ensemble. PLoS Genet 6:e1001074. https://doi.org/10.1371/journal.pgen.1001074
118. Salari R, Kimchi-Sarfaty C, Gottesman MM, Przytycka TM (2013) Sensitive measurement of single-nucleotide polymorphism-induced changes of RNA conformation: application to disease studies. Nucleic Acids Res 41:44–53. https://doi.org/10.1093/nar/gks1009
119. Seeburg PH, Higuchi M, Sprengel R (1998) RNA editing of brain glutamate receptor channels: mechanism and physiology. Brain Res Brain Res Rev 26:217–229
120. Fritz J, Strehblow A, Taschner A et al (2009) RNA-regulated interaction of transportin-1 and exportin-5 with the double-stranded RNA-binding domain regulates nucleocytoplasmic shuttling of ADAR1. Mol Cell Biol 29:1487–1497. https://doi.org/10.1128/MCB.01519-08
121. Brownawell AM, Macara IG (2002) Exportin-5, a novel karyopherin, mediates nuclear export of double-stranded RNA binding proteins. J Cell Biol 156:53–64. https://doi.org/10.1083/jcb.200110082
122. Lu KP, Liou YC, Vincent I (2003) Proline-directed phosphorylation and isomerization in mitotic regulation and in Alzheimer's disease. BioEssays 25:174–181. https://doi.org/10.1002/bies.10223
123. Marcucci R, Brindle J, Paro S et al (2011) Pin1 and WWP2 regulate GluR2 Q/R site RNA editing by ADAR2 with opposing effects. EMBO J 30:4211–4222. https://doi.org/10.1038/emboj.2011.303
124. Yaffe MB, Schutkowski M, Shen M et al (1997) Sequence-specific and phosphorylation-dependent proline isomerization: a potential mitotic regulatory mechanism. Science 278:1957–1960. https://doi.org/10.1126/science.278.5345.1957
125. Desterro JMP, Keegan LP, Jaffray E et al (2005) SUMO-1 modification alters

126. Garncarz W, Tariq A, Handl C et al (2013) A high-throughput screen to identify enhancers of ADAR-mediated RNA-editing. RNA Biol 10:192–204. https://doi.org/10.4161/rna.23208

127. Eidem TM, Kugel JF, Goodrich JA (2016) Noncoding RNAs: regulators of the mammalian transcription machinery. J Mol Biol 428:2652–2659. https://doi.org/10.1016/j.jmb.2016.02.019

128. Su Y, Wu H, Pavlosky A et al (2016) Regulatory non-coding RNA: new instruments in the orchestration of cell death. Cell Death Dis 7:e2333–e2333. https://doi.org/10.1038/cddis.2016.210

129. Nilsen TW, Graveley BR (2010) Expansion of the eukaryotic proteome by alternative splicing. Nature 463:457–463. https://doi.org/10.1038/nature08909

130. Khurana E, Fu Y, Chakravarty D et al (2016) Role of non-coding sequence variants in cancer. Nat Rev Genet 17:93–108. https://doi.org/10.1038/nrg.2015.17

131. Tilgner H, Knowles DG, Johnson R et al (2012) Deep sequencing of subcellular RNA fractions shows splicing to be predominantly co-transcriptional in the human genome but inefficient for lncRNAs. Genome Res 22:1616–1625. https://doi.org/10.1101/gr.134445.111

132. Roy SW, Gilbert W (2006) The evolution of spliceosomal introns: patterns, puzzles and progress. Nat Rev Genet 7:211–221. https://doi.org/10.1038/nrg1807

133. Steitz TA, Steitz JA (1993) A general two-metal-ion mechanism for catalytic RNA. Proc Natl Acad Sci U S A 90:6498–6502

134. Lim KH, Ferraris L, Filloux ME et al (2011) Using positional distribution to identify splicing elements and predict pre-mRNA processing defects in human genes. Proc Natl Acad Sci U S A 108:11093–11098. https://doi.org/10.1073/pnas.1101135108

135. Taggart AJ, DeSimone AM, Shih JS et al (2012) Large-scale mapping of branchpoints in human pre-mRNA transcripts in vivo. Nat Struct Mol Biol 19:719–721. https://doi.org/10.1038/nsmb.2327

136. Lee Y, Rio DC (2015) Mechanisms and regulation of alternative pre-mRNA splicing. Annu Rev Biochem 84:291–323. https://doi.org/10.1146/annurev-biochem-060614-034316

137. Fedorova L, Fedorov A (2003) Introns in gene evolution. In: Origin and evolution of new gene functions. Springer, Dordrecht, pp 123–131

138. Wong JJL, Au AYM, Ritchie W, Rasko JEJ (2015) Intron retention in mRNA: no longer nonsense. BioEssays 38:41–49. https://doi.org/10.1002/bies.201500117

139. Bicknell AA, Cenik C, Chua HN et al (2012) Introns in UTRs: why we should stop ignoring them. BioEssays 34:1025–1034. https://doi.org/10.1002/bies.201200073

140. Lykke-Andersen S, Jensen TH (2015) Nonsense-mediated mRNA decay: an intricate machinery that shapes transcriptomes. Nat Rev Mol Cell Biol 16:665–677. https://doi.org/10.1038/nrm4063

141. Hsiao Y-HE, Bahn JH, Yang Y et al (2018) RNA editing in nascent RNA affects pre-mRNA splicing. Genome Res 28:812–823. https://doi.org/10.1101/gr.231209.117

142. Rueter SM, Dawson TR, Emeson RB (1999) Regulation of alternative splicing by RNA editing. Nature 399:75–80. https://doi.org/10.1038/19992

143. Feng Y, Sansam CL, Singh M, Emeson RB (2006) Altered RNA editing in mice lacking ADAR2 autoregulation. Mol Cell Biol 26:480–488. https://doi.org/10.1128/MCB.26.2.480-488.2006

144. Maas S, Kawahara Y, Tamburro KM, Nishikura K (2006) A-to-I RNA editing and human disease. RNA Biol 3:1–9

145. Flomen R, Knight J, Sham P et al (2004) Evidence that RNA editing modulates splice site selection in the 5-HT2C receptor gene. Nucleic Acids Res 32:2113–2122. https://doi.org/10.1093/nar/gkh536

146. Chen Y-T, Chang IY-F, Liu H et al (2018) Tumor-associated intronic editing of HNRPLL generates a novel splicing variant linked to cell proliferation. J Biol Chem 293:10158–10171. https://doi.org/10.1074/jbc.RA117.001197

147. Siepel A, Bejerano G, Pedersen JS et al (2005) Evolutionarily conserved elements in vertebrate, insect, worm, and yeast genomes. Genome Res 15:1034–1050. https://doi.org/10.1101/gr.3715005

148. Gingerich TJ, Feige J-J, LaMarre J (2004) AU-rich elements and the control of gene expression through regulated mRNA stability. Anim Health Res Rev 5:49–63

149. Halees AS, Hitti E, Al-Saif M et al (2011) Global assessment of GU-rich regulatory content and function in the human

transcriptome. RNA Biol 8:681–691. https://doi.org/10.4161/rna.8.4.16283

150. Chen J-M, Férec C, Cooper DN (2006) A systematic analysis of disease-associated variants in the 3′ regulatory regions of human protein-coding genes I: general principles and overview. Hum Genet 120:1–21. https://doi.org/10.1007/s00439-006-0180-7

151. Sandberg R, Neilson JR, Sarma A et al (2008) Proliferating cells express mRNAs with shortened 3' untranslated regions and fewer MicroRNA target sites. Science 320:1643–1647. https://doi.org/10.1126/science.1155390

152. Chen J, Kastan MB (2010) 5′-3′-UTR interactions regulate p53 mRNA translation and provide a target for modulating p53 induction after DNA damage. Genes Dev 24:2146–2156. https://doi.org/10.1101/gad.1968910

153. Djuranovic S, Nahvi A, Green R (2012) miRNA-mediated gene silencing by translational repression followed by mRNA deadenylation and decay. Science 336:237–240. https://doi.org/10.1126/science.1215691

154. Vasudevan S, Tong Y, Steitz JA (2007) Switching from repression to activation: microRNAs can up-regulate translation. Science 318:1931–1934. https://doi.org/10.1126/science.1149460

155. Ghosh T, Soni K, Scaria V et al (2008) MicroRNA-mediated up-regulation of an alternatively polyadenylated variant of the mouse cytoplasmic β-actin gene. Nucleic Acids Res 36:6318–6332. https://doi.org/10.1093/nar/gkn624

156. Lee I, Ajay SS, Yook JI et al (2009) New class of microRNA targets containing simultaneous 5′-UTR and 3′-UTR interaction sites. Genome Res 19:1175–1183. https://doi.org/10.1101/gr.089367.108

157. Chen L-L, DeCerbo JN, Carmichael GG (2008) Alu element-mediated gene silencing. EMBO J 27:1694–1705. https://doi.org/10.1038/emboj.2008.94

158. Zhang Z, Carmichael GG (2001) The fate of dsRNA in the nucleus: a p54nrb-containing complex mediates the nuclear retention of promiscuously A-to-I edited RNAs. Cell 106:465–476. https://doi.org/10.1016/S0092-8674(01)00466-4

159. Wang Q, Hui H, Guo Z et al (2013) ADAR1 regulates ARHGAP26 gene expression through RNA editing by disrupting miR-30b-3p and miR-573 binding. RNA 19:1525–1536. https://doi.org/10.1261/rna.041533.113

160. Zhang L, Yang C-S, Varelas X, Monti S (2016) Altered RNA editing in 3′ UTR perturbs microRNA-mediated regulation of oncogenes and tumor-suppressors. Sci Rep 6:23226. https://doi.org/10.1038/srep23226

161. Borchert GM, Gilmore BL, Spengler RM et al (2009) Adenosine deamination in human transcripts generates novel microRNA binding sites. Hum Mol Genet 18:4801–4807. https://doi.org/10.1093/hmg/ddp443

162. Saxena S, Jónsson ZO, Dutta A (2003) Small RNAs with imperfect match to endogenous mRNA repress translation implications for off-target activity of small inhibitory RNA in mammalian cells. J Biol Chem 278:44312–44319. https://doi.org/10.1074/jbc.M307089200

163. Doench JG, Sharp PA (2004) Specificity of microRNA target selection in translational repression. Genes Dev 18:504–511. https://doi.org/10.1101/gad.1184404

164. Kertesz M, Iovino N, Unnerstall U et al (2007) The role of site accessibility in microRNA target recognition. Nat Publ Group 39:1278–1284. https://doi.org/10.1038/ng2135

165. Ha M, Kim VN (2014) Regulation of microRNA biogenesis. Nat Rev Mol Cell Biol 15:509–524. https://doi.org/10.1038/nrm3838

166. Kleinberger Y, Eisenberg E (2010) Large-scale analysis of structural, sequence and thermodynamic characteristics of A-to-I RNA editing sites in human Alu repeats. BMC Genomics 11:453. https://doi.org/10.1186/1471-2164-11-453

167. Lagos-Quintana M, Rauhut R, Lendeckel W, Tuschl T (2001) Identification of novel genes coding for small expressed RNAs. Science 294:853–858. https://doi.org/10.1126/science.1064921

168. Lau NC, Lim LP, Weinstein EG, Bartel DP (2001) An abundant class of tiny RNAs with probable regulatory roles in Caenorhabditis elegans. Science 294:858–862. https://doi.org/10.1126/science.1065062

169. Lee RC, Ambros V (2001) An extensive class of small RNAs in Caenorhabditis elegans. Science 294:862–864. https://doi.org/10.1126/science.1065329

170. Kozomara A, Griffiths-Jones S (2014) miRBase: annotating high confidence microRNAs using deep sequencing data. Nucleic Acids Res 42:D68–D73. https://doi.org/10.1093/nar/gkt1181

171. Kim Y-K, Kim VN (2007) Processing of intronic microRNAs. EMBO J 26:775–783. https://doi.org/10.1038/sj.emboj.7601512

172. Westholm JO, Lai EC (2011) Mirtrons: microRNA biogenesis via splicing. Biochimie 93:1897–1904. https://doi.org/10.1016/j.biochi.2011.06.017

173. Moretti F, Thermann R, Hentze MW (2010) Mechanism of translational regulation by miR-2 from sites in the 5′ untranslated region or the open reading frame. RNA 16:2493–2502. https://doi.org/10.1261/rna.2384610

174. Forman JJ, Legesse-Miller A, Coller HA (2008) A search for conserved sequences in coding regions reveals that the let-7 microRNA targets Dicer within its coding sequence. Proc Natl Acad Sci U S A 105:14879–14884. https://doi.org/10.1073/pnas.0803230105

175. Shin C, Nam J-W, Farh KK-H et al (2010) Expanding the MicroRNA targeting code: functional sites with centered pairing. Mol Cell 38:789–802. https://doi.org/10.1016/j.molcel.2010.06.005

176. Roberts JT, Borchert GM (2017) Computational prediction of MicroRNA target genes, target prediction databases, and web resources. In: Bioinformatics in MicroRNA research. Humana Press, New York, pp 109–122

177. Blow MJ, Grocock RJ, Van Dongen S et al (2006) RNA editing of human microRNAs. Genome Biol 7:R27. https://doi.org/10.1186/gb-2006-7-4-r27

178. Alon S, Mor E, Vigneault F et al (2012) Systematic identification of edited microRNAs in the human brain. Genome Res 22:1533–1540. https://doi.org/10.1101/gr.131573.111

179. Nigita G, Acunzo M, Romano G et al (2016) microRNA editing in seed region aligns with cellular changes in hypoxic conditions. Nucleic Acids Res 44:6298–6308. https://doi.org/10.1093/nar/gkw532

180. Pinto Y, Buchumenski I, Levanon EY, Eisenberg E (2018) Human cancer tissues exhibit reduced A-to-I editing of miRNAs coupled with elevated editing of their targets. Nucleic Acids Res 46:71–82. https://doi.org/10.1093/nar/gkx1176

181. Yang W, Chendrimada TP, Wang Q et al (2006) Modulation of microRNA processing and expression through RNA editing by ADAR deaminases. Nat Struct Mol Biol 13:13–21. https://doi.org/10.1038/nsmb1041

182. Liu Z, Wang J, Li G, Wang H-W (2014) Structure of precursor microRNA's terminal loop regulates human Dicer's dicing activity by switching DExH/D domain. Protein Cell 6:185–193. https://doi.org/10.1007/s13238-014-0124-2

183. Jankowsky E, Bowers H (2006) Remodeling of ribonucleoprotein complexes with DExH/D RNA helicases. Nucleic Acids Res 34:4181–4188. https://doi.org/10.1093/nar/gkl410

184. Yan KS, Yan S, Farooq A et al (2003) Structure and conserved RNA binding of the PAZ domain. Nature 426:469–474. https://doi.org/10.1038/nature02129

185. Park J-E, Heo I, Tian Y et al (2011) Dicer recognizes the 5′ end of RNA for efficient and accurate processing. Nature 475:201–205. https://doi.org/10.1038/nature10198

186. Tsutsumi A, Kawamata T, Izumi N et al (2011) Recognition of the pre-miRNA structure by *Drosophila* Dicer-1. Nat Struct Mol Biol 18:1153–1158. https://doi.org/10.1038/nsmb.2125

187. Kawahara Y, Zinshteyn B, Chendrimada TP et al (2007) RNA editing of the microRNA-151 precursor blocks cleavage by the Dicer-TRBP complex. EMBO Rep 8:763–769. https://doi.org/10.1038/sj.embor.7401011

188. Kawahara Y, Zinshteyn B, Sethupathy P et al (2007) Redirection of silencing targets by adenosine-to-inosine editing of miRNAs. Science 315:1137–1140. https://doi.org/10.1126/science.1138050

189. Choudhury Y, Tay FC, Lam DH et al (2012) Attenuated adenosine-to-inosine editing of microRNA-376a* promotes invasiveness of glioblastoma cells. J Clin Invest 122:4059–4076. https://doi.org/10.1172/JCI62925

190. Cesarini V, Silvestris DA, Tassinari V et al (2018) ADAR2/miR-589-3p axis controls glioblastoma cell migration/invasion. Nucleic Acids Res 46:2045–2059. https://doi.org/10.1093/nar/gkx1257

191. Gregory PA, Bert AG, Paterson EL et al (2008) The miR-200 family and miR-205 regulate epithelial to mesenchymal transition by targeting ZEB1 and SIP1. Nat Cell Biol 10:593–601. https://doi.org/10.1038/ncb1722

192. Park S-M, Gaur AB, Lengyel E, Peter ME (2008) The miR-200 family determines the epithelial phenotype of cancer cells by

targeting the E-cadherin repressors ZEB1 and ZEB2. Genes Dev 22:894–907. https://doi.org/10.1101/gad.1640608

193. Shoshan E, Mobley AK, Braeuer RR et al (2015) Reduced adenosine-to-inosine miR-455-5p editing promotes melanoma growth and metastasis. Nat Cell Biol 17:311–321. https://doi.org/10.1038/ncb3110

194. Velazquez-Torres G, Shoshan E, Ivan C et al (2018) A-to-I miR-378a-3p editing can prevent melanoma progression via regulation of PARVA expression. Nat Commun 9:461. https://doi.org/10.1038/s41467-018-02851-7

195. Moran VA, Perera RJ, Khalil AM (2012) Emerging functional and mechanistic paradigms of mammalian long non-coding RNAs. Nucleic Acids Res 40:6391–6400. https://doi.org/10.1093/nar/gks296

196. Harrow J, Frankish A, Gonzalez JM et al (2012) GENCODE: the reference human genome annotation for The ENCODE Project. Genome Res 22:1760–1774. https://doi.org/10.1101/gr.135350.111

197. Quek XC, Thomson DW, Maag JLV et al (2015) lncRNAdb v2.0: expanding the reference database for functional long noncoding RNAs. Nucleic Acids Res 43:D168–D173. https://doi.org/10.1093/nar/gku988

198. St Laurent G, Wahlestedt C, Kapranov P (2015) The landscape of long noncoding RNA classification. Trends Genet 31:239–251. https://doi.org/10.1016/j.tig.2015.03.007

199. Housman G, Ulitsky I (2016) Methods for distinguishing between protein-coding and long noncoding RNAs and the elusive biological purpose of translation of long non-coding RNAs. Biochim Biophys Acta 1859:31–40. https://doi.org/10.1016/j.bbagrm.2015.07.017

200. Nelson BR, Makarewich CA, Anderson DM et al (2016) A peptide encoded by a transcript annotated as long noncoding RNA enhances SERCA activity in muscle. Science 351:271–275. https://doi.org/10.1126/science.aad4076

201. Ruiz-Orera J, Verdaguer-Grau P, Villanueva-Cañas JL, et al (2017) Evidence for functional and non-functional classes of peptides translated from long non-coding RNAs. bioRxiv: 064915. doi: https://doi.org/10.1101/064915

202. Thomson DW, Dinger ME (2016) Endogenous microRNA sponges: evidence and controversy. Nat Rev Genet 17:272–283. https://doi.org/10.1038/nrg.2016.20

203. Wang X, Arai S, Song X et al (2008) Induced ncRNAs allosterically modify RNA-binding proteins in cis to inhibit transcription. Nature 454:126–130. https://doi.org/10.1038/nature06992

204. Schaukowitch K, Joo J-Y, Liu X et al (2014) Enhancer RNA facilitates NELF release from immediate early genes. Mol Cell 56:29–42. https://doi.org/10.1016/j.molcel.2014.08.023

205. Sigova AA, Abraham BJ, Ji X et al (2015) Transcription factor trapping by RNA in gene regulatory elements. Science 350:978–981. https://doi.org/10.1126/science.aad3346

206. Mondal T, Subhash S, Vaid R et al (2015) MEG3 long noncoding RNA regulates the TGF-β pathway genes through formation of RNA-DNA triplex structures. Nat Commun 6:7743. https://doi.org/10.1038/ncomms8743

207. Postepska-Igielska A, Giwojna A, Gasri-Plotnitsky L et al (2015) LncRNA Khps1 regulates expression of the proto-oncogene SPHK1 via triplex-mediated changes in chromatin structure. Mol Cell 60:626–636. https://doi.org/10.1016/j.molcel.2015.10.001

208. Gong J, Liu C, Liu W et al (2017) LNCediting: a database for functional effects of RNA editing in lncRNAs. Nucleic Acids Res 45:D79–D84. https://doi.org/10.1093/nar/gkw835

209. Picardi E, D'Erchia AM, Gallo A et al (2014) Uncovering RNA editing sites in long non-coding RNAs. Front Bioeng Biotechnol 2:64. https://doi.org/10.3389/fbioe.2014.00064

210. Luo H, Fang S, Sun L et al (2017) Comprehensive characterization of the RNA editomes in cancer development and progression. Front Genet 8:230. https://doi.org/10.3389/fgene.2017.00230

211. Novikova IV, Hennelly SP, Sanbonmatsu KY (2012) Structural architecture of the human long non-coding RNA, steroid receptor RNA activator. Nucleic Acids Res 40:5034–5051. https://doi.org/10.1093/nar/gks071

212. Mahlab S, Tuller T, Linial M (2012) Conservation of the relative tRNA composition in healthy and cancerous tissues. RNA 18:640–652. https://doi.org/10.1261/rna.030775.111

213. Itoh Y, Sekine S-I, Suetsugu S, Yokoyama S (2013) Tertiary structure of bacterial

selenocysteine tRNA. Nucleic Acids Res 41:6729–6738. https://doi.org/10.1093/nar/gkt321

214. Bunn CC, Mathews MB (1987) Autoreactive epitope defined as the anticodon region of alanine transfer RNA. Science 238:1116–1119

215. Becker HF, Corda Y, Mathews MB et al (1999) Inosine and N1-methylinosine within a synthetic oligomer mimicking the anticodon loop of human tRNA(Ala) are major epitopes for anti-PL-12 myositis autoantibodies. RNA 5:865–875

216. Torres AG, Piñeyro D, Filonava L et al (2014) A-to-I editing on tRNAs: biochemical, biological and evolutionary implications. FEBS Lett 588:4279–4286. https://doi.org/10.1016/j.febslet.2014.09.025

217. Gerber AP, Keller W (2001) RNA editing by base deamination: more enzymes, more targets, new mysteries. Trends Biochem Sci 26:376–384. https://doi.org/10.1016/S0968-0004(01)01827-8

218. Maas S, Gerber AP, Rich A (1999) Identification and characterization of a human tRNA-specific adenosine deaminase related to the ADAR family of pre-mRNA editing enzymes. Proc Natl Acad Sci U S A 96:8895–8900

219. Bjork GR, Jacobsson K, Nilsson K et al (2001) A primordial tRNA modification required for the evolution of life? EMBO J 20:231–239. https://doi.org/10.1093/emboj/20.1.231

220. Gerber AP, Keller W (1999) An adenosine deaminase that generates inosine at the wobble position of tRNAs. Science 286:1146–1149

221. Torres AG, Piñeyro D, Rodríguez-Escribà M et al (2015) Inosine modifications in human tRNAs are incorporated at the precursor tRNA level. Nucleic Acids Res 43:5145–5157. https://doi.org/10.1093/nar/gkv277

222. Alazami AM, Hijazi H, Al-Dosari MS et al (2013) Mutation in ADAT3, encoding adenosine deaminase acting on transfer RNA, causes intellectual disability and strabismus. J Med Genet 50:425–430. https://doi.org/10.1136/jmedgenet-2012-101378

223. Morse DP, Bass BL (1999) Long RNA hairpins that contain inosine are present in Caenorhabditis elegans poly(A)+ RNA. Proc Natl Acad Sci U S A 96:6048–6053

224. Hoopengardner B, Bhalla T, Staber C, Reenan R (2003) Nervous system targets of RNA editing identified by comparative genomics. Science 301:832–836. https://doi.org/10.1126/science.1086763

225. Boguski MS, Lowe T, Tolstoshev CM (1993) dbEST—database for "expressed sequence tags". Nat Genet 4:332–333. https://doi.org/10.1038/ng0893-332

226. Veneziano D, Di Bella S, Nigita G et al (2016) Noncoding RNA: current deep sequencing data analysis approaches and challenges. Hum Mutat 37:1283–1298. https://doi.org/10.1002/humu.23066

227. Li JB, Levanon EY, Yoon J-K et al (2009) Genome-wide identification of human RNA editing sites by parallel DNA capturing and sequencing. Science 324:1210–1213. https://doi.org/10.1126/science.1170995

228. Nigita G, Veneziano D, Ferro A (2015) A-to-I RNA editing: current knowledge sources and computational approaches with special emphasis on non-coding RNA molecules. Front Bioeng Biotechnol 3:37. https://doi.org/10.3389/fbioe.2015.00037

229. de Hoon MJL, Taft RJ, Hashimoto T et al (2010) Cross-mapping and the identification of editing sites in mature microRNAs in high-throughput sequencing libraries. Genome Res 20:257–264. https://doi.org/10.1101/gr.095273.109

230. Alon S, Eisenberg E (2013) Identifying RNA editing sites in miRNAs by deep sequencing. Methods Mol Biol 1038:159–170. https://doi.org/10.1007/978-1-62703-514-9_9

231. Alon S, Erew M, Eisenberg E (2015) DREAM: a webserver for the identification of editing sites in mature miRNAs using deep sequencing data. Bioinformatics 31:2568–2570. https://doi.org/10.1093/bioinformatics/btv187

232. Sakurai M, Yano T, Kawabata H et al (2010) Inosine cyanoethylation identifies A-to-I RNA editing sites in the human transcriptome. Nat Chem Biol 6:733–740. https://doi.org/10.1038/nchembio.434

233. Sakurai M, Ueda H, Yano T et al (2014) A biochemical landscape of A-to-I RNA editing in the human brain transcriptome. Genome Res 24:522–534. https://doi.org/10.1101/gr.162537.113

234. Kiran A, Baranov PV (2010) DARNED: a DAtabase of RNa EDiting in humans. Bioinformatics 26:1772–1776. https://doi.org/10.1093/bioinformatics/btq285

235. Laganà A, Paone A, Veneziano D et al (2012) miR-EdiTar: a database of predicted A-to-I edited miRNA target sites. Bioinformatics 28:3166–3168. https://doi.org/10.1093/bioinformatics/bts589

236. Kiran AM, O'Mahony JJ, Sanjeev K, Baranov PV (2013) Darned in 2013: inclusion of model organisms and linking with Wikipedia. Nucleic Acids Res 41:D258–D261. https://doi.org/10.1093/nar/gks961
237. Picardi E, D'Antonio M, Carrabino D et al (2011) ExpEdit: a webserver to explore human RNA editing in RNA-Seq experiments. Bioinformatics 27:1311–1312. https://doi.org/10.1093/bioinformatics/btr117
238. Distefano R, Nigita G, Macca V et al (2013) VIRGO: visualization of A-to-I RNA editing sites in genomic sequences. BMC Bioinformatics 14(Suppl 7):S5. https://doi.org/10.1186/1471-2105-14-S7-S5
239. Wahlstedt H, Luciano DJ, Enstero M, Öhman M (2009) Large-scale mRNA sequencing determines global regulation of RNA editing during brain development. Genome Res 19:978–986. https://doi.org/10.1101/gr.089409.108
240. Solomon O, Bazak L, Levanon EY et al (2014) Characterizing of functional human coding RNA editing from evolutionary, structural, and dynamic perspectives. Proteins 82:3117–3131. https://doi.org/10.1002/prot.24672
241. Ramaswami G, Li JB (2014) RADAR: a rigorously annotated database of A-to-I RNA editing. Nucleic Acids Res 42:D109–D113. https://doi.org/10.1093/nar/gkt996

Chapter 7

Computational Prediction of Functional MicroRNA–mRNA Interactions

Müşerref Duygu Saçar Demirci, Malik Yousef, and Jens Allmer

Abstract

Proteins have a strong influence on the phenotype and their aberrant expression leads to diseases. MicroRNAs (miRNAs) are short RNA sequences which posttranscriptionally regulate protein expression. This regulation is driven by miRNAs acting as recognition sequences for their target mRNAs within a larger regulatory machinery. A miRNA can have many target mRNAs and an mRNA can be targeted by many miRNAs which makes it difficult to experimentally discover all miRNA–mRNA interactions. Therefore, computational methods have been developed for miRNA detection and miRNA target prediction. An abundance of available computational tools makes selection difficult. Additionally, interactions are not currently the focus of investigation although they more accurately define the regulation than pre-miRNA detection or target prediction could perform alone. We define an interaction including the miRNA source and the mRNA target. We present computational methods allowing the investigation of these interactions as well as how they can be used to extend regulatory pathways. Finally, we present a list of points that should be taken into account when investigating miRNA–mRNA interactions. In the future, this may lead to better understanding of functional interactions which may pave the way for disease marker discovery and design of miRNA-based drugs.

Key words MicroRNA, Target, Regulation, Posttranscriptional regulation, Pathway extension, MiRNA–mRNA interaction

1 Introduction

The central dogma in biology describes a programmed flow of information from the genome to the phenotype [1]. However, as many dogmas before (e.g., the earth being the center of the universe), this viewpoint is challenged and a better explanation for reality can be gained by embracing relativity theory in biology [2]. This highlights the importance of regulation for the overall genetic programming which consists of many scopes including epigenetic, transcriptomic, and proteomic. Life, manifested at all scales, from metabolites to complete organisms, interplays and thereby implements the genetic program with the genome providing a parts list [2]. Posttranscriptional regulation via noncoding RNAs modifies protein abundance. Proteins can be transcription factors or parts of biological machines and pathways with direct involvement in transcription, closing the regulatory loop. Currently, evidence is accumulating supporting transcriptionally active miRNAs which are reimported into the nucleus [3], providing a direct involvement of microRNAs (miRNAs) in transcriptional control. Mature miRNAs are small (18–24 nt) noncoding RNAs derived from hairpins involving a complex molecular pathway [4]. These mature miRNAs act as recognition keys for their target genes within silencing complexes such as RISC. Either targeted mRNAs are silenced through endonucleolytic cleavage [5] or the translation process is modulated by either increasing [6] or decreasing protein abundance.

Regulation of gene expression is important and dysregulation often leads to disease. MiRNA dysregulation has been implicated in cancer, amyotrophic lateral sclerosis, and many other diseases. MiRNAs are also potentially instrumentalized by viruses [7, 8] and other organisms such as *Toxoplasma gondii* [9, 10] to modulate the host environment to their advantage. This may also work vice versa and there are accounts of human miRNAs shaping the gut microbiome [11] but in general the miRNA-based communication between host and microbiome is not yet well understood [12]. MiRNAs can be detected experimentally using a number of approaches such as degradome sequencing (Table 1).

Such approaches are time consuming and cannot investigate all possible states of a biological system which is why top-down computational ab initio prediction of miRNAs is important [16]. The same holds true for miRNA targets which can be experimentally detected using, for example, HITS-CLIP [17]. The multiplicity of miRNA targets further increases the challenge so that computational methods are prerogative [18]. MiRNAs and their target mRNAs can form a regulative network based on their source transcripts. It is, therefore, important to combine transcriptional and posttranscriptional regulation when investigating miRNAs and their targets.

The further text is structured into three parts. First we discuss miRNA detection including experimental and computational methods. Information on miRNA functionality follows with a focus on targeting. Finally, regulatory networks resulting from miRNAs and their targets are introduced.

2 MicroRNA Detection

2.1 Experimental miRNA Detection

Since the first discovery of miRNAs, many experimental and computational approaches have been developed for their detection and analysis. In earlier studies, such as the identification of let-7 in *Caenorhabditis elegans*, forward genetic approaches were applied

Table 1
Comparison of miRNA and miRNA target detection strategies

Method	Advantage	Disadvantage
Northern Blotting	New and known miRNAs Validation	Need lots of RNA MiRNAs with low abundance Low throughput Low sensitivity Time consuming
qRT-PCR	Highest dynamic range and accuracy Absolute quantification	Throughput problems Normalization and specificity
Microarray	Cheaper Screening tool	Not quantitative Lower sensitivity and dynamic range
NGS	High sensitivity Detection of sequence variation New miRNAs	Artifacts/contamination Time consuming Complex bioinformatics
HITS-CLIP	Stringent isolation of the miRNA-mRNA-Ago Analyzing miRNA–mRNA interactome	Inefficient UV cross-linking Requirement of large number of cells for library preparation
CLASH	High-throughput identification Independent of bioinformatic predictions	Stringent purification Low efficiency of RNA-RNA ligation

Adapted and updated from [13–15]

[19]. However, such methods are limited due to some characteristics of miRNAs, for example, their small size [20]. Today, Northern blotting, microarray, NGS, and qRT-PCR are popular experimental techniques used for detection and/or validation of miRNAs, each of these methods having some inherent strengths and weaknesses (Table 1).

Experiments will remain the golden standard for confirmation of miRNAs, but it has become clear that it is not possible to capture all miRNAs experimentally since they may only be expressed in low quantity, at specific developmental times, or in specific tissues. Therefore, computational detection of miRNAs has become important.

2.2 Computational miRNA Detection

Various tools have been designed for the in silico prediction of miRNAs (Table 2). Although they may vary in numerous parts, most of these programs rely on the secondary structure of the miRNA precursor [45]. Algorithms such as RNAfold [46] not only perform RNA secondary structure prediction but also calculate the thermodynamic stability of the proposed RNA hairpin structures. Available methods generally include (1) genome-wide

Table 2
Selected computational tools for miRNA prediction

Tool	Year	Conservation	Structure	Sequence	Machine learning	NGS application
miRscan [21]	2003	+	+			
miRAlign [22]	2005		+	+		
ProMiR [23]	2005	+	+	+	+	
Triplet-SVM [24]	2005		+	+	+	
miR-abela [25]	2005		+	+	+	
RNAmicro [26]	2006		+	+	+	
BayesMiRNAfind [27]	2006	+	+	+		
miRFinder [28]	2007	+	+		+	
miPred [29]	2007		+		+	
MiRRim [30]	2007	+	+		+	
miRDeep [31]	2008		+			+
miRanalyzer [32]	2009				+	+
SSCprofiler [33]	2009	+	+	+	+	+
HHMMiR [34]	2009		+	+	+	
MIReNA [35]	2010	+	+	+		+
miRPara [36]	2011		+	+	+	
miRNAFold [37]	2012		+	+		
miREval 2.0 [38]	2013	+	+	+	+	
miR-PREFeR [39]	2014		+			+
miRBoost [40]	2015		+	+	+	
iMiRNA-SSF [41]	2016		+	+	+	
izMiR [42]	2017		+	+	+	+
miRge2.0 [43]	2018		+	+	+	+

The list is updated version from [44]. The list is sorted by publication year

prediction of hairpin structures, (2) filtering and/or scoring of those hairpins, and (3) experimental confirmation [45].

MiRNA gene prediction algorithms can be divided into several categories. Usually, either homology modeling or ab initio methods are applied to extract possible miRNAs from a genome [16]. Homology modeling is based on the idea that if a miRNA is identified in one organism then it is possible that its homologs might be found in closely related species. Considering miRNA families, using homology-based approach in the same organism can be beneficial too. Since conservation is usually associated with function, most of the predictions made by tools using homology-based methods tend to be miRNAs. It has been shown that a miRNA prediction approach developed by using available genome sequences in castor bean (*Ricinus communis*) was able to detect 86.6% of miRNAs in Arabidopsis [47]. However, it is important to note that lack of conservation in miRNAs does not imply absence of function and that fast evolution of some miRNAs has been observed [48]. Although homology-based methods are quite helpful for initial screening of candidate miRNAs, they have several essential drawbacks. Most importantly, it is not possible to find novel and/or species-specific miRNAs. Therefore, various ab initio-based tools are developed for prediction of novel miRNAs.

Even though the ab initio approach does not directly rely on conservation, it still uses information obtained from known miRNAs. The majority of such approaches utilizes machine learning (ML) (Table 2) and specifically two-class classification. There are many factors that affect the performance of the prediction scheme but the most important one seems to be data quality [49]. Various tools have been designed for miRNA prediction. Some of them are constructed in a similar manner while others are using different approaches to achieve the task (Table 2). When comparing available tools and selecting one of them for analysis, the most important criteria should be accuracy of their results. Although some measurements like accuracy, sensitivity, and specificity are usually reported in papers about such tools, it is not possible to compare such values directly [42]. Based on a comparison of 15 tools in terms of sensitivity (min: 55, max: 98) and specificity (min: 40, max: 98), none of them provides sufficient confidence for experimental testing of all estimated 60 million miRNA-like structures found in the human genome [44]. Therefore, we recently suggested using consensus approaches for more reliable predictions rather than depending on any one tool [42].

Computational detection of miRNAs even in large eukaryotic genomes is now possible which can be used to answer questions such as about their locations within a genome and their multiplicity. Some miRNAs have large number of copies spread throughout the genome as for example for let-7. MiRNA copy number variation may lead to disease [50] which adds value to answering such questions. MiRNAs may vary in respect to their abundance with which they are encoded in a genome but their number of targets varies even more widely.

3 MicroRNA Targeting

MiRNAs follow a biogenesis pathway leading from their transcription to their incorporation into RISC [51]. Within RISC they help recognize their targets via sequence complementarity. The consensus is that RISC binding to their target mRNAs primes them for degradation or for translational repression when binding within the 3′-UTR region of the target mRNA. There have been accounts of translational activation when binding in the 5′-UTR and reimport of mature miRNAs into the nucleus leading to transcriptional control but this is not discussed here. Experimental approaches like PAR-CLIP [52], HITS-CLIP [53], and CLASH [54] are currently employed for the investigation of miRNA-based targeting (Table 1). Such techniques are limited to the availability of bound miRNA–mRNA in large enough quantity for detection. Competition among miRNA with multiple targets, miRNAs, or targets only expressed under specific conditions, and miRNAs or targets only expressed in low quantities, adds to this problem. MiRNA targets are stored in a number of databases such as miRTarBase [55] and TarBase [56]. Some databases contain computational predictions and others focus on experimentally validated targets (Table 3).

The experimental evidence points to the complementary binding of miRNA-loaded RISC to its targets as the most important factor for miRNA function. This is important for the design of computational methods for target prediction (Fig. 1). Investigation of the binding potential of miRNAs leads to a separation of the mature miRNA into seed region (nucleotides (1/2)-8 at 5′) and out region (3′ portion of the miRNA). The seed region generally forms perfect Watson-Crick complementarity with the target mRNA.

Computational approaches reflect the importance of complementary binding and heavily rely on it for detection of miRNA targets [18]. There are many means to classify computational tools for miRNA prediction and a recent survey grouped them into Web-based services, downloaded software, and R packages [73]. They conclude that Web-based tools are the most frequently used platform to predict miRNA–mRNA interactions and the top three tools from this category are TargetScan [74], miRanda [75], and DIANA Tools [76]. There has not been an independent assessment of prediction accuracy of the multitude of miRNA target prediction tools. Therefore, it cannot be judged whether preferential usage of these tools over standalone platforms and R scripts presents a problem. According to Riffo-Campos et al. TargetScan seems to be the most robust tool since its predictions have a higher probability of being biologically validated due to usage statistics [73].

Table 3
Databases containing miRNA targets and software for the prediction of miRNA targets

Name	Year	Type	Link
RNAhybrid [57]	2004	Webserver, predictions	https://bibiserv.cebitec.uni-bielefeld.de/rnahybrid;jsessionid=791293b42354681cb4afa2201b63
PicTar [58]	2005	Database, webserver, predictions	https://web.archive.org/web/20080724163022/http://pictar.bio.nyu.edu/
TargetScan [59]	2005	Database, webserver	http://www.targetscan.org/vert_72/
RNA22 [60]	2006	Webserver, predictions	https://cm.jefferson.edu/rna22/
TarBase [61]	2006	Database	http://diana.imis.athena-innovation.gr/DianaTools/index.php?r=tarbase/index
NBmiRtar [62]	2007	Webserver, predictions	http://wotan.wistar.upenn.edu/NBmiRTar/
PITA [63]	2007	Webserver, predictions	https://genie.weizmann.ac.il/pubs/mir07/mir07_data.html
Diana-microT [64]	2009	Webserver	http://diana.imis.athena-innovation.gr/DianaTools/index.php?r=tarbase/index
miRecords [65]	2009	Database	http://c1.accurascience.com/miRecords/
miRTarBase [66]	2011	Database	http://mirtarbase.mbc.nctu.edu.tw/php/index.php
miRwalk [67]	2011	Database, webserver	http://zmf.umm.uni-heidelberg.de/apps/zmf/mirwalk2/
RepTar [68]	2011	Database	http://bioinformatics.ekmd.huji.ac.il/reptar/
StarBase [69]	2014	Database	https://web.archive.org/web/20110222111721/http://starbase.sysu.edu.cn/
Cupid [70]	2015	Matlab script	http://cupidtool.sourceforge.net/
MBSTAR [71]	2015	Webserver, predictions	https://www.isical.ac.in/~bioinfo_miu/MBStar30.htm
StarScan [72]	2015	Web-based software	http://bioinformatics.psb.ugent.be/webtools/startscan/

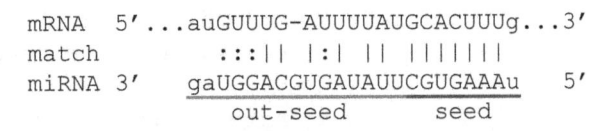

```
mRNA   5'...auGUUUG-AUUUUAUGCACUUUg...3'
match          :::||  |:| || |||||||
miRNA  3'    gaUGGACGUGAUAUUCGUGAAAu    5'
              out-seed         seed
```

Fig. 1 Example duplex structure of a miRNA and its target mRNA. Seed sequence (green) and out seed (blue) are indicated

Unfortunately, the molecular biology of target binding is not fully understood. Therefore, machine learning tools are used to automatically learn them from known examples. To perform machine learning, parameterization of the miRNA–mRNA duplex (Fig. 1) is an important step. Peterson et al. reviewed the features used by different computational tools for the prediction of miRNA targets [77]. They found that the four main aspects of the miRNA–mRNA target interaction modeled in the tools reviewed are seed match, evolutionary conservation, free energy, and site accessibility. We recently investigated feature selection for miRNA target prediction using machine learning and found large differences depending on the parameters used [78]. Among the first computational tools for miRNA target prediction, Diana-microT determined interaction rules using bioinformatics with coupled experimental validation and was able to predict all known *C. elegans* miRNA targets. Many more tools have been developed subsequently (Table 3). Another aspect of target site detection involves the fast and accurate detection of approximately complementary matches. The miRanda algorithm for example employs dynamic programming to optimally align miRNAs with their targets [79] but other approaches employ BLAST [80] or use different heuristics for sequence alignment. RNAhybrid further includes hybridization energy in its search for target sites [81]. Lai observed very little overlap among the predicted targets identified by several miRNA target prediction tools [82]. This could be explained by the utilization of different feature sets which capture distinct target sites. Sethupathy and colleagues [61] also compared miRNA target prediction tools and found that about 30% of experimentally validated target sites are non-conserved which can also partially explain the difference among tools found by Lai [82]. A large part of features used model sequence conservation. Such features may not generalize well since we were able to differentiate among target sites from different species using sequence-based features [83]. Yousef et al. developed a target-prediction method NBmiRTar using machine learning with a naïve Bayes classifier which does not incorporate sequence conservation but generates a model from sequence and miRNA–mRNA duplex information [62]. Training and testing examples were derived from validated target sequences and artificially generated negative data. NBmiRTar incorporates information from the seed and the "out-seed" segments of the miRNA–mRNA duplex (Fig. 1) and thereby produces fewer false-positive predictions and fewer target candidates to be tested than other methods.

Most recently, Riffo-Campos et al. reviewed miRNA target tools and described the fundamental biology which these prediction tools are based on. They also characterized the main sequence-based algorithms, and offered some insights into their uses by biologists [73]. Following miRNA target prediction, biomolecular validation is always necessary to confirm the miRNA-target gene interaction. Thus a protocol for validation of a miRNA target interaction is required. Different approaches are used for validation such as cloning of a dual-luciferase miRNA target expression vector, transfection of cells with this vector and a precursor miRNA (pre-miRNA), and subsequent luciferase assay [84]. Many other approaches such as degradome sequencing and methods in Table 1 are also used for validation. However the limitations of each validation approach need to be well understood. These limitations are well summarized in a recent review of experimental techniques for miRNA target identification [85]. Among the databases also giving supporting experimental evidence, miRTarBase [86] is the most updated resource for experimentally validated microRNA-target interactions.

While many miRNA targeting tools and several target databases are available, there still is a need to improve upon miRNA-targeting prediction in respect to prediction accuracy and toward quantifying the effect of the regulation. In the future, it would be beneficial for target prediction to include measures for target-site binding strength, target-site multiplicity on the target mRNA, and proximity to the stop of translation.

4 Detection of Gene: miRNA–Gene Interactions

MicroRNA detection and target prediction are important tasks. However, miRNAs can only convey function when co-expressed with their targets [87]. Therefore, it is important to refer to miRNA–mRNA interactions instead of analyzing miRNAs and their targets independently. Experimentally, such interactions can be analyzed using HITS-CLIP by cross-linking bound RNAs within protein complexes, isolating them, and sequencing the associated RNAs [53]. HITS-CLIP data was also instrumental in developing computational methods for the analysis of miRNA–mRNA regulatory interactions [88]. The HITS-CLIP methodology allows for the identification of functional interactions. Some limitations have been overcome with the CLASH protocol [54]. These approaches cannot replace computational methods, though, because it is not feasible to perform such experiments for all species, developmental stages, tissues, and external and internal stresses which affect regulation. Therefore, computational predictions of hairpins and their targets is an active area of research. A focus on miRNA–mRNA interactions will further shape the field in the future.

We here define a miRNA–mRNA interaction in terms of source gene interacting with its target gene, thereby abstracting all the detailed biological pathways and focusing only on the regulative role of the interaction. It needs to be noted that miRNAs can originate from anywhere in a genome. About 50% of the miRNAs are located within transcription units (40% within introns) and the other 50% are located in intergenic regions where they are usually clustered [89]. For example, miRNAs can form clusters in intergenic regions of a genome and be transcribed in a coordinated manner [90]. They are also co-transcribed with genes within their exons or noncoding parts. For example, the DiGeorge syndrome critical region gene 8 (DGCR8) contains a hairpin in its first exon which can be recognized and cleaved by the microprocessor complex leading to a truncated product and a miRNA [91]. We predicted more than 300 putative targets of the hairpin in human and the following six target genes seem to be affected most with two target-binding sites each: RASGRP1, LYNX1, TBC1D16, KLHL28, IPO8, and DPF1. This also exemplifies that miRNAs can have multiple targets. On the other hand, mRNAs can be targeted by multiple miRNAs and can have multiple binding sites per miRNA. Together, this can lead to a huge interaction network (Fig. 2).

It is beneficial to filter this large network by actually expressed interactions [92]. A complication exists for the acquisition of sequencing data for miRNAs, which requires special sequencing strategies since they are very short. Such expression data is often not available in public data from, for example, the sequence read archive [93]. Another approach, which is not dependent on short read sequences, uses the expression of the enclosing transcription unit as a measure for the expression of the miRNA. MiRNAs are often part of genes (Fig. 3) and are co-transcribed with them. Therefore, the expression of the gene can indicate the expression of miRNAs co-expressed within UTRs or introns. MiRNAs from exons can be treated in the same manner, but then it also needs to be taken into account that the resulting mRNA is also structurally affected and its sequence altered.

4.1 Databases Containing MicroRNA Targets

MiRNA and their targets are available in databases such as TarBase and miRTarBase. These are based on predictions or text mining. These databases also provide information about the evidence for the targeting such as sequencing support. Degradome sequencing has become a means of assessing functional miRNA–mRNA interactions and plant data is available in DPMIND [94]. Most of the target databases represent miRNA targets for the miRNAs available in miRBase. However, in most cases the source gene or miRNA cluster is not specified. Within the page source miRBase displays the overlapping transcripts with the miRNAs it hosts but does not visualize it in the web site, yet. Perhaps, this is a future feature which still needs further scrutiny. Clearly, it is an important piece of information facilitating the integration of miRNA and gene regulatory networks. Previously, we have constructed a database compiling the miRNA and target information from TarBase, miRTarBase, and miRBase [95] which also includes the

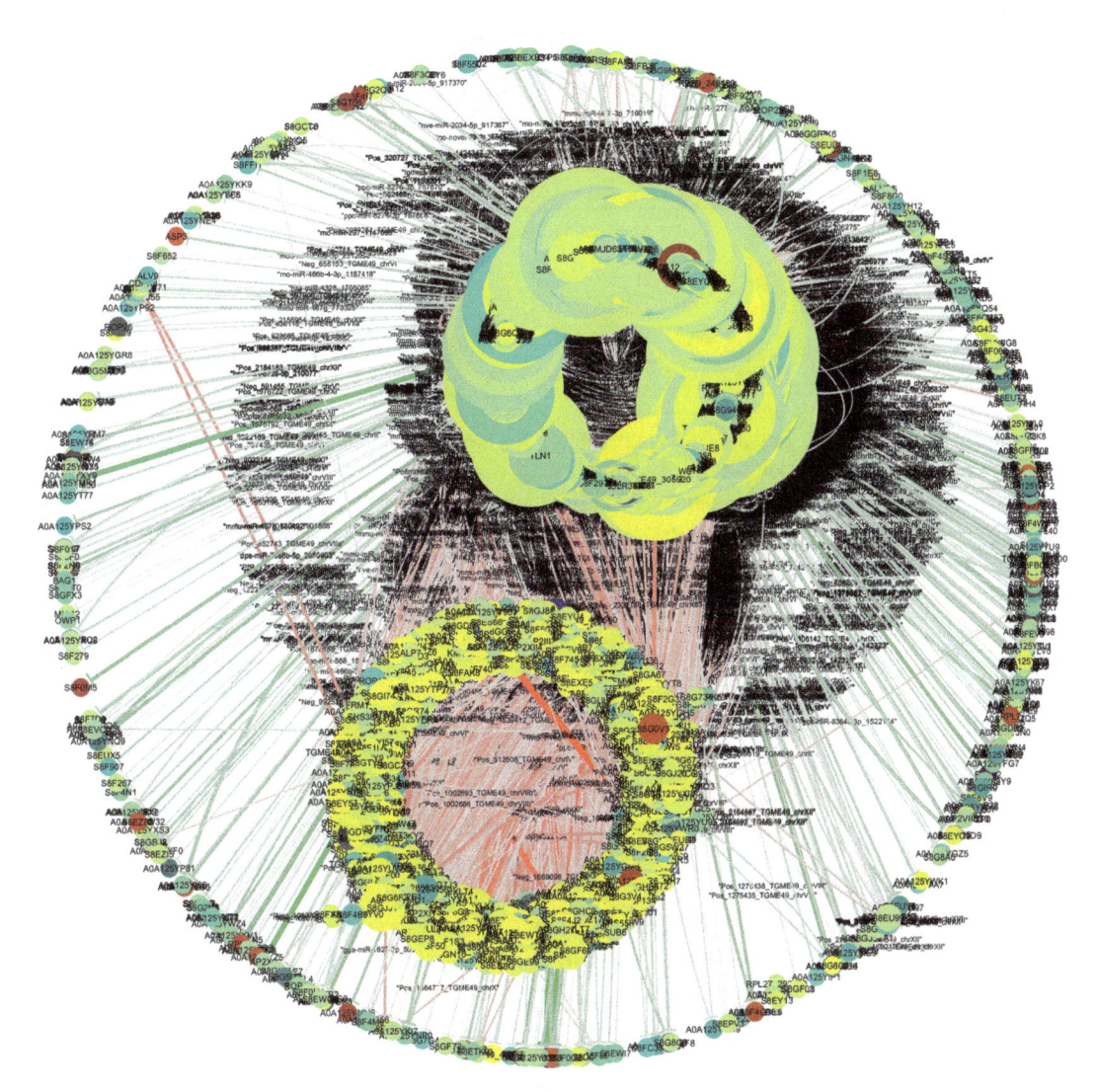

Fig. 2 Computationally predicted miRNA–mRNA interactions for *Toxoplasma gondii* forming a hairball that is difficult to interpret [92]

overlapping transcripts and can be queried using VANESA [96]. Figure 3 shows a regulatory network including genes, miRNAs, and proteins. Generally, pathway databases like Reactome [97] and KEGG [98] only include genes and their interactions. Implicitly, the interactions represent gene products which include transcription factors. VANESA facilitates the merging and enriching of KEGG pathways with miRNA interactions form the integrated database [99]. Thereby, pathways can be extended and can be investigated on multiple regulatory levels at the same time.

4.2 Other Approaches to Network Construction

The approach detailed above includes the origin of miRNAs in network construction in addition to targeting knowledge. To the best of our knowledge, no other approach takes this into account.

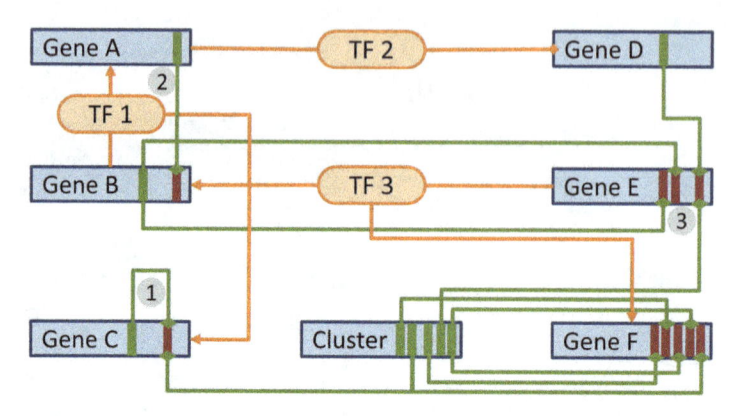

Fig. 3 A contrived example of a small regulatory network consisting of transcription factors (orange ovals), genes (blue rectangles), and miRNAs (green rectangles). Interactions are shown as lines with different arrowheads activating (arrows) or deactivating (diamonds). TFs are explicitly modeled here to show that three levels are cooperating to achieve regulation (genome, transcriptome, and proteome)

In general, networks are built from miRNAs (nodes) and their functional similarity (edges) to predict or extend disease networks. Le et al. refer to them as homogeneous networks and propose the use of bipartite graphs consisting of miRNAs and genes as nodes and their interactions as edges [100].

Regulatory networks that were established as pathways or discovered for various diseases or extended in ways as described above have also been employed for relating miRNAs to diseases. For this one of the two assumptions are generally used: (1) miRNAs that are associated with similar diseases must be similar, and (2) functionally similar miRNAs likely lead to a similar disease phenotype [101]. These assumptions lead to similarity measures or machine learning methods which suggest whether a miRNA is implicated in a particular pathway or disease [101].

In respect to (1) there are various methods to describe functional similarity. For instance, the shared targets among miRNAs can be used as a similarity measure [102]. Similarity for target gene regulation patterns has also been used [103]. Gene ontology enrichment of their targets was used to define similarity among miRNAs [104] and was further extended including protein interaction networks [105].

Considering (2), many machine learning approaches have been established. Different classifiers such as naïve Bayes [102] and support vector machines [103] were used. To remove the dependency on negative data without quality guarantee, a semi-supervised classifier was also used to prioritize candidate disease-related miRNAs [106].

Text mining of scientific publications to establish miRNA–disease associations has been employed to construct miRNA–disease networks. Kandhro et al. extracted miRNA–lipid disease associations from literature and extended the resulting interaction network with further miRNA target predictions [107]. Similarly, Honardoost et al. used literature mining combined with database (miRWalk and miRTarBase) extension for the investigation of autoimmune disease-related miRNAs deregulated in Th17 cells [108].

Databases containing miRNA interactions such as miRWalk [67] have been used to construct miRNA regulatory networks using methodologies based in social network analysis to decipher miRNA involvement in the regulation of intestinal epithelial cellular pathways [109].

4.3 Regulatory Networks

Regulation is of crucial importance for the survival of the organism. With thousands of genes and miRNAs and even more proteins and metabolites that can interact on the molecular level, a large network of interactions results. It is currently not feasible to construct a comprehensive network let alone analyze it and, therefore, the research focus is on smaller subnetworks (Fig. 4). An example for regulation involving few partners is the microprocessor self-regulation mentioned above: DGCR8 contains a hairpin in its first exon which leads to a truncated protein product when the hairpin is excised. The microprocessor complex posttranscriptionally regulates its own expression by cleaving the hairpin [111]. This presents a feedback loop where Drosha deactivates the microprocessor in a dose-dependent manner. Very short feedback loops are possible for example Fig. 3(1) where a miRNA co-expressed with gene C inhibits the translation of the gene product. Other examples not involving miRNAs are the myocyte-enhancing factor 2 and twist genes in Drosophila which are single-gene feedback loops

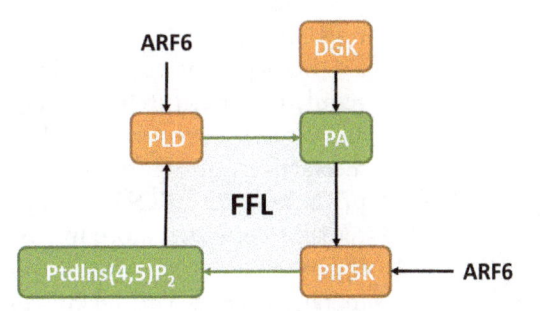

Fig. 4 Recreated after [110]. A feed-forward loop (FFL) regulates PA and PtdIns(4,5)P2 production. After ARF6 has activated PIP5K and/or PLD, a feed-forward loop is activated in which PLD-dependent PA production leads to the activation of PIP5K, PtdIns(4,5)P2, and PLD. Lipid enzymes are shown in orange and lipid products in green. Black arrows denote activation of a downstream protein or process, and green arrows denote conversion to a lipid product

[112]. Such very small regulatory motifs come together to form the overall regulatory network of an organism.

From the viewpoint of biological relativity, all scales of life ranging from the genome to the phenome partake in regulation and together represent the genetic program [2]. This entails a comprehensive regulatory network which allows a holistic view of an organism. Naturally, not all parts of the network interact at every time while other parts may be very active. These interactions need to be under tight time and space control and it is possible to find regulatory motifs as substructures in the overall network such as feedback loops and feed-forward loops (Fig. 4). Such motifs tie together via gene products or miRNAs which interact with multiple targets.

4.4 Regulatory Motifs

For example, a gene may co-produce a miRNA within an intron, which in turn downregulates the protein abundance of the same gene (Fig. 3). This represents the shortest possible feedback loop (FBL) including miRNAs. An equally short path would result from a gene coding for a TF which downregulates the gene itself. An example for this is PHOX2B [113]. Combined feedback loops consisting of miRNAs, TFs, and genes (Fig. 3) can also be envisioned. Such structures can represent molecular switches between cell states. A double-negative feedback loop between the miR-200 family and ZEB1-SIP1, for example, represents such a switch controlling the epithelial to mesenchymal transition [114]. Feed-forward loops (FFL) involving TFs and miRNAs are formed when a TF and an miRNA co-regulate a common target gene. This can, for example, be useful for signal noise buffering [115]. Another function could be the suppression of "leaky" transcription of target genes by reinforcing transcriptional control with posttranscriptional control [116]. Such small motifs can further be combined to larger regulatory motifs. Zhang et al. comprehensively summarized the possible regulatory motifs concerning TFs and miRNAs [117].

Perhaps not very intuitive at first glimpse, a very short path regulatory structure exists (Fig. 5). The miRNA targets (A–C) are all targeted by the same miRNA (for sake of simplicity, all outside interactions are ignored). When the transcription of one of the mRNAs (e.g.: mRNA A) increases while the miRNA expression remains constant, more miRNA target sites become available, acting as decoys. Thereby, the miRNA regulation is reduced for all of its targets (Fig. 5). The effect is most noticeable if mRNA expression levels, the target site-binding strengths, their accessibility, and multiplicity are similar for all targets [109].

In summary, miRNAs and their targets can be detected computationally. MicroRNA source information is important for integration with pathways and pathway extension. Since not all miRNAs or targets are known, other approaches to network

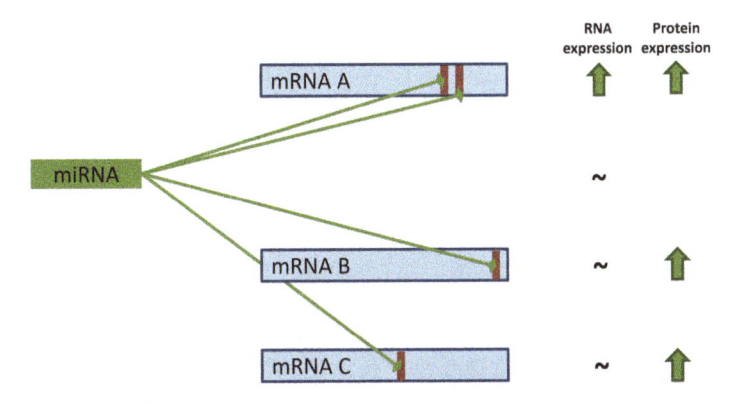

Fig. 5 Change in mRNA expression levels of one of the targets of one miRNA leads to increased protein expression for all its targets. For this example, assuming highly similar targets, mRNA A has two target sites for the miRNA, which would lead to inactivation of twice the increase in expression for mRNA A. Similarly, protein expression would increase by one-third of the expression change per mRNA

extension such as using literature mining are needed. Understanding regulatory pathways will be easier when regulatory motifs are annotated. Incorporation of (relative) expression levels for all players in such motifs will allow mathematical modeling of these substructures using for example Petri nets [118]. In the future, such motifs can be abstracted as circuits and combined into circuit diagrams.

5 Conclusion

Since the discovery of miRNAs in 1993 these small RNA molecules involved in posttranscriptional gene regulation have sparked a lot of research interest [19, 119]. They are generally thought to be involved in the downregulation of protein abundance, but have been shown to be involved in upregulation, as well. Recently, evidence is accumulating that they can be reimported into the nucleus where they are involved in transcriptional regulation. This versatility of miRNAs is not currently modeled in available pathway databases such as Reactome and KEGG. The methodology introduced at the beginning of the previous section which extends disease networks by miRNAs that are co-expressed with genes or target genes in known pathways can be extended with adding protein-protein interaction information. When that is done, all other methods described afterwards are implicitly contained in the solution if all miRNAs and their targets would be known or could be predicted. VANESA is a tool facilitating this approach [99, 120] including a database combining miRBase, miRTarBase, and TarBase as well as granting access to KEGG and the protein interaction databases Mint [121], IntAct [122], and HPRD [123]. The system further supports modeling using Petri nets. However, extension of networks with other means such as by disease association as well as detection of regulatory motifs and mapping of expression levels is not currently possible. Incorporation of expression information ensures that miRNA and targets are co-expressed and that regulation is possible. It also indicates whether regulation would lead to measurable effects. For example, if target levels and target-site abundance among all expressed targets are high and miRNA expression is comparably low, no regulatory effect should be expected (Fig. 5).

Future studies involving miRNA regulation should thus consider the following points:

1. Limitations of miRNA and target prediction algorithms or databases hosting them
2. The sources for miRNAs (intergenic or genic)
3. Integration of miRNAs with known pathways
4. Extension of such pathways using various methods
5. Incorporation of expression data
6. Analysis of miRNA expression levels together with its target(s) expression
7. Detection of regulatory motifs and their mathematical modeling

Taking into account these information and making the resulting models FAIR similarly to data publishing guidelines [124] will help assign function to miRNAs and ensure that miRNAs will be employed as biomarkers and drugs for precision medicine in the future.

References

1. Crick F (1970) Central dogma of molecular biology. Nature 227:561–563
2. Noble D (2012) A theory of biological relativity: no privileged level of causation. Interface Focus 2:55–64. https://doi.org/10.1098/rsfs.2011.0067
3. Liu H, Lei C, He Q, Pan Z, Xiao D, Tao Y (2018) Nuclear functions of mammalian MicroRNAs in gene regulation, immunity and cancer. Mol Cancer 17:64. https://doi.org/10.1186/s12943-018-0765-5
4. Yousef M, Allmer J (2014) miRNomics: microRNA biology and computational analysis. Humana Press, Totowa, NJ
5. Iwakawa H, Tomari Y (2015) The functions of microRNAs: mRNA decay and translational repression. Trends Cell Biol 25:651–665. https://doi.org/10.1016/j.tcb.2015.07.011
6. Ørom UA, Nielsen FC, Lund AH (2008) MicroRNA-10a binds the 5'UTR of ribosomal protein mRNAs and enhances their translation. Mol Cell 30:460–471. https://doi.org/10.1016/j.molcel.2008.05.001
7. Grundhoff A, Sullivan CS (2011) Virus-encoded microRNAs. Virology 411:325–343. https://doi.org/10.1016/j.virol.2011.01.002
8. Skalsky RL, Cullen BR (2010) Viruses, microRNAs, and host interactions. Annu Rev Microbiol 64:123–141. https://doi.org/10.1146/annurev.micro.112408.134243

9. Saçar Demirci MD, Bağcı C, Allmer J (2016) Differential expression of Toxoplasma gondii microRNAs in murine and human hosts. In: Non-coding RNAs and inter-kingdom communication. Springer International Publishing, Cham, pp 143–159
10. Saçar MD, Bağcı C, Allmer J (2014) Computational prediction of MicroRNAs from toxoplasma gondii potentially regulating the hosts' gene expression. Genomics, Proteomics Bioinformatics 12:228–238. https://doi.org/10.1016/j.gpb.2014.09.002
11. Liu S, Weiner HL (2016) Control of the gut microbiome by fecal microRNA. Microb cell (Graz, Austria) 3:176–177. https://doi.org/10.15698/mic2016.04.492
12. Williams MR, Stedtfeld RD, Tiedje JM, Hashsham SA (2017) MicroRNAs-based inter-domain communication between the host and members of the gut microbiome. Front Microbiol 8. https://doi.org/10.3389/fmicb.2017.01896
13. Baker M (2010) MicroRNA profiling: separating signal from noise. Nat Methods 7:687–692. https://doi.org/10.1038/nmeth0910-687
14. Chugh P, Dittmer DP (2012) Potential pitfalls in microRNA profiling. Wiley Interdiscip Rev RNA 3:601–616
15. Dong H, Lei J, Ding L, Wen Y, Ju H, Zhang X (2013) MicroRNA: function, detection, and bioanalysis. Chem Rev 113:6207–6233. https://doi.org/10.1021/cr300362f
16. Saçar MD, Allmer J (2014) Machine learning methods for microRNA gene prediction. In: Yousef M, Allmer J (eds) miRNomics: microRNA biology and computational analysis SE-10. Humana Press, pp 177–187
17. Licatalosi DD, Mele A, Fak JJ, Ule J, Kayikci M, Chi SW, Clark TA, Schweitzer AC, Blume JE, Wang X, Darnell JC, Darnell RB (2008) HITS-CLIP yields genome-wide insights into brain alternative RNA processing. Nature 456:464–469. https://doi.org/10.1038/nature07488
18. Hamzeiy H, Allmer J, Yousef M (2014) Computational methods for microRNA target prediction. Methods Mol Biol 1107:207–221. https://doi.org/10.1007/978-1-62703-748-8_12
19. Lee RC, Feinbaum RL, Ambros V (1993) The C. elegans heterochronic gene lin-4 encodes small RNAs with antisense complementarity to lin-14. Cell 75:843–854
20. Berezikov E, Cuppen E, RH P (2006) Approaches to microRNA discovery. Nat Genet 38(Suppl):S2–S7. https://doi.org/10.1038/ng1794
21. Lim LP, Lau NC, Weinstein EG, Abdelhakim A, Yekta S, Rhoades MW, Burge CB, Bartel DP (2003) The microRNAs of Caenorhabditis elegans. Genes Dev 17:991–1008. https://doi.org/10.1101/gad.1074403
22. Wang X, Zhang J, Li F, Gu J, He T, Zhang X, Li Y (2005) MicroRNA identification based on sequence and structure alignment. Bioinformatics 21:3610–3614. https://doi.org/10.1093/bioinformatics/bti562
23. Nam J-W, Kim J, Kim S-K, Zhang B-T (2006) ProMiR II: a web server for the probabilistic prediction of clustered, nonclustered, conserved and nonconserved microRNAs. Nucleic Acids Res 34:W455–W458. https://doi.org/10.1093/nar/gkl321
24. Xue C, Li F, He T, Liu G-P, Li Y, Zhang X (2005) Classification of real and pseudo microRNA precursors using local structure-sequence features and support vector machine. BMC Bioinformatics 6:310. https://doi.org/10.1186/1471-2105-6-310
25. Sewer A, Paul N, Landgraf P, Aravin A, Pfeffer S, Brownstein MJ, Tuschl T, van Nimwegen E, Zavolan M (2005) Identification of clustered microRNAs using an ab initio prediction method. BMC Bioinformatics 6:267. https://doi.org/10.1186/1471-2105-6-267
26. Hertel J, Stadler PF (2006) Hairpins in a Haystack: recognizing microRNA precursors in comparative genomics data. Bioinformatics 22:e197–e202. https://doi.org/10.1093/bioinformatics/btl257
27. Yousef M, Nebozhyn M, Shatkay H, Kanterakis S, Showe LC, Showe MK (2006) Combining multi-species genomic data for microRNA identification using a Naive Bayes classifier. Bioinformatics 22:1325–1334. https://doi.org/10.1093/bioinformatics/btl094
28. Huang T-H, Fan B, Rothschild MF, Hu Z-L, Li K, Zhao S-H (2007) MiRFinder: an improved approach and software implementation for genome-wide fast microRNA precursor scans. BMC Bioinformatics 8:341. https://doi.org/10.1186/1471-2105-8-341
29. Jiang P, Wu H, Wang W, Ma W, Sun X, Lu Z (2007) MiPred: classification of real and pseudo microRNA precursors using random forest prediction model with combined

features. Nucleic Acids Res 35:W339–W344. https://doi.org/10.1093/nar/gkm368
30. Terai G, Komori T, Asai K (2081–2090) Kin T (2007) miRRim: a novel system to find conserved miRNAs with high sensitivity and specificity. https://doi.org/10.1261/rna.655107.been
31. Friedländer MR, Chen W, Adamidi C, Maaskola J, Einspanier R, Knespel S, Rajewsky N (2008) Discovering microRNAs from deep sequencing data using miRDeep. Nat Biotechnol 26:407–415. https://doi.org/10.1038/nbt1394
32. Hackenberg M, Sturm M, Langenberger D, Falcón-Pérez JM, Aransay AM (2009) miRanalyzer: a microRNA detection and analysis tool for next-generation sequencing experiments. Nucleic Acids Res 37:W68–W76. https://doi.org/10.1093/nar/gkp347
33. Oulas A, Boutla A, Gkirtzou K, Reczko M, Kalantidis K, Poirazi P (2009) Prediction of novel microRNA genes in cancer-associated genomic regions—a combined computational and experimental approach. Nucleic Acids Res 37:3276–3287. https://doi.org/10.1093/nar/gkp120
34. Kadri S, Hinman V, Benos PV (2009) HHMMiR: efficient de novo prediction of microRNAs using hierarchical hidden Markov models. BMC Bioinformatics 10(Suppl 1):S35. https://doi.org/10.1186/1471-2105-10-S1-S35
35. Mathelier A, Carbone A (2010) MIReNA: finding microRNAs with high accuracy and no learning at genome scale and from deep sequencing data. Bioinformatics 26:2226–2234. https://doi.org/10.1093/bioinformatics/btq329
36. Wu Y, Wei B, Liu H, Li T, Rayner S (2011) MiRPara: a SVM-based software tool for prediction of most probable microRNA coding regions in genome scale sequences. BMC Bioinformatics 12:107. https://doi.org/10.1186/1471-2105-12-107
37. Tempel S, Tahi F (2012) A fast ab-initio method for predicting miRNA precursors in genomes. Nucleic Acids Res 40:e80. https://doi.org/10.1093/nar/gks146
38. Gao D, Middleton R, Rasko JEJ, Ritchie W (2013) miREval 2.0: a web tool for simple microRNA prediction in genome sequences. Bioinformatics 29:3225–3226. https://doi.org/10.1093/bioinformatics/btt545
39. Lei J, Sun Y (2014) miR-PREFeR: an accurate, fast and easy-to-use plant miRNA prediction tool using small RNA-Seq data. Bioinformatics 30:2837–2839. https://doi.org/10.1093/bioinformatics/btu380
40. Tran VDT, Tempel S, Zerath B, Zehraoui F, Tahi F (2015) miRBoost: boosting support vector machines for microRNA precursor classification. RNA 21:775–785. https://doi.org/10.1261/rna.043612.113
41. Chen J, Wang X, Liu B (2016) iMiRNA-SSF: improving the identification of microRNA precursors by combining negative sets with different distributions. Sci Rep 6:19062. https://doi.org/10.1038/srep19062
42. Saçar Demirci MD, Baumbach J, Allmer J (2017) On the performance of pre-microRNA detection algorithms. Nat Commun 8:330. https://doi.org/10.1038/s41467-017-00403-z
43. Lu Yi, Aras AS, Halushka MK (2018) miRge 2.0: an updated tool to comprehensively analyze microRNA sequencing data, bioRxiv, https://doi.org/10.1101/250779
44. Gomes CPC, Cho J-H, Hood L, Franco OL, Pereira RW, Wang K (2013) A review of computational tools in microRNA discovery. Front Genet 4:81. https://doi.org/10.3389/fgene.2013.00081
45. van der Burgt A, Fiers MWJE, Nap J-P, van Ham RCHJ (2009) In silico miRNA prediction in metazoan genomes: balancing between sensitivity and specificity. BMC Genomics 10:204. https://doi.org/10.1186/1471-2164-10-204
46. Hofacker IL (2003) Vienna RNA secondary structure server. Nucleic Acids Res 31:3429–3431. https://doi.org/10.1093/nar/gkg599
47. Zeng C, Wang W, Zheng Y, Chen X, Bo W, Song S, Zhang W, Peng M (2010) Conservation and divergence of microRNAs and their functions in Euphorbiaceous plants. Nucleic Acids Res 38:981–995. https://doi.org/10.1093/nar/gkp1035
48. Liang H, Li W-H (2009) Lowly expressed human microRNA genes evolve rapidly. Mol Biol Evol 26:1195–1198. https://doi.org/10.1093/molbev/msp053
49. Saçar Demirci MD, Allmer J (2017) Delineating the impact of machine learning elements in pre-microRNA detection. PeerJ 5:e3131. https://doi.org/10.7717/peerj.3131
50. Marcinkowska M, Szymanski M, Krzyzosiak WJ, Kozlowski P (2011) Copy number variation of microRNA genes in the human genome. BMC Genomics 12:183. https://doi.org/10.1186/1471-2164-12-183

51. Erson-Bensan AE (2014) Introduction to microRNAs in biological systems. Methods Mol Biol 1107:1–14. https://doi.org/10.1007/978-1-62703-748-8_1

52. Hafner M, Landthaler M, Burger L, Khorshid M, Hausser J, Berninger P, Rothballer A, Ascano M, Jungkamp A-C, Munschauer M, Ulrich A, Wardle GS, Dewell S, Zavolan M, Tuschl T (2010) Transcriptome-wide identification of RNA-binding protein and microRNA target sites by PAR-CLIP. Cell 141:129–141. https://doi.org/10.1016/j.cell.2010.03.009

53. Chi SW, Zang JB, Mele A, Darnell RB (2009) Argonaute HITS-CLIP decodes microRNA-mRNA interaction maps. Nature 460:479–486. https://doi.org/10.1038/nature08170.Ago

54. Helwak A, Kudla G, Dudnakova T, Tollervey D (2013) Mapping the human miRNA interactome by CLASH reveals frequent noncanonical binding. Cell 153:654–665. https://doi.org/10.1016/j.cell.2013.03.043

55. Hsu S-D, Tseng Y-T, Shrestha S, Lin Y-L, Khaleel A, Chou C-H, Chu C-F, Huang H-Y, Lin C-M, Ho S-Y, Jian T-Y, Lin F-M, Chang T-H, Weng S-L, Liao K-W, Liao I-E, Liu C-C, Huang H-D (2014) miRTarBase update 2014: an information resource for experimentally validated miRNA-target interactions. Nucleic Acids Res 42:D78–D85. https://doi.org/10.1093/nar/gkt1266

56. Vergoulis T, Vlachos IS, Alexiou P, Georgakilas G, Maragkakis M, Reczko M, Gerangelos S, Koziris N, Dalamagas T, Hatzigeorgiou AG (2012) TarBase 6.0: capturing the exponential growth of miRNA targets with experimental support. Nucleic Acids Res 40:D222–D229. https://doi.org/10.1093/nar/gkr1161

57. Krüger J, Rehmsmeier M (2006) RNAhybrid: microRNA target prediction easy, fast and flexible. Nucleic Acids Res 34:W451–W454. https://doi.org/10.1093/nar/gkl243

58. Krek A, Grün D, Poy MN, Wolf R, Rosenberg L, Epstein EJ, MacMenamin P, da Piedade I, Gunsalus KC, Stoffel M, Rajewsky N (2005) Combinatorial microRNA target predictions. Nat Genet 37:495–500. https://doi.org/10.1038/ng1536

59. Lewis BP, Burge CB, Bartel DP (2005) Conserved seed pairing, often flanked by adenosines, indicates that thousands of human genes are microRNA targets. Cell 120:15–20. https://doi.org/10.1016/j.cell.2004.12.035

60. Miranda KC, Huynh T, Tay Y, Ang Y-S, Tam W-L, Thomson AM, Lim B, Rigoutsos I (2006) A pattern-based method for the identification of MicroRNA binding sites and their corresponding heteroduplexes. Cell 126:1203–1217. https://doi.org/10.1016/j.cell.2006.07.031

61. Sethupathy P, Corda B, Hatzigeorgiou AG (2006) TarBase: a comprehensive database of experimentally supported animal microRNA targets. RNA 12:192–197. https://doi.org/10.1261/rna.2239606

62. Yousef M, Jung S, Kossenkov AV, Showe LC, Showe MK (2007) Naïve Bayes for microRNA target predictions—machine learning for microRNA targets. Bioinformatics 23:2987–2992. https://doi.org/10.1093/bioinformatics/btm484

63. Kertesz M, Iovino N, Unnerstall U, Gaul U, Segal E (2007) The role of site accessibility in microRNA target recognition. Nat Genet 39:1278–1284. https://doi.org/10.1038/ng2135

64. Maragkakis M, Alexiou P, Papadopoulos GL, Reczko M, Dalamagas T, Giannopoulos G, Goumas G, Koukis E, Kourtis K, Simossis VA, Sethupathy P, Vergoulis T, Koziris N, Sellis T, Tsanakas P, Hatzigeorgiou AG (2009) Accurate microRNA target prediction correlates with protein repression levels. BMC Bioinformatics 10:295. https://doi.org/10.1186/1471-2105-10-295

65. Xiao F, Zuo Z, Cai G, Kang S, Gao X, Li T (2009) miRecords: an integrated resource for microRNA-target interactions. Nucleic Acids Res 37:D105–D110. https://doi.org/10.1093/nar/gkn851

66. Hsu S-D, Lin F-M, Wu W-Y, Liang C, Huang W-C, Chan W-L, Tsai W-T, Chen G-Z, Lee C-J, Chiu C-M, Chien C-H, Wu M-C, Huang C-Y, Tsou A-P, Huang H-D (2011) miRTarBase: a database curates experimentally validated microRNA-target interactions. Nucleic Acids Res 39:D163–D169. https://doi.org/10.1093/nar/gkq1107

67. Dweep H, Sticht C, Pandey P, Gretz N (2011) miRWalk—database: prediction of possible miRNA binding sites by "walking" the genes of three genomes. J Biomed Inform 44:839–847. https://doi.org/10.1016/j.jbi.2011.05.002

68. Elefant N, Berger A, Shein H, Hofree M, Margalit H, Altuvia Y (2011) RepTar: a database of predicted cellular targets of host and viral miRNAs. Nucleic Acids Res 39:D188–D194. https://doi.org/10.1093/nar/gkq1233

69. Li J-H, Liu S, Zhou H, Qu L-H, Yang J-H (2014) starBase v2.0: decoding miRNA-ceRNA, miRNA-ncRNA and protein-RNA interaction networks from large-scale CLIP-Seq data. Nucleic Acids Res 42:D92–D97. https://doi.org/10.1093/nar/gkt1248
70. Chiu H-S, Llobet-Navas D, Yang X, Chung W-J, Ambesi-Impiombato A, Iyer A, Kim HR, Seviour EG, Luo Z, Sehgal V, Moss T, Lu Y, Ram P, Silva J, Mills GB, Califano A, Sumazin P (2015) Cupid: simultaneous reconstruction of microRNA-target and ceRNA networks. Genome Res 25:257–267. https://doi.org/10.1101/gr.178194.114
71. Bandyopadhyay S, Ghosh D, Mitra R, Zhao Z (2015) MBSTAR: multiple instance learning for predicting specific functional binding sites in microRNA targets. Sci Rep 5:8004. https://doi.org/10.1038/srep08004
72. Liu S, Li J-H, Wu J, Zhou K-R, Zhou H, Yang J-H, Qu L-H (2015) StarScan: a web server for scanning small RNA targets from degradome sequencing data. Nucleic Acids Res 43: W480–W486. https://doi.org/10.1093/nar/gkv524
73. Riffo-Campos Á, Riquelme I, Brebi-Mieville P (2016) Tools for sequence-based miRNA target prediction: what to choose? Int J Mol Sci 17:1987. https://doi.org/10.3390/ijms17121987
74. Lewis BP, Shih I, Jones-Rhoades MW, Bartel DP, Burge CB (2003) Prediction of mammalian microRNA targets. Cell 115:787–798
75. Enright AJ, John B, Gaul U, Tuschl T, Sander C, Marks DS (2003) MicroRNA targets in Drosophila. Genome Biol 5:R1. https://doi.org/10.1186/gb-2003-5-1-r1
76. Kiriakidou M, Nelson PT, Kouranov A, Fitziev P, Bouyioukos C, Mourelatos Z, Hatzigeorgiou A (2004) A combined computational-experimental approach predicts human microRNA targets. Genes Dev 18:1165–1178. https://doi.org/10.1101/gad.1184704
77. Peterson SM, JA T, Ufkin ML, Sathyanarayana P, Liaw L, Congdon CB (2014) Common features of microRNA target prediction tools. Front Genet 5:23. https://doi.org/10.3389/fgene.2014.00023
78. Yousef M, Allmer J, Khalifa W (2016) Feature selection for microRNA target prediction comparison of one-class feature selection methodologies. In: BIOINFORMATICS 2016—7th international conference on bioinformatics models, methods and algorithms, Proceedings; Part of 9th international joint conference on biomedical engineering systems and technologies, BIOSTEC 2016
79. John B, Enright AJ, Aravin A, Tuschl T, Sander C, Marks DS (2004) Human microRNA targets. PLoS Biol 2:e363. https://doi.org/10.1371/journal.pbio.0020363
80. Altschul SF, Gish W, Miller W, Myers EW, Lipman DJ (1990) Basic local alignment search tool. J Mol Biol 215:403–410. https://doi.org/10.1016/S0022-2836(05)80360-2
81. Rehmsmeier M, Steffen P, Hochsmann M, Giegerich R (2004) Fast and effective prediction of microRNA/target duplexes. RNA 10:1507–1517. https://doi.org/10.1261/rna.5248604
82. Lai EC (2004) Predicting and validating microRNA targets. Genome Biol 5:115. https://doi.org/10.1186/gb-2004-5-9-115
83. Yousef M, Nigatu D, Levy D, Allmer J, Henkel W (2017) Categorization of species based on their microRNAs employing sequence motifs, information-theoretic sequence feature extraction, and k-mers. EURASIP J Adv Signal Process 2017:70. https://doi.org/10.1186/s13634-017-0506-8
84. Heyn J, Hinske LC, Ledderose C, Limbeck E, Kreth S (2013) Experimental miRNA target validation. Methods Mol Biol 936:83–90. https://doi.org/10.1007/978-1-62703-083-0_7
85. Thomson DW, Bracken CP, Goodall GJ (2011) Experimental strategies for microRNA target identification. Nucleic Acids Res 39:6845–6853. https://doi.org/10.1093/nar/gkr330
86. Chou C-H, Shrestha S, Yang C-D, Chang N-W, Lin Y-L, Liao K-W, Huang W-C, Sun T-H, Tu S-J, Lee W-H, Chiew M-Y, Tai C-S, Wei T-Y, Tsai T-R, Huang H-T, Wang C-Y, Wu H-Y, Ho S-Y, Chen P-R, Chuang C-H, Hsieh P-J, Wu Y-S, Chen W-L, Li M-J, Wu Y-C, Huang X-Y, Ng FL, Buddhakosai W, Huang P-C, Lan K-C, Huang C-Y, Weng S-L, Cheng Y-N, Liang C, Hsu W-L, Huang H-D (2018) miRTarBase update 2018: a resource for experimentally validated microRNA-target interactions. Nucleic Acids Res 46:D296–D302. https://doi.org/10.1093/nar/gkx1067
87. Saçar MD, Allmer J (2013) Current limitations for computational analysis of miRNAs in cancer. Pakistan J Clin Biomed Res 1:3–5
88. Koo J, Zhang J, Chaterji S (2018) Tiresias: context-sensitive approach to decipher the presence and strength of microRNA

regulatory interactions. Theranostics 8:277–291. https://doi.org/10.7150/thno.22065

89. Kim VN, Han J, Siomi MC (2009) Biogenesis of small RNAs in animals. Nat Rev Mol Cell Biol 10:126–139. https://doi.org/10.1038/nrm2632

90. Altuvia Y, Landgraf P, Lithwick G, Elefant N, Pfeffer S, Aravin A, Brownstein MJ, Tuschl T, Margalit H (2005) Clustering and conservation patterns of human microRNAs. Nucleic Acids Res 33:2697–2706. https://doi.org/10.1093/nar/gki567

91. Mechtler P, Johnson S, Slabodkin H, Cohanim AB, Brodsky L, Kandel ES (2017) The evidence for a microRNA product of human DROSHA gene. RNA Biol 14:1508–1513. https://doi.org/10.1080/15476286.2017.1342934

92. Acar İE, Saçar Demirci MD, Groß U, Allmer J (2018) The expressed MicroRNA—mRNA interactions of Toxoplasma gondii. Front Microbiol 8. https://doi.org/10.3389/fmicb.2017.02630

93. Leinonen R, Sugawara H, Shumway M (2011) The sequence read archive. Nucleic Acids Res 39:D19–D21. https://doi.org/10.1093/nar/gkq1019

94. Fei Y, Wang R, Li H, Liu S, Zhang H, Huang J (2017) DPMIND: degradome-based Plant MiRNA-target interaction and network database. Bioinformatics. https://doi.org/10.1093/bioinformatics/btx824

95. Kozomara A, Griffiths-Jones S (2014) miRBase: annotating high confidence microRNAs using deep sequencing data. Nucleic Acids Res 42:D68–D73. https://doi.org/10.1093/nar/gkt1181

96. Brinkrolf C, Janowski SJ, Kormeier B, Lewinski M, Hippe K, Borck D, Hofestädt R (2014) VANESA—a software application for the visualization and analysis of networks in system biology applications. J Integr Bioinform 11:239. https://doi.org/10.2390/biecoll-jib-2014-239

97. Croft D, Mundo AF, Haw R, Milacic M, Weiser J, Wu G, Caudy M, Garapati P, Gillespie M, Kamdar MR, Jassal B, Jupe S, Matthews L, May B, Palatnik S, Rothfels K, Shamovsky V, Song H, Williams M, Birney E, Hermjakob H, Stein L, D'Eustachio P (2014) The Reactome pathway knowledgebase. Nucleic Acids Res 42:D472–D477. https://doi.org/10.1093/nar/gkt1102

98. Kanehisa M, Goto S (2000) KEGG: kyoto encyclopedia of genes and genomes. Nucleic Acids Res 28:27–30

99. Hamzeiy H, Suluyayla R, Brinkrolf C, Janowski SJ, Hofestaedt R, Allmer J (2017) Visualization and analysis of microRNAs within KEGG pathways using VANESA. J Integr Bioinform 14. https://doi.org/10.1515/jib-2016-0004

100. Le DH, Verbeke L, Son LH, Chu DT, Pham VH (2017) Random walks on mutual microRNA-target gene interaction network improve the prediction of disease-associated microRNAs. BMC Bioinformatics 18:1–13. https://doi.org/10.1186/s12859-017-1924-1

101. Zeng X, Zhang X, Zou Q (2016) Integrative approaches for predicting microRNA function and prioritizing disease-related microRNA using biological interaction networks. Brief Bioinform 17:193–203. https://doi.org/10.1093/bib/bbv033

102. Jiang Q, Hao Y, Wang G, Juan L, Zhang T, Teng M, Liu Y, Wang Y (2010) Prioritization of disease microRNAs through a human phenome-microRNAome network. BMC Syst Biol 4(Suppl 1):S2. https://doi.org/10.1186/1752-0509-4-S1-S2

103. Jiang Q, Hao Y, Wang G, Zhang T, Wang Y (2010) Weighted network-based inference of human microRNA-disease associations. In: 2010 Fifth international conference on frontier of computer science and technology. IEEE, pp 431–435

104. Wang D, Wang J, Lu M, Song F, Cui Q (2010) Inferring the human microRNA functional similarity and functional network based on microRNA-associated diseases. Bioinformatics 26:1644–1650. https://doi.org/10.1093/bioinformatics/btq241

105. Xu J, Li C-X, Li Y-S, Lv J-Y, Ma Y, Shao T-T, Xu L-D, Wang Y-Y, Du L, Zhang Y-P, Jiang W, Li C-Q, Xiao Y, Li X (2011) MiRNA-miRNA synergistic network: construction via co-regulating functional modules and disease miRNA topological features. Nucleic Acids Res 39:825–836. https://doi.org/10.1093/nar/gkq832

106. Chen X, Yan G-Y (2015) Semi-supervised learning for potential human microRNA-disease associations inference. Sci Rep 4:5501. https://doi.org/10.1038/srep05501

107. Kandhro AH, Shoombuatong W, Nantasenamat C, Prachayasittikul V, Nuchnoi P (2017) The microRNA interaction network of lipid diseases. Front Genet 8:1–14. https://doi.org/10.3389/fgene.2017.00116

108. Honardoost MA, Naghavian R, Ahmadinejad F, Hosseini A, Ghaedi K

(2015) Integrative computational mRNA-miRNA interaction analyses of the autoimmune-deregulated miRNAs and well-known Th17 differentiation regulators: an attempt to discover new potential miRNAs involved in Th17 differentiation. Gene 572:153–162. https://doi.org/10.1016/j.gene.2015.08.043

109. Robinson JM, Henderson WA (2018) Modelling the structure of a ceRNA-theoretical, bipartite microRNA-mRNA interaction network regulating intestinal epithelial cellular pathways using R programming. BMC Res Notes 11:1–7. https://doi.org/10.1186/s13104-018-3126-y

110. van den Bout I, Divecha N (2009) PIP5K-driven PtdIns(4,5)P2 synthesis: regulation and cellular functions. J Cell Sci 122:3837–3850. https://doi.org/10.1242/jcs.056127

111. Han J, Pedersen JS, Kwon SC, Belair CD, Kim Y, Yeom K, Yang W, Haussler D, Blelloch R, Kim VN (2009) Posttranscriptional crossregulation between Drosha and DGCR8. Cell 136:75–84. https://doi.org/10.1016/j.cell.2008.10.053

112. Crews ST, Pearson JC (2009) Transcriptional autoregulation in development. Curr Biol 19:R241–R246. https://doi.org/10.1016/j.cub.2009.01.015

113. Cargnin F, Flora A, Di Lascio S, Battaglioli E, Longhi R, Clementi F, Fornasari D (2005) PHOX2B regulates its own expression by a transcriptional auto-regulatory mechanism. J Biol Chem 280:37439–37448. https://doi.org/10.1074/jbc.M508368200

114. Bracken CP, Gregory PA, Kolesnikoff N, Bert AG, Wang J, Shannon MF, Goodall GJ (2008) A double-negative feedback loop between ZEB1-SIP1 and the microRNA-200 family regulates epithelial-mesenchymal transition. Cancer Res 68:7846–7854. https://doi.org/10.1158/0008-5472.CAN-08-1942

115. Osella M, Bosia C, Corá D, Caselle M (2011) The role of incoherent microRNA-mediated feedforward loops in noise buffering. PLoS Comput Biol 7. https://doi.org/10.1371/journal.pcbi.1001101

116. Tsang J, Zhu J, van Oudenaarden A (2007) MicroRNA-mediated feedback and feedforward loops are recurrent network motifs in mammals. Mol Cell 26:753–767. https://doi.org/10.1016/j.molcel.2007.05.018

117. Zhang HM, Kuang S, Xiong X, Gao T, Liu C, Guo AY (2013) Transcription factor and microRNA co-regulatory loops: Important regulatory motifs in biological processes and diseases. Brief Bioinform 16:45–58. https://doi.org/10.1093/bib/bbt085

118. Yousef M, Trinh HV, Allmer J (2014) Intersection of microRNA and gene regulatory networks and their implication in cancer. Curr Pharm Biotechnol 15:445–454. https://doi.org/10.2174/1389201015666140519120855

119. Wightman B, Ha I, Ruvkun G (1993) Posttranscriptional regulation of the heterochronic gene lin-14 by lin-4 mediates temporal pattern formation in C. elegans. Cell 75:855–862

120. Hamzeiy H, Suluyayla R, Brinkrolf C, Janowski SJ, Hofestädt R, Allmer J (2018) Visualization and analysis of miRNAs implicated in amyotrophic lateral sclerosis within gene regulatory pathways. Stud Heal Technol Inform 253:183–187

121. Licata L, Briganti L, Peluso D, Perfetto L, Iannuccelli M, Galeota E, Sacco F, Palma A, Nardozza AP, Santonico E, Castagnoli L, Cesareni G (2012) MINT, the molecular interaction database: 2012 Update. Nucleic Acids Res 40

122. Kerrien S, Aranda B, Breuza L, Bridge A, Broackes-Carter F, Chen C, Duesbury M, Dumousseau M, Feuermann M, Hinz U, Jandrasits C, Jimenez RC, Khadake J, Mahadevan U, Masson P, Pedruzzi I, Pfeiffenberger E, Porras P, Raghunath A, Roechert B, Orchard S, Hermjakob H (2012) The IntAct molecular interaction database in 2012. Nucleic Acids Res 40

123. Liu B, Hu B (2010) HPRD: a high performance RDF database. Int J Parallel Emergent Distrib Syst 25:123–133

124. Wilkinson MD, Dumontier M, Aalbersberg IJ, Appleton G, Axton M, Baak A, Blomberg N, Boiten J-W, da Silva Santos LB, Bourne PE, Bouwman J, Brookes AJ, Clark T, Crosas M, Dillo I, Dumon O, Edmunds S, Evelo CT, Finkers R, Gonzalez-Beltran A, Gray AJG, Groth P, Goble C, Grethe JS, Heringa J, 't Hoen PA, Hooft R, Kuhn T, Kok R, Kok J, Lusher SJ, Martone ME, Mons A, Packer AL, Persson B, Rocca-Serra P, Roos M, van Schaik R, Sansone S-A, Schultes E, Sengstag T, Slater T, Strawn G, Swertz MA, Thompson M, van der Lei J, van Mulligen E, Velterop J, Waagmeester A, Wittenburg P, Wolstencroft K, Zhao J, Mons B (2016) The FAIR Guiding Principles for scientific data management and stewardship. Sci Data 3:160018. https://doi.org/10.1038/sdata.2016.18

Part III

Bioinformatics Tools and Databases for ncRNA Analyses

Chapter 8

Tools for Understanding miRNA–mRNA Interactions for Reproducible RNA Analysis

Andrea Bagnacani, Markus Wolfien, and Olaf Wolkenhauer

Abstract

MicroRNAs (miRNAs) are an integral part of gene regulation at the post-transcriptional level. The use of RNA data in gene expression analysis has become increasingly important to gain insights into the regulatory mechanisms behind miRNA–mRNA interactions. As a result, we are confronted with a growing landscape of tools, while standards for reproducibility and benchmarking lag behind. This work identifies the challenges for reproducible RNA analysis, and highlights best practices on the processing and dissemination of scientific results. We found that the success of a tool does not solely depend on its performances: equally important is how a tool is received, and then supported within a community. This leads us to a detailed presentation of the RNA workbench, a community effort for sharing workflows and processing tools, built on top of the Galaxy framework. Here, we follow the community guidelines to extend its portfolio of RNA tools with the integration of the TriplexRNA (https://triplexrna.org). Our findings provide the basis for the development of a recommendation system, to guide users in the choice of tools and workflows.

Key words miRNA–mRNA interactions, Gene regulation, RNA workbench, Galaxy, Database

1 Introduction to Computational Data Analysis Challenges in the Life Sciences

MicroRNAs (miRNAs) play a key role in gene regulation at the post-transcriptional level. Their presence can be correlated with the progression of diseases [1, 2], and are therefore used for the design of diagnostic and prognostic markers [3–5]. For these reasons, transcript quantification and discovery by RNA sequencing (RNA-Seq) is at the basis of diverse experiments in life science research [6]. RNA-Seq is a high-throughput technique that does not require predetermined DNA probes to known genes, and is therefore a key technology for the discovery of new exons, splice variants, and small RNAs [7].

The functional characterization of miRNA–mRNA interactions implies the use of specialized computational tools for RNA-Seq data analysis. This, from both users and developers' sides, is a process entailing challenges whose solutions do not yet leverage on a unique corpus of standards, but rather a set of community-based best practices [8, 9].

From a developer's perspective, the need for designing and implementing a new computational tool for RNA analysis is dictated by the lack of the desired functions within all available software tools. This might be the case when the investigation is novel, or when available tools miss the necessary parametrization that allows room for testing the hypothesis under scrutiny. Another scenario can be that of achieving the desired computational approach by chaining existing tools in a *workflow*. Nonetheless, developers first look for tools already implementing part of the devised strategy, not to reinvent the wheel. In turn, choosing to implement new tools by leveraging on third-party modules, workflows, and frameworks can narrow down the types of input data format that are processed throughout the analysis. However, the use of a restricted set of data formats should not be seen as a limit, but rather an effort to provide de facto standards for reproducible analyses [8].

From a user's perspective, the need for adopting a computational tool for RNA analysis is dictated by the high availability of RNA-centric data, and by the applicability and usability of software tools. While it has become increasingly cheap to produce next-generation sequencing (NGS) data, the required efforts to gain insights by mining it have likewise increased in time and resources [10]. Furthermore, such tasks entail the adoption of statistical and computational methods, which are rarely the domain of a life science curriculum. As a result, users are often confronted with a new technical jargon, spanning from software interface tutorials to operative system-dependent installation instructions. Under such light, software solutions look like seas of alternatives to choose from.

Not surprisingly, community forums such as BioStars (https://www.biostars.org) and Stack Overflow (https://stackoverflow.com) have become hubs for sharing expertise across users of diverse expertise. Multiple RNA-centric tools in fact endorsed lively communities as support platforms, or started their very own development right from decentralized repositories such as GitHub (https://github.com) or BitBucket (https://bitbucket.org).

Although these platforms represent a valuable starting point to approach data analysis problems, there is still a high demand for providing guidance, and ultimately hands-on training on how to face such tasks. As a response, in recent years multiple training initiatives have started delivering topic- and/or tool-specific hands-on sessions and workshops to life scientists [11]. The European Life Sciences Infrastructure for Biological Information (ELIXIR, https://www.elixir-europe.org) training program TeSS (https://tess.elixir-europe.org), the Global Organisation for Bioinformatics Learning, Education & Training (Goblet, https://www.mygoblet.org), the German Network for Bioinformatics Infrastructure (de.NBI, https://www.denbi.de), the Software Carpentry

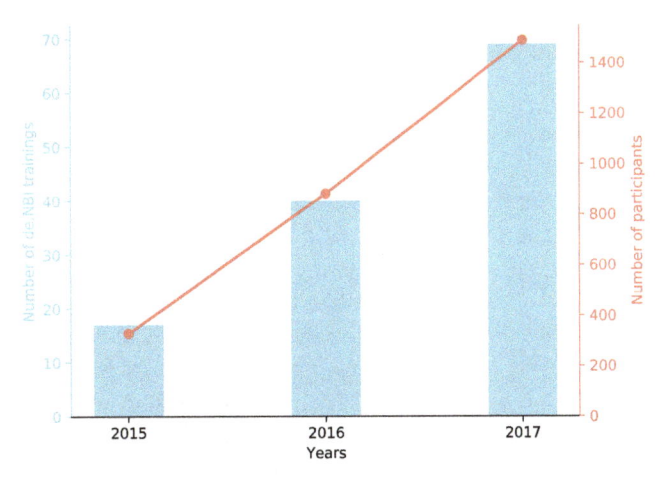

Fig. 1 Attendance to training courses organized by the German Network for Bioinformatics Infrastructure (de.NBI) between 2015 and 2017. The graph shows how the number of participants grows with the number of provided training courses. Trainings provided by de.NBI include topic- and/or tool-specific courses, focusing on computational approaches for life science data analyses. Further information is available at https:/www.denbi.de/training

(https://software-carpentry.org) initiative, and the Galaxy Training Network (https://galaxyproject.org/teach/gtn) are examples of active training providers across the globe. Figure 1 highlights how the attendance to topic- and tool-specific training events has increased in the last years within the de.NBI network.

However, the role of trainers becomes twofold: on the one hand they propose a biological problem and a computational solution that allows for tailored parametric options, while on the other they make sense of the whole plethora of software solutions, and provide a structured overview of tools, workflows, libraries, and frameworks to users. This interaction brings diverse technical jargons together, establishing a community where disparate expertise converge.

From a community's perspective, the adoption of a selected tool, workflow, or framework consolidates a computational approach into an established practice, paving the path for robust and reproducible data analysis [8]. Indeed, the success of a tool does not solely depend on how its algorithm performs: usability, as well as interoperability to different computing environments, plays a key role on the dissemination of the tool itself, and the community around it finally shapes its acceptance.

In this section, we overviewed the challenges that developers and users face when approaching life science problems computationally. Within this framework, the investigation of miRNA–mRNA interactions, for the acquisition of a better understanding of the phenomena and implications, represents a subproblem holding the same traits: the need for usable software solutions developing around a set of standards, and a lively community able to share its diverse expertise, and provide guidance and best practices for reproducible science.

2 Online Catalogs of Bioinformatic Tools

In response to the growing amount of software solutions for downstream analysis in the life sciences, tools have been organized into registries: online indexes of tools, collected and categorized on the basis of their functions and target analyses.

Tool categorization has been implemented with diverse approaches and for different communities: for bioinformatic tools, established examples are the *Molecular Biology Database List* published by *Nucleic Acids Research*'s *Database Issues* [12, 13], the *Bioinformatics Links Directory* [14, 15], the *EMBRACE Registry* [16, 17], and *BioCatalogue Web services* [18].

In recent years, two new players have entered the scene: *OMICtools* [19] (https://omictools.com) and *bio.tools* [20] (https://bio.tools).

OMICtools is a Web-based search engine for biomedical resources. Its inventory is manually curated, and organized under a tailor-made taxonomy. Tools can be browsed by submitting keywords in a user-friendly input form. The whole inventory's descriptions are meant for human readers, and are compiled by both users and developers. Coupled with a catalog of tools, the service additionally integrates an editor where users can provide their review and feedback. Such a feature makes OMICtools not just a tool registry, but also a community hub.

Bio.tools is ELIXIR's effort of providing a manually curated registry of bioinformatic tools. By offering as well a user-friendly Web-based search engine to browse for computational resources, it furthermore provides the consultation of its catalog to both human and software agents. This is made possible by its Application Program Interface (API), which serves tool descriptions organized by means of the EMBRACE Data And Methods ontology (EDAM) [21]: a structured vocabulary of terms and their relations within the domain of bioinformatic (a) operations, (b) identifiers and data types, (c) data formats, and (d) topics.

From a developer's perspective, such catalogs represent a channel for the dissemination of software solutions, while users can directly benefit from the structured overview they offer for all available tools pertaining a specific biological problem or analysis.

Indeed, if a community of users promotes the adoption of an already implemented tool, this becomes established, and future users will save time and resources by channeling their research in the direction of testing hypotheses, rather than implementing new prototypes apt at addressing the same set of questions.

Tools however might be modified to accommodate new features. For this reason, new *similar* tools might be implemented to make room for specialized parameters and functionalities.

Tool similarity and redundant functionalities are not synonyms of tool *duplication*. Similar tools can be used by different communities to investigate diverse aspects of a biological problem, as well as to benchmark their combined outcomes. For example, several algorithms have been devised and implemented for aligning individual sequence reads against a reference genome. This is the case of genomic alignment tools such as TopHat2 [22], HiSat2 [23], STAR [24], or Segemehl [25]. Indeed, the reason behind the development of all these tools is not to merely rebrand the problem of sequence alignment with different names, but rather to address specific applications, approaches, and parameter features such as accuracy, sensitivity, and splice-site detection.

A benchmark of the aforementioned tools has been in fact provided in terms of their compared performances [26]. Under this light, tool similarity represents a feature, rather than a problem of redundancy.

3 A Platform for Understanding miRNA–mRNA Interactions: The RNA Workbench

In a computational analysis, each tool represents a logical step toward the final result. Here, tools are *modules*, whose outputs become the inputs for the next modules at each successive iteration. Complex biological questions are addressed *computationally* by organizing tools into *workflows* which, in turn, comprise the minimal set of combinable modules whose functions are tailored for a specific analysis. As a consequence, each tool should be designed to be simple, extensible, easy to be maintained, and repurposed by other developers [27, 28]. Such design principles are necessary, albeit not sufficient, to make research reproducible. Indeed, even when sharing resources such as code, data, and parameter settings, the variability of computing environments makes it difficult to reproduce results. During workflow design, it is in fact common to leverage on specific module versions, and one or more software libraries.

These requirements constitute the set of *dependencies* that have to be satisfied at every run of the analysis. Moreover, reproducibility is tempered by the lack of standards, and the use of large dataset or multiple data sources (e.g., integrative omics experiments) [29].

For RNA analyses, all the aforementioned challenges have been addressed by the RNA workbench [8], a Galaxy [30] instance tailored for RNA-centric data analyses. The workbench combines a comprehensive set of tools for the analysis of RNA structures,

Table 1
An overview of the most prominent tools provided by the RNA Workbench, for the understanding of miRNA–mRNA interactions

Tool name	Tool application
ViennaRNA [31]	A suite of tools aiming at the prediction of RNA secondary structures. The ViennaRNA package covers predictions for optimal and suboptimal structures from single sequences, as well as sequence alignments, predictions of ensemble base-pair probabilities, and RNA–RNA interactions. Furthermore, it enables hard and soft constraint parametrizations for the prediction of RNA structures
LocARNA [32, 33]	Realizes a variant of the Sankoff algorithm [34] for comparative analysis of unaligned RNAs by simultaneously folding and aligning
PARalyzer [35]	Provides high-resolution maps of the interaction sites occurring between RNA-binding proteins and their targets
RNAz [36]	Predicts structurally conserved and thermodynamically stable RNA secondary structure
doRiNA [37]	A database for the investigation of the regulatory pattern occurring between RNA-binding proteins (RBPs) and miRNAs. Information about this phenomenon is accessible directly through Galaxy for the incorporation of doRiNA-based queries in custom computational pipelines

genomic aligners, RNA–RNA and RNA–protein interactions, RNA-Seq, ribosome profiling, genome annotation, and more. Table 1 provides an overview of the most prominent computational approaches for understanding miRNA–mRNA interactions offered by the RNA workbench. An up-to-date overview of its growing set of tools can be found online, at https://bgruening.github.io/galaxy-rna-workbench.

The key reasons for the development of a comprehensive RNA analysis framework relying on Galaxy are its scalability, which enables the RNA workbench to run on single-CPU installations as well as on large multi-node high-performance computing environments, and workflow reproducibility, which provides researchers with a means to share their own analyses with colleagues, as well as import third-party workflows.

Tool and tool-version dependencies are resolved via BioConda (https://bioconda.github.io): the bioinformatics channel for the Conda (https://conda.io) package manager. BioConda facilitates version-aware software packaging, enabling installation at user level. Finally, the workbench is containerized with Docker (https://www.docker.com) to deal with different installation environments and provide a platform agnostic suite for RNA analysis. This layer of virtualization also allows the handling of user-defined input data in a secured manner, which represents a crucial requirement for the analysis of sensitive data (e.g., patient data in clinics).

In this section we illustrated how the dissemination of tools and best practices can benefit from a comprehensive and system-agnostic environment, able to cope with software-dependency resolution. For these reasons, the Galaxy community provides technical guidance on how to bring *stand-alone* services to their ample portfolio of topic-specific tools. This is achieved through Planemo (https://planemo.readthedocs.io): a set of command-line utilities to help developers building and publishing tools in Galaxy.

4 From Stand-Alone RNA Tool to a Comprehensive Environment: The TriplexRNA

Among the RNA tools that investigate miRNA-mediated gene regulation is the TriplexRNA database. The aim of this database is consistent with the study and characterization of our understanding of miRNA–mRNA interactions. However, as with plenty of RNA-centric software tools available, the TriplexRNA is *stand-alone*, and not integrated with the growing community around the RNA workbench.

The TriplexRNA focuses on investigating a regulatory pattern involving a pair of miRNAs, cooperatively inhibiting the expression of a mutual target gene [38, 39]. The *triplex* model is not a computational artifact: Sætrom et al. experimentally validated this complex, further characterizing its structural constraints in 2007 [40]; additional experimental evidence confirmed this regulatory mechanism, whose knowledge has been harvested from the literature and computationally simulated in 2012 [38]. The results of this study were finally gathered in a dedicated database in 2014 [39], the TriplexRNA, accessible at https://triplexrna.org.

The database contains all cooperating miRNA pairs in human and mouse, as well as their target genes, graphical illustrations of triplex secondary structures, their Gibbs free energies, and predicted equilibrium concentrations. A link with known human diseases is provided by a dedicated interactive interface, to search for cooperative miRNA pairs linked to KEGG pathways [41]. These features allow the identification and testing of disease-specific miRNAs that could be used as diagnostic and prognostic markers.

In this section, we implement a TriplexRNA *wrapper* and show its usage with Planemo, to provide all functionalities of the original database within the Galaxy Web interface.

4.1 Queries for Understanding Cooperative miRNA Regulation

In its online stand-alone version, the TriplexRNA organizes all its functionalities behind a unique *entry point*: a form, consisting of a sentence which is adapted to reflect the desired interrogation (Fig. 2). Upon user submission, the results are returned within an interactive table, with per-column filters and selectors for easy consultation of the retrieved results (Fig. 3). Moreover, for promoting data reuse, and enabling researchers to integrate database

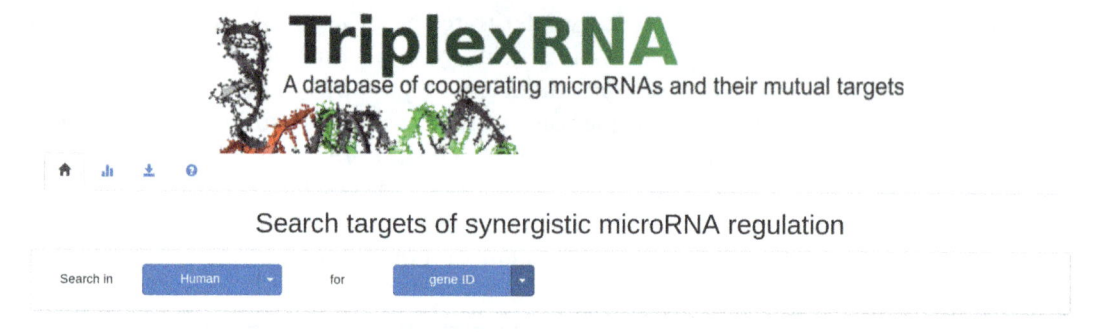

Fig. 2 The TriplexRNA form. The form constitutes a unique entry point, whereby queries are launched depending on the composed sentence. The default operation is to look for genes subjects of concerted miRNA regulation in humans. This query can be modified by selecting the desired organism, and gene or miRNA of interest

Gene ID	RefSeq ID	miRNA1 ID	miRNA2 ID	Seed distance (nt)	Free energy (Kcal/mol)	Energy gain (Kcal/mol)	Triplex details
CDKN1A	NM_000389	hsa-miR-224	hsa-miR-370	27	−35.46	−14.48	more >
CDKN1A	NM_000389	hsa-miR-370	hsa-miR-708	25	−35.06	−12.68	more >
CDKN1A	NM_000389	hsa-miR-93	hsa-miR-186	30	−34.66	−8.18	more >
CDKN1A	NM_000389	hsa-miR-132	hsa-miR-708	23	−33.76	−15.28	more >
CDKN1A	NM_000389	hsa-miR-132	hsa-miR-873	25	−33.46	−14.48	more >
CDKN1A	NM_000389	hsa-miR-186	hsa-miR-519d	30	−32.66	−10.18	more >
CDKN1A	NM_000389	hsa-miR-212	hsa-miR-708	23	−32.36	−14.68	more >
CDKN1A	NM_000389	hsa-miR-186	hsa-miR-20b	30	−32.06	−8.98	more >
CDKN1A	NM_000389	hsa-miR-212	hsa-miR-873	25	−32.06	−13.18	more >
CDKN1A	NM_000389	hsa-miR-28-5p	hsa-miR-101	21	−31.96	−9.68	more >

Showing 1 to 10 of 19 entries

Fig. 3 The TriplexRNA result table. In this example, the table presents all putative triplexes involving the target gene CDKN1A. The table provides per-column filters and selectors to narrow down the entries to the results of interest. More information on the meaning of each column can be found in the help section at https:/triplexrna.org

queries within their computational pipelines, the TriplexRNA implements an API for retrieving results as HTML (https://www.w3.org/html), CSV (https://tools.ietf.org/html/rfc4180), and JSON (https://www.json.org). Each of these *standard*

Table 2
The TriplexRNA queries, implemented to investigate cooperative miRNA regulation. All functionalities are available through the standard Web interface, as well as API requests. API requests are launched from user pipelines, using an interrogation path which encodes the query parameters in a URI, for programmatic database retrieval. The table describes all TriplexRNA queries, and their API counterparts for retrieving data in JSON format

Query function	Interrogation path
Single gene query: retrieve all RNA triplexes of organism O, involving target gene X	triplexrna.org/JSON/O/gene/X
Multiple gene query: retrieve all RNA triplexes of organism O, involving the genes X, Y, Z	triplexrna.org/JSON/O/genes/$X/Y/Z$
Triplex query: retrieve all details of organism O's RNA triplex T	triplexrna.org/JSON/O/triplex/T
Pathway query: retrieve all RNA triplexes of organism O, involved in KEGG pathway K	triplexrna.org/JSON/O/pathway/K
Single miRNA query: retrieve all RNA triplexes of organism O, involving miRNA M	triplexrna.org/JSON/O/mirna/M
miRNA pair query: retrieve all RNA triplexes of organism O, involving miRNAs M and N	triplexrna.org/JSON/O/mirna/M/N
Targets of cooperative miRNA pair query: retrieve all RNA triplexes of organism O, involving miRNA pair M and N, and targeting genes X, Y, Z	triplexrna.org/JSON/O/mirna/M/N/targeting/$X/Y/$

representations is returned depending on the selected *interrogation path*, i.e., a URI (https://www.w3.org/Addressing) which replaces the original user-operated Web form (entry point), for programmatic database access. Table 2 offers an overview of the TriplexRNA functionalities, their meaning, and the corresponding interrogation path that users shall adopt to automate data retrieval.

The integration of the TriplexRNA database functionalities within Galaxy leverages on the interrogation paths from the aforementioned table. These are implemented in a Python 2.7 (https://www.python.org) wrapper, available at https://github.com/bagnacan/triplexrna-planemo. The wrapper maps each database query to a specific command-line invocation, which is triggered once the corresponding arguments have been called.

Galaxy users do not interact directly with tool command lines; conversely, they invoke their functions by means of their dedicated Web interfaces. The tool wrapper helps developers in moving toward this solution: the next step toward the integration of the TriplexRNA within Galaxy leverages in fact on the wrapper's options and parameter semantics, to define the tool's very own Galaxy Web interface.

4.2 Integrating RNA Cooperativity Investigations in Galaxy

The integration of novel tools within Galaxy is carried out through Planemo.

In the first part of this section we explained the aim of its set of command-line utilities. Here, we overview their practical application, in relation to the TriplexRNA wrapper functions outlined in Table 2.

Planemo semi-automates the creation of a tool's Web page through the *planemo tool_init* function. This creates a draft XML (https://www.w3.org/XML) file, which includes the mandatory sections describing the tool in terms of its package dependencies, input/output formats, and mode of operations. For the creation of the TriplexRNA XML file, we used Planemo version 0.48.0, which generates the code shown in Fig. 4. All sections are then parsed by Galaxy, and rendered together as a dedicated Web interface that users can adopt to interact with the underlying tool.

The rendering is performed after a validation phase. This operation is carried out by *planemo lint*, which parses the developer-provided descriptions, to test for inconsistencies within the XML content. The linting phase has to be repeated at every successive editing of the XML file.

The TriplexRNA XML descriptor file is available at https://github.com/bagnacan/triplexrna-planemo.

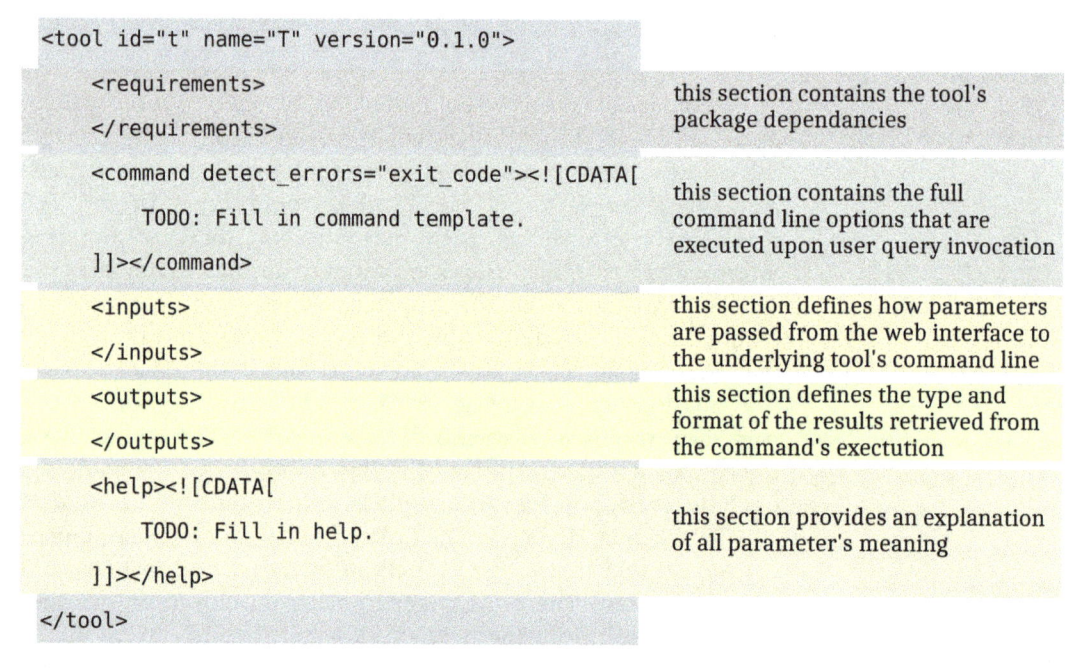

```
<tool id="t" name="T" version="0.1.0">

    <requirements>
                                                    this section contains the tool's
    </requirements>                                 package dependancies

    <command detect_errors="exit_code"><![CDATA[
                                                    this section contains the full
        TODO: Fill in command template.             command line options that are
                                                    executed upon user query invocation
    ]]></command>

    <inputs>                                        this section defines how parameters
                                                    are passed from the web interface to
    </inputs>                                       the underlying tool's command line

    <outputs>                                       this section defines the type and
                                                    format of the results retrieved from
    </outputs>                                      the command's exectution

    <help><![CDATA[
                                                    this section provides an explanation
        TODO: Fill in help.                         of all parameter's meaning

    ]]></help>

</tool>
```

Fig. 4 The Planemo draft XML file. This includes the mandatory descriptions for a tool's integration in Galaxy. Each description is enclosed in a dedicated XML section, which provides a mean for the characterization of the underlying command-line tool, in terms of its environment requirements, input/output formats, and modes of operation. An overview of additional XML section is provided on the Planemo documentation, available at https:/planemo.readthedocs.io

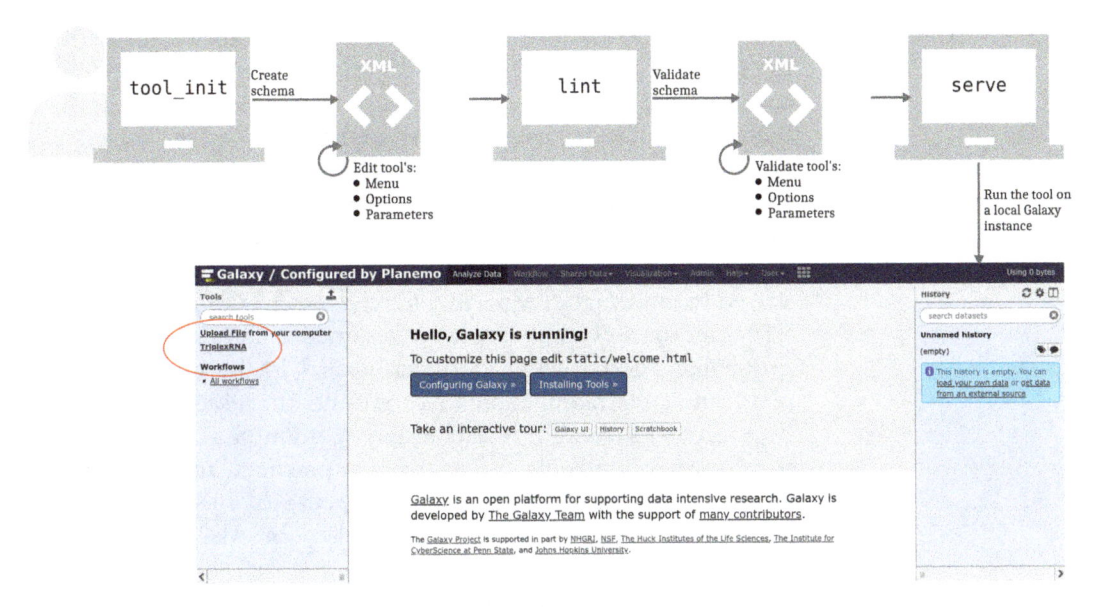

Fig. 5 The first steps toward integrating a novel tool within Galaxy using Planemo. (1) Planemo *tool_init* creates a scaffold XML file, filled by the user to accommodate each parameter option. (2) The provided XML is parsed and validated by Planemo *lint*. (3) Ultimately, Planemo *serve* includes the tool's wrapper and XML file for execution within a local Galaxy instance

Once all tool dependencies, input/output formats, and modes of operation are defined, the tool can be tested for running within Galaxy, by using the *planemo serve* command. This phase runs a local Galaxy instance, whose list of tools includes the one that underwent the previous edit and linting phases.

Figure 5 shows the aforementioned tool_init, lint, and serve steps we took to (1) create and edit the TriplexRNA XML descriptor for the underlying tool wrapper, (2) iteratively edit and validate the tools input/output options, and (3) launch the tool within a local Galaxy instance.

Planemo also provides utilities to test the new tool against predefined input datasets, assess the correct execution of each mode of operation, as well as publish the tool in the Galaxy Tool Shed (https://toolshed.g2.bx.psu.edu), for a later inclusion within dedicated Galaxy instances. These phases are not within the scope of the present work; however, they are mandatory for the development of new Galaxy tools. Further documentation on these topics can be found at https://planemo.readthedocs.io.

5   Discussion

The investigation and functional characterization of miRNA–mRNA interactions can be assessed computationally. A plethora of RNA software tools have been implemented for analyzing their mechanisms; however, their solutions leverage on mathematical and statistical foundations: subjects that are rarely within the focus of life science curricula. As a result, before answering a biological question, users are required to look for available solutions by delving into online catalogs of tools, and make sense of their data formats, parametrization semantics, and software dependencies, ultimately facing a novel technical jargon.

This scenario has highlighted the need for creating a corpus of shared bioinformatics expertise, for the dissemination of computational approaches and best practices, aimed at addressing biological problems. The challenge has been dealt with by numerous Web platforms and training initiatives around the globe, which have been proved successful in establishing communities for discussing topic-specific problems, discussing best practices, and creating frameworks for the harmonization and reuse of shared processing tools and workflows [11]. In particular, for RNA analyses, the coherence of computational environments, data formats, interoperability, and reproducibility has been addressed by the Galaxy [30] community with the creation of the RNA workbench [8]. Such platform acts as a hub for developers as well as users: on the one hand to easily integrate and deploy novel or existing tools and workflows, and on the other to readily address biological questions, regardless of the background experience in administering computing environments. Following their community guidelines, we showcased Planemo and Galaxy, by realizing a wrapper for the TriplexRNA database [38, 39]. Finally, we provided a starting point for the inclusion of this stand-alone RNA tool as a new building block of the broad portfolio of tools within the RNA workbench.

The RNA workbench is growing, incorporating further tools for RNA analysis, and enriching the platform of diverse interoperable methods for understanding miRNA–mRNA interactions. However, this is coming at a cost, because as the list of tools increases, it is necessary to accommodate them in a structured catalog that must be both *clear* and *usable*: it should group the tools under pertinent and agreed-upon terms, to avoid forcing its users open each category before finding the desired tool, and abstract enough to accommodate novel entries, therefore not forcing its users to endlessly browse through an overwhelming list of categories. Failing to meet these trade-offs will systematically penalize the overall user experience, hindering tools below a long list of browsable fuzzy terms. Albeit on a smaller scale, the problem of organizing tools into catalogs [19, 20], as we highlighted in Section 2, reappears in the Galaxy framework.

In practice, both *Galaxy Main* (https://usegalaxy.org) and the *Galaxy EU* (https://usegalaxy.eu) public instances solve the problem of ambiguous terms by organizing their tools under pertinent and agreed-upon terms; however, due to the large number of tools,

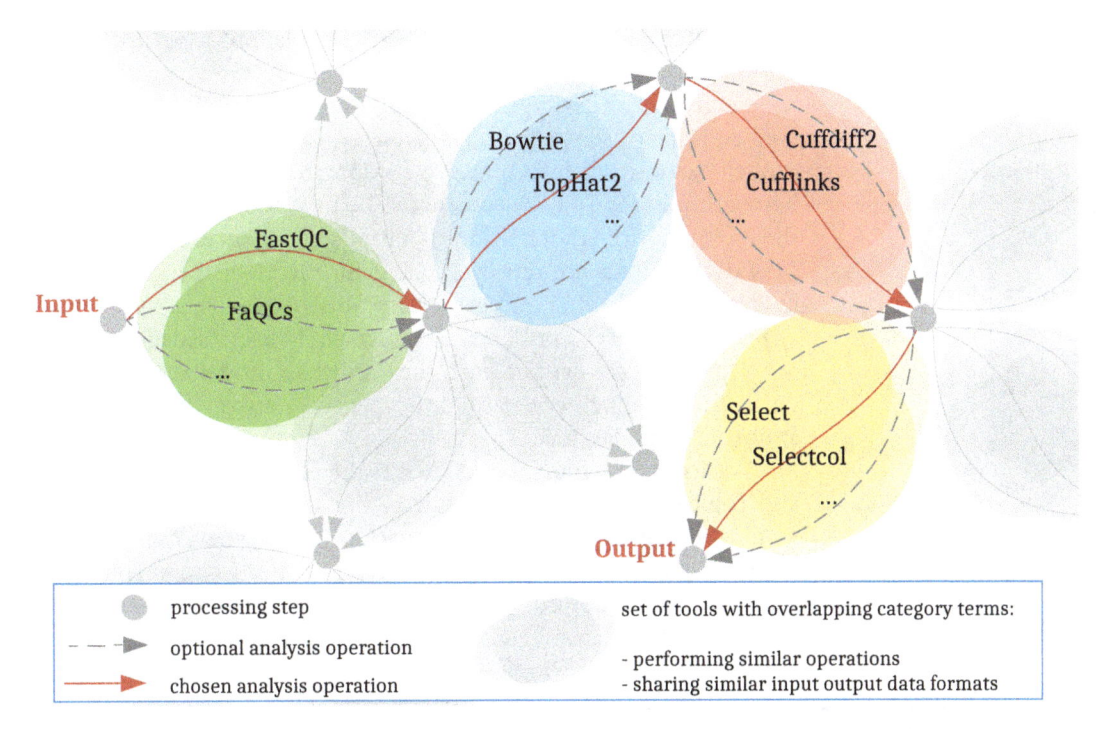

Fig. 6 Design of the Galaxy recommendation system. Workflow analyses are represented as paths over a graph comprising all known analysis pipelines. Here, each node represents a state of the processed data, and each edge a tool's function, whose allowed inputs and generated outputs correspond to the states of data it connects. In this framework, the next recommended tool must be pertinent with the scope of the desired analysis

novice users are left with no choice other than to browse most of the categories before finding the desired tool. This usability problem is partly overcome with the addition of a *search bar*; however, this feature requires the users to already know the name of the desired tool, which, especially for novice users, is not always the case. Moreover, such a solution does not advertise newly incorporated tools, therefore missing the promotion of possible alternatives and benchmarking.

To address the usability problem, we propose the integration of a *recommendation system*, aiming at providing guidance toward the completion of the desired analysis, while at the same time hiding all tools and categories behind a unique interface, where suggestions for next tools are loaded at each iteration. Figure 6 shows a candidate approach for the design of such a system. Here, we define a node as a state of the processed data, representing the starting or landing point of an edge: a tool's function, whose input and output (I/O) formats correspond to the specific nodes it connects. Within this framework, we define the next candidate tool as *pertinent* in terms of (a) allowed I/O data formats, accepted and generated by the tool, and (b) the original tool's categorization, attributed by the instance's administrator to organize tools into a list of agreed-upon terms.

Such system leverages on a knowledge base built from anonymized Galaxy user-generated analyses, on top of which a predictive model *chains* tools into putative computational workflows on the basis of tool usage and occurrence. The model is further refined by the aforementioned definition of a tool's pertinence: here, allowed I/O data formats are retrieved using BioBlend (https://bioblend.readthedocs.io), while categories are obtained using a *web crawler* against Galaxy EU's public Web interface (https://usegalaxy.eu).

A proof of concept of both the predictive model and the tool's pertinence refinement are available at https://github.com/anuprulez/similar_galaxy_workflow.

We argue that frameworks such as the Galaxy Main, Galaxy EU, and topic-specific instances such as the Galaxy RNA workbench would benefit from a recommendation system, because it would improve the overall user experience, and therefore increase its acceptance and adoption across diverse communities, as a reference framework for shared best practices and reproducible analyses.

Acknowledgments

The authors would like to thank the de.NBI and ELIXIR initiatives, for their support in the bioinformatics infrastructure. Thanks also to the Galaxy community, for developing, maintaining, and providing guidance on the use of this comprehensive framework. A warm thank you goes to the RBC Freiburg group, in particular to Anup Kumar, Björn Grüning, and Rolf Backofen for their efforts and commitment in improving the Galaxy framework.

References

1. Lu J, Getz G, Miska EA et al (2005) MicroRNA expression profiles classify human cancers. Nature 435:834–838
2. Croce CM, Calin GA (2005) miRNAs, cancer, and stem cell division. Cell 122:6–7
3. Mitchell PS, Parkin RK, Kroh EM et al (2008) Circulating microRNAs as stable blood-based markers for cancer detection. PNAS 105:10513–10518
4. Chen X, Ba Y, Ma L et al (2008) Characterization of microRNAs in serum: a novel class of biomarkers for diagnosis of cancer and other diseases. Cell Res 18:997–1006
5. Cho WCS (2010) MicroRNAs: potential biomarkers for cancer diagnosis, prognosis and targets for therapy. Int J Biochem Cell Biol 42:1273–1281
6. Linsen SEV, de Wit E, Janssens G et al (2009) Limitations and possibilities of small RNA digital gene expression profiling. Nat Methods 6:474–476
7. Wang Z, Gerstein M, Snyder M (2009) RNA-Seq: a revolutionary tool for transcriptomics. Nat Rev Genet 10:57–63
8. Grüning BA, Fallmann J, Yusuf D et al (2017) The RNA workbench: best practices for RNA and high-throughput sequencing bioinformatics in Galaxy. Nucleic Acids Res 45:W560–W566. https://doi.org/10.1093/nar/gkx409
9. Conesa A, Madrigal P, Tarazona S et al (2016) A survey of best practices for RNA-seq data analysis. Genome Biol 17:13
10. Sboner A, Mu XJ, Greenbaum D, Auerbach RK, Gerstein MB (2011) The real cost of

sequencing: higher than you think! Genome Biol 12:125
11. Batut B, Hiltemann S, Bagnacani A, et al (2017) Community-driven data analysis training for biology. bioRxiv: 225680
12. Burks C (1999) Molecular biology database list. Nucleic Acids Res 27:1–9
13. Galperin MY, Rigden DJ, Fernández-Suárez XM (2015) The 2015 nucleic acids research database issue and molecular biology database collection. Nucleic Acids Res 43:D1–D5
14. Fox JA, Butland SL, McMillan S, Campbell G, Ouellette BFF (2005) The bioinformatics links directory: a compilation of molecular biology web servers. Nucleic Acids Res 33:W3–W24
15. Brazas MD, Yim D, Yeung W, Ouellette BFF (2012) A decade of web server updates at the bioinformatics links directory: 2003–2012. Nucleic Acids Res 40:W3–W12
16. Pettifer S, Thorne D, McDermott P, Attwood T, Baran J, Bryne JC, Hupponen T, Mowbray D, Vriend G (2009) An active registry for bioinformatics web services. Bioinformatics 25:2090–2091
17. Pettifer S, Ison J, Kalaš M et al (2010) The EMBRACE web service collection. Nucleic Acids Res 38:W683–W688
18. Bhagat J, Tanoh F, Nzuobontane E et al (2010) BioCatalogue: a universal catalogue of web services for the life sciences. Nucleic Acids Res 38:W689–W694
19. Henry VJ, Bandrowski AE, Pepin A-S, Gonzalez BJ, Desfeux A (2014) OMICtools: an informative directory for multi-omic data analysis. Database (Oxford). https://doi.org/10.1093/database/bau069
20. Ison J, Rapacki K, Ménager H et al (2016) Tools and data services registry: a community effort to document bioinformatics resources. Nucleic Acids Res 44:D38–D47
21. Ison J, Kalaš M, Jonassen I, Bolser D, Uludag M, McWilliam H, Malone J, Lopez R, Pettifer S, Rice P (2013) EDAM: an ontology of bioinformatics operations, types of data and identifiers, topics and formats. Bioinformatics 29:1325–1332
22. Kim D, Pertea G, Trapnell C, Pimentel H, Kelley R, Salzberg SL (2013) TopHat2: accurate alignment of transcriptomes in the presence of insertions, deletions and gene fusions. Genome Biol 14:R36
23. Kim D, Langmead B, Salzberg SL (2015) HISAT: a fast spliced aligner with low memory requirements. Nat Methods 12:357–360
24. Dobin A, Davis CA, Schlesinger F, Drenkow J, Zaleski C, Jha S, Batut P, Chaisson M, Gingeras TR (2013) STAR: ultrafast universal RNA-seq aligner. Bioinformatics 29:15–21
25. Hoffmann S, Otto C, Doose G et al (2014) A multi-split mapping algorithm for circular RNA, splicing, trans-splicing and fusion detection. Genome Biol 15:R34
26. Engström PG, Steijger T, Sipos B et al (2013) Systematic evaluation of spliced alignment programs for RNA-seq data. Nat Methods 10:1185–1191
27. Möller S, Prescott SW, Wirzenius L et al (2017) Robust cross-platform workflows: how technical and scientific communities collaborate to develop, test and share best practices for data analysis. Data Sci Eng 2:232–244
28. Sandve GK, Nekrutenko A, Taylor J, Hovig E (2013) Ten simple rules for reproducible computational research. PLoS Comput Biol 9: e1003285
29. Goecks J, Nekrutenko A, Taylor J (2010) Galaxy: a comprehensive approach for supporting accessible, reproducible, and transparent computational research in the life sciences. Genome Biol 11:R86
30. Afgan E, Baker D, van den Beek M et al (2016) The Galaxy platform for accessible, reproducible and collaborative biomedical analyses: 2016 update. Nucleic Acids Res 44:W3–W10
31. Lorenz R, Bernhart SH, Höner zu Siederdissen C, Tafer H, Flamm C, Stadler PF, Hofacker IL (2011) ViennaRNA Package 2.0. Algorithms Mol Biol 6:26
32. Will S, Joshi T, Hofacker IL, Stadler PF, Backofen R (2012) LocARNA-P: accurate boundary prediction and improved detection of structural RNAs. RNA 18:900–914
33. Will S, Reiche K, Hofacker IL, Stadler PF, Backofen R (2007) Inferring noncoding RNA families and classes by means of genome-scale structure-based clustering. PLoS Comput Biol 3:e65
34. Zuker M, Sankoff D (1984) RNA secondary structures and their prediction. Bltn Mathcal Biol 46:591–621
35. Corcoran DL, Georgiev S, Mukherjee N, Gottwein E, Skalsky RL, Keene JD, Ohler U (2011) PARalyzer: definition of RNA binding sites from PAR-CLIP short-read sequence data. Genome Biol 12:R79
36. Gruber AR, Findeiß S, Washietl S, Hofacker IL, Stadler PF (2010) Rnaz 2.0: improved noncoding RNA detection. Pac Symp Biocomput 15:69–79
37. Blin K, Dieterich C, Wurmus R, Rajewsky N, Landthaler M, Akalin A (2015) DoRiNA 2.0—upgrading the doRiNA database of RNA

interactions in post-transcriptional regulation. Nucleic Acids Res 43:D160–D167

38. Lai X, Schmitz U, Gupta SK, Bhattacharya A, Kunz M, Wolkenhauer O, Vera J (2012) Computational analysis of target hub gene repression regulated by multiple and cooperative miRNAs. Nucleic Acids Res 40:8818–8834

39. Schmitz U, Lai X, Winter F, Wolkenhauer O, Vera J, Gupta SK (2014) Cooperative gene regulation by microRNA pairs and their identification using a computational workflow. Nucleic Acids Res 42:7539–7552

40. Sætrom P, Heale BSE, Snøve O, Aagaard L, Alluin J, Rossi JJ (2007) Distance constraints between microRNA target sites dictate efficacy and cooperativity. Nucleic Acids Res 35:2333–2342

41. Kanehisa M, Goto S (2000) KEGG: kyoto encyclopedia of genes and genomes. Nucleic Acids Res 28:27–30

Chapter 9

Computational Resources for Prediction and Analysis of Functional miRNA and Their Targetome

Isha Monga and Manoj Kumar

Abstract

microRNAs are evolutionarily conserved, endogenously produced, noncoding RNAs (ncRNAs) of approximately 19–24 nucleotides (nts) in length known to exhibit gene silencing of complementary target sequence. Their deregulated expression is reported in various disease conditions and thus has therapeutic implications. In the last decade, various computational resources are published in this field. In this chapter, we have reviewed bioinformatics resources, i.e., miRNA-centered databases, algorithms, and tools to predict miRNA targets. First section has enlisted more than 75 databases, which mainly covers information regarding miRNA registries, targets, disease associations, differential expression, interactions with other noncoding RNAs, and all-in-one resources. In the algorithms section, we have compiled about 140 algorithms from eight subcategories, *viz.* for the prediction of precursor (pre-) and mature miRNAs. These algorithms are developed on various sequence, structure, and thermodynamic based features incorporated into different machine learning techniques (MLTs). In addition, computational identification of miRNAs from high-throughput next generation sequencing (NGS) data and their variants, *viz.* isomiRs, differential expression, miR-SNPs, and functional annotation, are discussed. Prediction and analysis of miRNAs and their associated targets are also evaluated under miR-targets section providing knowledge regarding novel miRNA targets and complex host-pathogen interactions. In conclusion, we have provided comprehensive review of in silico resources published in miRNA research to help scientific community be updated and choose the appropriate tool according to their needs.

Key words microRNA, Transcription factor, Database, Algorithm, Analysis tools, Machine learning tools

1 Introduction

1.1 miRNA Definition   The microRNAs (miRNAs) are one of the important RNA molecules involved in RNA interference (RNAi), which are produced endogenously in cells. They are evolutionarily preserved, double stranded, approximately 21–23 nucleotides (nts) long with two nucleotide overhangs at 3′ end, hairpin-derived small non-coding RNAs (ncRNAs). They function as "dimmer switch" by binding to perfect or imperfect complementary target gene sequences and cause mRNA degradation or translational repression on the basis of the degree of complementarity [1].

1.2 Discovery

In 1993, Victor Ambros and colleagues from Harvard University discovered that lin-4, a small RNA of 22nts, is implicated in the postembryonic development of a nematode *Caenorhabditis elegans* [2]. It negatively represses the lin-14 gene expression by binding at the 3′ untranslated region (UTR) of its messenger RNA (mRNA) [2]. It was a new landmark in the post transcriptional gene silencing field. After this breakthrough discovery, numerous miRNA genes were testified in *C. elegans* and other species [3–5]. Since then, miRNAs have emerged as vital gene expression regulators. Further, keeping their importance, miRNA genes from numerous species have been reported in the last decade resulting in enormous expansion of the miRBase, a central miRNA sequence repository [6]. Currently, miRBase release version 21 reports a total of 1881 pre- and 2588 mature miRNAs in the human genome (GRCh38). Further, miRNAs were also discovered in viruses. In 2004, Tuschl et al. have discovered viral encoded miRNAs in Epstein-Barr virus (EBV) [7]. After this breakthrough discovery, there were numerous reports of viral miRNAs in many other viruses [8, 9]. Presently, VIRmiRNA, a comprehensive resource for experimentally validated viral miRNAs, has reported a total of 1308 viral miRNA sequences [10].

miRNAs intensify complexity in eukaryotic gene regulation mechanisms by playing their role in multifarious and multifactorial ways. One of the characteristic features of miRNA-mediated gene regulation is fully matching of 2nd–8th nts from the 5′ end of the mature miRNA molecule with the target sequence, termed as seed region [11]. Hence, this nominal prerequisite results in the targeting of many mRNAs by a single miRNA and targeting of one mRNA by several miRNAs designated as multiplicity and cooperativity respectively [12, 13]. Additionally, miRNAs are reported to have length and sequence variability termed as isomiRs, which are related in changed targeting repertoire.

1.3 Molecular Mechanism

Biogenesis of miRNA is a highly coordinated mechanism of enzymatic chemical reactions (*see* Fig. 1). Mature miRNA sequence is generated by a two-step processing pathway, *viz.* one in nucleus and the other one in cytoplasm, respectively [14]. In nucleus, usually RNA polymerase II (pol II) transcribes long sequence with a hairpin shape of several kilobases (kbs) in length called primary miRNA (pri-miRNA) [15]. However, miRNA genes are reported to be transcribed by RNA pol III, *viz.* a miRNA cluster in human chromosome 19 (C19MC) [16] and miRNAs of murine γ-herpesvirus 68 [17]. Involvement of pol III is also reported in *Neurospora crassa* [18].

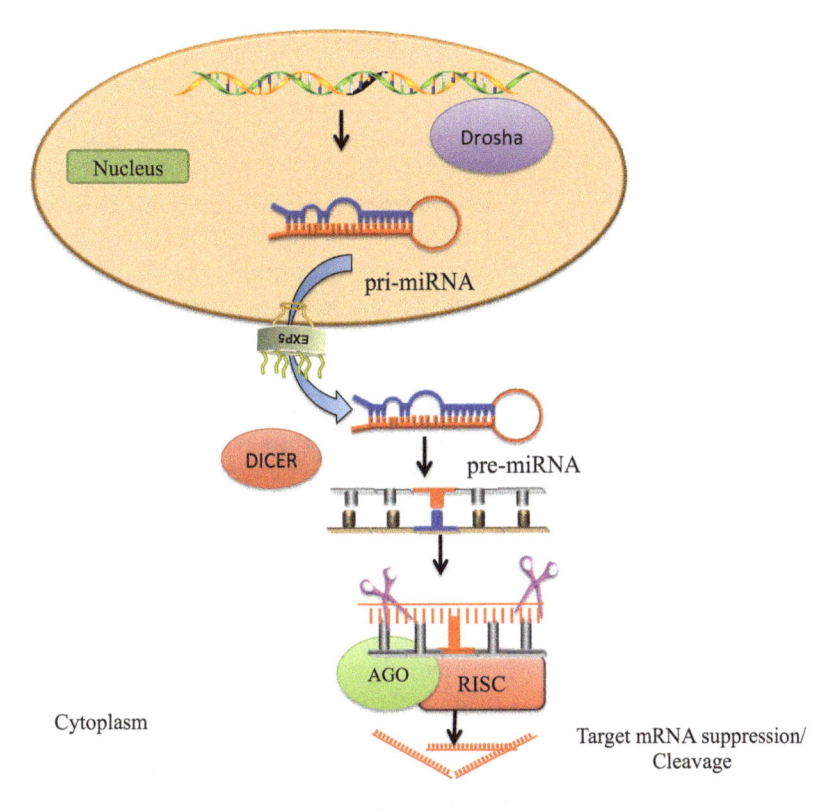

Fig. 1 Molecular mechanism of miRNA biogenesis

After pri-miRNA transcription, these underwent slicing and processing into pre-miRNA, a ~70 nt long hairpin structure by a microprocessor complex (Drosha and DGCR8). This pre-miRNA has 2 nt overhangs at 3' end of the hairpin sequence. Then, it is transferred to the cytoplasm by Exportin-5. In cytoplasm, Dicer (DCR) cleaves off the pre-miRNA hairpin to generate nascent double-stranded (ds) RNA species. Nascent ds miRNA-miRNA* duplex is loaded on RISC. One of the strand binds with cognate target mRNA called guide strand. The strand, which was degraded by endonuclease domain of Argonaute (AGO) protein of RISC complex, is called as passenger strand [15]. This binding is assisted first with the help of seed sequence, a stretch of nucleotides having positions 2 to 8 from 5' end of guide strand. Guide strand sequence undergoes mRNA cleavage or inhibition of translational initiation of the target gene depending upon the sequence complementarity between miRNA and mRNA. The terms guide and passenger strand have been revised to 5p and 3p in recent literature [19].

1.4 Role in Biological Processes

miRNAs are critical components of gene silencing mechanism in eukaryotic organisms. They play a immense regulatory roles in various cellular pathways ranging among cell division, differentiation, and growth. Their abnormal expression is associated with the various diseases including cancer and viral infections, which makes them potential diagnostic and prognostic biomarkers.

Further, recent studies have revealed a landscape of novel and non canonical discoveries of miRNAs apart from conventional miRNA functions. These include their presence in the exons apart from intergenic regions, participation of passenger strand in the gene regulation, and miRNA functions in nucleus.

Currently discovered animal miRNAs are mapped to either the intergenic regions in the form of single or clustered miRNA genes or the intronic regions of the coding genes [20, 21]. However, major portion (68%) from currently discovered human miRNAs belongs to intergenic origin, while rest of the proportion is reported to be intragenic. Additionally, out of intragenic miRNAs, mostly miRNA genes are transcribed from intronic regions followed by repeats, long ncRNAs, and UTRs [22]. It is reported that intergenic miRNA genes that are located together termed as miRNA cluster are often co-transcribed and under the regulation of host gene promoter.

Similarly, intragenic miRNAs are transcribed together with gene and dependent on the gene promoter. This is further supported by good correlation in their expression patterns [23]. On the contrary, miRNA genes in introns can be transcribed self-sufficiently from host gene promoter [24]. Similarly, miRNAs in polycistronic transcribed sequence can be alternatively spliced to generate specific miRNA [25]. Apart from long studied intergenic or intronic miRNA genes, they have been surprisingly also described in exons of protein-coding regions. In mouse testes, miRNA genes are reported in exon regions. In intergenic regions, they are found to present as solo or clustered miRNA-coding genes [24].

It was postulated that pre-miRNA sequence processing through DCR1 liberates a miRNA duplex termed as miRNA-miRNA* or guide-passenger strand complex, respectively [26]. While miRNA arm is known to get incorporated in AGO2 of RISC complex, miRNA* arm is only considered as a carrier of miRNA arm. However, recent studies conducted on different organisms like *Drosophila*, *C. elegans*, mouse, and human revealed regulatory roles of so-called passenger arm. These miRNA* arms are found to be highly conserved around seed sequence region in humans [27], Drosophila [28], and other vertebrates [29]. Therefore, all these evidence-based studies indicate that all miRNA locations can potentially give two distinct mature miRNA sequences evenly important in function [28, 30].

Further, there are experimental studies, which report nuclear import motif in miRNAs [31]. miRNAs are not only reported to be involved in posttranscriptional gene silencing; instead they have also been reported in the nucleus suggesting their role in transcriptional silencing [32, 33], chromatin remodeling, and many more [34].

Table 1
microRNA-based therapeutics in preclinical studies

S. No.	miRNAs with tumor-suppressive function (miRNA mimics as therapeutics)		
	miRNA name	Disease	In vivo delivery systems used
1	let-7	B-cell lymphoma	Neutral lipid emulsions
2	miR-34a	Myeloma, B-cell lymphoma	Lipid nanoparticles
3	miR-143	Lymphoid leukemia	Liposomes
4	miR-200	Solid tumors (breast, ovarian)	Liposomes
	OncomiRs (antimiRs as therapeutic agents)		
5	miR-10b	Solid tumors (breast and glioma)	LNA antimiRs
6	miR-155	B cell lymphoma, Lymphoid leukemia	pHLIP-conjugated antimiR
7	miR-221	Solid tumors (liver, pancreas)	Cholesterol-conjugated antimiR
	Other		
8	miR-122	HCV infection	PS DNA–LNA antimiR
9	miR-33	Atherosclerosis	LNA antimiR
10	miR-208	Cardiac disease, myocardial infarction	LNA antimiR
11	miR-21	Kidney fibrosis	LNA antimiR
12	miR-192	Diabetes-related kidney complications	LNA miRNA mimic
13	miR-29c	Diabetes-related kidney complications	Naked antagomiRs
14	miR-103	Diabetes	LNA antimiR
15	miR-15	Myocardial infarction	LNA antimiR

Note: Abbreviations used: *PS* phosphorothioate, *LNA* locked nucleic acid, *pHLIP* pH low-insertion peptide

1.5 Therapeutic Importance

Discovery of miRNAs two decades ago has made them important tools in biomedicine. miRNAs hold many attractive features, which makes them suitable therapeutic modality like having small size, specificity, and known sequence with conserved nature [35]. They can be exploited in the drug market via two ways, *viz.* miRNA mimics and anti-miRNAs (antimiRs), respectively [36]. miRNA mimics have been reported in cases where disease is caused due to lack or reduced expression of a particular miRNA [37]. Chemically synthesized miRNA mimics are incorporated to overcome the underexpression of miRNA causing the disease [38] (Tables 1 and 2).

Contrariwise, targeting with antimiR compounds treats those diseases, which are caused due to overexpression of miRNA. Recently there are some reports in animals and humans indicating the importance of antimiR compounds that inhibit specific miRNAs to treat cardiac remodeling [39], hepatitis C virus (HCV) [40], and neoangiogenesis [41] (Tables 1 and 2).

Table 2
microRNA-based therapeutics in clinical trials

Therapeutic	Company	Therapeutic type	Target disease	Clinical trial stage	Reference
Miravirsen	SantarisPharma	AntimiR-122	Chronic Hepatitis C	Phase I, completed	NCT01646489
				Phase II, completed	NCT01200420
				Phase II, ongoing	NCT01872936
				Phase II, ongoing	NCT02031133
				Phase II, ongoing	NCT02508090
RG-101	Regulus	AntimiR-122	Chronic hepatitis C	Phase I, completed	–
				Phase II, ongoing	–
RG-125	Regulus	AntimiR-103	Type 2 diabetes	Phase I, ongoing	NCT02612662
				Phase I/IIa, ongoing	NCT02826525
MRG-106	miRagen	AntimiR-155	Cutaneous T-cell lymphoma	Phase I, ongoing	NCT02580552
MRG-201	miRagen	miR-29 mimic	Scleroderma	Phase I, ongoing	NCT02603224
MesomiR-1	EnGeneIC	miR-16 mimic	Mesothelioma	Phase I, ongoing	NCT02369198
MRX34	Mirna	miR-34 mimic	Multiple solid tumors	Phase I, terminated	NCT01829971

1.6 Current Status of Computational Resources

Due to emergence of next-generation sequencing (NGS) technologies as a powerful platform to discover miRNAs in a high-throughput manner, there has been a wealth of data regarding miRNAs in various organisms in recent past. Various databases were reported for collection of either experimentally validated or predicted miRNAs. However, experimental methods still remain laborious because it requires the identification of miRNAs from millions of reads by matching them with reference database and still it is dependent on detection of novel miRNAs with the help of any computational algorithm [42]. Hence, genome-wide computational prediction of novel miRNAs has emerged as a faster alternative for miRNA identification. Thus, it will further widen our understanding toward complex interactions at posttranscriptional level during various infections.

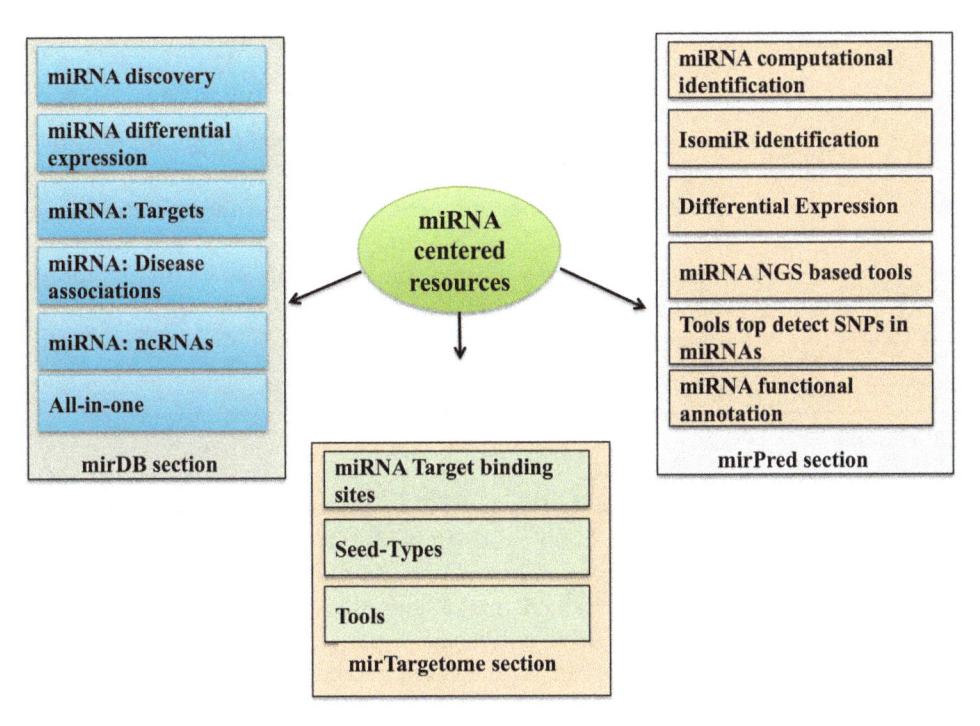

Fig. 2 Schematic representation of the miRNA-centered resources in mirDB, miRPred, and miR-Targetome sections

This chapter provides a concise overview of the state-of-the-art bioinformatics resources developed in miRNA-based research till now (Fig. 2). The first section comprehends databases in the present field, which includes archives from the following subgroups miRNA registries, targets, disease associations, differential expression, interactions with other noncoding RNAs, and all-in-one resources. Further, their comparative discussion on the basis of various aspects like number of records, organisms covered, miRNA target sequence, and disease-related information is done. In the second section, computational tools and algorithms are assembled from eight different classes, *viz.* prediction of precursor (pre-) and mature miRNA, and computational identification of miRNAs from next-generation sequencing (NGS) data and their variants, *viz.* isomiRs, differential expression, miR-SNPs, and functional annotation. These algorithms are developed on various sequence, structure, and thermodynamic based features incorporated into different machine learning techniques (MLTs). Additionally, under isomiRs and miR-SNPs sections, we have highlighted the consequence of genetic variants on the miRNA expression and functioning.

Computational tools developed for the prediction of miRNA targets were evaluated under *miR-Targetome* section (Fig. 2). These tools are helpful in delivering knowledge regarding target repertoire of novel miRNA and hence helping to elucidate complex host-pathogen interactions.

Therefore, we have summarized and evaluated tools and databases published in miRNA research field. We have also briefly discussed current therapeutic advancements in miRNA field with an emphasis on upcoming challenges for computational biology for identifying novel miRNA and targets. Thus, we hope that this chapter will help scientific community in choosing the appropriate tools according to their research needs along with instructions to use a specific resource.

2 Databases

In recent years, the number of experimentally validated miRNAs is accelerated in an exponential rate resulting in the need of repositories to store and organize them in a structured way. Therefore, many databases encompassing miRNAs and miRNA-target associations have been published. Further, availability of experimentally testified miRNA sequences in databases and technological advances in the NGS-based technologies has made computational identification of miRNAs and miRNA-target associations comparatively easier. Thus, various archives were published in literature holding predicted miRNA sequences. Further, some databases not only archive predicted miRNAs, but also encompass predicted targets and their disease associations. Therefore, current portion represents databases from six different subcategories like miRNA registries, differential expression, targets, disease associations, all-in-one resources, and noncoding RNA archives. In total, we have provided comprehensive list of more than 75 miRNA-centered databases. Some repositories are catering the need of all species, while other ones are from specialized domains like encompassing species- or tissue-specific miRNAs.

2.1 miRNA Registries

Experimental identification of miRNAs is done by approaches like microarray, cloning, northern blotting, and in situ hybridization, which were time consuming and costly. However, NGS has enabled miRNA identification easy and comparatively cost effective in recent times. Therefore, thousands of miRNA sequences were reported for various organisms like animals, plants, and viruses. Several algorithms were developed to discover novel miRNAs from high-throughput sequencing data. These tools are developed on sequence-structure features characteristic to miRNA like sequence composition, conservation among various species, position and number of hairpin loops, number of bulges and base pairs, and structural features like secondary structure and Gibbs free energy.

Hence, to cater the storage need of these miRNA sequences, a microRNA Registry was established in 2004 [43]. It was taken over by miRBase in 2006, the central online depository of miRNA sequences. Presently, miRBase comprehends over 24,000 miRNA genes reported to have more than 30,000 mature miRNAs along with their comprehensive annotations from over 200 species [44]. Current version contains >1800 human pre-miRNAs and over 2500 mature miRNAs. Apart from miRBase, species-specific databases provide more comprehensive coverage of experimentally validated miRNA sequences, *viz.* PMRD, a repository encompassing miRNAs from plants. Similarly, VIRmiRNA is the first specialized resource encompassing three sub-databases archiving experimentally verified virus-encoded miRNAs, their targets, and antiviral miRNAs (*see* Table 3 for all databases covered in miRNA discovery section).

Table 3
Repositories for miRNA discovery

S. No.	Database	Description	Organism	Source of data	Reference
1	mirBase	Central mirBase repository	All species	E	[190, 191]
2	PMRD	Plant miRNAs	Plants	E	[192]
3	EpimiR	Epigenetic changes and miRNAs	Huamn	E	[45]
4	AvirmiR	Antiviral miRNAs	Virus	E	[10]
5	VIRmiRNA	Viral miRNAs	Virus	E	[10]
6	MirGeneDB	miRNA-miRNA* duplex annotation	All species	E	[48]
7	miRviewer	miRNA genes and candidate homolog identification	Human	E, P	[46]
8	miRBase Tracker	miRNA annotation	All species	E, P	[47]
9	mirPub	Literature searching Web interface	All species	E, P	[54]
10	YM500v2	miRNA quantification, isomiR identification	All species	E, P	[52]
11	CoGemiR	Conservation of microRNAs during evolution in different animal species	Human	E, P	[50]
12	mESAdb	Multivariate analysis of miRNAs	Human, mouse, etc.	E, P	[51]
13	miRNEST	miRNA storage and predictions from deep sequencing data	All species	E, P	[53]
14	Vir-Mir db	Predicted viral miRNA hairpins	Virus	P	[193]
15	miROrtho	Predictions of pre-miRNA genes combining orthology and a support vector machine	All species	P	[49]

Further, repositories holding miRNAs according to certain specific condition or experimental platforms utilized to extract them are also reported (see Table 3 for more details), viz. EpimiR, a comprehensive resource of mutual regulations between epigenetic modifications and miRNAs [45], and AvirmiR, a sub-database of VIRmiRNA encompassing host encoding miRNAs reported to have antiviral effect (antiviral miRNAs) [10].

Some databases deal with miRNA annotations along with visualization tools like miRviewer, Web server providing a visualization of miRNAs from miRBase and their corresponding homologs classified using miRNAminer [46]. Similarly, miRBase Tracker, as the name indicates, updates users regarding changes in miRNA annotation and simplifies their annotation [47]. Correspondingly, MirGeneDB provides comprehensive annotations of the miRNA-miRNA* duplex by imposing an evolutionary hierarchy [48]. However others focus on the evolutionary aspect of miRNA along with their storage like miROrtho, an online service that provides pre-miRNA prediction from numerous animal genomes in combination with orthology [49]. CoGemiR is an online database proposing miRNA conservation during evolution [50]. mESAdb is a database encompassing miRNA expression from multiple taxa [51].

With the advancements in NGS-based techniques, various depositories were developed archiving predicted miRNAs and various other features from NGS-based data like YM500v2—archive for prediction and quantification of miRNA and their isomiRs from small RNA sequencing datasets (sRNA-seq). Cancer miRNAome studies are also provided in its current version [52]. miRNEST is an integrated database harboring experimental predicted miRNAs from various species [53]. Literature search-enhancing depositories in the miRNA field are also available, viz. mirPub, which is a comprehensive database along with search facility to provide researchers publications pertaining to particular miRNA [54].

2.2 miRNA Differential Expression

It is already proven that miRNAs are gene expression regulators; hence their perturbed expression is documented in various diseases. In recent times, differential expression of miRNAs has been reported via various miRNA-profiling methods like RT-PCR, southern blot, and NGS-based techniques. NGS-based methods have become popular due to accuracy and relative fastness. Thus, methods providing insight into *differentially expressed miRNAs (DEmiRs)* will be very helpful to elucidate biomarkers for various diseases. Therefore, we have kept a separate category of resources dealing with storage and analysis of *DEmiRs* (see Table 4 for more details).

Computational resources harboring miRNA expression profiles include mirEX 2.0 [55] and HMED (human miRNA expression database) [56]. mirEX and HMED are comprehensive programs

Table 4
Computational resources catering the miRNA differential expression data

S. No.	Database	Description	Organism	Source of data	Reference	Year
1	bloodmiRs	Cell-specific microRNA catalog of human peripheral blood	Human	E	[57]	2017
2	mirEX 2.0	Comparative analysis of pri-miRNA expression data	Plants	E	[55]	2015
3	PmiRExAt	Plant miRNA expression in multiple tissues	Plants	E	[61]	2016
4	ExcellmiRDB	miRNA expression levels in biofluids	Human	E	[58]	2015
5	miRandola	Extracellular circulating miRNAs	Human	E	[59]	2012
6	miREnvironment	miRNA and environmental factors playing a role in diseases	Human	E/P	[60]	2011
7	HMED	Human miRNA expression	Human	E	[56]	2014

for analysis of miRNA expression datasets. Further, many databases store the DEmiRs from different conditions; for example bloodmiRs encompasses cell-specific miRNAs from human peripheral blood [57]. Similarly, ExcellmiRDB is the first specialized database providing curated knowledge regarding DEmiRs in biofluids [58]. Correspondingly, miRandola provides comprehensive knowledge about various extracellular circulating miRNAs [59].

Environmental factors also play a vital role along with miRNAs during any disease. To store experimentally supported data showing associations of environmental factor, miRNAs, and their related phenotypes, a catalog namely miREnvironment was developed [60]. Some resources are specialized to store the DE miRNAs from plant tissues; PmiRExAt is a repository with an inclusive view of plant miRNA expression profiles from several tissues, in developing stages. They have also performed further examination of expression patterns of miRNAs from the freely accessible small RNA datasets [61]. HPVbase comprehensively covers expression profiles of differentially regulated miRNAs implicated in HPV infection [62].

2.3 miRNA: Targets

The discovery of abundant number of miRNAs encoded in various organisms raised an important point regarding their function. To answer this, cloning and computational methods were developed to find their targets. Hence, growing number of miRNAs and their targets from various organisms suggested the need of both types of repositories in this direction. Repositories encompassing predicted miRNA targets are predicted by exploiting features from miRNA: gene sequence like evolutionary conservation of miRNAs, degree of complementarity between miRNA and target gene sequence,

Gibbs free energy to access the level of binding between the two, target-site accessibility and abundance for miRNA, nucleotide composition (AU/GU/GC), number of base pairs and bulges in seed sequence, etc. We have compiled a list of archives harboring either experientially validated or predicted miRNA targets or both (*see* Table 5).

Table 5
Bioinformatics tools and databases encompassing targets of miRNAs

S. No.	Database	Description	Organism	Source of data	Reference	Year
1	TarBase	Catalog of miRNA targets	All species	E	[63]	2012
2	miRTarBase	miRNA-target interactions	Human, mouse, viruses	E	[64, 65]	2016
3	miRGate	Human, mouse, and rat miRNAs/mRNA targets	Human, rat, mouse	E	[66]	2015
4	VIRmirTar	Targets for viral miRNAs	Virus	E	[10]	2014
5	MtiBase	miRNA-binding sites by CLIP-seq	All species	E	[82]	2015
6	miRdSNP	Manually curated dSNPs on the 3' UTRs of human genes	Human	E	[79]	2012
7	MirSNP	Human SNPs in predicted miRNA-mRNA-binding sites	Human	E	[194]	2015
8	PNRD	11 different types of ncRNAs from 150 plant species	Plants	E	[67]	2015
9	PolymiRTS Database	Functional impact of genetic polymorphisms in miRNA seed regions and target sites	Human, mouse	E	[80, 81]	2014
10	TargetScanS	microRNA targets conserved in five vertebrates	Vertebrates	E	[195]	
11	VIRmiRNA	Experimentally proven virus-encoded miRNAs and their associated targets	Virus	E	[10]	2014
12	CSmiRTar	Condition-specific microRNA targets	Human, mouse	E	[78]	2017
13	miRecords	Animal miRNA-target interactions	Human, rat, mouse, fly, worm, chicken	E/P	[68]	2009
14	miRNA—Target Gene Prediction at EMBL	Access to the 2003 and 2005 miRNA-target predictions for Drosophila miRNAs	Fly	E/P	[196]	2005

(continued)

Table 5
(continued)

S. No.	Database	Description	Organism	Source of data	Reference	Year
15	miRSel	Automated extraction of associations between miRNAs and genes	Human, mouse	E/P	[84]	2010
16	miRSystem	A database which integrates seven well-known miRNA target gene prediction programs	Human, mouse	E/P	[70]	2012
17	miRWalk	Predicted as well as the experimentally validated miRNA-target interactions	Human, mouse	E/P	[69]	2011
18	targetHub	Programmer-friendly interface to access miRNA targets	Human	E/P	[197]	2013
19	miRPathDB	miRNAs, different miRNA target sets, and functional biochemical categories	Human, mouse	E/P	[198]	2017
20	multiMiR	A miRNA-target interaction R package	Human, mouse	E/P	[77]	2014
21	DIANA-microT Web server v5.0	miRNA functional analysis workflows	Human	E/P	[74]	2013
22	HOCTARdb	Predicted target genes for 290 intragenic miRNAs annotated in human	Human	P	[83]	2011
23	ViTa	Prediction of host microRNA targets on viruses	Virus	P	[72]	2007
24	miRTar	Resource for miRNA target interactions	Human	P	[73]	2011
25	DIANA-microT-CDS	Web server for miRNA:target interactions	All species	P	[74]	2013
26	MicroCosm Targets	microRNAs across many species	Animals	P	[199]	
27	microPIR2	Database containing over 80 and 40 million predicted microRNA target sites located within human and mouse promoter sequences	Human, mouse	P	[71]	2014
28	miRDB	2.1 million predicted gene targets regulated by 6709 miRNAs	Human, rat, mouse, dog, chicken	P	[75, 76]	2015
29	ViTa	Predicted host miRNA targets for viral miRNAs	Human, mouse	P	[72]	2007

We will discuss the depositories harboring information about experimentally validated miRNA targets followed by predicted ones. TarBase is the main representative catalog encompassing experimentally validated miRNA targets from various species [63]. miRTarBase deals with miRNA-target interactions [64, 65]. miRGate is a manually curated database to examine miRNA and gene targets along with isoforms from 3' UTRs of human, mouse, and rat [66]. TargetScanS provides miRNA targets conserved in five vertebrates. PNRD is a specialized resource for miR:targets in plant species along with other ncRNAs. Presently it encompasses more than 25,000 records of approximately 10 kinds of ncRNAs from more than 150 plant species [67]. VIRmirTar is a sub-repository under VIRmiRNA integrated resource providing extensive information about cellular as well as viral targets of miRNAs encoded by viruses [10].

Some well-known archives hold both experimental and computationally predicted data. miRecords is an all-inclusive source for miRNA-target interactions. It has two sub-database constituents. The former one consists of manually curated, experimentally validated miRNA targets collected from comprehensive literature search, while the latter one encompasses computationally predicted targets. Eleven well-known prediction programs govern these predictions [68]. Similarly, miRWalk is a freely accessible all-inclusive database providing experimentally testified and predicted miRNA-target interaction pairs [69]. miRSystem incorporates predictions from seven well-known miRNA target gene prediction algorithms: DIANA, miRanda, miRBridge, PicTar, PITA, rna22, and TargetScan [70].

There are in silico predictions of miRNA-binding sites in the promoter sequences of representative sequences; for example microPIR2 is an archive comprehending computationally identified miRNA target sites in promoter sequences of human and mouse genomes. It also contains investigational information pertaining to predicted targets [71]. Similarly, ViTa hosts experimentally established host miRNA targets on virus from miRBase [72].

On the other hand, depositories holding predicted miR:target interactions are of two types either holding data from multiple species or species specific; for example miRTar is a store for miRNA-target interactions [73], and DIANA-microT-CDS is an online platform for miRNA:target interactions [74]. MicroCosm provides computationally predicted miRNA targets from numerous taxa. mirDB comprises more than two million predicted gene targets regulated by >6500 miRNAs [75, 76] (*see* Table 5). *miRNA—Target Gene Prediction at EMBL* is an archive providing information about miRNA-target predictions for fruit fly.

Some computational resources provide target prediction by utilizing available algorithms along with other associated functionalities; for example multiMiR is a catalog-cum-miRNA-target R package. This resource delivers collection of approximately 50 million entries from human and mouse from numerous databases [77]. Similarly, CSmiRTar, namely condition-specific miRNA target database, provides predictions based upon different conditions [78].

It is well known that SNPs in the miRNA-binding sites of genes resulted in the changed miRNA target repertoire. Hence, to deal with the available biological information in this context MirSNP, miRdSNP, and PolymiRTS were developed. MirSNP is an assembly of predicted SNPs in human miRNAs. Similarly, miRdSNP is a comprehensive repository of SNPs on the 3′ UTRs of human genes [79]. PolymiRTS is an integrated database-cum-Web service for investigating the effect of SNPs in miRNA seed sequence. It enables users to explore the associations of SNPs with gene expression traits, pathways, and physiological and disease phenotypes [80, 81].

Some archives hold miRNA targets from NGS datasets, e.g., MtiBase Database harboring miRNA-binding sites by CLIP-seq [82]. HOCTARdb is an integrated database providing predicted target genes from transcriptomic data. Currently, it contains information for 290 intragenic miRNAs annotated in human [83]. Further, some resources are developed based on automated extraction, *viz.* miRSel. It is based upon automated extraction of miR targets from literature on a daily basis [84]. DIANA-microT Web server provides miRNA functional analysis workflows [74].

2.4 miRNA: Disease Associations

Further, archives holding miRNAs according to disease-specific conditions or experimental platforms utilized to extract them are also reported (Table 6). Some instances are HMED that is an archive incorporating human miRNA expression [56]. EpimiR is a wide-ranging source of correlations between epigenetic modifications and miRNAs. CoGemiR is an online platform offering conservation of miRNAs during evolution in various animal taxa.

Some computational resources focus on a specialized category of disease/s; for example dbDEMC, miRCancer, and OncomiRDB are catalogs of DEmiRs from various types of cancers [85, 86]. PhenomiR is an online service providing information about DEmiRs in various disorders and other biological pathways [87]. PMTED is developed to analyze expression profiles of miRNA and their associated targets from published plant microarray studies [88]. Further, miRStress is a catalog hosting miRNAs associated with stress conditions [89]. Dietary microRNA Database (DMD) holds published miRNA testified in human food [90]. PASmiR was designed to develop a program for collection, standardization, and searching of these miRNA-stress regulation data in plants. EpimiRBase is a user-friendly archive providing information on miRNAs implicated in epilepsy [91]. Similarly, AVIRmiR, a sub-database of VIRmiRNA [10], encompasses antiviral miRNAs during viral infections.

Table 6
Tools dedicated for miRNA disease associations

S. No.	Database	Description	Organism	Source of data	Reference	Year
1	dbDEMC	Differentially expressed miRNAs in human cancers	Human	E	[85, 200]	2010, 2017
2	miRCancer	miRNA expression profiles in various human cancers	Human	E	[201]	2013
3	EpimiRBase	miRNAs in epilepsy	Human	E	[91]	2016
4	miRStress	miRNA deregulation implicated in various stress stimuli	Human, mouse	E	[89]	2013
5	DIANA miRPath v.2.0	Combinatorial effect of microRNAs in pathways	Human	E/P	[202, 203]	2015
6	HMDD	Human MicroRNA and Disease Database	Human	E	[204]	2014
7	OncomiRDB	Oncogenic and tumor-suppressive miRNAs	Human	E	[86]	2014

2.5 Non-coding RNA Interactions

There are some published repositories providing information on miRNA and other ncRNA interactions. We have compiled a list of some representative archives here; for example DIANA-LncBase is the first broad catalog encompassing miRNA and lncRNA interactions. They have also provided experimentally testified and computationally predicted miRNA response elements (MREs) on lncRNAs [92]. AntagomirBase, as name indicates, delivers information on antagomirs and miRNA heterodimers along with the tool to design the putative antagomirs [93]. CircuitsDB2 is a comprehensive platform linking miRNAs with other regulatory modules like other ncRNAs, genes, and transcription factors [94].

2.6 All-in-One Resources

There are several computational resources, which have hosted more than one aspect of miRNA-based research. We have enlisted these kinds of reserves in all-in-one resources category in Table 7. These resources are discussed in detail below. For ease of comprehension, these have been further subcategorized based upon several features like organism specificity, experimental platform employed, and disease specificity.

The microRNA.org is an all-inclusive catalog for predicted targets of miRNA and corresponding available expression profile studies. In silico determination of miRNA targets involves use of miRanda algorithm, which is based upon known biological rules [95]. miRNAMap harbors verified miRNAs and target genes from PubMed in metazoan organisms. Apart from experimental miRNA:

Table 7
All-in-one resources

S. No.	Database	Description	Organism	Source of data	Reference	Year
1	MicroRNA.org	miRNA target predictions	Human, mouse	P	[95]	2008
2	miRNAMap	Experimentally verified miRNAs and target genes	Human, rat, mouse, fly, worm, chicken	E/P	[96, 205]	2008
3	MtiBase	CDS/5′ UTR-located miRNA-binding sites	Human, mouse	E	[82]	2015
4	PMTED	Expression profiles of miRNA targets from microarray studies	Plants	E	[88]	2013
5	SomamiR DB 2.0	Cancer somatic mutations in miRNAs and their target sites	Human	E	[99]	2016
6	miRGator	miRNA diversity, expression profiles, target relationships	Human	E/P	[100, 101]	2011
7	DIANA-miRGen v3.0	Cell-line-specific miRNA gene transcription start sites (TSSs)	Human, rat, mouse	E	[104]	2016
8	PASmiR	A literature-curated and Web-accessible database	Plants	E	[206]	2013
9	PhenomiR	Differentially regulated miRNA expression in diseases	Human	E	[207]	2012
10	DIANA-TarBase	Experimentally validated miRNA: gene interactions	Animals, plants, virus	E	[105]	2016
11	mimiRNA	miRNA expression profile data across different tissues and cell lines	Human	E/P	[97]	2010
12	miR2Disease	miRNA deregulation in various human diseases	Human	E/P	[98]	2009
13	starBase	Pan-Cancer and Interaction Networks of lncRNAs, miRNAs, competing endogenous RNAs (ceRNAs), and mRNAs from large-scale CLIP-Seq data	Human, mouse	E	[102, 103]	2014

target hosting, computational prediction of miR:targets is achieved employing three software, *viz.* miRanda, RNAhybrid, and TargetScan. Furthermore, miRNA expression profiles are provided to shed valued evidences on the DEmiRs w.r.t tissue specificity [96]. mimiRNA is a reserve of miRNA profiling studies along with algorithm, namely ExParser. It is used for clustering identical miRNA and mRNA pairs from different articles into one group [97].

There are some resources, encompassing miRNA and gene interactions from high-throughput platforms. MtiBase is one instance in this category. It harbors targets on the basis of miRNA-binding sites on CDS and 5′ UTR from various experimental platforms like cross-linked immunoprecipitation (CLIP). Thus, the organized information helps to widen the miRNA regulatory effects on mRNA stability and translation [82].

Computational resources from all-in-one category also include disease-specific data, e.g., miR2Disease, which is a comprehensive catalog delivering manually curated information on miRNA alterations implicated in numerous disorders [98]. Similarly, SomamiRDB is a collection of somatic and germinal alterations in miRNAs [99]. miRGator is a database comprehending miRNA: target profiling data along with analysis tools on miRNA diversity [100, 101]. Further, starBase is intended for deciphering complex interaction networks of various ncRNAs like lncRNAs, miRNAs, ceRNAs with RNA-binding proteins (RBPs), and mRNAs from NGS data of cancer patients [102, 103].

All-in-one category has some reserves from DIANA-suite too. DIANA-miRGen furnishes manually curated cell-line-specific miRNAs, their transcription start sites (TSSs), and information on transcriptional factors (TF) as their targets. It furnishes comprehensive experimental data on more than 270 miRNA TSSs resulting in ~430 precursors and >19M binding sites of ~200 TFs reported in studies involving more than eight cell lines and six tissues from human and mouse origin [104]. DIANA-TarBase delivers thousands of manually curated experimentally certified miRNA: gene interactions along with details of reported study, *viz.* experimental technique, conditions, and cell line/tissue employed [105]. PMTED is aimed to obtain and investigate plant miRNA: target profiles collected from published microarray data along with gene ontology [88].

2.7 How to Use a miRNA-Centered Database

miRNA-centered depositories are well structured into exploratory options, *viz.* search, browse, and other analysis tools, to allow their users to fetch useful biological information from them. Therefore, we have provided a generalized overview on the usage of these databases by taking an illustrative example of *hhv6b-miR-ro6-4* encoded by human herpes virus 6 (HHV-6) from VIRmiRNA (*see* Fig. 3).

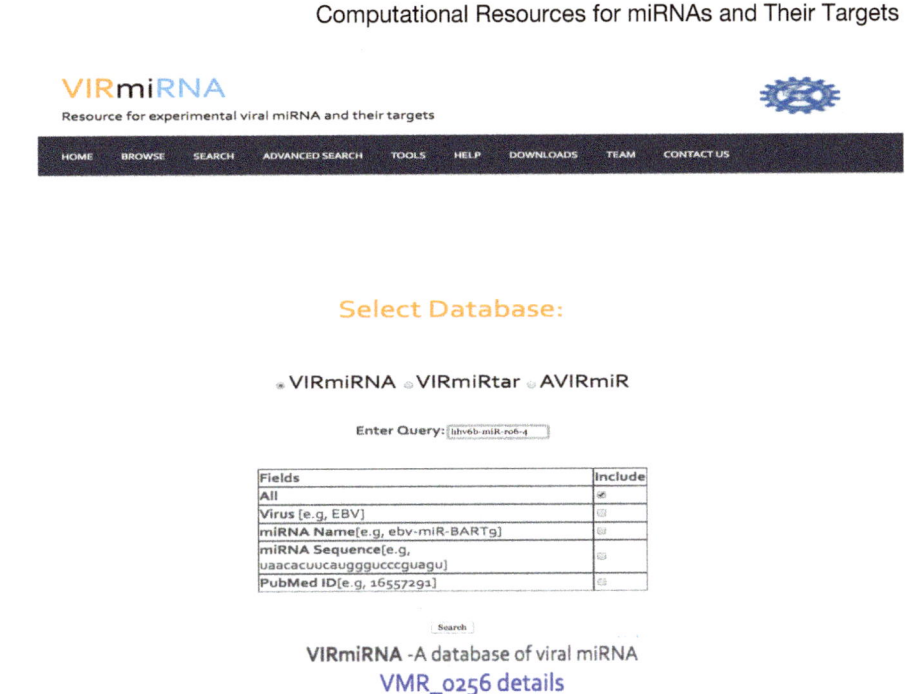

Fig. 3 A screenshot of the miRNA-centered resource VIRmiRNA to illustrate how to use a computational resource

3 Tools/Algorithms

3.1 miRNA Computational Identification

There are some tools for computational identification of miRNAs from genomes, *viz.* miRdentify [106], MIReNA [107], miREval [108], miRPlant [109], DIANA-mirExTra 2.0 [110], etc.; details of individual tool are provided below.

MIReNA identifies miRNA sequences by using five parameters specific and known for true pre-miRNAs. It can detect novel miR-NAs by homology-based approach. Further, miREval 2.0 identifies novel miRNAs from user-provided sequence using numerous bioinformatics methods. It is specialized for identification in mammals, plants, and viruses [108]. miRPlant performs similarly to identify novel plant miRNA [109].

DIANA-mirExTra 2.0 performs the sequence-based identification of miRNA-enriched motifs between user-provided sets of genes [110]. miRQuest is a comprehensible algorithm designed for pre-miRNA prediction. It also offers benchmarking facility to assess which tool is performing better on user-provided query [111].

3.1.1 Precursor miRNA Prediction

While precursor miRNA prediction involves finding true miRNA-coding hairpins from the genome, mature miRNA is determined from within pre-miRNA sequences on the basis of location information about mature sequence [42, 112]. Various computational approaches have been published over the past few years to predict precursor (*see* Table 8) and mature miRNAs (*see* Table 9). Thus, computational tools developed for identification of miRNAs and their targets incorporating new feature or methods represent an

Table 8
List of available algorithms for prediction of precursor miRNA sequences along with details of the features incorporated in their predictive model

S. No.	Name	Classifier used	Reference
1	Triplet-SVM	SVM	[208]
2	microPred	SVM	[135]
3	MiPred	RF	[127]
4	miPred	SVM	[209]
5	miR-BAG	SVM, BF tree, and naive Bayes	[128]
6	ViralmiR	SVM, RF	[210]
7	MiRFinder	SVM	[124]
8	PMirP	SVM	[211]

Table 9
List of available algorithms for the prediction of mature miRNAs

S. No.	Name	Classifier used	Reference
1	MaturePred	SVM	[141]
2	MatureBayes	Naive Bayes	[138]
3	MiRFinder	SVM	[124]
4	miRDup	Random forest	[140]
5	miRLocator	SVM	[42]
6	MatPred	SVM	[112]
7	miRanalyzer	RF	[139]
8	MiRmat	RF	[142]

opportunity to predict novel miRNA candidates including nonclassical ones.

The reserves in the area of miRNAs would be helpful in expanding our current understanding toward the role of miRNAs in disease and using it as a start point of therapeutic development has the potential to generate a whole new class of drugs. Further, novel miRNAs identified by computational prediction in viruses will not only help in the development of miRNA-based therapeutics, but also be useful for translating that bench discovery into a therapeutically useful modality [113].

In earlier tools, performing global alignment of miRNA genes among various species discovered miRNA features. These studies led to the inference that miRNA genes are usually conserved and exhibit secondary hairpin structure. Therefore, numerous computational approaches utilizing homology-based miRNA identification came into existence, *viz.* srnaloop [114], MiRscan [115, 116], miRseeker [117], and RNAmicro [118]. These methods are dependent on the conserved miRNA features, *viz.* secondary structure and conservedness adopted by miRNA structures. These tools firstly search for conserved regions in the genome, which can form hairpin structure. However, these methods are not able to identify novel miRNAs, which do not exhibit the miRNA-enriched features from above-mentioned studies.

To improve this, some tools were developed which take evolutionary conservedness as secondary structure and sequence properties as input for developing algorithm [119]. ProMiR II and MiRRim are the examples of this category. ProMiR II was developed on probabilistic model, where features generated from comparative methods started as input to machine learning techniques (MLTs). It incorporates features like Gibbs free energy data, entropy, G/C content, and conservation score apart from the candidate miRNA sequence from mouse and human genomes. Further, this method was also applicable for miRNA identification in viral genomes. MiRRim has also exploited similar features as an input using hidden Markov models (HMMs) [119].

Further, machine learning-based predictors were developed using various features, such as minimum free energy (MFE), evolutionary preservedness, and length of various sections of miRNA sequence like loop, stem, and base pairs [120, 121]. Xue et al. have developed a tool for the classification of true pre-miRNAs from pseudo hairpins using local structure-sequence features and support vector machine (SVM). It was the first ab initio method for distinguishing pre-miRNAs from sequence segments with pre-miRNA-like hairpin structures. miRabela has used similar approach with improved features [115, 116]. ProMir has incorporated a probabilistic co-learning method; ProMiR II has utilized naive Bayes classifier [122].

BayesMiRNA finder predicts mature-miRNA position in pre-miRNA using naive Bayes classifier [123]. Similarly, MiRFinder is a software for genome-wide pre-miRNA precursor identification [124]. PMirP is a prediction tool in this area based on structure-sequence hybrid features [125]. Couplet Syntax has developed a new notation to describe local continuous structure-sequence information for recognizing new pre-miRNAs [122]. MiRPara is meant for prediction of putative miRNA-coding regions in genome-scale sequences [126]. Similarly, miR-BAG is developed using Bagging-based method to create an ensemble of complementary trees. It encompasses six unalike species, *viz.* human, mouse, rat, dog, nematode, and fruit fly [127]. Recently, a Web server is published, *viz.* iMiRNA-SSF, to improve the identification of miRNA precursors using negative dataset with distinctive statistical distributions [124].

Some Web servers use heterogeneous approach to computationally identify miRNAs. HeteroMirPred is one example. It has utilized discriminative features and modified SMOTE bagging for pre-miRNA classification [128]. Similarly, iMiRNA-PseDPC is developed with a pseudo distance-pair composition approach [129]. miRNAFold is an ab initio method for predicting pre-miRNAs in genomes [130]. miRBoost utilizes support vector machines for pre-miRNA classification [131]. omiRas predicts DEmiRs from sRNA-Seq experimental studies [132].

Further, some tools are dedicated for computational prediction of miRNAs in plants like PlantMiRNAPred [133]. Similarly, tools for viral miRNA prediction are VMiR [134], ViralmiR, etc. However, some are dedicated for more than one species, *viz.* microPred [135], CSHMM [136], miR-KDE [137], etc. Apart from SVM, various other MLTs have been utilized to develop algorithms, *viz.* miPred and HHMMiR.

3.1.2 Mature miRNA Prediction

MatPred involves identification of mature miRNAs within novel pre-miRNAs [112]; Mature Bayes integrates a naive Bayes classifier to identify mature miRNA sequences [138]. Further, miRanalyzer [139], miRDup [140], and MiRFinder are similar examples of this category. MaturePred performs efficient identification of mature miRNAs in plants [141]. Furthermore, miRLocator is a machine learning algorithm utilizing 440 miRNA-based features from miRNA duplexes [124]. MiRmat is a tool to predict the mature miRNA sequence. It is mainly comprised of two segments: the estimation of sites involving Drosha Dicer processing cleavage [142].

3.2 isomiR Identification

This section describes computational resources meant for identification, annotation, and analysis of isomiRs. IsomiRage provides detailed depiction of miRNAs and their isomiRs from NGS data [143]. isomiReX is a freely available tool to detect and visualize isomiRs [144]. DeAnnIso and miRge are designed for identification of isomiR from small RNA sequencing data [145, 146]. MIRPIPE is a pipeline for the quantification of miRNA based on small RNA sequencing reads [147]. mirPRo is a tool for identification of known as well as novel miRNAs from RNA-seq data [148].

Further, isomiR2Function performs high-throughput identification of plant isomiRs from any miRNA-seq profiling study [149]. miRSeqNovel is an integrated pipeline for identification of novel miRNAs from NGS-based data [150]. sRNAtoolbox is developed to provide numerous other tools including small RNA profiling from NGS-based experiments [151].

3.3 Differential Expression

This section covers resources encompassing information about differential expression of miRNAs, *viz.* CAP-miRSeq [152], MAGI [153], miRNAkey [154], mirnaTA [155], miRNet [156], miRSeqNovel [150], MMIA [157], MTide [158], Oasis 2.0 [159], and Omics Pipe [160]. Some cover both plants and animals, *viz.* UEA sRNA Workbench [161]. shortran [162], and wapRNA [163]. For plants, there are some tools like plant-DARIO [164].

3.4 miRNA Sequencing Tools

This section harbors computational resources having NGS-based data. General-purpose tools, which cover more than one type of organism, are BioVLAB-MMIA-NGS [165]. CPSS entails data about human, mouse, rat, chicken, and fish [166]. Some Web servers are exclusive for animals like eRNA [167], iMir [168], MIREAP [169], miRge [146], miRDeep2 [170], miRExpress [171], miRNAkey [154], MIRPIPE [147], mirPRo [148], miRSeqNovel [150], and ncPRO-seq [172].

Some are plant-centric resources, *viz.* miR-PREFeR [173], miRDeep-P [174], and PIPmir [175]. miR-PREFeR provides miRNA annotation by employing expression profiles [173]. miRDeep-P predicts miRNA genes in plants [174]. MTide identifies miRNA and their cognate targets based on RNA-seq datasets [158]. Further, for novel miRNA gene prediction, an ab initio algorithm is developed, namely PIPmir [175].

3.5 miR-SNP Analysis

Since the advent of NGS techniques, sequence variation in miRNA genes and their impact on regulatory network have been studied a lot. Therefore, tools have been published in this arena. We have covered these tools in miR-SNP analysis section. These are miRVaS [176], MicroSNiPer [177], miRNA-SniPer [178], MirSNP [179], and submiRine [180]. miRVaS and MicroSNiPer take into account the sequence variation in miRNA hairpin and predict its functional outcome [176, 177].

miRNA-SniPer is exclusively used for identification of miRNA SNPs in vertebrates [178]. Similarly, MirSNP is a repertoire of predicted SNPs in the binding sites of miRNA from human [179]. Similarly, submiRine provides information about SNPs from clinical data [180].

3.6 miRNA Functional Annotation

There are some resources for functional annotation of miRNAs, viz. CHNmiRD [181], ActMiR [182], etc. CHNmiRDis developed for identification of miRNA and disease interactions utilizing complex heterogeneous network (CHN) [181]. Similarly, ActMiR assigns miRNA function and activity using miRNA expression level studies [182]. miR2GO analyzes miRNA functions in a comparative way using two main components, miRmut2GO and miRpair2GO, respectively [183].

3.7 Others

miRiadne is an algorithm for re-annotating miRNAs according to their standard and updated nomenclature [184]. MirCompare is an advanced method for miRNA comparison from different classifications. It offers comparative assessment based on sequential comparing filtering, and also provides phylogenetic investigation of provided miRNAs [185].

miRprimer designs primers for the purpose of miRNA PCR amplification [186]. miRseqViewer provides visualization of the user-provided sequences in the form of alignment [187]. Some tools are dedicated for identifying the modifications in miRNAs; miTRATA is one of them. It predicts 3′ modifications in miRNAs [188]. miRTex is a server based on text mining, which mines miRNA, gene, and target gene associations from literature. Currently, it harbors more than 12,000 PubMed indexed articles [189].

4 Conclusion

In conclusion, we have summarized and evaluated tools and databases published and utilized in miRNA research field, to help scientific community in choosing the appropriate tools according to their research needs along with instructions to use a specific resource. We have also briefly discussed current therapeutic advancements in miRNA field with an emphasis on upcoming challenges for computational biology for identifying novel miRNA and targets.

References

1. Castel SE, Martienssen RA (2013) RNA interference in the nucleus: roles for small RNAs in transcription, epigenetics and beyond. Nat Rev Genet 14(2):100–112. https://doi.org/10.1038/nrg3355
2. Lee RC, Feinbaum RL, Ambros V (1993) The C. elegans heterochronic gene lin-4 encodes small RNAs with antisense complementarity to lin-14. Cell 75(5):843–854
3. Pasquinelli AE, Reinhart BJ, Slack F, Martindale MQ, Kuroda MI, Maller B, Hayward DC, Ball EE, Degnan B, Muller P, Spring J, Srinivasan A, Fishman M, Finnerty J, Corbo J, Levine M, Leahy P, Davidson E, Ruvkun G (2000) Conservation of the sequence and temporal expression of let-7 heterochronic regulatory RNA. Nature 408(6808):86–89. https://doi.org/10.1038/35040556
4. Lagos-Quintana M, Rauhut R, Lendeckel W, Tuschl T (2001) Identification of novel genes coding for small expressed RNAs. Science 294(5543):853–858. https://doi.org/10.1126/science.1064921
5. Lee RC, Ambros V (2001) An extensive class of small RNAs in Caenorhabditis elegans. Science 294(5543):862–864. https://doi.org/10.1126/science.1065329
6. Lau NC, Lim LP, Weinstein EG, Bartel DP (2001) An abundant class of tiny RNAs with probable regulatory roles in Caenorhabditis elegans. Science 294(5543):858–862. https://doi.org/10.1126/science.1065062
7. Pfeffer S, Zavolan M, Grasser FA, Chien M, Russo JJ, Ju J, John B, Enright AJ, Marks D, Sander C, Tuschl T (2004) Identification of virus-encoded microRNAs. Science 304(5671):734–736. https://doi.org/10.1126/science.1096781
8. Flores O, Nakayama S, Whisnant AW, Javanbakht H, Cullen BR, Bloom DC (2013) Mutational inactivation of herpes simplex virus 1 microRNAs identifies viral mRNA targets and reveals phenotypic effects in culture. J Virol 87(12):6589–6603. https://doi.org/10.1128/jvi.00504-13
9. Glazov EA, Horwood PF, Assavalapsakul W, Kongsuwan K, Mitchell RW, Mitter N, Mahony TJ (2010) Characterization of microRNAs encoded by the bovine herpesvirus 1 genome. J Gen Virol 91(Pt 1):32–41. https://doi.org/10.1099/vir.0.014290-0
10. Qureshi A, Thakur N, Monga I, Thakur A, Kumar M (2014) VIRmiRNA: a comprehensive resource for experimentally validated viral miRNAs and their targets. Database (Oxford) 2014:bau103. https://doi.org/10.1093/database/bau103
11. Bartel DP (2009) MicroRNAs: target recognition and regulatory functions. Cell 136(2):215–233. https://doi.org/10.1016/j.cell.2009.01.002
12. Lim LP, Lau NC, Garrett-Engele P, Grimson A, Schelter JM, Castle J, Bartel DP, Linsley PS, Johnson JM (2005) Microarray analysis shows that some microRNAs downregulate large numbers of target mRNAs. Nature 433(7027):769–773. https://doi.org/10.1038/nature03315
13. Peter ME (2010) Targeting of mRNAs by multiple miRNAs: the next step. Oncogene 29(15):2161–2164. https://doi.org/10.1038/onc.2010.59
14. Shalgi R, Pilpel Y, Oren M (2010) Repression of transposable-elements – a microRNA anti-cancer defense mechanism? Trends Genet 26(6):253–259. https://doi.org/10.1016/j.tig.2010.03.006
15. Denli AM, Tops BB, Plasterk RH, Ketting RF, Hannon GJ (2004) Processing of primary microRNAs by the microprocessor complex. Nature 432(7014):231–235. https://doi.org/10.1038/nature03049
16. Zhang R, Wang YQ, Su B (2008) Molecular evolution of a primate-specific microRNA family. Mol Biol Evol 25(7):1493–1502. https://doi.org/10.1093/molbev/msn094
17. Pfeffer S, Sewer A, Lagos-Quintana M, Sheridan R, Sander C, Grasser FA, van Dyk LF, Ho CK, Shuman S, Chien M, Russo JJ, Ju J, Randall G, Lindenbach BD, Rice CM, Simon V, Ho DD, Zavolan M, Tuschl T (2005) Identification of microRNAs of the herpesvirus family. Nat Methods 2(4):269–276. https://doi.org/10.1038/nmeth746

18. Yang Q, Li L, Xue Z, Ye Q, Zhang L, Li S, Liu Y (2013) Transcription of the major neurospora crassa microRNA-like small RNAs relies on RNA polymerase III. PLoS Genet 9(1): e1003227. https://doi.org/10.1371/journal.pgen.1003227

19. Jayaraj GG, Nahar S, Maiti S (2015) Nonconventional chemical inhibitors of microRNA: therapeutic scope. Chem Commun (Camb) 51(5):820–831. https://doi.org/10.1039/c4cc04514a

20. Bartel DP (2004) MicroRNAs: genomics, biogenesis, mechanism, and function. Cell 116(2):281–297

21. Kim VN, Nam JW (2006) Genomics of microRNA. Trends Genet 22(3):165–173. https://doi.org/10.1016/j.tig.2006.01.003

22. Londin E, Loher P, Telonis AG, Quann K, Clark P, Jing Y, Hatzimichael E, Kirino Y, Honda S, Lally M, Ramratnam B, Comstock CE, Knudsen KE, Gomella L, Spaeth GL, Hark L, Katz LJ, Witkiewicz A, Rostami A, Jimenez SA, Hollingsworth MA, Yeh JJ, Shaw CA, McKenzie SE, Bray P, Nelson PT, Zupo S, Van Roosbroeck K, Keating MJ, Calin GA, Yeo C, Jimbo M, Cozzitorto J, Brody JR, Delgrosso K, Mattick JS, Fortina P, Rigoutsos I (2015) Analysis of 13 cell types reveals evidence for the expression of numerous novel primate- and tissue-specific microRNAs. Proc Natl Acad Sci U S A 112(10):E1106–E1115. https://doi.org/10.1073/pnas.1420955112

23. Baskerville S, Bartel DP (2005) Microarray profiling of microRNAs reveals frequent coexpression with neighboring miRNAs and host genes. RNA 11(3):241–247. https://doi.org/10.1261/rna.7240905

24. Isik M, Korswagen HC, Berezikov E (2010) Expression patterns of intronic microRNAs in Caenorhabditis elegans. Silence 1(1):5. https://doi.org/10.1186/1758-907X-1-5

25. Ramalingam P, Palanichamy JK, Singh A, Das P, Bhagat M, Kassab MA, Sinha S, Chattopadhyay P (2014) Biogenesis of intronic miRNAs located in clusters by independent transcription and alternative splicing. RNA 20(1):76–87. https://doi.org/10.1261/rna.041814.113

26. Mah SM, Buske C, Humphries RK, Kuchenbauer F (2010) miRNA*: a passenger stranded in RNA-induced silencing complex? Crit Rev Eukaryot Gene Expr 20(2):141–148

27. Yang JS, Phillips MD, Betel D, Mu P, Ventura A, Siepel AC, Chen KC, Lai EC (2011) Widespread regulatory activity of vertebrate microRNA* species. RNA 17 (2):312–326. https://doi.org/10.1261/rna.2537911

28. Okamura K, Phillips MD, Tyler DM, Duan H, Chou YT, Lai EC (2008) The regulatory activity of microRNA* species has substantial influence on microRNA and 3' UTR evolution. Nat Struct Mol Biol 15 (4):354–363. https://doi.org/10.1038/nsmb.1409

29. Guo L, Lu Z (2010) The fate of miRNA* strand through evolutionary analysis: implication for degradation as merely carrier strand or potential regulatory molecule? PLoS One 5 (6):e11387. https://doi.org/10.1371/journal.pone.0011387

30. Ogata A, Furukawa C, Sakurai K, Iba H, Kitade Y, Ueno Y (2010) Biaryl modification of the 5'-terminus of one strand of a microRNA duplex induces strand specificity. Bioorg Med Chem Lett 20(24):7299–7302. https://doi.org/10.1016/j.bmcl.2010.10.077

31. Hwang HW, Wentzel EA, Mendell JT (2007) A hexanucleotide element directs microRNA nuclear import. Science 315(5808):97–100. https://doi.org/10.1126/science.1136235

32. Marcon E, Babak T, Chua G, Hughes T, Moens PB (2008) miRNA and piRNA localization in the male mammalian meiotic nucleus. Chromosom Res 16(2):243–260. https://doi.org/10.1007/s10577-007-1190-6

33. Kim DH, Saetrom P, Snove O Jr, Rossi JJ (2008) MicroRNA-directed transcriptional gene silencing in mammalian cells. Proc Natl Acad Sci U S A 105(42):16230–16235. https://doi.org/10.1073/pnas.0808830105

34. Salmanidis M, Pillman K, Goodall G, Bracken C (2014) Direct transcriptional regulation by nuclear microRNAs. Int J Biochem Cell Biol 54:304–311. https://doi.org/10.1016/j.biocel.2014.03.010

35. van Rooij E, Purcell AL, Levin AA (2012) Developing microRNA therapeutics. Circ Res 110(3):496–507. https://doi.org/10.1161/circresaha.111.247916

36. Stenvang J, Kauppinen S (2008) MicroRNAs as targets for antisense-based therapeutics. Expert Opin Biol Ther 8(1):59–81. https://doi.org/10.1517/14712598.8.1.59

37. Nishimura M, Jung EJ, Shah MY, Lu C, Spizzo R, Shimizu M, Han HD, Ivan C, Rossi S, Zhang X, Nicoloso MS, Wu SY, Almeida MI, Bottsford-Miller J, Pecot CV, Zand B, Matsuo K, Shahzad MM, Jennings NB, Rodriguez-Aguayo C, Lopez-Berestein G, Sood AK, Calin GA (2013) Therapeutic

38. Prakash TP, Bhat B (2007) 2′-Modified oligonucleotides for antisense therapeutics. Curr Top Med Chem 7(7):641–649
39. van Rooij E, Sutherland LB, Qi X, Richardson JA, Hill J, Olson EN (2007) Control of stress-dependent cardiac growth and gene expression by a microRNA. Science (New York, NY) 316(5824):575–579. https://doi.org/10.1126/science.1139089
40. Jopling CL, Schutz S, Sarnow P (2008) Position-dependent function for a tandem microRNA miR-122-binding site located in the hepatitis C virus RNA genome. Cell Host Microbe 4(1):77–85. https://doi.org/10.1016/j.chom.2008.05.013
41. Bonauer A, Carmona G, Iwasaki M, Mione M, Koyanagi M, Fischer A, Burchfield J, Fox H, Doebele C, Ohtani K, Chavakis E, Potente M, Tjwa M, Urbich C, Zeiher AM, Dimmeler S (2009) MicroRNA-92a controls angiogenesis and functional recovery of ischemic tissues in mice. Science (New York, NY) 324(5935):1710–1713. https://doi.org/10.1126/science.1174381
42. Cui H, Zhai J, Ma C (2015) miRLocator: machine learning-based prediction of mature MicroRNAs within plant pre-miRNA sequences. PLoS One 10(11):e0142753. https://doi.org/10.1371/journal.pone.0142753
43. Griffiths-Jones S (2004) The microRNA registry. Nucleic Acids Res 32(Database issue):D109–D111. https://doi.org/10.1093/nar/gkh023
44. Kozomara A, Griffiths-Jones S (2014) miRBase: annotating high confidence microRNAs using deep sequencing data. Nucleic Acids Res 42(Database issue):D68–D73. https://doi.org/10.1093/nar/gkt1181
45. Dai E, Yu X, Zhang Y, Meng F, Wang S, Liu X, Liu D, Wang J, Li X, Jiang W (2014) EpimiR: a database of curated mutual regulation between miRNAs and epigenetic modifications. Database (Oxford) 2014:bau023. https://doi.org/10.1093/database/bau023
46. Kiezun A, Artzi S, Modai S, Volk N, Isakov O, Shomron N (2012) miRviewer: a multispecies microRNA homologous viewer. BMC Res Notes 5:92. https://doi.org/10.1186/1756-0500-5-92
47. Van Peer G, Lefever S, Anckaert J, Beckers A, Rihani A, Van Goethem A, Volders PJ, Zeka F, Ongenaert M, Mestdagh P, Vandesompele J (2014) miRBase Tracker: keeping track of microRNA annotation changes. Database (Oxford) 2014:bau080. https://doi.org/10.1093/database/bau080
48. Fromm B, Billipp T, Peck LE, Johansen M, Tarver JE, King BL, Newcomb JM, Sempere LF, Flatmark K, Hovig E, Peterson KJ (2015) A uniform system for the annotation of vertebrate microRNA genes and the evolution of the human microRNAome. Annu Rev Genet 49:213–242. https://doi.org/10.1146/annurev-genet-120213-092023
49. Gerlach D, Kriventseva EV, Rahman N, Vejnar CE, Zdobnov EM (2009) miROrtho: computational survey of microRNA genes. Nucleic Acids Res 37(Database issue):D111–D117. https://doi.org/10.1093/nar/gkn707
50. Maselli V, Di Bernardo D, Banfi S (2008) CoGemiR: a comparative genomics microRNA database. BMC Genomics 9:457. https://doi.org/10.1186/1471-2164-9-457
51. Kaya KD, Karakulah G, Yakicier CM, Acar AC, Konu O (2011) mESAdb: microRNA expression and sequence analysis database. Nucleic Acids Res 39(Database issue):D170–D180. https://doi.org/10.1093/nar/gkq1256
52. Cheng WC, Chung IF, Tsai CF, Huang TS, Chen CY, Wang SC, Chang TY, Sun HJ, Chao JY, Cheng CC, Wu CW, Wang HW (2015) YM500v2: a small RNA sequencing (smRNA-seq) database for human cancer miRNome research. Nucleic Acids Res 43(Database issue):D862–D867. https://doi.org/10.1093/nar/gku1156
53. Szczesniak MW, Makalowska I (2014) miRNEST 2.0: a database of plant and animal microRNAs. Nucleic Acids Res 42(Database issue):D74–D77. https://doi.org/10.1093/nar/gkt1156
54. Vergoulis T, Kanellos I, Kostoulas N, Georgakilas G, Sellis T, Hatzigeorgiou A, Dalamagas T (2015) mirPub: a database for searching microRNA publications. Bioinformatics 31(9):1502–1504. https://doi.org/10.1093/bioinformatics/btu819
55. Zielezinski A, Dolata J, Alaba S, Kruszka K, Pacak A, Swida-Barteczka A, Knop K, Stepien A, Bielewicz D, Pietrykowska H, Sierocka I, Sobkowiak L, Lakomiak A, Jarmolowski A, Szweykowska-Kulinska Z, Karlowski WM (2015) mirEX 2.0 – an integrated environment for expression profiling of plant microRNAs. BMC Plant Biol 15:144. https://doi.org/10.1186/s12870-015-0533-2

56. Gong J, Wu Y, Zhang X, Liao Y, Sibanda VL, Liu W, Guo AY (2014) Comprehensive analysis of human small RNA sequencing data provides insights into expression profiles and miRNA editing. RNA Biol 11(11):1375–1385. https://doi.org/10.1080/15476286.2014.996465

57. Juzenas S, Venkatesh G, Hubenthal M, Hoeppner MP, Du ZG, Paulsen M, Rosenstiel P, Senger P, Hofmann-Apitius M, Keller A, Kupcinskas L, Franke A, Hemmrich-Stanisak G (2017) A comprehensive, cell specific microRNA catalogue of human peripheral blood. Nucleic Acids Res 45(16):9290–9301. https://doi.org/10.1093/nar/gkx706

58. Barupal JK, Saini AK, Chand T, Meena A, Beniwal S, Suthar JR, Meena N, Kachhwaha S, Kothari SL (2015) ExcellmiRDB for translational genomics: a curated online resource for extracellular microRNAs. OMICS 19(1):24–30. https://doi.org/10.1089/omi.2014.0106

59. Russo F, Di Bella S, Nigita G, Macca V, Lagana A, Giugno R, Pulvirenti A, Ferro A (2012) miRandola: extracellular circulating microRNAs database. PLoS One 7(10):e47786. https://doi.org/10.1371/journal.pone.0047786

60. Yang Q, Qiu C, Yang J, Wu Q, Cui Q (2011) miREnvironment database: providing a bridge for microRNAs, environmental factors and phenotypes. Bioinformatics 27(23):3329–3330. https://doi.org/10.1093/bioinformatics/btr556

61. Gurjar AK, Panwar AS, Gupta R, Mantri SS (2016) PmiRExAt: plant miRNA expression atlas database and web applications. Database (Oxford) 2016:baw060. https://doi.org/10.1093/database/baw060

62. Kumar Gupta A, Kumar M (2015) HPVbase—a knowledgebase of viral integrations, methylation patterns and microRNAs aberrant expression: as potential biomarkers for Human papillomaviruses mediated carcinomas. Sci Rep 5:12522. https://doi.org/10.1038/srep12522

63. Vergoulis T, Vlachos IS, Alexiou P, Georgakilas G, Maragkakis M, Reczko M, Gerangelos S, Koziris N, Dalamagas T, Hatzigeorgiou AG (2012) TarBase 6.0: capturing the exponential growth of miRNA targets with experimental support. Nucleic Acids Res 40(Database issue):D222–D229. https://doi.org/10.1093/nar/gkr1161

64. Hsu SD, Lin FM, Wu WY, Liang C, Huang WC, Chan WL, Tsai WT, Chen GZ, Lee CJ, Chiu CM, Chien CH, Wu MC, Huang CY, Tsou AP, Huang HD (2011) miRTarBase: a database curates experimentally validated microRNA-target interactions. Nucleic Acids Res 39(Database issue):D163–D169. https://doi.org/10.1093/nar/gkq1107

65. Chou CH, Chang NW, Shrestha S, Hsu SD, Lin YL, Lee WH, Yang CD, Hong HC, Wei TY, Tu SJ, Tsai TR, Ho SY, Jian TY, Wu HY, Chen PR, Lin NC, Huang HT, Yang TL, Pai CY, Tai CS, Chen WL, Huang CY, Liu CC, Weng SL, Liao KW, Hsu WL, Huang HD (2016) miRTarBase 2016: updates to the experimentally validated miRNA-target interactions database. Nucleic Acids Res 44(D1):D239–D247. https://doi.org/10.1093/nar/gkv1258

66. Andres-Leon E, Gonzalez Pena D, Gomez-Lopez G, Pisano DG (2015) miRGate: a curated database of human, mouse and rat miRNA-mRNA targets. Database (Oxford) 2015:bav035. https://doi.org/10.1093/database/bav035

67. Yi X, Zhang Z, Ling Y, Xu W, Su Z (2015) PNRD: a plant non-coding RNA database. Nucleic Acids Res 43(Database issue):D982–D989. https://doi.org/10.1093/nar/gku1162

68. Xiao F, Zuo Z, Cai G, Kang S, Gao X, Li T (2009) miRecords: an integrated resource for microRNA-target interactions. Nucleic Acids Res 37(Database issue):D105–D110. https://doi.org/10.1093/nar/gkn851

69. Dweep H, Sticht C, Pandey P, Gretz N (2011) miRWalk—database: prediction of possible miRNA binding sites by "walking" the genes of three genomes. J Biomed Inform 44(5):839–847. https://doi.org/10.1016/j.jbi.2011.05.002

70. Lu TP, Lee CY, Tsai MH, Chiu YC, Hsiao CK, Lai LC, Chuang EY (2012) miRSystem: an integrated system for characterizing enriched functions and pathways of microRNA targets. PLoS One 7(8):e42390. https://doi.org/10.1371/journal.pone.0042390

71. Piriyapongsa J, Bootchai C, Ngamphiw C, Tongsima S (2014) microPIR2: a comprehensive database for human-mouse comparative study of microRNA-promoter interactions. Database (Oxford) 2014:bau115. https://doi.org/10.1093/database/bau115

72. Hsu PW, Lin LZ, Hsu SD, Hsu JB, Huang HD (2007) ViTa: prediction of host microRNAs targets on viruses. Nucleic Acids Res 35(Database issue):D381–D385. https://doi.org/10.1093/nar/gkl1009

73. Hsu JB, Chiu CM, Hsu SD, Huang WY, Chien CH, Lee TY, Huang HD (2011) miR-Tar: an integrated system for identifying miRNA-target interactions in human. BMC Bioinformatics 12:300. https://doi.org/10.1186/1471-2105-12-300
74. Paraskevopoulou MD, Georgakilas G, Kostoulas N, Vlachos IS, Vergoulis T, Reczko M, Filippidis C, Dalamagas T, Hatzigeorgiou AG (2013) DIANA-microT web server v5.0: service integration into miRNA functional analysis workflows. Nucleic Acids Res 41(Web Server issue):W169–W173. https://doi.org/10.1093/nar/gkt393
75. Wang X (2008) miRDB: a microRNA target prediction and functional annotation database with a wiki interface. RNA 14(6):1012–1017. https://doi.org/10.1261/rna.965408
76. Wong N, Wang X (2015) miRDB: an online resource for microRNA target prediction and functional annotations. Nucleic Acids Res 43 (Database issue):D146–D152. https://doi.org/10.1093/nar/gku1104
77. Ru Y, Kechris KJ, Tabakoff B, Hoffman P, Radcliffe RA, Bowler R, Mahaffey S, Rossi S, Calin GA, Bemis L, Theodorescu D (2014) The multiMiR R package and database: integration of microRNA-target interactions along with their disease and drug associations. Nucleic Acids Res 42(17):e133. https://doi.org/10.1093/nar/gku631
78. Wu WS, Tu BW, Chen TT, Hou SW, Tseng JT (2017) CSmiRTar: condition-specific microRNA targets database. PLoS One 12(7): e0181231. https://doi.org/10.1371/journal.pone.0181231
79. Bruno AE, Li L, Kalabus JL, Pan Y, Yu A, Hu Z (2012) miRdSNP: a database of disease-associated SNPs and microRNA target sites on 3′UTRs of human genes. BMC Genomics 13:44. https://doi.org/10.1186/1471-2164-13-44
80. Bhattacharya A, Ziebarth JD, Cui Y (2014) PolymiRTS Database 3.0: linking polymorphisms in microRNAs and their target sites with human diseases and biological pathways. Nucleic Acids Res 42(Database issue): D86–D91. https://doi.org/10.1093/nar/gkt1028
81. Ziebarth JD, Bhattacharya A, Chen A, Cui Y (2012) PolymiRTS Database 2.0: linking polymorphisms in microRNA target sites with human diseases and complex traits. Nucleic Acids Res 40(Database issue): D216–D221. https://doi.org/10.1093/nar/gkr1026
82. Guo ZW, Xie C, Yang JR, Li JH, Yang JH, Zheng L (2015) MtiBase: a database for decoding microRNA target sites located within CDS and 5′UTR regions from CLIP-Seq and expression profile datasets. Database (Oxford) 2015:bav102. https://doi.org/10.1093/database/bav102
83. Gennarino VA, Sardiello M, Mutarelli M, Dharmalingam G, Maselli V, Lago G, Banfi S (2011) HOCTAR database: a unique resource for microRNA target prediction. Gene 480(1–2):51–58. https://doi.org/10.1016/j.gene.2011.03.005
84. Naeem H, Kuffner R, Csaba G, Zimmer R (2010) miRSel: automated extraction of associations between microRNAs and genes from the biomedical literature. BMC Bioinformatics 11:135. https://doi.org/10.1186/1471-2105-11-135
85. Yang Z, Ren F, Liu C, He S, Sun G, Gao Q, Yao L, Zhang Y, Miao R, Cao Y, Zhao Y, Zhong Y, Zhao H (2010) dbDEMC: a database of differentially expressed miRNAs in human cancers. BMC Genomics 11(Suppl 4):S5. https://doi.org/10.1186/1471-2164-11-S4-S5
86. Wang D, Gu J, Wang T, Ding Z (2014) OncomiRDB: a database for the experimentally verified oncogenic and tumor-suppressive microRNAs. Bioinformatics 30 (15):2237–2238. https://doi.org/10.1093/bioinformatics/btu155
87. Ruepp A, Kowarsch A, Schmidl D, Buggenthin F, Brauner B, Dunger I, Fobo G, Frishman G, Montrone C, Theis FJ (2010) PhenomiR: a knowledgebase for microRNA expression in diseases and biological processes. Genome Biol 11(1):R6. https://doi.org/10.1186/gb-2010-11-1-r6
88. Sun X, Dong B, Yin L, Zhang R, Du W, Liu D, Shi N, Li A, Liang Y, Mao L (2013) PMTED: a plant microRNA target expression database. BMC Bioinformatics 14:174. https://doi.org/10.1186/1471-2105-14-174
89. Jacobs LA, Bewicke-Copley F, Poolman MG, Pink RC, Mulcahy LA, Baker I, Beaman EM, Brooks T, Caley DP, Cowling W, Currie JM, Horsburgh J, Kenehan L, Keyes E, Leite D, Massa D, McDermott-Rouse A, Samuel P, Wood H, Kadhim M, Carter DR (2013) Meta-analysis using a novel database, miRStress, reveals miRNAs that are frequently associated with the radiation and hypoxia stress-responses. PLoS One 8(11):e80844. https://doi.org/10.1371/journal.pone.0080844
90. Chiang K, Shu J, Zempleni J, Cui J (2015) Dietary MicroRNA Database (DMD): an archive database and analytic tool for food-borne microRNAs. PLoS One 10(6):

e0128089. https://doi.org/10.1371/journal.pone.0128089
91. Mooney C, Becker BA, Raoof R, Henshall DC (2016) EpimiRBase: a comprehensive database of microRNA-epilepsy associations. Bioinformatics 32(9):1436–1438. https://doi.org/10.1093/bioinformatics/btw008
92. Paraskevopoulou MD, Vlachos IS, Karagkouni D, Georgakilas G, Kanellos I, Vergoulis T, Zagganas K, Tsanakas P, Floros E, Dalamagas T, Hatzigeorgiou AG (2016) DIANA-LncBase v2: indexing microRNA targets on non-coding transcripts. Nucleic Acids Res 44(D1):D231–D238. https://doi.org/10.1093/nar/gkv1270
93. Ganguli S, Mitra S, Datta A (2011) Antagomirbase- a putative antagomir database. Bioinformation 7(1):41–43
94. Friard O, Re A, Taverna D, De Bortoli M, Cora D (2010) CircuitsDB: a database of mixed microRNA/transcription factor feedforward regulatory circuits in human and mouse. BMC Bioinformatics 11:435. https://doi.org/10.1186/1471-2105-11-435
95. Betel D, Wilson M, Gabow A, Marks DS, Sander C (2008) The microRNA.org resource: targets and expression. Nucleic Acids Res 36(Database issue):D149–D153. https://doi.org/10.1093/nar/gkm995
96. Hsu PW, Huang HD, Hsu SD, Lin LZ, Tsou AP, Tseng CP, Stadler PF, Washietl S, Hofacker IL (2006) miRNAMap: genomic maps of microRNA genes and their target genes in mammalian genomes. Nucleic Acids Res 34(Database issue):D135–D139. https://doi.org/10.1093/nar/gkj135
97. Ritchie W, Flamant S, Rasko JE (2010) mimiRNA: a microRNA expression profiler and classification resource designed to identify functional correlations between microRNAs and their targets. Bioinformatics 26(2):223–227. https://doi.org/10.1093/bioinformatics/btp649
98. Jiang Q, Wang Y, Hao Y, Juan L, Teng M, Zhang X, Li M, Wang G, Liu Y (2009) miR2-Disease: a manually curated database for microRNA deregulation in human disease. Nucleic Acids Res 37(Database issue):D98–D104. https://doi.org/10.1093/nar/gkn714
99. Bhattacharya A, Cui Y (2016) SomamiR 2.0: a database of cancer somatic mutations altering microRNA-ceRNA interactions. Nucleic Acids Res 44(D1):D1005–D1010. https://doi.org/10.1093/nar/gkv1220
100. Nam S, Kim B, Shin S, Lee S (2008) miRGator: an integrated system for functional annotation of microRNAs. Nucleic Acids Res 36(Database issue):D159–D164. https://doi.org/10.1093/nar/gkm829
101. Cho S, Jun Y, Lee S, Choi HS, Jung S, Jang Y, Park C, Kim S, Lee S, Kim W (2011) miRGator v2.0: an integrated system for functional investigation of microRNAs. Nucleic Acids Res 39(Database issue):D158–D162. https://doi.org/10.1093/nar/gkq1094
102. Yang JH, Li JH, Shao P, Zhou H, Chen YQ, Qu LH (2011) starBase: a database for exploring microRNA-mRNA interaction maps from Argonaute CLIP-Seq and Degradome-Seq data. Nucleic Acids Res 39(Database issue):D202–D209. https://doi.org/10.1093/nar/gkq1056
103. Li JH, Liu S, Zhou H, Qu LH, Yang JH (2014) starBase v2.0: decoding miRNA-ceRNA, miRNA-ncRNA and protein-RNA interaction networks from large-scale CLIP-Seq data. Nucleic Acids Res 42(Database issue):D92–D97. https://doi.org/10.1093/nar/gkt1248
104. Georgakilas G, Vlachos IS, Zagganas K, Vergoulis T, Paraskevopoulou MD, Kanellos I, Tsanakas P, Dellis D, Fevgas A, Dalamagas T, Hatzigeorgiou AG (2016) DIANA-miRGen v3.0: accurate characterization of microRNA promoters and their regulators. Nucleic Acids Res 44(D1):D190–D195. https://doi.org/10.1093/nar/gkv1254
105. Paraskevopoulou MD, Vlachos IS, Hatzigeorgiou AG (2016) DIANA-TarBase and DIANA suite tools: studying experimentally supported microRNA targets. Curr Protoc Bioinformatics 55:12.14.11–12.14.18. https://doi.org/10.1002/cpbi.12
106. Hansen TB, Veno MT, Kjems J, Damgaard CK (2014) miRdentify: high stringency miRNA predictor identifies several novel animal miRNAs. Nucleic Acids Res 42(16):e124. https://doi.org/10.1093/nar/gku598
107. Mathelier A, Carbone A (2010) MIReNA: finding microRNAs with high accuracy and no learning at genome scale and from deep sequencing data. Bioinformatics 26(18):2226–2234. https://doi.org/10.1093/bioinformatics/btq329
108. Gao D, Middleton R, Rasko JE, Ritchie W (2013) miREval 2.0: a web tool for simple microRNA prediction in genome sequences. Bioinformatics 29(24):3225–3226. https://doi.org/10.1093/bioinformatics/btt545

109. An J, Lai J, Sajjanhar A, Lehman ML, Nelson CC (2014) miRPlant: an integrated tool for identification of plant miRNA from RNA sequencing data. BMC Bioinformatics 15:275. https://doi.org/10.1186/1471-2105-15-275

110. Vlachos IS, Vergoulis T, Paraskevopoulou MD, Lykokanellos F, Georgakilas G, Georgiou P, Chatzopoulos S, Karagkouni D, Christodoulou F, Dalamagas T, Hatzigeorgiou AG (2016) DIANA-mirExTra v2.0: Uncovering microRNAs and transcription factors with crucial roles in NGS expression data. Nucleic Acids Res 44(W1):W128–W134. https://doi.org/10.1093/nar/gkw455

111. Aguiar RR, Ambrosio LA, Sepulveda-Hermosilla G, Maracaja-Coutinho V, Paschoal AR (2016) miRQuest: integration of tools on a Web server for microRNA research. Genet Mol Res 15:1. https://doi.org/10.4238/gmr.15016861

112. Li J, Wang Y, Wang L, Feng W, Luan K, Dai X, Xu C, Meng X, Zhang Q, Liang H (2015) MatPred: computational identification of mature MicroRNAs within novel Pre-MicroRNAs. Biomed Res Int 2015:546763. https://doi.org/10.1155/2015/546763

113. Jamal S, Periwal V, Scaria V (2012) Computational analysis and predictive modeling of small molecule modulators of microRNA. J Cheminform 4(1):16. https://doi.org/10.1186/1758-2946-4-16

114. Grad Y, Aach J, Hayes GD, Reinhart BJ, Church GM, Ruvkun G, Kim J (2003) Computational and experimental identification of C. elegans microRNAs. Mol Cell 11 (5):1253–1263

115. Lim LP, Lau NC, Weinstein EG, Abdelhakim A, Yekta S, Rhoades MW, Burge CB, Bartel DP (2003) The microRNAs of Caenorhabditis elegans. Genes Dev 17 (8):991–1008. https://doi.org/10.1101/gad.1074403

116. Lim LP, Glasner ME, Yekta S, Burge CB, Bartel DP (2003) Vertebrate microRNA genes. Science 299(5612):1540. https://doi.org/10.1126/science.1080372

117. Lai EC, Tomancak P, Williams RW, Rubin GM (2003) Computational identification of Drosophila microRNA genes. Genome Biol 4 (7):R42. https://doi.org/10.1186/gb-2003-4-7-r42

118. Hu LL, Huang Y, Wang QC, Zou Q, Jiang Y (2012) Benchmark comparison of ab initio microRNA identification methods and software. Genet Mol Res 11(4):4525–4538. https://doi.org/10.4238/2012.October.17.4

119. Terai G, Komori T, Asai K, Kin T (2007) miRRim: a novel system to find conserved miRNAs with high sensitivity and specificity. RNA 13(12):2081–2090. https://doi.org/10.1261/rna.655107

120. Bentwich I (2005) Prediction and validation of microRNAs and their targets. FEBS Lett 579(26):5904–5910. https://doi.org/10.1016/j.febslet.2005.09.040

121. Oulas A, Boutla A, Gkirtzou K, Reczko M, Kalantidis K, Poirazi P (2009) Prediction of novel microRNA genes in cancer-associated genomic regions—a combined computational and experimental approach. Nucleic Acids Res 37(10):3276–3287. https://doi.org/10.1093/nar/gkp120

122. Nam JW, Kim J, Kim SK, Zhang BT (2006) ProMiR II: a web server for the probabilistic prediction of clustered, nonclustered, conserved and nonconserved microRNAs. Nucleic Acids Res 34(Web Server issue):W455–W458. https://doi.org/10.1093/nar/gkl321

123. Berezikov E, Guryev V, van de Belt J, Wienholds E, Plasterk RH, Cuppen E (2005) Phylogenetic shadowing and computational identification of human microRNA genes. Cell 120(1):21–24. https://doi.org/10.1016/j.cell.2004.12.031

124. Huang TH, Fan B, Rothschild MF, Hu ZL, Li K, Zhao SH (2007) MiRFinder: an improved approach and software implementation for genome-wide fast microRNA precursor scans. BMC Bioinformatics 8:341. https://doi.org/10.1186/1471-2105-8-341

125. Nam JW, Shin KR, Han J, Lee Y, Kim VN, Zhang BT (2005) Human microRNA prediction through a probabilistic co-learning model of sequence and structure. Nucleic Acids Res 33(11):3570–3581. https://doi.org/10.1093/nar/gki668

126. Yousef M, Nebozhyn M, Shatkay H, Kanterakis S, Showe LC, Showe MK (2006) Combining multi-species genomic data for microRNA identification using a Naive Bayes classifier. Bioinformatics 22(11):1325–1334. https://doi.org/10.1093/bioinformatics/btl094

127. Jiang P, Wu H, Wang W, Ma W, Sun X, Lu Z (2007) MiPred: classification of real and pseudo microRNA precursors using random forest prediction model with combined features. Nucleic Acids Res 35(Web Server issue):W339–W344. https://doi.org/10.1093/nar/gkm368

128. Jha A, Chauhan R, Mehra M, Singh HR, Shankar R (2012) miR-BAG: bagging based identification of microRNA precursors. PLoS One 7(9):e45782. https://doi.org/10.1371/journal.pone.0045782

129. Chen J, Wang X, Liu B (2016) iMiRNA-SSF: improving the identification of MicroRNA precursors by combining negative sets with different distributions. Sci Rep 6:19062. https://doi.org/10.1038/srep19062

130. Friedlander MR, Chen W, Adamidi C, Maaskola J, Einspanier R, Knespel S, Rajewsky N (2008) Discovering microRNAs from deep sequencing data using miRDeep. Nat Biotechnol 26(4):407–415. https://doi.org/10.1038/nbt1394

131. Tran Vdu T, Tempel S, Zerath B, Zehraoui F, Tahi F (2015) miRBoost: boosting support vector machines for microRNA precursor classification. RNA 21(5):775–785. https://doi.org/10.1261/rna.043612.113

132. Muller S, Rycak L, Winter P, Kahl G, Koch I, Rotter B (2013) omiRas: a Web server for differential expression analysis of miRNAs derived from small RNA-Seq data. Bioinformatics 29(20):2651–2652. https://doi.org/10.1093/bioinformatics/btt457

133. Lertampaiporn S, Thammarongtham C, Nukoolkit C, Kaewkamnerdpong B, Ruengjitchatchawalya M (2013) Heterogeneous ensemble approach with discriminative features and modified-SMOTEbagging for pre-miRNA classification. Nucleic Acids Res 41(1):e21. https://doi.org/10.1093/nar/gks878

134. Song X, Wang M, Chen YP, Wang H, Han P, Sun H (2013) Prediction of pre-miRNA with multiple stem-loops using pruning algorithm. Comput Biol Med 43(5):409–416. https://doi.org/10.1016/j.compbiomed.2013.02.003

135. Batuwita R, Palade V (2009) microPred: effective classification of pre-miRNAs for human miRNA gene prediction. Bioinformatics 25(8):989–995. https://doi.org/10.1093/bioinformatics/btp107

136. Agarwal S, Vaz C, Bhattacharya A, Srinivasan A (2010) Prediction of novel precursor miRNAs using a context-sensitive hidden Markov model (CSHMM). BMC Bioinformatics 11 (Suppl 1):S29. https://doi.org/10.1186/1471-2105-11-S1-S29

137. Chang DT, Wang CC, Chen JW (2008) Using a kernel density estimation based classifier to predict species-specific microRNA precursors. BMC Bioinformatics 9(Suppl 12):S2. https://doi.org/10.1186/1471-2105-9-s12-s2

138. Gkirtzou K, Tsamardinos I, Tsakalides P, Poirazi P (2010) MatureBayes: a probabilistic algorithm for identifying the mature miRNA within novel precursors. PLoS One 5(8):e11843. https://doi.org/10.1371/journal.pone.0011843

139. Hackenberg M, Sturm M, Langenberger D, Falcon-Perez JM, Aransay AM (2009) miRanalyzer: a microRNA detection and analysis tool for next-generation sequencing experiments. Nucleic Acids Res 37(Web Server issue):W68–W76. https://doi.org/10.1093/nar/gkp347

140. Leclercq M, Diallo AB, Blanchette M (2013) Computational prediction of the localization of microRNAs within their pre-miRNA. Nucleic Acids Res 41(15):7200–7211. https://doi.org/10.1093/nar/gkt466

141. Xuan P, Guo M, Huang Y, Li W, Huang Y (2011) MaturePred: efficient identification of microRNAs within novel plant pre-miRNAs. PLoS One 6(11):e27422. https://doi.org/10.1371/journal.pone.0027422

142. He C, Li YX, Zhang G, Gu Z, Yang R, Li J, Lu ZJ, Zhou ZH, Zhang C, Wang J (2012) MiRmat: mature microRNA sequence prediction. PLoS One 7(12):e51673. https://doi.org/10.1371/journal.pone.0051673

143. Muller H, Marzi MJ, Nicassio F (2014) IsomiRage: from functional classification to differential expression of miRNA isoforms. Front Bioeng Biotechnol 2:38. https://doi.org/10.3389/fbioe.2014.00038

144. Sablok G, Milev I, Minkov G, Minkov I, Varotto C, Yahubyan G, Baev V (2013) isomiRex: web-based identification of microRNAs, isomiR variations and differential expression using next-generation sequencing datasets. FEBS Lett 587(16):2629–2634. https://doi.org/10.1016/j.febslet.2013.06.047

145. Zhang Y, Zang Q, Zhang H, Ban R, Yang Y, Iqbal F, Li A, Shi Q (2016) DeAnnIso: a tool for online detection and annotation of isomiRs from small RNA sequencing data. Nucleic Acids Res 44(W1):W166–W175. https://doi.org/10.1093/nar/gkw427

146. Baras AS, Mitchell CJ, Myers JR, Gupta S, Weng LC, Ashton JM, Cornish TC, Pandey A, Halushka MK (2015) miRge – a multiplexed method of processing small RNA-Seq data to determine MicroRNA entropy. PLoS One 10(11):e0143066. https://doi.org/10.1371/journal.pone.0143066

147. Kuenne C, Preussner J, Herzog M, Braun T, Looso M (2014) MIRPIPE: quantification of microRNAs in niche model organisms.

Bioinformatics 30(23):3412–3413. https://doi.org/10.1093/bioinformatics/btu573
148. Shi J, Dong M, Li L, Liu L, Luz-Madrigal A, Tsonis PA, Del Rio-Tsonis K, Liang C (2015) mirPRo-a novel standalone program for differential expression and variation analysis of miRNAs. Sci Rep 5:14617. https://doi.org/10.1038/srep14617
149. Yang K, Sablok G, Qiao G, Nie Q, Wen X (2017) isomiR2Function: an integrated workflow for identifying MicroRNA variants in plants. Front Plant Sci 8:322. https://doi.org/10.3389/fpls.2017.00322
150. Qian K, Auvinen E, Greco D, Auvinen P (2012) miRSeqNovel: an R based workflow for analyzing miRNA sequencing data. Mol Cell Probes 26(5):208–211. https://doi.org/10.1016/j.mcp.2012.05.002
151. Rueda A, Barturen G, Lebron R, Gomez-Martin C, Alganza A, Oliver JL, Hackenberg M (2015) sRNAtoolbox: an integrated collection of small RNA research tools. Nucleic Acids Res 43(W1):W467–W473. https://doi.org/10.1093/nar/gkv555
152. Sun Z, Evans J, Bhagwate A, Middha S, Bockol M, Yan H, Kocher JP (2014) CAP-miRSeq: a comprehensive analysis pipeline for microRNA sequencing data. BMC Genomics 15:423. https://doi.org/10.1186/1471-2164-15-423
153. Kozakai T, Takahashi M, Higuchi M, Hara T, Saito K, Tanaka Y, Masuko M, Takizawa J, Sone H, Fujii M (2018) MAGI-1 expression is decreased in several types of human T-cell leukemia cell lines, including adult T-cell leukemia. Int J Hematol 107(3):337–344. https://doi.org/10.1007/s12185-017-2359-1
154. Ronen R, Gan I, Modai S, Sukacheov A, Dror G, Halperin E, Shomron N (2010) miRNAkey: a software for microRNA deep sequencing analysis. Bioinformatics 26(20):2615–2616. https://doi.org/10.1093/bioinformatics/btq493
155. Cer RZ, Herrera-Galeano JE, Anderson JJ, Bishop-Lilly KA, Mokashi VP (2014) miRNA Temporal Analyzer (mirnaTA): a bioinformatics tool for identifying differentially expressed microRNAs in temporal studies using normal quantile transformation. Gigascience 3:20. https://doi.org/10.1186/2047-217X-3-20
156. Fan Y, Siklenka K, Arora SK, Ribeiro P, Kimmins S, Xia J (2016) miRNet – dissecting miRNA-target interactions and functional associations through network-based visual analysis. Nucleic Acids Res 44(W1): W135–W141. https://doi.org/10.1093/nar/gkw288
157. Nam S, Li M, Choi K, Balch C, Kim S, Nephew KP (2009) MicroRNA and mRNA integrated analysis (MMIA): a web tool for examining biological functions of microRNA expression. Nucleic Acids Res 37(Web Server issue):W356–W362. https://doi.org/10.1093/nar/gkp294
158. Zhang Z, Jiang L, Wang J, Gu P, Chen M (2015) MTide: an integrated tool for the identification of miRNA-target interaction in plants. Bioinformatics 31(2):290–291. https://doi.org/10.1093/bioinformatics/btu633
159. Capece V, Garcia Vizcaino JC, Vidal R, Rahman RU, Pena Centeno T, Shomroni O, Suberviola I, Fischer A, Bonn S (2015) Oasis: online analysis of small RNA deep sequencing data. Bioinformatics 31(13):2205–2207. https://doi.org/10.1093/bioinformatics/btv113
160. Fisch KM, Meissner T, Gioia L, Ducom JC, Carland TM, Loguercio S, Su AI (2015) Omics Pipe: a community-based framework for reproducible multi-omics data analysis. Bioinformatics 31(11):1724–1728. https://doi.org/10.1093/bioinformatics/btv061
161. Stocks MB, Moxon S, Mapleson D, Woolfenden HC, Mohorianu I, Folkes L, Schwach F, Dalmay T, Moulton V (2012) The UEA sRNA workbench: a suite of tools for analysing and visualizing next generation sequencing microRNA and small RNA datasets. Bioinformatics 28(15):2059–2061. https://doi.org/10.1093/bioinformatics/bts311
162. Gupta V, Markmann K, Pedersen CN, Stougaard J, Andersen SU (2012) shortran: a pipeline for small RNA-seq data analysis. Bioinformatics 28(20):2698–2700. https://doi.org/10.1093/bioinformatics/bts496
163. Zhao W, Liu W, Tian D, Tang B, Wang Y, Yu C, Li R, Ling Y, Wu J, Song S, Hu S (2011) wapRNA: a web-based application for the processing of RNA sequences. Bioinformatics 27(21):3076–3077. https://doi.org/10.1093/bioinformatics/btr504
164. Patra D, Fasold M, Langenberger D, Steger G, Grosse I, Stadler PF (2014) plantDARIO: web based quantitative and qualitative analysis of small RNA-seq data in plants. Front Plant Sci 5:708. https://doi.org/10.3389/fpls.2014.00708
165. Chae H, Rhee S, Nephew KP, Kim S (2015) BioVLAB-MMIA-NGS: microRNA-mRNA integrated analysis using high-throughput sequencing data. Bioinformatics 31

166. Zhang Y, Xu B, Yang Y, Ban R, Zhang H, Jiang X, Cooke HJ, Xue Y, Shi Q (2012) CPSS: a computational platform for the analysis of small RNA deep sequencing data. Bioinformatics 28(14):1925–1927. https://doi.org/10.1093/bioinformatics/bts282

167. Yuan T, Huang X, Dittmar RL, Du M, Kohli M, Boardman L, Thibodeau SN, Wang L (2014) eRNA: a graphic user interface-based tool optimized for large data analysis from high-throughput RNA sequencing. BMC Genomics 15:176. https://doi.org/10.1186/1471-2164-15-176

168. Giurato G, De Filippo MR, Rinaldi A, Hashim A, Nassa G, Ravo M, Rizzo F, Tarallo R, Weisz A (2013) iMir: an integrated pipeline for high-throughput analysis of small non-coding RNA data obtained by smallRNA-Seq. BMC Bioinformatics 14:362. https://doi.org/10.1186/1471-2105-14-362

169. Williamson V, Kim A, Xie B, McMichael GO, Gao Y, Vladimirov V (2013) Detecting miRNAs in deep-sequencing data: a software performance comparison and evaluation. Brief Bioinform 14(1):36–45. https://doi.org/10.1093/bib/bbs010

170. Friedlander MR, Mackowiak SD, Li N, Chen W, Rajewsky N (2012) miRDeep2 accurately identifies known and hundreds of novel microRNA genes in seven animal clades. Nucleic Acids Res 40(1):37–52. https://doi.org/10.1093/nar/gkr688

171. Wang WC, Lin FM, Chang WC, Lin KY, Huang HD, Lin NS (2009) miRExpress: analyzing high-throughput sequencing data for profiling microRNA expression. BMC Bioinformatics 10:328. https://doi.org/10.1186/1471-2105-10-328

172. Chen CJ, Servant N, Toedling J, Sarazin A, Marchais A, Duvernois-Berthet E, Cognat V, Colot V, Voinnet O, Heard E, Ciaudo C, Barillot E (2012) ncPRO-seq: a tool for annotation and profiling of ncRNAs in sRNA-seq data. Bioinformatics 28(23):3147–3149. https://doi.org/10.1093/bioinformatics/bts587

173. Lei J, Sun Y (2014) miR-PREFeR: an accurate, fast and easy-to-use plant miRNA prediction tool using small RNA-Seq data. Bioinformatics 30(19):2837–2839. https://doi.org/10.1093/bioinformatics/btu380

174. Yang X, Li L (2011) miRDeep-P: a computational tool for analyzing the microRNA transcriptome in plants. Bioinformatics 27(18):2614–2615. https://doi.org/10.1093/bioinformatics/btr430

175. Breakfield NW, Corcoran DL, Petricka JJ, Shen J, Sae-Seaw J, Rubio-Somoza I, Weigel D, Ohler U, Benfey PN (2012) High-resolution experimental and computational profiling of tissue-specific known and novel miRNAs in Arabidopsis. Genome Res 22(1):163–176. https://doi.org/10.1101/gr.123547.111

176. Cammaerts S, Strazisar M, Dierckx J, Del Favero J, De Rijk P (2016) miRVaS: a tool to predict the impact of genetic variants on miRNAs. Nucleic Acids Res 44(3):e23. https://doi.org/10.1093/nar/gkv921

177. Barenboim M, Zoltick BJ, Guo Y, Weinberger DR (2010) MicroSNiPer: a web tool for prediction of SNP effects on putative microRNA targets. Hum Mutat 31(11):1223–1232. https://doi.org/10.1002/humu.21349

178. Zorc M, Skok DJ, Godnic I, Calin GA, Horvat S, Jiang Z, Dovc P, Kunej T (2012) Catalog of microRNA seed polymorphisms in vertebrates. PLoS One 7(1):e30737. https://doi.org/10.1371/journal.pone.0030737

179. Liu C, Zhang F, Li T, Lu M, Wang L, Yue W, Zhang D (2012) MirSNP, a database of polymorphisms altering miRNA target sites, identifies miRNA-related SNPs in GWAS SNPs and eQTLs. BMC Genomics 13:661. https://doi.org/10.1186/1471-2164-13-661

180. Maxwell EK, Campbell JD, Spira A, Baxevanis AD (2015) SubmiRine: assessing variants in microRNA targets using clinical genomic data sets. Nucleic Acids Res 43(8):3886–3898. https://doi.org/10.1093/nar/gkv256

181. Shi H, Zhang G, Zhou M, Cheng L, Yang H, Wang J, Sun J, Wang Z (2016) Integration of multiple genomic and phenotype data to infer novel miRNA-disease associations. PLoS One 11(2):e0148521. https://doi.org/10.1371/journal.pone.0148521

182. Lee E, Ito K, Zhao Y, Schadt EE, Irie HY, Zhu J (2016) Inferred miRNA activity identifies miRNA-mediated regulatory networks underlying multiple cancers. Bioinformatics 32(1):96–105. https://doi.org/10.1093/bioinformatics/btv531

183. Bhattacharya A, Cui Y (2015) miR2GO: comparative functional analysis for microRNAs. Bioinformatics 31(14):2403–2405. https://doi.org/10.1093/bioinformatics/btv140

184. Bonnal RJ, Rossi RL, Carpi D, Ranzani V, Abrignani S, Pagani M (2015) miRiadne: a web tool for consistent integration of

miRNA nomenclature. Nucleic Acids Res 43 (W1):W487–W492. https://doi.org/10.1093/nar/gkv381

185. Pirro S, Minutolo A, Galgani A, Potesta M, Colizzi V, Montesano C (2016) Bioinformatics prediction and experimental validation of MicroRNAs involved in cross-kingdom interaction. J Comput Biol 23(12):976–989. https://doi.org/10.1089/cmb.2016.0059

186. Busk PK (2014) A tool for design of primers for microRNA-specific quantitative RT-qPCR. BMC Bioinformatics 15:29. https://doi.org/10.1186/1471-2105-15-29

187. Jang I, Chang H, Jun Y, Park S, Yang JO, Lee B, Kim W, Kim VN, Lee S (2015) miRseqViewer: multi-panel visualization of sequence, structure and expression for analysis of microRNA sequencing data. Bioinformatics 31(4):596–598. https://doi.org/10.1093/bioinformatics/btu676

188. Patel P, Ramachandruni SD, Kakrana A, Nakano M, Meyers BC (2016) miTRATA: a web-based tool for microRNA Truncation and Tailing Analysis. Bioinformatics 32(3):450–452. https://doi.org/10.1093/bioinformatics/btv583

189. Li G, Ross KE, Arighi CN, Peng Y, Wu CH, Vijay-Shanker K (2015) miRTex: a text mining system for miRNA-gene relation extraction. PLoS Comput Biol 11(9): e1004391. https://doi.org/10.1371/journal.pcbi.1004391

190. Kozomara A, Griffiths-Jones S (2011) miRBase: integrating microRNA annotation and deep-sequencing data. Nucleic Acids Res 39 (Database issue):D152–D157. https://doi.org/10.1093/nar/gkq1027

191. Griffiths-Jones S, Grocock RJ, van Dongen S, Bateman A, Enright AJ (2006) miRBase: microRNA sequences, targets and gene nomenclature. Nucleic Acids Res 34(Database issue):D140–D144. https://doi.org/10.1093/nar/gkj112

192. Zhang Z, Yu J, Li D, Zhang Z, Liu F, Zhou X, Wang T, Ling Y, Su Z (2010) PMRD: plant microRNA database. Nucleic Acids Res 38 (Database issue):D806–D813. https://doi.org/10.1093/nar/gkp818

193. Li SC, Shiau CK, Lin WC (2008) Vir-Mir db: prediction of viral microRNA candidate hairpins. Nucleic Acids Res 36(Database issue): D184–D189. https://doi.org/10.1093/nar/gkm610

194. Yousef GM (2015) miRSNP-based approach identifies a miRNA that regulates prostate-specific antigen in an allele-specific manner.
Cancer Discov 5(4):351–352. https://doi.org/10.1158/2159-8290.CD-15-0230

195. Lewis BP, Burge CB, Bartel DP (2005) Conserved seed pairing, often flanked by adenosines, indicates that thousands of human genes are microRNA targets. Cell 120 (1):15–20. https://doi.org/10.1016/j.cell.2004.12.035

196. Brennecke J, Stark A, Russell RB, Cohen SM (2005) Principles of microRNA-target recognition. PLoS Biol 3(3):e85. https://doi.org/10.1371/journal.pbio.0030085

197. Manyam G, Ivan C, Calin GA, Coombes KR (2013) targetHub: a programmable interface for miRNA-gene interactions. Bioinformatics 29(20):2657–2658. https://doi.org/10.1093/bioinformatics/btt439

198. Backes C, Kehl T, Stockel D, Fehlmann T, Schneider L, Meese E, Lenhof HP, Keller A (2017) miRPathDB: a new dictionary on microRNAs and target pathways. Nucleic Acids Res 45(D1):D90–D96. https://doi.org/10.1093/nar/gkw926

199. Griffiths-Jones S, Saini HK, van Dongen S, Enright AJ (2008) miRBase: tools for microRNA genomics. Nucleic Acids Res 36(Database issue):D154–D158. https://doi.org/10.1093/nar/gkm952

200. Yang Z, Wu L, Wang A, Tang W, Zhao Y, Zhao H, Teschendorff AE (2017) dbDEMC 2.0: updated database of differentially expressed miRNAs in human cancers. Nucleic Acids Res 45(D1):D812–D818. https://doi.org/10.1093/nar/gkw1079

201. Xie B, Ding Q, Han H, Wu D (2013) miRCancer: a microRNA-cancer association database constructed by text mining on literature. Bioinformatics 29(5):638–644. https://doi.org/10.1093/bioinformatics/btt014

202. Vlachos IS, Kostoulas N, Vergoulis T, Georgakilas G, Reczko M, Maragkakis M, Paraskevopoulou MD, Prionidis K, Dalamagas T, Hatzigeorgiou AG (2012) DIANA miRPath v.2.0: investigating the combinatorial effect of microRNAs in pathways. Nucleic Acids Res 40(Web Server issue):W498–W504. https://doi.org/10.1093/nar/gks494

203. Vlachos IS, Zagganas K, Paraskevopoulou MD, Georgakilas G, Karagkouni D, Vergoulis T, Dalamagas T, Hatzigeorgiou AG (2015) DIANA-miRPath v3.0: deciphering microRNA function with experimental support. Nucleic Acids Res 43(W1): W460–W466. https://doi.org/10.1093/nar/gkv403

204. Li Y, Qiu C, Tu J, Geng B, Yang J, Jiang T, Cui Q (2014) HMDD v2.0: a database for experimentally supported human microRNA and disease associations. Nucleic Acids Res 42 (Database issue):D1070–D1074. https://doi.org/10.1093/nar/gkt1023

205. Hsu SD, Chu CH, Tsou AP, Chen SJ, Chen HC, Hsu PW, Wong YH, Chen YH, Chen GH, Huang HD (2008) miRNAMap 2.0: genomic maps of microRNAs in metazoan genomes. Nucleic Acids Res 36(Database issue):D165–D169. https://doi.org/10.1093/nar/gkm1012

206. Zhang S, Yue Y, Sheng L, Wu Y, Fan G, Li A, Hu X, Shangguan M, Wei C (2013) PASmiR: a literature-curated database for miRNA molecular regulation in plant response to abiotic stress. BMC Plant Biol 13:33. https://doi.org/10.1186/1471-2229-13-33

207. Ruepp A, Kowarsch A, Theis F (2012) PhenomiR: microRNAs in human diseases and biological processes. Methods Mol Biol 822:249–260. https://doi.org/10.1007/978-1-61779-427-8_17

208. Xue C, Li F, He T, Liu GP, Li Y, Zhang X (2005) Classification of real and pseudo microRNA precursors using local structure-sequence features and support vector machine. BMC Bioinformatics 6:310. https://doi.org/10.1186/1471-2105-6-310

209. Ng KL, Mishra SK (2007) De novo SVM classification of precursor microRNAs from genomic pseudo hairpins using global and intrinsic folding measures. Bioinformatics 23 (11):1321–1330. https://doi.org/10.1093/bioinformatics/btm026

210. Huang KY, Lee TY, Teng YC, Chang TH (2015) ViralmiR: a support-vector-machine-based method for predicting viral microRNA precursors. BMC Bioinformatics 16(Suppl 1):S9. https://doi.org/10.1186/1471-2105-16-s1-s9

211. Zhao D, Wang Y, Luo D, Shi X, Wang L, Xu D, Yu J, Liang Y (2010) PMirP: a pre-microRNA prediction method based on structure-sequence hybrid features. Artif Intell Med 49(2):127–132. https://doi.org/10.1016/j.artmed.2010.03.004

Chapter 10

Noncoding RNAs Databases: Current Status and Trends

Vinicius Maracaja-Coutinho, Alexandre Rossi Paschoal,
José Carlos Caris-Maldonado, Pedro Vinícius Borges, Almir José Ferreira,
and Alan Mitchell Durham

Abstract

One of the most important resources for researchers of noncoding RNAs is the information available in public databases spread over the internet. However, the effective exploration of this data can represent a daunting task, given the large amount of databases available and the variety of stored data. This chapter describes a classification of databases based on information source, type of RNA, source organisms, data formats, and the mechanisms for information retrieval, detailing the relevance of each of these classifications and its usability by researchers. This classification is used to update a 2012 review, indexing now more than 229 public databases. This review will include an assessment of the new trends for ncRNA research based on the information that is being offered by the databases. Additionally, we will expand the previous analysis focusing on the usability and application of these databases in pathogen and disease research. Finally, this chapter will analyze how currently available database schemas can help the development of new and improved web resources.

Key words Bioinformatics, Databases, Noncoding RNAs, Biomedicine, Biomedical, Disease, MicroRNA, lncRNA, Circulating RNAs, Review

1 Introduction

Noncoding RNAs (ncRNAs) have been a hot research topic for more than a decade. This was prompted by comprehensive genome-wide transcriptional efforts from international collaborative initiatives like The Encyclopedia of DNA Elements (ENCODE) [1] and The Functional Annotation of the Mammalian Genome (FANTOM) [2]. Their existence is ubiquitous to all domains of life [3], and their mode of action varies according to the RNA family it belongs [4]. Dysregulation of different types of ncRNAs can lead to neurological, cardiovascular, developmental, and other diseases [5].

One of the most important resources for ncRNA research is the information available in public databases spread over the internet.

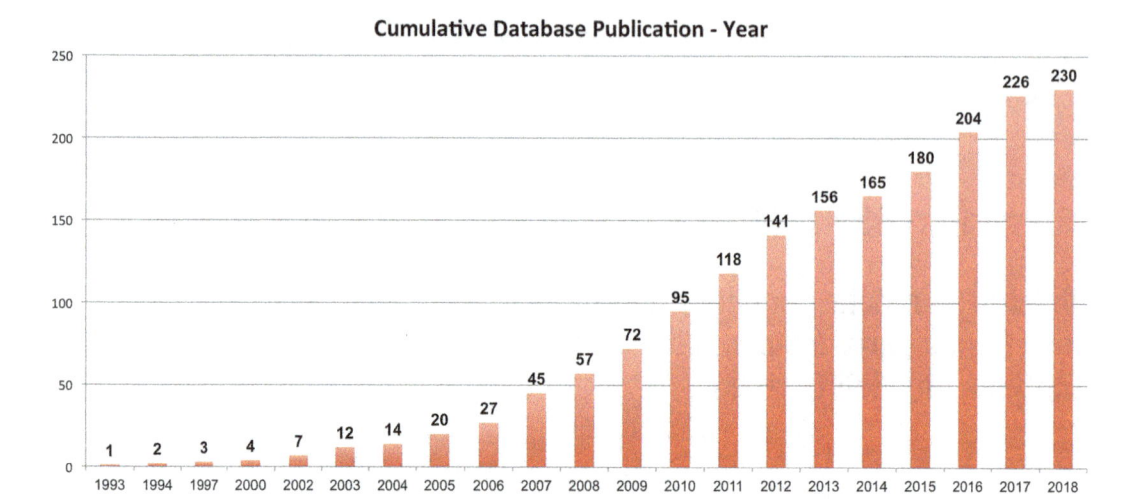

Fig. 1 The cumulative database publication by year

With the progression of the field, an increasing number of public databases have been made publicly available to the research community. These databases cover a wide range of information such as sequences, target genes or proteins, genomic locations, organisms, genomic variants, and disease associations. The number of databases focused on ncRNAs has been increasing steadily in the last 25 years, with 229 public repositories currently available on the subject (Fig. 1). However, the effective exploration of this data can represent a daunting task, given the large amount of databases available and the variety of data stored.

We have manually reviewed each of these databases and catalogued their information (i.e., websites, original publication, search methods, target organisms, disease relationship, target transcripts, information source), along with a small overview or description in an integrated repository called "The Non-coding RNA Database Resource" (NRDR), originally published in 2012 [6]. NRDR is a public web portal that provides a search interface where users can retrieve a list of databases filtered by user-defined criteria, providing a tool for researchers to quickly locate repositories that will be relevant for their research. Details on the cataloged information and on the classification criteria can be found on NRDR original publication.

In this chapter we analyze and update the result of this survey and try to portray a picture of the current status of ncRNA research, based on the contents of the publicly available databases and its evolution since our first survey in 2012 [6]. We give particular emphasis on the databases related to biomedicine, obtaining a general overview of the repositories containing information associated with cancers, cardiovascular and nervous systems diseases, as well as human pathogens, drugs, and other disorders. All data researched for this chapter is being made publicly available to the research community at the NRDR database website (http://ncrnadatabases.org).

2 Current Status of Available Noncoding RNA Databases

In this section we analyze the current status of the ncRNA databases based on three criteria: information source, search methods, and RNA families. For each of these criteria we analyzed the distribution of the databases. These distributions that cannot be obtained directly from NRDR are given explicitly by the enumeration of the databases in lists. Larger lists are given in the form of tables; smaller lists are embedded in the text.

2.1 Information Source: How Is Data Obtained/Generated?

Information on ncRNA databases can come from multiple sources: in silico annotation, literature, manual curation, and experimental. Experimental confirmation is clearly the most reliable source, followed by manual curation. However, both are time-consuming processes of obtaining information. Experimentation, in particular, can involve a very high financial cost. In silico annotation is the most time and cost-efficient way of producing information on ncRNAs, but automatic annotation processes are in general subject to nontrivial error rates and are very sensitive to propagation of errors when it is based on existing database information. Literature can be a reliable source of information, and new artificial intelligence approaches can help with the automatic processing of scientific databases. This probably explains why the information sources that have increased more in ncRNA databases in the last seven years are in silico annotation (141% increase, from 81 to 196 databases) and literature (143% increase, from 37 to 90 databases). Databases with experimental and manual curation sources have increased in roughly the same proportion (respectively 86% and 85%) and clearly are lagging behind the duplication of the total number of ncRNA databases available. In spite of the relative increase, in silico annotation and experimental information still dominate as information sources (196 databases, 85%, use in silico annotation; 125, 54%, use experimental data). We expect the above trend to continue in the foreseeable future unless new low-cost biological experimentation protocols are developed to produce experimentally confirmed information. Information on the databases that use particular sources can be easily obtained in NRDR in the *statistics* page (see http://ncrnadatabases.org). Figure 2 shows the distribution of information sources in all surveyed repositories.

2.2 Search Methods: Data Retrieving and Exploration

Databases can be searched and data retrieved in many ways. Availability of a search method can define the usability of a database. We categorize search methods in six types: keyword, TAG, similarity, genome location, density of ncRNAs, and Tabular. The most

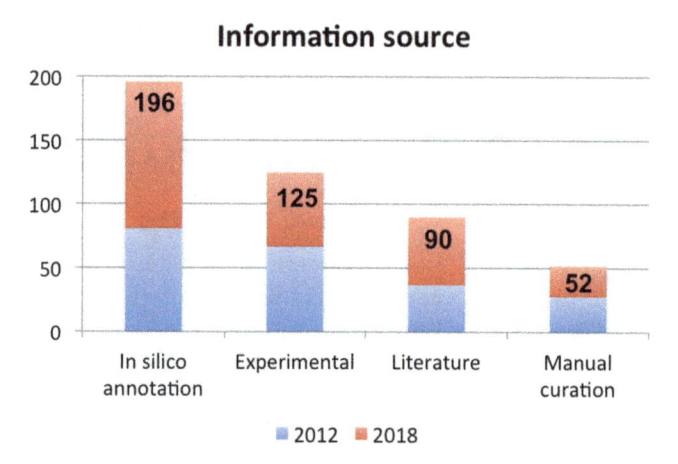

Fig. 2 The distribution of information sources in ncRNA databases. The blue color bars represent databases available in NRDR 2012, while where the red color bars represent databases included in NRDR 2018

popular search method is the keyword, where users can retrieve information based on arbitrary user-defined words. This method can be really effective for exploratory searches in databases and can be easily implemented with an automatic word indexing of the terms present in the record descriptions. Probably this explains its widespread availability: 183 (~80%) databases include this search method, a percentage that has been maintained in the last seven years. The second most popular search method is TAGs. This type of search is based on a preselected set of terms, and its usability heavily depends on the selection of a set of terms predefined by database maintainers. TAG searches are available in ~60% (137) of the ncRNA databases.

The third most widely available search method is Tabular search. This is not exactly a search method; it means that data is viewable in a tabular form, spreadsheet style. The user needs to browse the tables and find the desired content manually. Sometimes the tables can be ordered by a particular field or downloaded to be used directly in spreadsheet software. Despite the fact that tabular search is still widely used, it is not very effective for scanning large datasets and its popularity has decreased in the last years. Currently, 82 databases (~36%) implement tabular displays of data.

Sequence similarity search is present in 60 (26%) databases. In a similarity search, the user has a subject RNA or DNA sequence that is compared to the set of sequences stored in the database. This type of search is very effective to find information on sequences that are closely related to the subject sequences. Similarity search is not as effective in ncRNA molecules as it is for protein-coding DNA sequences, since many times RNA function is defined mostly by the structure and not by sequence and tends to be less conserved than coding gene sequences. Therefore, its usability can be impaired when comparing more distantly related ncRNAs. One

Search Mechanism

[Bar chart showing distribution of search mechanisms with values: Keyword 183, TAG 137, Similarity 60, Genomic location 30, Density of ncRNAs 6, Tabular 82. Blue bars = 2012, Red bars = 2018.]

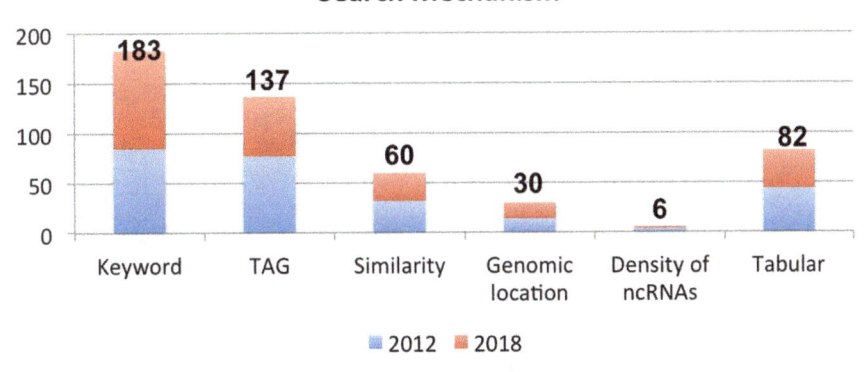

Fig. 3 The distribution of search mechanisms across ncRNAs databases. The blue color bars indicate number of databases in NRDR 2012 where the red color bars indicate number of databases in NRDR 2018

noteworthy alternative is the use of covariance models, a probabilistic model to represent ncRNA families that includes structure and/or sequence information. Covariance model search is available only on the database Rfam. However, it can be very useful for finding ncRNA molecules that are not evolutionarily close to the query sequence, being a powerful source for de novo annotation of transcriptome projects of non-model organisms (i.e., human pathogens, disease vectors) [4]. In particular, coupled with other computational approaches, similarity search is commonly used for mapping miRNAs or natural antisense transcripts (NATs) in target transcripts.

The genomic location search retrieves information on ncRNA molecules present in user-specified regions of the genome. Even though being offered by only 30 databases (13%), genomic location is one of the search methods that has shown the highest increase, 114% in the last seven years. Some of the applications of genomic location search are the identification of RNA hotspots, transcribed potentially by the same transcriptional unit; the characterization of ncRNAs according to neighboring genes; and the characterization of regulatory regions located upstream to the set of ncRNAs of interest. Figure 3 shows the distribution and relative increase in the last seven years of each of the search methods.

Finally, six databases offer search by density of ncRNAs. This search can be useful to locate ncRNA hotspots in the genome corresponding to a single family of transcriptional regulation [7]. The list of databases offering each search method can be easily obtained in the *statistics* page of NRDR.

2.3 RNA Classes

Analyzing the evolution of the ncRNA databases in the last five years can shed light on some of the new trends on ncRNA research and on the panorama of the field today. Figure 4 shows a chart representing the absolute number of databases for each ncRNA class in 2011 and 2018. Clearly miRNAs dominate the field with

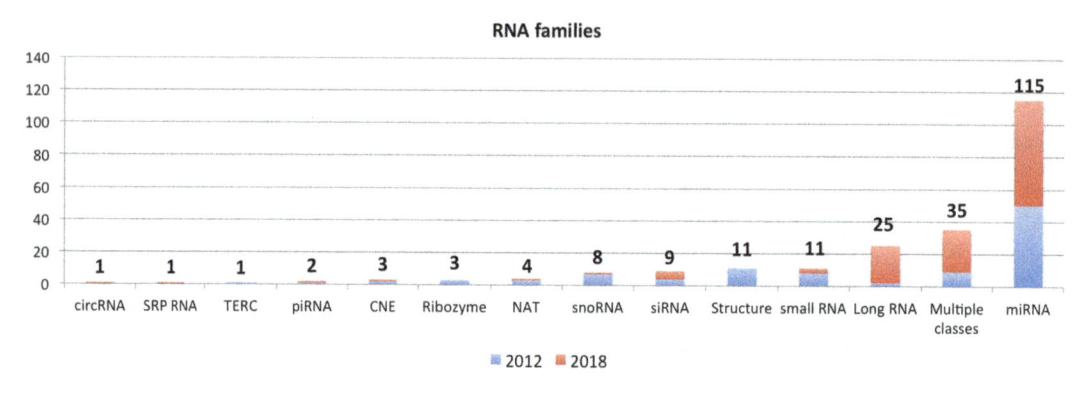

Fig. 4 The evolution of the number of databases for each ncRNA class in the last 6 years. The blue color bars represent the number of databases in 2012 and the red color bars represent the additional number of databases in 2018

roughly 50% (115 in total) of all databases dedicated to them, and representing an increase of 130% in the number of repositories for this particular RNA class. If we consider small ncRNAs without a particular RNA class associated, they constitute the subject of ~5% of all databases. This can be partially explained by the set of other small RNA families obtained from high-throughput RNA sequencing assays, in which several other small RNAs yet without a functional classification are obtained. In general, these nonannotated short transcripts are simply classified as small RNAs in databases and publications. However, the most dramatic shift in the interest of the ncRNA community seems to be the increasing attention given to long ncRNAs. In 2011 we could locate only two databases dedicated to the subject. This number increased to 25 (a growth of 1150%), constituting now about 11% of all databases researched. Figure 4 shows the evolution on the number of databases covering each ncRNA class in the last 6 years. In the rest of this section we analyze the ncRNA classes with more than eight databases.

2.3.1 microRNAs (miRNAs)

MicroRNAs are small ncRNAs (normally containing 22 nucleotides in its mature form) that, by interacting with messenger RNAs via complementarity of their nucleotides, participate in RNA silencing and posttranscriptional regulation of gene expression, inducing mRNA degradation or translational repression. They have been found in different kingdoms (animals, plants, and viruses) and are key regulators of important biological processes [8]. miRNA dysfunctions may lead to the development of different human chronic diseases (i.e., cancers, neurological disorders, diabetes, cardiovascular diseases). Furthermore, during viral infection, some viruses use miRNAs to regulate viral and host gene expression to escape the immune system of host cells, facilitating viral latency [9]. The continuous increase in the number of publicly available databases focused on miRNAs reinforced the tendency that most of the ncRNA research has concentrated on miRNAs. Since the original publication of the NRDR repository in 2012, the number of miRNA databases increased by 130% (50 in 2012, 115 in 2018) in 6 years.

Apart from general repositories covering multiple RNA classes (i.e., Rfam, RNAcentral, NONCODE), our survey identified a total of 115 public databases related exclusively to miRNA research, of which 35 cover a wide range of species. The most important microRNA database is miRBase, the state-of-the-art repository of publicly available microRNA information. It was established in 2002 and aims to contain all published microRNA sequences and associated annotation, with the primary aim of assigning stable and consistent names to newly discovered microRNAs [10]. The importance of miRNAs for human health may explain the fact that the majority of publicly available databases of miRNAs include human sequences: 34 databases include only human sequences, 11 cover data from humans and mice, five of them store information from humans, mice, and rats, and two of them include also data from other vertebrates. There are also 6 databases dedicated to plants and seven covering other species-specific miRNAs, most of which are related to model or economically important organisms (i.e., *Arabidopsis thaliana*, grape, tomato, rice, trout, salmon). Finally, we have seven covering human virus miRNAs, which is of particular interest to biomedical research, since it is dedicated to matching miRNAs to their targets host genes. This increase in the number of plants and species-specific miRNA databases may be the result of the increasing use of RNA next-generation sequencing in non-model organisms. Table 1 provides a list of databases for each species or combination of species.

Table 1
MicroRNA databases and organisms: list of databases for each species or combination of species

Species	Databases
Human	CCGD, ceRDB, ComiRNet, CSCdb, dbDEMC, dbSMR, dPORE, ExcellmiRDB, ExprTargetDB, HMDD, HOCTARdb, IGDB.NSCLC, Immune-miR, microPIR, mimiRNA, miR-EdiTar, miR2Disease, miRCancer, mirCoX, miRdSNP, miReg, miRGator, miRmine, miRnalyze, miRvar, NCG, OCDB, PhenomiR, Psmir, PuTmiR, RCGD, S-MED, SMiR-NBI, SomamiR, TMREC
Human+Mouse	ChIPBase (+vertebrates), CircuitsDB, CSmiRTar, ExoCarta (+vertebrates), IntmiR, Isomirs, miRGate (+rat), mirGen, miRNA Body Map(+rat), miRPathDB, miRSel (+rat), miRWalk (+rat), miTALOS, OncomiRdbB, PolymiRTS, ViTa (+rat), virusTarget finding
Plant	AtmiRNET, comTAR, MicroPC, miRFANs, PmiRExAt, PmiRKB
Other Species Specific	*A. thaliana* (AtmiRNET,mirEX), Trout (MicroTrout), *Vitis vinifera* (GrapeMiRNA), *Mus musculus* (miRNeye), Tomato (miSolRNA), Rice (RiceATM), Salmon (Ssa miRNAs DB)
Viruses	ViTa, VIRsiRNAdb, Vir-Mir, VIRmiRNA, vHoT DB, ViRBase, ZIKV-DB

The vast majority (104 in total) of miRNA databases contain information generated by in silico annotation protocols. Also, 54 databases include information obtained from the literature search. Probably due to the relevance of miRNA research and to the low cost of RNA-seq experiments, almost half of the databases (63 in total) contain information based on experimental sources. However, only 71 of the databases make the data available for download, of which only 14 offer the data directly in FASTA format or similar sequence file. Sequence data is crucial for functional annotation and prediction of novel miRNAs in non-model organisms through sequence similarity searches, or for the development of novel tools for miRNA prediction [11]. A surprisingly small number of just four databases provide genomic coordinate data for download (i.e., BED, GFF, GTF, or PSL). Genomic location is important for the functional annotation of reconstructed small transcripts in RNA-seq pipelines. Data in sequence or genomic coordinates formats retrieved from public databases has been widely used as reference for high-throughput expression experiments, commonly implemented to functionally characterize miRNAs [12] or to associate them as potential biomarkers for human diseases [13]. The complete list of databases offering each data type can be easily obtained in the *statistics* page of NRDR.

Expression profiles of miRNAs from different cultured cell lines, tissues, specific conditions, or diseases are made available in 31 databases. Other 62 databases possess information related to the interaction between microRNA and their target transcripts. Data from target coding genes, noncoding RNAs, or transposable elements are made available in these databases. Single nucleotide polymorphisms (SNPs) are the most common type of genetic variation. SNPs can affect human health or development, being important for human ancestry and admixture studies [14, 15], drug resistance [16], or evolutionary research in other species [17]. SNPs available in microRNA sequences or target regions are depicted in nine databases (dPORE, IGDB.NSCLC, miRdSNP, mirGen, miRNA SNiPer, miRNASNP, OCDB, PmiRKB, PolymiRTS). This information is summarized in Table 2.

Other relevant information in miRNA databases includes structure (microTranspoGene, mirEX, miRTarBase), epigenetic modifications (EpimiR), function (CCGD, mESAdb, MicroPC, miRDB, miRFANs, GrapeMiRNA, miRNA Body, miRPathDB, miRSponge, miRTarBase, miTALOS, NCG, RiceATM), promoter characterization (AtmiRNET, dPORE, CircuitsDB, mirGen), RNA-binding proteins (doRiNA), and other phenotype data (miRò, miReg, miREnvironment, RiceATM).

Table 2
MicroRNA databases and their information content

Information content	Databases
Expression profiles	ChIPBase, CoGemiR, dbDEMC, ExcellmiRDB, exoRbase, ExprTargetDB, IGDB.NSCLC, Isomirs, mESAdb, microRNA.org, mimiRNA, miRandb, miRò, miReg, *mirEX*, miRFANs, mirGen, miRmine, miRNA Body, miRNAMap, miRNApath, miRNASNP, *miRNeye*, miRSel, *miSolRNA*, NCG, PhenomiR, PmiRExAt, PmiRKB, RiceATM, S-MED
miRNA/Target interactions	AtmiRNET, CircuitsDB, ComiRNet, comTAR, CIRC2TRAITS, CSmiRTar, CSCdb, DIANA miRPath, DIANA-TarBase, ExcellmiRDB, ExprTargetDB, HOCTARdb, Immune-miR, IntmiR, mESAdb, MicroPC, microPIR, microRNA.org, *MicroTrout*, mimiRNA, miR-EdiTar, miR2Disease, miRandb, miRBase, mirCoX, miRDB, miRDeathDB, mirDIP, miRdSNP, miRecords, miReg, miRFANs, miRGate, miRGator, miRnalyze, miRNAMap, miRNApath, miRNASNP, miRNA_Targets, miRNEST, miRPathDB, miRSel, miRSponge, miRTarBase, miRWalk, *miSolRNA*, miTALOS, MNDR, OCDB, OncomiRdbB, PmiRKB, RCGD, SMiR-NBI, SomamiR, Ssa miRNAs DB, vHoT, ViTa, Vir-Mir, VIRBase, VIRmiRNA, ZIKV-DB
SNPs	dPORE, IGDB.NSCLC, miRdSNP, mirGen, miRNA SNiPer, miRNASNP, OCDB, PmiRKB, PolymiRTS, ChIPBase, CIRC2TRAITS
Phylogeny	CCGD, CoGemiR, comTAR, Isomirs, MDTE DB, MicroPC, microRNA.org, miRTarBase, NCG, SomamiR, ZooMir

2.3.2 Long Noncoding RNAs (lncRNAs)

Since our first survey of public ncRNA databases in 2012, a class that has gained increasing attention from research community is long noncoding RNA (lncRNA). The advent of high-throughput expression profiling technologies resulted in the identification of thousands of lncRNAs in eukaryotes [1, 18]. For instance, lncRNAs compose the main transcriptional repertoire of the mammalian genome, with over 100,000 human lncRNAs cataloged in public databases [19]. Although lncRNAs are known by their participation in the fine-tuning regulation of different cellular processes (i.e., transcriptional regulation, chromatin modification, cell differentiation, epigenetic regulation, and immune responses), the functions of only a small fraction have been experimentally tested [20]. The attention of the scientific community to this particular RNA type is clearly observed in the evolution of lncRNA exclusive databases, which increased from 2 to 25 since the first publication of NRDR. Of these, 4 are dedicated to plant lncRNAs (CANTATAdb, GreeC, PeTMbase, Plncdb), 12 are dedicated just to humans (CSG, GermlncRNA, LNCat, LncSNP, Lnc2Cancer, lnCeDB, LNCMap, Lnc2Meth, lncATLAS, lncPedia, lncRNAdisease, lncRNome), 5 cover a wider range of vertebrates (ANGIO-GENES, DIANA-LncBase, lncEditing, lncRNACNP, NRED), and 4 cover a wide range of organisms (EVLnRNAs, GreeNC, LNCediting, lncRNAdb).

Despite the significant increase in the number of lncRNA databases, there are still a small number of them. Of these, 80% (20 in total) contains reliable experimental data and 96% (24 in total) includes in silico annotation. Finally, 76% (19 in total) present information based on literature surveys, and only 32% (8 databases) contain manually curated information. Many lncRNA databases provide data for downloading useful for downstream analyses: in total, four databases provide the nucleotides sequences for downloading in FASTA format; 4 provide their genomic coordinates in standardized formats such as BED, GFF, GTF, or PSL; and 17 provide other kind of data for downloading in nonstandardized formats, containing useful information regarding their association with disease, expression profiles, or interaction with other biomolecules such as DNA, RNA, or proteins. The list of databases in each category can be easily obtained with a search on NRDR.

Expression profiles of lncRNAs are made available in seven databases (ANGIOGENES, CANTATA, ChIPBase, exoRbase, GermLncRNA, NRED, Plncdb), covering tissue or cell lines specificity, as well as disease expression data. Information related to the biogenesis and regulation of these lncRNAs is made available in three databases, with two of them (LincSNP and LNCEditing) covering information associated with transcriptional binding sites (TFBSs) and only one specific for differentially methylated lncRNAs in diverse human diseases (Lnc2Meth).

As we previously mentioned, the majority of lncRNAs do not have their function uncovered so far. However, functional mechanisms of action for lncRNAs are made available in 5 databases (lncRNAdb, LNCat, lncRNome, The Functional lncRNA Database, VIRBase). Their interaction with other RNAs, DNA, or proteins is covered in five databases (ChIPBase, EVLncRNAs, lncRInter, lncRNAdisease, lncRNome). RNA secondary structure information, important to characterize stability and function, is present in five databases (GreeNC, LNCipedia, LNCat, LncRNASNP, NRED). Interestingly, other five databases include information about microRNA-lncRNA interaction (DIANA-LncBase, lnCeDB, LncRNASNP, LNCipedia, PeTMbase). This interaction may cover lncRNAs potentially regulated by microRNAs, or lncRNAs that act as miRNA mimics, which bind to the miRNA molecules, avoiding these miRNAs to regulate the expression of their normal target gene. Unfortunately, even with the broad range information associating lncRNAs with disease, scarcer information is made available related to SNPs located in these long transcripts. Only two databases (LncRNASNP, LincSNP) cover SNPs information. Finally, MNDR is a database specific for mammal ncRNA diseases.

2.3.3 Other RNA Classes   We observed, in our first survey of publicly available in noncoding RNA databases, that a limited amount of long ncRNA data. Now, this changed. On the other side, for the research of specific small RNA classes (i.e., tRNA-derived small RNAs, rRNA-derived small RNA, piwi-interacting RNA), only piRNAs were categorized in two novel specific databases (piRNA Cluster database, piRNA-Bank). Apart from small RNAs without functional associations, databases for small RNA also cover other classes such as the well-studied snoRNAs, siRNAs/RNAi, and specific RNAs from bacteria. Bacterial RNAs are stored in three databases (SRD, sRNAmap, Rfam), with information related to a wide range of small RNAs with a variety of specific functional roles in regulating important cellular processes.

Of the databases dedicated to other small RNA classes, 10 cover a wider range of organisms (snoRNAdb, snOPY, snoRNP, Sno/scaRNAbase, tasiRNAdb, VIRsiRNAdb, RNAi Codex, piRNA Cluster database, piRNABank, starBase), 6 are dedicated to human (HuSiDa, snoRNA-LBME-db, tasiRNAdb, RNAimmuno, Autism Genetic Database, DASHR), 4 to plants (Plant snoRNA database, ASRP, CSRDB, The Tomato small RNA Database), one to mammals in general (The MIT/ICBP siRNA Database), one to viruses (VIRsiRNAdb), one to yeast (Yeast snoRNA database), and one specific to Arabidopsis (ASRP). More specific repositories are the few databases for snoRNAs from plants (ASRP, CSRDB, Plant snRNA database, The Tomato small RNA Database), yeast (Yeast snoRNA database), and from multiple species including human and organisms from different kingdoms (Methylation Guide snoRNA, Sno/scaRNAbase, snOPY, U12DB). Databases for small interfering RNAs (siRNAs/RNAi) are available for plants (tasiRNAdb), pathogens (VIRsiRNAdb), and model organisms (GenomeRNAi, HuSiDo, RNAi Codez, RNAiAtlas, siRecords, siRNAdb, The MIT/ICBP siRNA database). Only 4 databases covering different classes of small RNAs are related to human disease (The MIT/ICBP siRNA Database, Autism Genetic Database, starBase), with only one associated with different cancer types (The MIT/ICBP siRNA database). In total, 17 databases (ASRP, CSRDB, DASHR, HuSiDa, RNAimmuno, snOPY, snoRNA-LBME-db, snoRNP, sRNAMap, starBase, tasiRNAdb, The MIT/ICBP siRNA Database, VIRsiRNAdb, RNAi Codex, The Tomato small RNA Database, Yeast snoRNA database) cover information related to any type of interaction between these small RNAs and other molecular types (other RNAs, DNAs, or proteins). Other relevant information available in small RNA databases include tissue and stage specificity (DASHR), genomic location (snOPY, piRNABank), function (snoRNA-LBME-db), expression (DASHR, sRNAMap, The Tomato small RNA Database), methylation (snoRNAdb), and structure (snoRNA-LBME-db, RNAi Codex, sRNAMap).

2.3.4 Multiple-Class RNA Databases

Apart from all long and small RNA-specific databases mentioned before, there is a set of other 28 databases covering multiple RNA classes. We observed a considerable increase of 211% (up from nine databases) in the number of multiple RNA classes databases. It is known that different classes of RNAs interact each other in order to develop their functional roles in the cells. This increase in the number of multiple classes databases might be related to the evolution of computational and experimental methodologies to retrieve information related to RNA-RNA interactions. The most important group is certainly that of general repositories that cover a wide range of species from all different kingdoms (Rfam, RNACentral, NONCODE, UCSC Genome Browser, NCBI RefSeq, Ensembl, deepBase, fRNAdb, NAPP, ncRNAdb, NPInter, NSDNA, RNArchitecture). However, it is worth noting that the UCSC Genome Browser, NCBI RefSeq, and Ensembl are not specific for noncoding RNAs. In general, these repositories provide annotation information from coding genes, ncRNAs, and pseudogenes. The remaining databases cover a more limited range of RNA classes with variable annotation information, such as genomic imprinting associations (ncRNAdb, ncRNAImprint, NONCODE), phylogeny (NAPP, RNArchitecture), and SNPs (rVarBase). Some databases cover general RNA classes but are specific for certain species or evolutionary groups, such as plants (PNRD contains information from 11 different types of ncRNAs from 150 plant species); human (CircNet, decodeRNA, NRDTD, rVarBase, YM500—with mouse); mammals (ncRNAimprint, RNADb, SpermBase, MnbcRNAdb); and Leishmania (Leishdb). Finally, another interesting case is the databases associated with RNAs circulating in body fluids (miRandola, exoRBase). These RNAs have been described as important biomarkers for different diseases and belong to different classes such as circular RNAs (circRNAs), lncRNAs, and microRNAs.

3 An Overview of Public Databases Applied to Biomedicine

The increased attention of the scientific community to noncoding RNAs is also reflected in the amount of public databases with data relevant to human health. The various types of data stored in the databases with potential associations with biomedicine cover information related to expression patterns of ncRNAs in disease-associated cell lines, tissues, or other types of blood samples (i.e., urine, blood, plasma); the response of ncRNAs to drugs and other chemical compounds administration; the response of human or model organisms to a potential pathogen exposition; and the repertoire of ncRNAs available in human pathogens genomes. An analysis of this information will paint a general picture of the results of ncRNA research associated with human health and biomedicine available in public databases and can serve as a guide for further investigation in the role of ncRNAs in different diseases.

To identify and categorize all biomedicine-related ncRNA databases, we manually examined the ones indexed in NRDR, by mapping the repositories that contain information related to diseases, pathogens, drugs, or association with chemical compounds. We examined the terms associated with these three categories and used this information to summarize the recent progress and breakthroughs of ncRNA research on human health.

We retrieved a total of 1526 terms related to disease from 47 databases, and 1393 to drugs or chemical compounds from 10 databases. In general, data available in these databases was obtained from literature survey or using different experimental and computational approaches using human and model organisms (i.e., *Mus musculus, Rattus norvegicus, Drosophila melanogaster*). The vast majority of the databases cover mainly three RNA types: microRNAs, circular RNAs, or long noncoding RNAs (including lincRNAs); but other four types (piwiRNAs, shRNAs/siRNAs, Telomerase RNA, other small RNAs) are also observed. There is scarce information from other RNA classes, which is available in databases containing data from human pathogens like *Leishmania braziliensis* (LeishDB) and bacteria (Rfam).

There is a wide range of information available in these databases, containing data related to expression profiles; interaction between ncRNA and their target transcripts in an RNA:RNA, RNA:protein, or RNA:DNA relationship; biogenesis regulation (i.e., transcription factor binding sites, epigenetic markers); and putative regulated networks and pathways. We categorized the diseases available in public databases in five main categories: "cardiovascular system diseases," "cancers," "nervous system diseases," "viruses and other pathogens," and "other diseases," which will be addressed in separate subsections.

3.1 Cardiovascular Disease Databases

The advent of high-throughput expression measuring technologies revealed different types of noncoding RNAs as important components of the fine-tuning regulation of heart development and cardiovascular systems diseases [21, 22]. For instance, *circANRIL* is a circRNA transcribed from the *locus* of *INK4*—a genomic region phenotypically associated with atherosclerosis, which has been proven to possess protective roles by controlling and modulating atherogenic pathways [23]. *Myheart* is a lncRNA that acts by remodeling chromatin structure, antagonizing the effects of Brg1 and preventing a decontrolled transcription of genes associated with cardiac myopathies. Its repression is directly associated with cardiomyopathies, while the restoration of initial levels of expression protects the heart from hypertrophy and cardiac insufficiency [24]. Our survey revealed a total of 12 databases (CIRC2TRAITS,

PhenomiR, HMDD, miRdSNP, LncRNADisease, lncRNAdb, miR2Disease, MNDR, EVLncRNAs, lncRnome, ImmunemiR, BioM2MetDisease) covering 223 terms associated with cardiovascular diseases. The most representative disease is diabetes, represented in all databases. Although there is no specific repository for ncRNAs associated with cardiovascular system diseases, they are strongly represented in more general repositories, covering a wide range of diseases. Noncoding RNA classes represented in these databases are circRNAs, lncRNAs, miRNAs, piRNAs, and snoRNAs.

3.2 Cancer-Related Databases

Similar to cardiovascular diseases, ncRNAs has been widely associated with cancers. They are relevant oncogenic drivers and tumor suppressors in every major cancer type [25]. The cells during their cycle of life express a large amount of ncRNA types varying according to their normal functions. However, cancer cells alter their expression and adjust genetics at various levels. The changes may occur in a profile of RNAs other than those circulating in a person not affected by the disease. This approach was used for the application of microRNAs in new experimental protocols. Their importance to cancers is reflected in the abundance of ncRNAs related to a large number of cancers available in the surveyed databases, with 629 terms associated with different cancer types. They are represented in almost all databases, with exception to those specific for viruses, leishmaniasis, and neurological disorders. The most representative cancer type is leukemia, with 58 associated terms (i.e., acute myeloid leukemia, T-cell acute lymphoblastic leukemia, monocytic leukemia) appearing 112 times in 16 databases (exoCarta, miRandola, CIRC2TRAITS, Phenomir, miReg, HMDD, VirmiRNA, VIRBase, miRSNP, lncRNADiesease, lncRNAdb, miR2Disease, MNDR, ImmunemiR, EVLncRNAs, dbDEMC). There are also 13 databases specific for cancer research (CCG, CCDB, HMDD, IGDB.NSCLC, YM500, miREnvironment, CSCdb, NCG, SomamiR, dbDEMC, miRCancer, decodeRNA, RCDB). CCDB, IGDB.NSCLC, CSCdb, and RCDB are specific repositories for microRNAs associated with cervix, lung, stem cells, and renal cancer, respectively. Associations of the other RNA classes (circRNAs, lncRNAs, snoRNAs, snRNAs) to different cancer types are spread over the remaining databases.

3.3 Nervous System-Related Databases

Different ncRNAs have been associated with the human nervous system, having essential roles in brain evolution, development, plasticity, and disease [26]; as well as in diseases affecting the central [27] and peripheral nervous system [28], with some of them with essential roles in neurodegenerative disorders [29]. There are four databases specific for neurological diseases. OCDB (miRNAs) is specific for obsessive-compulsive disorder, the Autism Genetic Database (piRNAs, miRNAs, snoRNAs, siRNAs) for autism,

EpimiRBase (miRNAs) for epilepsy, and NSDNA (miRNAs, lncRNAs siRNAs, snoRNAs, piRNAs) for 16 nervous systems disorders. OCDB also contains information related to the potential associations of 48 drugs administration and the set of miRNAs stored there. Despite the rising importance of circulating RNAs as biomarkers for neurodegenerative diseases and other neurologic pathologies [13, 30], only NSDNA contains data related to circulating miRNAs. We found 284 terms associated with nervous system-related diseases, with multiple sclerosis, the most representative, present in 11 databases (miRandola, CIRC2TRAITS, PhenomiR, miReg, HMDD, LncRNADisease, miR2Disease, MNDR, ImmunemiR, EVLncRNAs, NSDNA).

3.4 Viruses and Other Pathogens

Viral ncRNAs are known to participate in diverse biological processes important for the regulation of viral replication, viral persistence, host immune evasion, and cellular transformation [31]. It is known that some miRNAs may even participate in the oncogenic process during viral infection [9]. Our survey revealed seven databases specific for virus-related ncRNAs (ViTa, VIRsiRNAdb, Vir-Mir, VIRmiRNA, vHoT DB, ViRBase, ZIKV—DB), covering mainly data associated with host target genes and functional annotation. ViRBase and ViTa databases store data from both viruses and host-affected ncRNAs, while the remaining are specific for viral ncRNAs. There are five databases specific for miRNAs (ViTa, Vir-Mir, VIRmiRNA, vHoT DB, ZIKV—DB), one specific for siRNA/shRNA (VIRsiRNAdb), and one containing data related to miRNAs, siRNAs/shRNAs, lncRNAs, and snoRNAs (ViRBase). Information from thousands of human viruses is stored in ViTa and Vir-Mir databases. The remaining databases cover dozens of viral species, with only ZIKV—DB specific for Zika virus. There are also other non-viral-specific databases covering viral ncRNAs (miRBase, for miRNAs) or host-affected ncRNAs (LncRNADisease, lncRNAdb, and EVLncRNAs for lncRNAs; miReg, miR2Disease, miRandola, miRdSNP, and HMDD for miRNAs; and HMDD for miRNAs, lncRNAs, piRNAs, and snoRNAs).

Another pathogen-specific database is LeishDB, which covers the repertory of ncRNAs available in *Leishmania braziliensis* genome, a protozoan parasite with an infection rate estimated in 0.7–1.3 million new cases yearly [32]. Data stored in LeishDB was predicted using sequence similarity searches and covariance models, validated by RNA-seq experiments. Human lncRNAs affected by leishmaniasis infection can be found at LncRNADisease.

There is not a specific database for bacterial pathogen species, but ncRNAs available in their genomes (Rfam) or identified as affected by a bacterial infection (lncRNAdb, LncRNADisease, MNDR, EVLncRNAs) can be found in other databases.

3.5 Other Diseases Similar to cancer, other hundreds of immunological, metabolic, and chronic diseases are included in almost all nonspecific disease databases, with all RNA types described in this book chapter represented. The only specific database associated with other than cardiovascular, cancers, or neurological diseases is Immune-miR. This database provides data related to miRNA-disease association through interactome network, obtained by the integration of 2702 papers and different miRNA, target genes, and immunological databases. It covers a total of 92 diseases associated with inflammation, cancer, and autoimmune disorders.

3.6 Drugs and Other Chemical Compounds The effects of drugs and other chemical compounds have been recently modeled in different systems biology approaches, in order to obtain the transcriptional response to particular treatments and vaccines [33, 34]. There are four databases specific for the effects of drugs in different transcripts and gene networks. Pharmaco-miR provides associations of miRNAs, genes they regulate, and a set of 941 drugs annotated in literature as dependent on these genes. SMiR-NBI provides miRNAs with expression affected (up- or downregulated) by the administration of 154 small molecules. GEAR provides comprehensive information about genomic pattern associated with drug resistance for 66 miRNAs and 29 drugs, including SNP information related to that potential resistance. NRDTD provides a repertoire of clinically or experimentally curated data obtained from literature for 94 drugs, and its association with miRNAs and lncRNAs. Other drugs and chemical compounds associations with ncRNAs are made available in more general databases (i.e., miRandola, miReg, BioM2MetDisease, lncRNAdb, EVLncRNAs, NSDNA).

3.7 Circulating RNAs Finally, an emerging field in RNA biomedicine research is related to circulating ncRNAs available in body fluids. These RNAs are stable and protected from degradation and can be used as noninvasive biomarkers in a strategy for chronic and infectious diseases prognostics or diagnosis [13, 35]. They can be found encapsulated in large lipoprotein complexes, such as exosomes or microvesicles (MVs), which are actively released by living cells [36]. Fluid samples such as blood or urine can replace invasive biopsies, increasing patients' quality of life and costs. Our survey revealed three databases related to circulating RNAs detected in exosomes body fluids: exoRbase, ExoCarta, and miRandola. The first, exoRBase, is a repository of circRNAs, lncRNAs, and messenger RNAs (mRNAs) derived from RNA-seq data analyses of healthy and diseased (colorectal cancer, coronary heart disease, hepatocellular carcinoma, pancreatic adenocarcinoma) human blood exosomes. The second, ExoCarta, is a compendium of exosome datasets generated by the integration of 64 studies focused in identifying and characterizing proteins, mRNA, and miRNAs available in exosomes derived from different body fluids, reporting the their association with 28 different diseases (27 cancers and ascites). The third, miRandola, is a comprehensive, manually curated, repository of circulating miRNAs, lncRNAs, and circRNAs, built through the integration of data retrieved from 314 papers which analyzed the expression of these RNAs in serum, plasma, saliva, urine, and other body fluids in 197 diseases and cell lines. Finally, NSDNA (The Nervous System Disease NcRNAome Atlas) is not specific for circulating RNAs but includes information from microRNAs detected in body fluid obtained from diverse nervous system diseases.

4 Conclusion

Noncoding RNA is a thriving research topic. The last 25 years have seen an increasing number of articles and databases dedicated to these molecules. We have presented an overview of the information available in public databases related to the theme, which give us a snapshot of the current status of their research. The new sequencing technologies, in silico annotation, and big data protocols are enabling scientists to add an immense amount of information and the field should continue to grow in the following years. In particular, miRNA will most likely continue to be one of the hot topics, given its important role in the regulation of protein synthesis in living organisms. Another clear trend seems to be the research on long noncoding RNAs, which presented a spectacular growth in the number of databases in the last 6 years. A third potentially hot topic to be observed is circulating RNAs, due to its application as biomarkers for diseases diagnosis and prognosis. These three ncRNA classes are of particular biomedical interest, given their already reported relationship to different chronic and infectious diseases. However, so far biological validation protocols are still expensive. Most information regarding experimentally validated ncRNAs, with the exception of expression data, will continue to constitute a small percentage of the total information available on the field. Given the costs involved in RNA sequencing we expect this information to present a steady, but less drastic, growth most likely housed either in large wide-spectrum repositories or in smaller, more specialized databases. All information about each database mentioned in this chapter can be retrieved using NRDR (http://ncrnadtabases.org), a web resource that constitutes the companion to this work.

5 List of Databases, Their Web Address, and Original Reference

Table 3 lists all the ncRNAs databases surveyed for this chapter, cross referencing them with the http address and reference for the original article.

Table 3
Noncoding RNA public databases: database name, http address, and article reference

Name	URL	Reference
Condor	http://condor.nimr.mrc.ac.uk/	[37]
TFCONES	http://tfcones.fugu-sg.org/	[38]
deepBase	http://deepbase.sysu.edu.cn/	[39]
fRNAdb	http://www.ncrna.org/frnadb	[40]
ncRNAdb	http://ncrnadb.trna.ibch.poznan.pl/	[41]
ncRNAimprint	http://rnaqueen.sysu.edu.cn/ncRNAimprint/	[42]
NONCODE	http://www.noncode.org/	[43]
Rfam	http://rfam.sanger.ac.uk/	[44]
RNAdb	http://research.imb.uq.edu.au/rnadb/	[45]
GISSD	http://www.rna.whu.edu.cn/gissd/	[46]
Mobile Group II Introns	http://webapps2.ucalgary.ca/~groupii/	[47]
lncRNAdb	http://lncrnadb.com/	[20]
NRED	http://jsm-research.imb.uq.edu.au/nred/	[48]
CircuitsDB	http://biocluster.di.unito.it/circuits/	[49]
CoGemiR	http://cogemir.tigem.it/	[50]
dbDEMC	http://159.226.118.44/dbDEMC/	[51]
dbSMR	http://miracle.igib.res.in/polyreg/	[52]
dPORE	http://apps.sanbi.ac.za/dpore/	[53]
ExprTargetDB	http://www.scandb.org/apps/microrna/	[54]
HMDD	http://202.38.126.151/hmdd/mirna/md/	[55]
HOCTARdb	http://hoctar.tigem.it/	[56]
IntmiR	http://rgcb.res.in/intmir/	[57]
mESAdb	http://konulab.fen.bilkent.edu.tr/mirna/	[58]
MicroPC	http://konulab.fen.bilkent.edu.tr/mirna/	[59]
microRNA.org	http://www.microrna.org/	[60]
microTranspoGene	http://transpogene.tau.ac.il/microTranspoGene.html	[61]

(continued)

Table 3
(continued)

Name	URL	Reference
mimiRNA	http://mimirna.centenary.org.au/	[62]
MIR@NT@N	http://mironton.uni.lu/	[63]
miR2Disease	http://www.mir2disease.org/	[64]
miRBase	http://www.mirbase.org/	[65]
miRDB	http://www.mirdb.org/	[66]
mirDIP	http://ophid.utoronto.ca/mirDIP/	[67]
miRecords	http://mirecords.biolead.org/	[68]
miReg	http://www.iioab-mireg.webs.com/	[69]
miRGator	http://mirgator.kobic.re.kr/	[70]
mirGen	http://www.microrna.gr/mirgen/	[71]
miRHrt	http://sysbio.suda.edu.cn/mirhrt/	[72]
miRNAMap	http://mirnamap.mbc.nctu.edu.tw/	[73]
miRNApath	http://lgmb.fmrp.usp.br/mirnapath/	[74]
miRNeye	http://mirneye.tigem.it/	[75]
miROrtho	http://cegg.unige.ch/mirortho/	[76]
miRSel	http://services.bio.ifi.lmu.de/mirsel/	[77]
miRTarBase	http://mirtarbase.mbc.nctu.edu.tw/	[78]
miRvar	http://genome.igib.res.in/mirlovd	[79]
miRWalk	http://www.ma.uni-heidelberg.de/apps/zmf/mirwalk/	[80]
miSolRNA	http://www.misolrna.org/	[81]
miTALOS	http://mips.helmholtz-muenchen.de/mitalos/	[82]
NetAge	http://netage-project.org/	[83]
Patrocles	http://www.patrocles.org/	[84]
PhenomiR	http://mips.helmholtz-muenchen.de/phenomir/	[85]
PmiRKB	http://bis.zju.edu.cn/pmirkb/	[86]
PMRD	http://bioinformatics.cau.edu.cn/PMRD/	[87]
PolymiRTS	http://compbio.uthsc.edu/miRSNP/	[88]
PuTmiR	http://www.isical.ac.in/~bioinfo_miu/TF-miRNA.php	[89]
RepTar	http://bioinformatics.ekmd.huji.ac.il/reptar/	[90]
S-MED	http://www.oncomir.umn.edu/SMED/	[91]
starBase	http://starbase.sysu.edu.cn/	[92]

(continued)

Table 3
(continued)

Name	URL	Reference
TarBase	http://diana.cslab.ece.ntua.gr/tarbase/	[93]
TransmiR	http://202.38.126.151/hmdd/mirna/tf/	[94]
UCbase & miRfunc	http://microrna.osu.edu/.UCbase4/	[95]
ZooMir	http://insr.ibms.sinica.edu.tw/ZooMir/	[96]
antiCODE	http://bioinfo.ibp.ac.cn/ANTICODE	[97]
NATsDB	http://natsdb.cbi.pku.edu.cn/	[98]
Trans-SAMap	http://trans.cbi.pku.edu.cn/	[99]
NPInter	http://www.bioinfo.org.cn/NPInter/	[100]
piRNABank	http://pirnabank.ibab.ac.in/	[101]
The ribonuclease P database	http://www.mbio.ncsu.edu/rnasep/	[102]
HuSiDa	http://itb.biologie.hu-berlin.de/~nebulus/sirna/	[103]
siRecords	http://sirecords.biolead.org/	[104]
siRNAdb	http://sirna.sbc.su.se/	[105]
The MIT/ICBP siRNA Database	http://web.mit.edu/sirna/	–
ASRP	http://asrp.cgrb.oregonstate.edu/	[106]
Autism Genetic Database	http://wren.bcf.ku.edu/	[107]
cre-siRNA	http://cresirna.cmp.uea.ac.uk/	[108]
CSRDB	http://sundarlab.ucdavis.edu/smrnas/	[109]
SiLoDb	http://silodb.cmp.uea.ac.uk/	–
smiRNAdb	http://www.mirz.unibas.ch/cloningprofiles/	[110]
sRNAmap	http://srnamap.mbc.nctu.edu.tw/	[111]
sRNATarBase	http://ccb.bmi.ac.cn/srnatarbase/	[112]
The Tomato small RNA Database	http://ted.bti.cornell.edu/	[113]
Sno/scaRNAbase	http://gene.fudan.sh.cn/snoRNAbase.nsf	[114]
snoRNA-LBME-db	http://www-snorna.biotoul.fr/	[115]
Methylation Guide snoRNA	http://lowelab.ucsc.edu/snoRNAdb/	[116]
Plant snoRNA database	http://bioinf.scri.sari.ac.uk/cgi-bin/plant_snorna/home	[117]
snoRNP database	http://www.evolveathome.com/snoRNA/snoRNA.php	[118]
U12DB	http://genome.imim.es/datasets/u12/	[119]
Yeast snoRNA database	http://people.biochem.umass.edu/fournierlab/snornadb/	[120]

(continued)

Table 3
(continued)

Name	URL	Reference
SRPDB	http://rnp.uthscsa.edu/rnp/SRPDB/SRPDB.html	[121]
BPS	http://bps.rutgers.edu/	[122]
BRAliBase	http://projects.binf.ku.dk/pgardner/bralibase/	[123]
CRW	http://www.rna.ccbb.utexas.edu/	[124]
MeRNA	http://merna.lbl.gov/	[125]
NCIR	http://prion.bchs.uh.edu/bp_type/	[126]
NNDB	http://rna.urmc.rochester.edu/NNDB/	[127]
PseudoBase	http://pseudobaseplusplus.utep.edu/	[128]
RNA FRABASE	http://rnafrabase.ibch.poznan.pl/	[129]
RNA STRAND	http://www.rnasoft.ca/strand/	[130]
RNAJunction	http://rnajunction.abcc.ncifcrf.gov/	[131]
SCOR	http://scor.berkeley.edu/	[132]
Telomerase Database	http://telomerase.asu.edu/	[133]
The UCSC Genome Browser database	http://genome.ucsc.edu/	[134]
miRNA Body Map	http://www.mirnabodymap.org	[135]
IGDB.NSCLC	http://igdb.nsclc.ibms.sinica.edu.tw	[136]
miRNA SNiPer	http://www.integratomics-time.com/miRNA-SNiPer	[137]
microRNAviewer	http://people.csail.mit.edu/akiezun/microRNAviewer	[138]
miREnvironment	http://cmbi.bjmu.edu.cn/miren	[139]
NAPP	http://napp.u-psud.fr/	[140]
PlantNATsDB	http://bis.zju.edu.cn/pnatdb/	[141]
doRiNA	http://dorina.mdc-berlin.de	[142]
Antagomirbase	http://bioinfopresidencycollegekolkata.edu.in/antagomirs.html	[143]
Pharmaco-miR	http://www.pharmaco-mir.org/	[144]
Plncdb	http://chualab.rockefeller.edu/gbrowse2/homepage.html	[145]
LNCipedia	http://www.lncipedia.org	[146]
OCDB	http://ocdb.di.univr.it	[147]
miRdSNP	http://mirdsnp.ccr.buffalo.edu/	[148]
SomamiR	http://compbio.uthsc.edu/SomamiR/	[149]

(continued)

Table 3
(continued)

Name	URL	Reference
lncRNAdisease	http://202.38.126.151/hmdd/html/tools/lncrnadisease.html	[150]
miRandola	http://atlas.dmi.unict.it/mirandola/index.html	[151]
lncRNome	http://genome.igib.res.in/lncRNome	[152]
GenomeRNAi	http://www.genomernai.org	[153]
mirEX	http://bioinfo.amu.edu.pl/mirex	[154]
miRFANs	http://www.cassava-genome.cn/mirfans	[155]
sRNAdb	http://bioinfo.mikrobio.med.uni-giessen.de/sRNAdb	[156]
Ssa miRNAs DB	http://www.molgenv.com/ssa_mirnas_db_home.php	[157]
The Functional lncRNA Database	http://valadkhanlab.org/database	[158]
UCNEbase	http://ccg.vital-it.ch/UCNEbase	[159]
vHoT	http://best.snu.ac.kr/vhot/	[160]
YM500	http://ngs.ym.edu.tw/ym500/	[161]
miRCancer	http://mircancer.ecu.edu/	[162]
miRDeathDB	http://rna-world.org/mirdeathdb/	[163]
DIANA miRPath v.2.0	http://www.microrna.gr/miRPathv2	[164]
DIANA-LncBase	www.microrna.gr/LncBase	[165]
HNOCDB	http://gyanxet.com/hno.html	[166]
microPIR	http://www4a.biotec.or.th/micropir	[167]
miR-EdiTar	http://microrna.osumc.edu/mireditar	[168]
AthaMap	http://www.athamap.de/	[169]
RNAiAtlas	http://www.rnaiatlas.ethz.ch	[170]
TRiP	http://www.flyrnai.org/TRiP-HOME.html	[171]
VIRsiRNAdb	http://crdd.osdd.net/servers/virsirnadb	[172]
GtRNAdb	http://gtrnadb.ucsc.edu/	[173]
tmRNA Website	http://bioinformatics.sandia.gov/tmrna/	[174]
snOPY	http://snoopy.med.miyazaki-u.ac.jp	[175]
tRNAdb	http://trnadb.bioinf.uni-leipzig.de	[176]
PeTMbase	http://petmbase.org	[177]
EVLncRNAs	http://biophy.dzu.edu.cn/EVLncRNAs/	[178]
lncATLAS	http://lncatlas.crg.eu	[179]

(continued)

Non-coding RNA Databases   273

Table 3
(continued)

Name	URL	Reference
ANGIOGENES	http://angiogenes.uni-frankfurt.de	[180]
lncRInter	http://bioinfo.life.hust.edu.cn/lncRInter	[181]
BmncRNAdb	http://gene.cqu.edu.cn/BmncRNAdb/index.php	[182]
LincSNP	http://bioinfo.hrbmu.edu.cn/LincSNP	[183]
NRDTD	http://chengroup.cumt.edu.cn/NRDTD	[184]
CCG	http://www.xingene.net/ccg/	[185]
NSDNA	http://www.bio-bigdata.net/nsdna/	[186]
GreeNC	http://greenc.sciencedesigners.com	[187]
LNCediting	http://bioinfo.life.hust.edu.cn/LNCediting/	[188]
decodeRNA	http://www.decoderna.org/index.php	[189]
LncRNASNP	http://bioinfo.life.hust.edu.cn/lncRNASNP2#!/	[190]
Lnc2Meth	http://www.bio-bigdata.com/Lnc2Meth	[191]
RNArchitecture	http://iimcb.genesilico.pl/RNArchitecture/	[192]
SRD	http://srd.genouest.org	[193]
piRNA cluster database	http://www.smallrnagroup.uni-mainz.de/piRNAclusterDB.html	[194]
Isomirs	http://hood.systemsbiology.net/cgi-bin/isomir/find.pl	[195]
ARN	http://210.27.80.93/arn/	[196]
CANTATAdb	http://yeti.amu.edu.pl/CANTATA/	[197]
miRnalyze	http://www.mirnalyze.in	[198]
comTAR	http://rnabiology.ibr-conicet.gov.ar/comtar/	[199]
MicroTrout	http://www.mennigen-lab.com/microtrout.html	[200]
ComiRNet	http://comirnet.di.uniba.it:8080	[201]
GermlncRNA	http://germlncrna.cbiit.cuhk.edu.hk	[202]
AtmiRNET	http://atmirnet.itps.ncku.edu.tw	[203]
ceRDB	http://www.oncomir.umn.edu/cefinder/	[204]
tasiRNAdb	http://bioinfo.jit.edu.cn/tasiRNADatabase/	[205]
mirCoX	http://210.212.254.116/mircox/pages/index.php	[206]
ERISdb	http://lemur.amu.edu.pl/share/ERISdb/home.html	[207]
Psmir	http://www.bio-bigdata.com/Psmir/index.jsp	[208]
SpermBase	http://www.spermbase.org	[209]
LeishDB	http://leishdb.com/	[32]

(continued)

Table 3
(continued)

Name	URL	Reference
miRSponge	http://www.bio-bigdata.com/miRSponge/	[210]
miRGate	http://mirgate.bioinfo.cnio.es/miRGate/	[211]
miRNEST	http://rhesus.amu.edu.pl/mirnest/copy/	[212]
RNAcentral	http://rnacentral.org	[213]
miRNA	http://www.itb.cnr.it/ptp/grapemirna/	[214]
RiceATM	http://syslab3.nchu.edu.tw/rice/	[215]
miRNASNP	http://bioinfo.life.hust.edu.cn/miRNASNP2/	[216]
OncomiRdbB	http://tdb.ccmb.res.in/OncomiRdbB/index.htm	[217]
CCDB	http://crdd.osdd.net/raghava/ccdb	[218]
PNRD	http://structuralbiology.cau.edu.cn/PNRD/	[219]
ViTa	http://vita.mbc.nctu.edu.tw/	[220]
ExcellmiRDB	http://www.excellmirdb.brfjaisalmer.com	[221]
RCGD	http://www.juit.ac.in/attachments/jsr/rcdb/homenew.html	[222]
CircNet	http://circnet.mbc.nctu.edu.tw	[223]
MDTE DB	http://bioinf.njnu.edu.cn/MDTE/MDTE.php	[224]
miRandb	http://mirandb.ir	[225]
miRNA_Targets	http://mamsap.it.deakin.edu.au/~amitkuma/mirna_targetsnew/find_mirna.html	[226]
RNAimmuno	http://rnaimmuno.ibch.poznan.pl/	[227]
rVarBase	http://rv.psych.ac.cn/index.do	[228]
BioM2MetDisease	http://www.bio-bigdata.com/BioM2MetDisease/	[229]
PmiRExAt	http://pmirexat.nabi.res.in	[230]
Immune-miR	http://biominingbu.org/immunemir/index.html	[231]
miRPathDB	https://mpd.bioinf.uni-sb.de/overview.html	[232]
EpimiR	http://210.46.85.180:8080/EpimiR/index.jsp	[233]
GEAR	http://gear.comp-sysbio.org	[234]
CSmiRTar	http://cosbi.ee.ncku.edu.tw/CSmiRTar/	[235]
SM2miR	http://210.46.85.180:8080/sm2mir/index.jsp	[236]
EpimiRBase	http://www.epimirbase.eu	[237]
NCG 5.0	http://ncg.kcl.ac.uk/index.php	[238]
RNAi Codex	http://cancan.cshl.edu/cgi-bin/Codex/Codex.cgi	[239]

(continued)

**Table 3
(continued)**

Name	URL	Reference
SorghumFDB	http://structuralbiology.cau.edu.cn/sorghum/index.html	[240]
Lnc2Cancer	http://www.bio-bigdata.com/lnc2cancer/	[241]
Isomir Bank	http://mcg.ustc.edu.cn/bsc/isomir/	[242]
TMREC	http://210.46.85.180:8080/TMREC/	[243]
DASHR	http://www.lisanwanglab.org/DASHR/smdb.php	[244]
STarMirDB	http://sfold.wadsworth.org/starmirDB.php	[245]
lnCeDB	http://gyanxet-beta.com/lncedb/	[246]
LNCmap	http://www.bio-bigdata.com/LNCmap/	[247]
miRmine	http://guanlab.ccmb.med.umich.edu/mirmine/	[248]
LNCat	http://biocc.hrbmu.edu.cn/LNCat/	[249]
ZIKV-DB	http://zikadb.cpqrr.fiocruz.br/	[250]
exoRbase	http://www.exorbase.org/	[251]
ExoCarta	http://www.exocarta.org/	[252]
CIRC2TRAITS	http://gyanxet-beta.com/circdb/	[253]
Vir-Mir	http://alk.ibms.sinica.edu.tw/	[254]
VIRmiRNA	http://crdd.osdd.net/servers/virmirna/	[255]
VIRBase	http://www.rna-society.org/virbase/	[256]
MNDR	http://www.bioinformatics.ac.cn/mndr/	[257]
ChIPBase	http://rna.sysu.edu.cn/chipbase/	[258]
miRò2	http://microrna.osumc.edu/miro/	[259]
SMiR-NBI	http://lmmd.ecust.edu.cn/database/smir-nbi/	[260]
CSCdb	http://bioinformatics.ustc.edu.cn/cscdb/	[261]
TriplexRNA	https://triplexrna.org/	[262]

Acknowledgments

VMC was funded by grants from Comisión Nacional de Investigación Científica y Tecnológica (CONICYT): grants FONDECYT (11161020), PAI (PAI79170021), and FONDAP (15130011), Chile. ARP was funded by Conselho Nacional de Desenvolvimento Científico e Tecnológico (CNPq), grant 00454505/2014-0, Brazil. AMD was funded by CNPq (309566/2015-0) and Coordenação de Aperfeiçoamento de Pessoal de Nível Superior (CAPES) (25/2013), Brazil.

References

1. Djebali S, Davis CA, Merkel A et al (2012) Landscape of transcription in human cells. Nature 489(7414):101–108
2. Okazaki Y, Furuno M, Kasukawa T et al (2002) Analysis of the mouse transcriptome based on functional annotation of 60,770 full-length cDNAs. Nature 420 (6915):563–573
3. Mattick JS (2010) The central role of RNA in the genetic programming of complex organisms. An Acad Bras Cienc 82(4):933–939
4. Arias-Carrasco R, Vásquez-Morán Y, Nakaya HI, Maracaja-Coutinho V (2018) StructRNAfinder: an automated pipeline and web server for RNA families prediction. BMC Bioinformatics 19(1):55
5. Esteller M (2011) Non-coding RNAs in human disease. Nat Rev Genet 12 (12):861–874
6. Paschoal AR, Maracaja-Coutinho V, Setubal JC et al (2012) Non-coding transcription characterization and annotation: a guide and web resource for non-coding RNA databases. RNA Biol 9(3):274–282
7. Wang Y, Luo J, Zhang H, Lu J (2016) microRNAs in the same clusters evolve to coordinately regulate functionally related genes. Mol Biol Evol 33(9):2232–2247
8. Guo L, Liang T (2016) MicroRNAs and their variants in an RNA world: implications for complex interactions and diverse roles in an RNA regulatory network. Brief Bioinform 19 (2):245–253
9. Piedade D, Azevedo-Pereira J (2016) The role of microRNAs in the pathogenesis of Herpesvirus infection. Viruses 8(6):156
10. Kozomara A, Griffiths-Jones S (2014) miRBase: annotating high confidence microRNAs using deep sequencing data. Nucleic Acids Res 42(D1):D68–D73
11. Aguiar RR, Ambrosio LA, Sepúlveda-Hermosilla G et al (2016) miRQuest: integration of tools on a Web server for microRNA research. Genet Mol Res 15(1):gmr6861. https://doi.org/10.4238/gmr.15016861
12. Orell A, Tripp V, Aliaga-Tobar V et al (2018) A regulatory RNA is involved in RNA duplex formation and biofilm regulation in Sulfolobus acidocaldarius. Nucleic Acids Res 46 (9):4794–4806
13. Matamala JM, Arias-Carrasco R, Sanchez C et al (2018) Genome-wide circulating microRNA expression profiling reveals potential biomarkers for amyotrophic lateral sclerosis. Neurobiol Aging 64:123–138
14. Naslavsky MS, Yamamoto GL, de Almeida TF et al (2017) Exomic variants of an elderly cohort of Brazilians in the ABraOM database. Hum Mutat 38(7):751–763
15. Eyheramendy S, Martinez FI, Manevy F et al (2015) Genetic structure characterization of Chileans reflects historical immigration patterns. Nat Commun 6:6472
16. Brimacombe M, Hazbon M, Motiwala AS, Alland D (2007) Antibiotic resistance and Single-Nucleotide Polymorphism cluster grouping type in a multinational sample of resistant Mycobacterium tuberculosis isolates. Antimicrob Agents Chemother 51 (11):4157–4159
17. Morin PA, Luikart G, Wayne RK et al (2004) SNPs in ecology, evolution and conservation. Trends Ecol Evol 19(4):208–216
18. Amaral PP, Leonardi T, Han N, Viré E et al (2018) Genomic positional conservation identifies topological anchor point RNAs linked to developmental loci. Genome Biol 19(1):32
19. Zhao Y, Li H, Fang S et al (2016) NONCODE 2016: an informative and valuable data source of long non-coding RNAs. Nucleic Acids Res 44(D1):D203–D208
20. Amaral PP, Clark MB, Gascoigne DK et al (2011) lncRNAdb: a reference database for long noncoding RNAs. Nucleic Acids Res 39:D146–D151
21. Korostowski L, Sedlak N, Engel N (2012) The Kcnq1ot1 long non-coding RNA affects chromatin conformation and expression of Kcnq1, but does not regulate its imprinting in the developing heart. PLoS Genet 8(9):e1002956
22. Wang J, Geng Z, Weng J et al (2016) Microarray analysis reveals a potential role of lncRNAs expression in cardiac cell proliferation. BMC Dev Biol 16(1):41
23. Holdt LM, Stahringer A, Sass K et al (2016) Circular non-coding RNA ANRIL modulates ribosomal RNA maturation and atherosclerosis in humans. Nat Commun 7:12429
24. Han P, Li W, Lin C-H et al (2014) A long noncoding RNA protects the heart from pathological hypertrophy. Nature 514 (7520):102–106
25. Anastasiadou E, Jacob LS, Slack FJ (2018) Non-coding RNA networks in cancer. Nat Rev Cancer 18(1):5–18
26. Qureshi IA, Mehler MF (2012) Emerging roles of non-coding RNAs in brain evolution,

26. development, plasticity and disease. Nat Rev Neurosci 13(8):528–541
27. Pastori C, Wahlestedt C (2012) Involvement of long noncoding RNAs in diseases affecting the central nervous system. RNA Biol 9 (6):860–870
28. Mehler MF, Mattick JS (2006) Non-coding RNAs in the nervous system. J Physiol 575 (2):333–341
29. Salta E, De Strooper B (2012) Non-coding RNAs with essential roles in neurodegenerative disorders. Lancet Neurol 11(2):189–200
30. Sheinerman KS, Umansky SR (2013) Circulating cell-free microRNA as biomarkers for screening, diagnosis and monitoring of neurodegenerative diseases and other neurologic pathologies. Front Cell Neurosci 7:150
31. Tycowski KT, Guo YE, Lee N et al (2015) Viral noncoding RNAs: more surprises. Genes Dev 29(6):567–584
32. Torres F, Arias-Carrasco R, Caris-Maldonado JC et al (2017) LeishDB: a database of coding gene annotation and non-coding RNAs in *Leishmania braziliensis*. Database 2017: bax047
33. Hagan T, Nakaya HI, Subramaniam S, Pulendran B (2015) Systems vaccinology: enabling rational vaccine design with systems biological approaches. Vaccine 33(40):5294–5301
34. Delhalle S, Bode SFN, Balling R et al (2018) A roadmap towards personalized immunology. NPJ Syst Biol Appl 4:9
35. Leung RK-K, Wu Y-K (2015) Circulating microbial RNA and health. Sci Rep 5:16814
36. Kishikawa T, Otsuka M, Ohno M et al (2015) Circulating RNAs as new biomarkers for detecting pancreatic cancer. World J Gastroenterol 21(28):8527–8540
37. Woolfe A, Goode DK, Cooke J et al (2007) CONDOR: a database resource of developmentally associated conserved non-coding elements. BMC Dev Biol 7(1):100
38. Lee AP, Yang Y, Brenner S, Venkatesh B (2007) TFCONES: a database of vertebrate transcription factor-encoding genes and their associated conserved noncoding elements. BMC Genomics 8:441
39. Yang J-H, Shao P, Zhou H et al (2010) deepBase: a database for deeply annotating and mining deep sequencing data. Nucleic Acids Res 38:D123–D130
40. Kin T, Yamada K, Terai G et al (2007) fRNAdb: a platform for mining/annotating functional RNA candidates from non-coding RNA sequences. Nucleic Acids Res 35: D145–D148
41. Szymanski M, Erdmann VA, Barciszewski J (2007) Noncoding RNAs database (ncRNAdb). Nucleic Acids Res 35: D162–D164
42. Zhang Y, Guan D-G, Yang J-H et al (2010) ncRNAimprint: a comprehensive database of mammalian imprinted noncoding RNAs. RNA 16(10):1889–1901
43. He S, Liu C, Skogerbo G et al (2007) NONCODE v2.0: decoding the non-coding. Nucleic Acids Res 36:D170–D172
44. Gardner PP, Daub J, Tate J et al (2011) Rfam: Wikipedia, clans and the "decimal" release. Nucleic Acids Res 39:D141–D145
45. Pang KC (2004) RNAdb--a comprehensive mammalian noncoding RNA database. Nucleic Acids Res 33:D125–D130
46. Zhou Y, Lu C, Wu Q-J, Wang Y, Sun Z-T, Deng J-C et al (2007) GISSD: Group I intron sequence and structure database. Nucleic Acids Res 36(Suppl 1):D31–D37
47. Dai L, Toor N, Olson R et al (2003) Database for mobile group II introns. Nucleic Acids Res 31(1):424–426
48. Dinger ME, Pang KC, Mercer TR et al (2009) NRED: a database of long noncoding RNA expression. Nucleic Acids Res 37: D122–D126
49. Friard O, Re A, Taverna D et al (2010) CircuitsDB: a database of mixed microRNA/ transcription factor feed-forward regulatory circuits in human and mouse. BMC Bioinformatics 11:435
50. Maselli V, Di Bernardo D, Banfi S (2008) CoGemiR: a comparative genomics microRNA database. BMC Genomics 9:457
51. Yang Z, Ren F, Liu C et al (2010) dbDEMC: a database of differentially expressed miRNAs in human cancers. BMC Genomics 11(Suppl 4):S5
52. Hariharan M, Scaria V, Brahmachari SK (2009) dbSMR: a novel resource of genome-wide SNPs affecting microRNA mediated regulation. BMC Bioinformatics 10:108
53. Schmeier S, Schaefer U, MacPherson CR, Bajic VB (2011) dPORE-miRNA: polymorphic regulation of microRNA genes. PLoS One 6(2):e16657
54. Gamazon ER, Im H-K, Duan S et al (2010) Exprtarget: an integrative approach to predicting human microRNA targets. PLoS One 5(10):e13534
55. Lu M, Zhang Q, Deng M et al (2008) An analysis of human microRNA and disease associations. PLoS One 3(10):e3420

56. Gennarino VA, Sardiello M, Mutarelli M et al (2011) HOCTAR database: a unique resource for microRNA target prediction. Gene 480(1–2):51–58
57. Girijadevi R, Sreedevi VCS, Sreedharan JV, Pillai MR (2011) IntmiR: a complete catalogue of intronic miRNAs of human and mouse. Bioinformation 5(10):458–459
58. Kaya KD, Karakülah G, Yakicier CM et al (2011) mESAdb: microRNA expression and sequence analysis database. Nucleic Acids Res 39:D170–D180
59. Mhuantong W, Wichadakul D (2009) MicroPC (microPC): A comprehensive resource for predicting and comparing plant microRNAs. BMC Genomics 10:366
60. Betel D, Wilson M, Gabow A et al (2008) The microRNA.org resource: targets and expression. Nucleic Acids Res 36:D149–D153
61. Levy A, Sela N, Ast G (2008) TranspoGene and microTranspoGene: transposed elements influence on the transcriptome of seven vertebrates and invertebrates. Nucleic Acids Res 36:D47–D52
62. Ritchie W, Flamant S, Rasko JEJ (2010) mimiRNA: a microRNA expression profiler and classification resource designed to identify functional correlations between microRNAs and their targets. Bioinformatics 26(2):223–227
63. Le Béchec A, Portales-Casamar E, Vetter G et al (2011) MIR@NT@N: a framework integrating transcription factors, microRNAs and their targets to identify sub-network motifs in a meta-regulation network model. BMC Bioinformatics 12(1):67
64. Jiang Q, Wang Y, Hao Y et al (2009) miR2-Disease: a manually curated database for microRNA deregulation in human disease. Nucleic Acids Res 37:D98–D104
65. Griffiths-Jones S, Saini HK, van Dongen S, Enright AJ (2008) miRBase: tools for microRNA genomics. Nucleic Acids Res 36:D154–D158
66. Wang X (2008) miRDB: a microRNA target prediction and functional annotation database with a wiki interface. RNA 14(6):1012–1017
67. Tokar T, Pastrello C, Rossos AEM et al (2017) mirDIP 4.1—integrative database of human microRNA target predictions. Nucleic Acids Res 46(D1):D360–D370
68. Xiao F, Zuo Z, Cai G et al (2009) miRecords: an integrated resource for microRNA-target interactions. Nucleic Acids Res 37:D105–D110
69. Barh D, Bhat D, Viero C (2010) miReg: a resource for microRNA regulation. J Integr Bioinform 7:1
70. Cho S, Jun Y, Lee S et al (2011) miRGator v2.0 : an integrated system for functional investigation of microRNAs. Nucleic Acids Res 39:D158–D162
71. Alexiou P, Vergoulis T, Gleditzsch M et al (2009) miRGen 2.0: a database of microRNA genomic information and regulation. Nucleic Acids Res 38:D137–D141
72. Liu G, Ding M, Chen J et al (2010) Computational analysis of microRNA function in heart development. Acta Biochim Biophys Sin 42(9):662–670
73. Hsu S-D, Chu C-H, Tsou A-P et al (2007) miRNAMap 2.0: genomic maps of microRNAs in metazoan genomes. Nucleic Acids Res 36:D165–D169
74. Chiromatzo AO, Oliveira TYK, Pereira G et al (2007) miRNApath: a database of miRNAs, target genes and metabolic pathways. Genet Mol Res 6(4):859–865
75. Karali M, Peluso I, Gennarino VA et al (2010) miRNeye: a microRNA expression atlas of the mouse eye. BMC Genomics 11:715
76. Gerlach D, Kriventseva EV, Rahman N et al (2009) miROrtho: computational survey of microRNA genes. Nucleic Acids Res 37:D111–D117
77. Naeem H, Küffner R, Csaba G, Zimmer R (2010) miRSel: automated extraction of associations between microRNAs and genes from the biomedical literature. BMC Bioinformatics 11:135
78. Chou C-H, Shrestha S, Yang C-D et al (2018) miRTarBase update 2018: a resource for experimentally validated microRNA-target interactions. Nucleic Acids Res 46(D1):D296–D302
79. Bhartiya D, Laddha SV, Mukhopadhyay A, Scaria V (2011) miRvar: A comprehensive database for genomic variations in microRNAs. Hum Mutat 32(6):E2226–E2245
80. Dweep H, Sticht C, Pandey P, Gretz N (2011) miRWalk – Database: Prediction of possible miRNA binding sites by "walking" the genes of three genomes. J Biomed Inform 44(5):839–847
81. Bazzini AA, Asís R, González V et al (2010) miSolRNA: A tomato micro RNA relational database. BMC Plant Biol 10:240
82. Preusse M, Theis FJ, Mueller NS (2016) miTALOS v2: analyzing tissue specific microRNA function. PLoS One 11(3):e0151771

83. Tacutu R, Budovsky A, Fraifeld VE (2010) The NetAge database: a compendium of networks for longevity, age-related diseases and associated processes. Biogerontology 11(4):513–522
84. Hiard S, Charlier C, Coppieters W et al (2010) Patrocles: a database of polymorphic miRNA-mediated gene regulation in vertebrates. Nucleic Acids Res 38:D640–D651
85. Ruepp A, Kowarsch A, Schmidl D et al (2010) PhenomiR: a knowledgebase for microRNA expression in diseases and biological processes. Genome Biol 11(1):R6
86. Meng Y, Gou L, Chen D et al (2011) PmiRKB: a plant microRNA knowledge base. Nucleic Acids Res 39:D181–D187
87. Cui X, Wang Q, Yin W et al (2012) PMRD: a curated database for genes and mutants involved in plant male reproduction. BMC Plant Biol 12:215
88. Ziebarth JD, Bhattacharya A, Chen A, Cui Y (2011) PolymiRTS Database 2.0: linking polymorphisms in microRNA target sites with human diseases and complex traits. Nucleic Acids Res 40(D1):D216–D221
89. Bandyopadhyay S, Bhattacharyya M (2010) PuTmiR: a database for extracting neighboring transcription factors of human microRNAs. BMC Bioinformatics 11:190
90. Elefant N, Berger A, Shein H (2011) RepTar: a database of predicted cellular targets of host and viral miRNAs. Nucleic Acids Res 39:D188–D194
91. Sarver AL, Phalak R, Thayanithy V, Subramanian S (2010) S-MED: sarcoma microRNA expression database. Lab Investig 90(5):753–761
92. Yang J-H, Li J-H, Shao P (2010) starBase: a database for exploring microRNA–mRNA interaction maps from Argonaute CLIP-Seq and Degradome-Seq data. Nucleic Acids Res 39:D202–D209
93. Papadopoulos GL, Reczko M, Simossis VA et al (2009) The database of experimentally supported targets: a functional update of TarBase. Nucleic Acids Res 37:D155–D158
94. Wang J, Lu M, Qiu C, Cui Q (2010) TransmiR: a transcription factor-microRNA regulation database. Nucleic Acids Res 38:D119–D122
95. Taccioli C, Fabbri E, Visone R et al (2009) UCbase & miRfunc: a database of ultraconserved sequences and microRNA function. Nucleic Acids Res 37:D41–D48
96. Li S-C, Chan W-C, Hu L-Y et al (2010) Identification of homologous microRNAs in 56 animal genomes. Genomics 96(1):1–9
97. Yin Y, Zhao Y, Wang J et al (2007) antiCODE: a natural sense-antisense transcripts database. BMC Bioinformatics 8:319
98. Zhang Y, Li J, Kong L et al (2007) NATsDB: Natural Antisense Transcripts DataBase. Nucleic Acids Res 35:D156–D161
99. Li J-T, Zhang Y, Kong L et al (2008) Trans-natural antisense transcripts including non-coding RNAs in 10 species: implications for expression regulation. Nucleic Acids Res 36(15):4833–4844
100. Wu T, Wang J, Liu C et al (2006) NPInter: the noncoding RNAs and protein related biomacromolecules interaction database. Nucleic Acids Res 34:D150–D152
101. Sai Lakshmi S, Agrawal S (2008) piRNABank: a web resource on classified and clustered Piwi-interacting RNAs. Nucleic Acids Res 36:D173–D177
102. Brown JW (1997) The ribonuclease P database. Nucleic Acids Res 25(1):263–264
103. Truss M, Swat M, Kielbasa SM et al (2005) HuSiDa--the human siRNA database: an open-access database for published functional siRNA sequences and technical details of efficient transfer into recipient cells. Nucleic Acids Res 33:D108–D111
104. Ren Y, Gong W, Zhou H et al (2009) siRecords: a database of mammalian RNAi experiments and efficacies. Nucleic Acids Res 37:D146–D149
105. Chalk AM, Warfinge RE, Georgii-Hemming P, Sonnhammer ELL (2005) siRNAdb: a database of siRNA sequences. Nucleic Acids Res 33:D131–D134
106. Gustafson AM, Allen E, Givan S (2005) ASRP: the Arabidopsis Small RNA Project Database. Nucleic Acids Res 33:D637–D640
107. Matuszek G, Talebizadeh Z (2009) Autism Genetic Database (AGD): a comprehensive database including autism susceptibility gene-CNVs integrated with known noncoding RNAs and fragile sites. BMC Med Genet 10:102
108. Molnár A, Schwach F, Studholme DJ (2007) miRNAs control gene expression in the single-cell alga Chlamydomonas reinhardtii. Nature 447(7148):1126–1129
109. Johnson C, Bowman L, Adai AT (2007) CSRDB: a small RNA integrated database and browser resource for cereals. Nucleic Acids Res 35:D829–D833
110. Landgraf P, Rusu M, Sheridan R et al (2007) A mammalian microRNA expression atlas based on small RNA library sequencing. Cell 129(7):1401–1414

111. Huang H-Y, Chang H-Y, Chou C-H et al (2009) sRNAMap: genomic maps for small non-coding RNAs, their regulators and their targets in microbial genomes. Nucleic Acids Res 37:D150–D154

112. Cao Y, Wu J, Liu Q et al (2010) sRNATarBase: a comprehensive database of bacterial sRNA targets verified by experiments. RNA 16(11):2051–2057

113. Fei Z, Joung J-G, Tang X et al (2011) Tomato Functional Genomics Database: a comprehensive resource and analysis package for tomato functional genomics. Nucleic Acids Res 39:D1156–D1163

114. Xie J, Zhang M, Zhou T et al (2007) Sno/scaRNAbase: a curated database for small nucleolar RNAs and cajal body-specific RNAs. Nucleic Acids Res 35:D183–D187

115. Lestrade L (2006) snoRNA-LBME-db, a comprehensive database of human H/ACA and C/D box snoRNAs. Nucleic Acids Res 34:D158–D162

116. Dennis PP, Omer A, Lowe T (2001) A guided tour: small RNA function in Archaea. Mol Microbiol 40(3):509–519

117. Brown JWS (2003) Plant snoRNA database. Nucleic Acids Res 31(1):432–435

118. Ellis JC, Brown DD, Brown JW (2010) The small nucleolar ribonucleoprotein (snoRNP) database. RNA 16(4):664–666

119. Alioto TS (2007) U12DB: a database of orthologous U12-type spliceosomal introns. Nucleic Acids Res 35:D110–D115

120. Piekna-Przybylska D, Decatur WA, Fournier MJ (2007) New bioinformatic tools for analysis of nucleotide modifications in eukaryotic rRNA. RNA 13(3):305–312

121. Andersen ES, Rosenblad MA, Larsen N et al (2006) The tmRDB and SRPDB resources. Nucleic Acids Res 34:D163–D168

122. Xin Y, Olson WK (2009) BPS: a database of RNA base-pair structures. Nucleic Acids Res 37:D83–D88

123. Löwes B, Chauve C, Ponty Y, Giegerich R (2016) The BRaliBase dent—a tale of benchmark design and interpretation. Brief Bioinform 2016:bbw022

124. Cannone JJ, Subramanian S, Schnare MN et al (2002) The comparative RNA web (CRW) site: an online database of comparative sequence and structure information for ribosomal, intron, and other RNAs. BMC Bioinformatics 3:2

125. Stefan LR, Zhang R, Levitan AG et al (2006) MeRNA: a database of metal ion binding sites in RNA structures. Nucleic Acids Res 34:D131–D134

126. Nagaswamy U, Larios-Sanz M, Hury J et al (2002) NCIR: a database of non-canonical interactions in known RNA structures. Nucleic Acids Res 30(1):395–397

127. Turner DH, Mathews DH (2010) NNDB: the nearest neighbor parameter database for predicting stability of nucleic acid secondary structure. Nucleic Acids Res 38:D280–D282

128. Taufer M, Licon A, Araiza R et al (2009) PseudoBase++: an extension of PseudoBase for easy searching, formatting and visualization of pseudoknots. Nucleic Acids Res 37:D127–D135

129. Popenda M, Błażewicz M, Szachniuk M, Adamiak RW (2007) RNA FRABASE version 1.0: an engine with a database to search for the three-dimensional fragments within RNA structures. Nucleic Acids Res 36:D386–D391

130. Andronescu M, Bereg V, Hoos HH, Condon A (2008) RNA STRAND: the RNA secondary structure and statistical analysis database. BMC Bioinformatics 9:340

131. Bindewald E, Hayes R, Yingling YG et al (2008) RNAJunction: a database of RNA junctions and kissing loops for three-dimensional structural analysis and nanodesign. Nucleic Acids Res 36:D392–D397

132. Klosterman PS, Tamura M, Holbrook SR, Brenner SE (2002) SCOR: a Structural Classification of RNA database. Nucleic Acids Res 30(1):392–394

133. Podlevsky JD, Bley CJ, Omana RV et al (2007) The telomerase database. Nucleic Acids Res 36:D339–D343

134. Casper J, Zweig AS, Villarreal C et al (2018) The UCSC Genome Browser database: 2018 update. Nucleic Acids Res 46(D1):D762–D769

135. Mestdagh P, Lefever S, Pattyn F et al (2011) The microRNA body map: dissecting microRNA function through integrative genomics. Nucleic Acids Res 39(20):e136

136. Kao S, Shiau C-K, Gu D-L et al (2012) IGDB.NSCLC: integrated genomic database of non-small cell lung cancer. Nucleic Acids Res 40:D972–D977

137. Zorc M, Skok DJ, Godnic I et al (2012) Catalog of microRNA seed polymorphisms in vertebrates. PLoS One 7(1):e30737

138. Kiezun A, Artzi S, Modai S (2012) miRviewer: a multispecies microRNA homologous viewer. BMC Res Notes 5:92

139. Yang Q, Qiu C, Yang J et al (2011) miREnvironment database: providing a bridge for microRNAs, environmental factors and phenotypes. Bioinformatics 27(23):3329–3330

140. Ott A, Idali A, Marchais A, Gautheret D (2012) NAPP: the Nucleic Acid Phylogenetic Profile Database. Nucleic Acids Res 40: D205–D209

141. Chen D, Yuan C, Zhang J et al (2012) PlantNATsDB: a comprehensive database of plant natural antisense transcripts. Nucleic Acids Res 40:D1187–D1193

142. Blin K, Dieterich C, Wurmus R et al (2014) DoRiNA 2.0—upgrading the doRiNA database of RNA interactions in post-transcriptional regulation. Nucleic Acids Res 43(D1):D160–D167

143. Ganguli S, Mitra S, Datta A (2011) Antagomirbase- a putative antagomir database. Bioinformation 7(1):41–43

144. Rukov JL, Wilentzik R, Jaffe I et al (2014) Pharmaco-miR: linking microRNAs and drug effects. Brief Bioinform 15(4):648–659

145. Jin J, Liu J, Wang H et al (2013) PLncDB: plant long non-coding RNA database. Bioinformatics 29(8):1068–1071

146. Volders PJ, Verheggen K, Menschaert G et al (2015) An update on LNCipedia: a database for annotated human lncRNA sequences. Nucleic Acids Res 43(8):4363–4364

147. Privitera AP, Distefano R, Wefer HA et al (2015) OCDB: a database collecting genes, miRNAs and drugs for obsessive-compulsive disorder. Database 2015:bav069

148. Bruno AE, Li L, Kalabus JL et al (2012) miRdSNP: a database of disease-associated SNPs and microRNA target sites on 3'UTRs of human genes. BMC Genomics 13:44

149. Bhattacharya A, Cui Y (2015) SomamiR 2.0: a database of cancer somatic mutations altering microRNA–ceRNA interactions. Nucleic Acids Res 44(D1):D1005–D1010

150. Chen G, Wang Z, Wang D et al (2013) LncRNADisease: a database for long-non-coding RNA-associated diseases. Nucleic Acids Res 41:D983–D986

151. Russo F, Di Bella S, Nigita G et al (2012) miRandola: extracellular circulating microRNAs database. PLoS One 7(10):e47786

152. Bhartiya D, Pal K, Ghosh S et al (2013) lncRNome: a comprehensive knowledgebase of human long noncoding RNAs. Database 2013:bat034

153. Schmidt EE, Pelz O, Buhlmann S et al (2013) GenomeRNAi: a database for cell-based and in vivo RNAi phenotypes, 2013 update. Nucleic Acids Res 41:D1021–D1026

154. Bielewicz D, Dolata J, Zielezinski A et al (2012) mirEX: a platform for comparative exploration of plant pri-miRNA expression data. Nucleic Acids Res 40:D191–D197

155. Liu H, Jin T, Liao R, Wan L, Xu B, Zhou S et al (2012) miRFANs: an integrated database for Arabidopsis thaliana microRNA function annotations. BMC Plant Biol 12:68

156. Pischimarov J, Kuenne C, Billion A et al (2012) sRNAdb: a small non-coding RNA database for gram-positive bacteria. BMC Genomics 13:384

157. Reyes D, Cepeda V, González R, Vidal R (2012) Ssa miRNAs DB: Online repository of in silico predicted miRNAs in Salmo salar. Bioinformation 8(6):284–286

158. Niazi F, Valadkhan S (2012) Computational analysis of functional long noncoding RNAs reveals lack of peptide-coding capacity and parallels with 3' UTRs. RNA 18(4):825–843

159. Dimitrieva S, Bucher P (2013) UCNEbase--a database of ultraconserved non-coding elements and genomic regulatory blocks. Nucleic Acids Res 41:D101–D109

160. Kim H, Park S, Min H, Yoon S (2012) vHoT: a database for predicting interspecies interactions between viral microRNA and host genomes. Arch Virol 157(3):497–501

161. Cheng W-C, Chung I-F, Huang T-S et al (2013) YM500: a small RNA sequencing (smRNA-seq) database for microRNA research. Nucleic Acids Res 41:D285–D294

162. Xie B, Ding Q, Han H, Wu D (2013) miRCancer: a microRNA-cancer association database constructed by text mining on literature. Bioinformatics 29(5):638–644

163. Xu J, Li Y-H (2012) miRDeathDB: a database bridging microRNAs and the programmed cell death. Cell Death Differ 19(9):1571

164. Vlachos IS, Kostoulas N, Vergoulis T et al (2012) DIANA miRPath v.2.0: investigating the combinatorial effect of microRNAs in pathways. Nucleic Acids Res 40(W1): W498–W504

165. Paraskevopoulou MD, Vlachos IS, Karagkouni D et al (2016) DIANA-LncBase v2: indexing microRNA targets on non-coding transcripts. Nucleic Acids Res 44(D1): D231–D238

166. Mitra S, Das S, Das S et al (2012) HNOCDB: a comprehensive database of genes and miRNAs relevant to head and neck and oral cancer. Oral Oncol 48(2):117–119

167. Piriyapongsa J, Bootchai C, Ngamphiw C, Tongsima S (2012) microPIR: an integrated database of microRNA target sites within human promoter sequences. PLoS One 7 (3):e33888

168. Laganà A, Paone A, Veneziano D et al (2012) miR-EdiTar: a database of predicted A-to-I

edited miRNA target sites. Bioinformatics 28(23):3166–3168
169. Bülow L, Bolívar JC, Ruhe J et al (2012) "MicroRNA Targets", a new AthaMap web-tool for genome-wide identification of miRNA targets in Arabidopsis thaliana. BioData Min 5(1):7
170. Mazur S, Csucs G, Kozak K (2012) RNAiAtlas: a database for RNAi (siRNA) libraries and their specificity. Database 2012:bas027
171. Dietzl G, Chen D, Schnorrer F et al (2007) A genome-wide transgenic RNAi library for conditional gene inactivation in Drosophila. Nature 448(7150):151–156
172. Thakur N, Qureshi A, Kumar M (2012) VIRsiRNAdb: a curated database of experimentally validated viral siRNA/shRNA. Nucleic Acids Res 40:D230–D236
173. Chan PP, Lowe TM (2009) GtRNAdb: a database of transfer RNA genes detected in genomic sequence. Nucleic Acids Res 37:D93–D97
174. Williams KP, Bartel DP (1998) The tmRNA Website. Nucleic Acids Res 26(1):163–165
175. Yoshihama M, Nakao A, Kenmochi N (2013) snOPY: a small nucleolar RNA orthological gene database. BMC Res Notes 6:426
176. Jühling F, Mörl M, Hartmann RK et al (2009) tRNAdb 2009: compilation of tRNA sequences and tRNA genes. Nucleic Acids Res 37:D159–D162
177. Karakülah G, Yücebilgili Kurtoğlu K, Unver T (2016) PeTMbase: A Database of Plant Endogenous Target Mimics (eTMs). PLoS One 11(12):e0167698
178. Zhou B, Zhao H, Yu J et al (2018) EVLncRNAs: a manually curated database for long non-coding RNAs validated by low-throughput experiments. Nucleic Acids Res 46(D1):D100–D105
179. Mas-Ponte D, Carlevaro-Fita J, Palumbo E et al (2017) LncATLAS database for subcellular localization of long noncoding RNAs. RNA 23(7):1080–1087
180. Müller R, Weirick T, John D et al (2016) ANGIOGENES: knowledge database for protein-coding and noncoding RNA genes in endothelial cells. Sci Rep 6:32475
181. Liu C-J, Gao C, Ma Z et al (2017) lncRInter: a database of experimentally validated long non-coding RNA interaction. J Genet Genomics 44(5):265–268
182. Zhou Q-Z, Zhang B, Yu Q-Y, Zhang Z (2016) BmncRNAdb: a comprehensive database of non-coding RNAs in the silkworm, Bombyx mori. BMC Bioinformatics 17(1):370

183. Ning S, Yue M, Wang P et al (2016) LincSNP 2.0: an updated database for linking disease-associated SNPs to human long non-coding RNAs and their TFBSs. Nucleic Acids Res 45(D1):D74–D78
184. Chen X, Sun Y-Z, Zhang D-H et al (2017) NRDTD: a database for clinically or experimentally supported non-coding RNAs and drug targets associations. Database 2017:bax057
185. Liu M, Yang Y-CT, Xu G et al (2016) CCG: an integrative resource of cancer protein-coding genes and long noncoding RNAs. Discov Med 22(123):351–359
186. Wang J, Cao Y, Zhang H et al (2017) NSDNA: a manually curated database of experimentally supported ncRNAs associated with nervous system diseases. Nucleic Acids Res 45(D1):D902–D907
187. Paytuví Gallart A, Hermoso Pulido A, Anzar Martínez de Lagrán I et al (2016) GREENC: a Wiki-based database of plant lncRNAs. Nucleic Acids Res 44(D1):D1161–D1166
188. Gong J, Liu C, Liu W et al (2017) LNCediting: a database for functional effects of RNA editing in lncRNAs. Nucleic Acids Res 45(D1):D79–D84
189. Lefever S, Anckaert J, Volders P-J et al (2017) decodeRNA- predicting non-coding RNA functions using guilt-by-association. Database 2017:bax042
190. Miao Y-R, Liu W, Zhang Q, Guo A-Y (2018) lncRNASNP2: an updated database of functional SNPs and mutations in human and mouse lncRNAs. Nucleic Acids Res 46(D1):D276–D280
191. Zhi H, Li X, Wang P et al (2018) Lnc2Meth: a manually curated database of regulatory relationships between long non-coding RNAs and DNA methylation associated with human disease. Nucleic Acids Res 46(D1):D133–D138
192. Boccaletto P, Magnus M, Almeida C et al (2017) RNArchitecture: a database and a classification system of RNA families, with a focus on structural information. Nucleic Acids Res 46(D1):D202–D205
193. Sassi M, Augagneur Y, Mauro T et al (2015) SRD: a Staphylococcus regulatory RNA database. RNA 21(5):1005–1017
194. Rosenkranz D (2016) piRNA cluster database: a web resource for piRNA producing loci. Nucleic Acids Res 44(D1):D223–D230
195. Lee LW, Zhang S, Etheridge A et al (2010) Complexity of the microRNA repertoire revealed by next-generation sequencing. RNA 16(11):2170–2180

196. Huang Y, Wang L, Zan L-S (2016) ARN: Analysis and Visualization System for adipogenic regulation network information. Sci Rep 6:39347
197. Szcześniak MW, Rosikiewicz W, Makałowska I (2016) CANTATAdb: A collection of plant long non-coding RNAs. Plant Cell Physiol 57 (1):e8
198. Subhra Das S, James M, Paul S, Chakravorty N (2017) miRnalyze: an interactive database linking tool to unlock intuitive microRNA regulation of cell signaling pathways. Database 2017:bax015
199. Chorostecki U, Palatnik JF (2014) comTAR: a web tool for the prediction and characterization of conserved microRNA targets in plants. Bioinformatics 30(14):2066–2067
200. Mennigen JA, Zhang D (2016) MicroTrout: a comprehensive, genome-wide miRNA target prediction framework for rainbow trout, Oncorhynchus mykiss. Comp Biochem Physiol Part D Genomics Proteomics 20:19–26
201. Pio G, Ceci M, Malerba D, D'Elia D (2015) ComiRNet: a web-based system for the analysis of miRNA-gene regulatory networks. BMC Bioinformatics 16(Suppl 9):S7
202. Luk AC-S, Gao H, Xiao S et al (2015) GermlncRNA: a unique catalogue of long non-coding RNAs and associated regulations in male germ cell development. Database 2015:bav044
203. Chien C-H, Chiang-Hsieh Y-F, Chen Y-A et al (2015) AtmiRNET: a web-based resource for reconstructing regulatory networks of Arabidopsis microRNAs. Database 2015:bav042
204. Sarver AL, Subramanian S (2012) Competing endogenous RNA database. Bioinformation 8 (15):731–733
205. Zhang C, Li G, Zhu S et al (2014) tasiRNAdb: a database of ta-siRNA regulatory pathways. Bioinformatics 30(7):1045–1046
206. Giles CB, Girija-Devi R, Dozmorov MG, Wren JD (2013) mirCoX: a database of miRNA-mRNA expression correlations derived from RNA-seq meta-analysis. BMC Bioinformatics 14(Suppl 14):S17
207. Szcześniak MW, Kabza M, Pokrzywa R et al (2013) ERISdb: a database of plant splice sites and splicing signals. Plant Cell Physiol 54(2): e10
208. Meng F, Wang J, Dai E et al (2016) Psmir: a database of potential associations between small molecules and miRNAs. Sci Rep 6:19264
209. Schuster A, Tang C, Xie Y (2016) SpermBase: a database for Sperm-Borne RNA contents. Biol Reprod 95(5):99
210. Wang P, Zhi H, Zhang Y et al (2015) miRSponge: a manually curated database for experimentally supported miRNA sponges and ceRNAs. Database 2015:bav098
211. Andrés-León E, González Peña D, Gómez-López G, Pisano DG (2015) miRGate: a curated database of human, mouse and rat miRNA-mRNA targets. Database 2015: bav035
212. Szcześniak MW, Makałowska I (2013) miRNEST 2.0: a database of plant and animal microRNAs. Nucleic Acids Res 42(D1): D74–D77
213. The RNAcentral Consortium (2016) RNAcentral: a comprehensive database of non-coding RNA sequences. Nucleic Acids Res 45(D1):D128–D134
214. Lazzari B, Caprera A, Cestaro A et al (2009) Ontology-oriented retrieval of putative microRNAs in Vitis vinifera via GrapeMiRNA: a web database of de novo predicted grape microRNAs. BMC Plant Biol 9:82
215. Liu W-T, Yang C-C, Chen R-K et al (2016) RiceATM: a platform for identifying the association between rice agronomic traits and miRNA expression. Database 2016:baw151
216. Gong J, Liu C, Liu W et al (2015) An update of miRNASNP database for better SNP selection by GWAS data, miRNA expression and online tools. Database 2015:bav029
217. Khurana R, Verma VK, Rawoof A et al (2014) OncomiRdbB: a comprehensive database of microRNAs and their targets in breast cancer. BMC Bioinformatics 15:15
218. Agarwal SM, Raghav D, Singh H, Raghava GPS (2011) CCDB: a curated database of genes involved in cervix cancer. Nucleic Acids Res 39:D975–D979
219. Yi X, Zhang Z, Ling Y et al (2015) PNRD: a plant non-coding RNA database. Nucleic Acids Res 43:D982–D989
220. Hsu PW-C, Lin L-Z, Hsu S-D et al (2007) ViTa: prediction of host microRNAs targets on viruses. Nucleic Acids Res 35:D381–D385
221. Barupal JK, Saini AK, Chand T et al (2015) ExcellmiRDB for translational genomics: a curated online resource for extracellular microRNAs. OMICS 19(1):24–30
222. Ramana J (2012) RCDB: Renal Cancer Gene Database. BMC Res Notes 5:246
223. Liu Y-C, Li J-R, Sun C-H et al (2016) CircNet: a database of circular RNAs derived from transcriptome sequencing data. Nucleic Acids Res 44(D1):D209–D215

224. Wei G, Qin S, Li W et al (2015) MDTE DB: a database for microRNAs derived from Transposable element. IEEE/ACM Trans Comput Biol Bioinform 3(6):1155–1160
225. Aghaee-Bakhtiari SH, Arefian E, Lau P (2018) miRandb: a resource of online services for miRNA research. Brief Bioinform 19(2):254–262
226. Kumar A, Wong AK-L, Tizard ML et al (2012) miRNA_Targets: a database for miRNA target predictions in coding and non-coding regions of mRNAs. Genomics 100(6):352–356
227. Olejniczak M, Galka-Marciniak P, Polak K et al (2012) RNAimmuno: a database of the nonspecific immunological effects of RNA interference and microRNA reagents. RNA 18(5):930–935
228. Guo L, Du Y, Qu S, Wang J (2016) rVarBase: an updated database for regulatory features of human variants. Nucleic Acids Res 44(D1):D888–D893
229. Xu Y, Yang H, Wu T et al (2017) BioM2MetDisease: a manually curated database for associations between microRNAs, metabolites, small molecules and metabolic diseases. Database 2017:bax037
230. Gurjar AKS, Panwar AS, Gupta R, Mantri SS (2016) PmiRExAt: plant miRNA expression atlas database and web applications. Database 2016:baw060
231. Prabahar A, Natarajan J (2017) ImmunemiR – A Database of Prioritized Immune miRNA Disease Associations and its Interactome. Microrna 6(1):71–78
232. Backes C, Kehl T, Stöckel D et al (2017) miRPathDB: a new dictionary on microRNAs and target pathways. Nucleic Acids Res 45(D1):D90–D96
233. Dai E, Yu X, Zhang Y et al (2014) EpimiR: a database of curated mutual regulation between miRNAs and epigenetic modifications. Database 2014:bau023
234. Wang Y-Y, Chen W-H, Xiao P-P et al (2017) GEAR: A database of Genomic Elements Associated with drug Resistance. Sci Rep 7:44085
235. Wu W-S, Tu B-W, Chen T-T et al (2017) CSmiRTar: Condition-Specific microRNA targets database. PLoS One 12(7):e0181231
236. Liu X, Wang S, Meng F et al (2013) SM2miR: a database of the experimentally validated small molecules' effects on microRNA expression. Bioinformatics 29(3):409–411
237. Mooney C, Becker BA, Raoof R, Henshall DC (2016) EpimiRBase: a comprehensive database of microRNA-epilepsy associations. Bioinformatics 32(9):1436–1438
238. An O, Dall'Olio GM, Mourikis TP, Ciccarelli FD (2015) NCG 5.0: updates of a manually curated repository of cancer genes and associated properties from cancer mutational screenings. Nucleic Acids Res 44(D1):D992–D999
239. Olson A, Sheth N, Lee JS et al (2006) RNAi Codex: a portal/database for short-hairpin RNA (shRNA) gene-silencing constructs. Nucleic Acids Res 34:D153–D157
240. Tian T, You Q, Zhang L et al (2016) SorghumFDB: sorghum functional genomics database with multidimensional network analysis. Database 2016:baw099
241. Ning S, Zhang J, Wang P et al (2016) Lnc2Cancer: a manually curated database of experimentally supported lncRNAs associated with various human cancers. Nucleic Acids Res 44(D1):D980–D985
242. Zhang Y, Zang Q, Xu B et al (2016) IsomiR Bank: a research resource for tracking IsomiRs. Bioinformatics 32(13):2069–2071
243. Wang S, Li W, Lian B et al (2015) TMREC: A database of transcription factor and miRNA regulatory cascades in human diseases. PLoS One 10(5):e0125222
244. Leung YY, Kuksa PP, Amlie-Wolf A et al (2016) DASHR: database of small human noncoding RNAs. Nucleic Acids Res 44(D1):D216–D222
245. Rennie W, Kanoria S, Liu C et al (2016) STarMirDB: A database of microRNA binding sites. RNA Biol 13(6):554–560
246. Das S, Ghosal S, Sen R, Chakrabarti J (2014) lnCeDB: database of human long noncoding RNA acting as competing endogenous RNA. PLoS One 9(6):e98965
247. Yang H, Shang D, Xu Y et al (2017) The lncRNA connectivity map: using lncRNA signatures to connect small molecules, lncRNAs, and diseases. Sci Rep 7(1):6655
248. Panwar B, Omenn GS, Guan Y (2017) miRmine: a database of human miRNA expression profiles. Bioinformatics 33(10):1554–1560
249. Xu J, Bai J, Zhang X et al (2017) A comprehensive overview of lncRNA annotation resources. Brief Bioinform 18(2):236–249
250. Pylro VS, Oliveira FS, Morais DK et al (2016) ZIKV – CDB: A collaborative database to guide research linking sncRNAs and ZIKA virus disease symptoms. PLoS Negl Trop Dis 10(6):e0004817
251. Li S, Li Y, Chen B et al (2017) exoRBase: a database of circRNA, lncRNA and mRNA in

human blood exosomes. Nucleic Acids Res 46 (D1):D106–D112
252. Mathivanan S, Simpson RJ (2009) ExoCarta: A compendium of exosomal proteins and RNA. Proteomics 9(21):4997–5000
253. Ghosal S, Das S, Sen R et al (2013) Circ2Traits: a comprehensive database for circular RNA potentially associated with disease and traits. Front Genet 4:283
254. Li S-C, Shiau C-K, Lin W-C (2008) Vir-Mir db: prediction of viral microRNA candidate hairpins. Nucleic Acids Res 36:D184–D189
255. Qureshi A, Thakur N, Monga I et al (2014) VIRmiRNA: a comprehensive resource for experimentally validated viral miRNAs and their targets. Database 2014:bau103
256. Li Y, Wang C, Miao Z et al (2014) ViRBase: a resource for virus–host ncRNA-associated interactions. Nucleic Acids Res 43(D1):D578–D582
257. Cui T, Zhang L, Huang Y et al (2017) MNDR v2.0: an updated resource of ncRNA–disease associations in mammals. Nucleic Acids Res 46(D1):D371–D374
258. Zhou K-R, Liu S, Sun W-J et al (2016) ChIPBase v2.0: decoding transcriptional regulatory networks of non-coding RNAs and protein-coding genes from ChIP-seq data. Nucleic Acids Res 45(D1):D43–D50
259. Lagana A, Forte S, Giudice A et al (2009) miRo: a miRNA knowledge base. Database 2009:bap008
260. Li J, Lei K, Wu Z et al (2016) Network-based identification of microRNAs as potential pharmacogenomic biomarkers for anticancer drugs. Oncotarget 7(29):45584–45596
261. Shen Y, Yao H, Li A, Wang M (2016) CSCdb: a cancer stem cells portal for markers, related genes and functional information. Database 2016:baw023
262. Schmitz U, Lai X, Winter F, Wolkenhauer O, Vera J, Gupta S (2016) Cooperative gene regulation by microRNA pairs and their identification using a computational workflow. Nucleic Acids Res 42(12):7539–7552

Part IV

Network-Based Methods for Characterizing ncRNA Function

Chapter 11

Controllability Methods for Identifying Associations Between Critical Control ncRNAs and Human Diseases

Jose C. Nacher and Tatsuya Akutsu

Abstract

Human diseases are not only associated to mutations in protein-coding genes. Contrary to what was thought decades ago, the human genome is largely transcribed which generates a large amount of nonprotein-coding RNAs (ncRNAs). Interestingly, these ncRNAs are not only able to perform biological functions and interact with other molecules such as proteins, but also have been reported involved in human diseases. In this book chapter, we review the recent research done on controllability methods related to associations between ncRNAs and human diseases. First, we introduce the bipartite complex network resulting from the interactions of ncRNAs and proteins. We then explain the theoretical background of controllability algorithms and apply these methods to the problem of identifying ncRNAs with critical roles in network control. Then, by performing statistical analyses we can answer the question on whether the subset of critical control ncRNAs is also enriched by human diseases. In addition, we review three-layer network models for prediction of ncRNA-disease associations.

Key words Network controllability, Noncoding RNA, ncRNA-protein interactions, Disease associations, Bipartite networks, Minimum dominating sets

1 Introduction

RNA infrastructure is a term referred to all cellular network processes that involve RNAs [1, 2]. Although most of the human genome is not directly translated into proteins, a large number of infrastructural ncRNAs, such as tRNAs, rRNAs, spliceosomal uRNAs (snRNAs), and small nucleolar RNAs (snoRNAs), have been associated to specific cellular functions and processes [3, 4]. Recent experimental work has demonstrated that a fraction of these ncRNAs are not only actively involved in regulatory processes but also play a role in human disorders such as cancers [5, 6]. Indeed, small ncRNA molecules expressed in cells, such as snoRNAs, miRNAs, and siRNAs, are reported to target and regulate the expression levels of different life molecules, such as proteins, rather than coding proteins themselves [3, 7]. Moreover, several works have shown that ncRNAs may play a major role in epigenetics by participating in cellular network processes as long ncRNAs (e.g., XIST and HOTAIR) [8, 9], and as short ncRNAs such as miRNAs, siRNAs, and piRNAs. In particular, miRNAs have been reported playing roles in RNA networks behind stem-cell self-renewal and differentiation [10, 11]. Oncogenesis and viral infections have also been increasingly associated to miRNAs [12–17]. Precisely, miRNA-related mutations and dysregulations are responsible for specific human disorders [18–21], which have motivated some miRNA-based therapeutic targeting approaches [22, 23].

On the other hand, the large-scale network interactions of infrastructure ncRNA and small RNAs such as miRNAs had not been assembled in large numbers. Recently, NPInter database released an updated version which included almost 93,000 interactions between ncRNAs and proteins in human organisms [24]. The assembled network provides a framework to investigate for the first time associations between ncRNAs and network controllability roles.

Network controllability has captured attention of multidisciplinary researchers because it offers a framework to integrate control theory analysis with large-scale networks that already exist in natural and engineering systems. Liu et al. proposed the maximum matching (MM) approach to control directed complex networks approximated by an underlying linear dynamics [25]. Their results show that random networks are very easy to control, but many driver nodes (more than 95%) are required to control natural systems such as transcriptional regulatory networks which follow power-law degree distributions with the functional form $k^{-\gamma}$, where k indicates the node degree and γ is a characteristic degree exponent. In order to address nonlinear dynamics, and to decrease the number of driver nodes required to control a real-world network, Nacher and Akutsu proposed a minimum dominating set (MDS)-based model [26]. The proposed MDS model allows us to address controllability not only in undirected and directed unipartite networks [27] but also, as we will show later, in bipartite networks [28], by solving relatively simple integer linear programming (ILP) problems (see Fig. 1 for a comparison of MM and MDS models). The MDS methodology has been further adopted by several research groups to investigate domination problem in scale-free networks [29] and controllability in biological networks [30–36]. In particular, Wuchty identified a statistically significant association between an MDS set of proteins and oncogenes using the undirected version of the MDS-based algorithm [26, 30].

To investigate ncRNA-protein network control features, a more complex framework is necessary. In particular, a bipartite network-based MDS may allow us to identify specific driver nodes (ncRNAs) associated to control and regulatory functions in a large

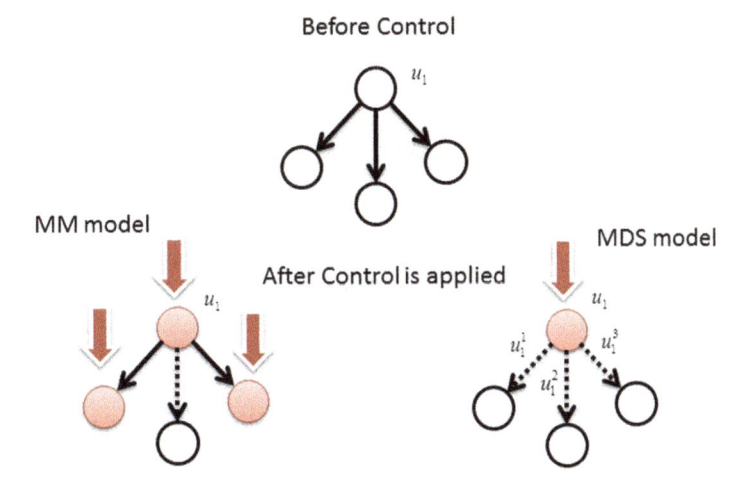

Fig. 1 Comparison between the maximum matching (MM) model [25] and the minimum dominating set (MDS) model [26]. In the MM model, after a matching link is selected we still need to control three nodes (colored nodes) with external signal (*see* red arrows). In contrast, the MDS model only requires one node as driver node u_1 (colored node). The rest of the nodes are controlled by the driver node u_1 using an independent signal for each outgoing link (*see* dashed arrows denoted as signal variables u_1^1, u_1^2, u_1^3). *See* also Ref. [26] for details of the model and mathematical proofs

network. Moreover, the MDS does not provide a unique solution; therefore, all nodes in a network have to be classified in separate control categories such as critical, intermittent, and redundant classes [27, 37]. The details of the methodology will be explained in the following Subheading 2. By combining the identified critical ncRNA molecules with human ncRNA disease associations collected from the HMDD and OMiR databases [38, 39], we could further investigate whether ncRNAs with control roles are also statistically significantly involved in human disorders [40].

Recently, bipartite network modeling was extended to three-layer network modeling in order to add information on diseases, which was effectively utilized for prediction of ncRNA-disease associations. Therefore, we also briefly review these three-layer models to infer ncRNA-disease associations [41, 42].

2 Methods

Here, we will explain the main methodology underlying the ncRNA-protein interaction network analysis shown in [40] that led to the identification of critical control ncRNAs. First, we introduce the details of the datasets used in our study in Subheading 2.1. We then present the bipartite network MDS methodology to determine critical driver nodes in Subheading. 2.2.

2.1 Collected Datasets

Here, we used the NPInter v2.0 database to collect interactions between noncoding RNAs and other biomolecules, including proteins, RNAs, and genomic DNAs. We extracted information related to human organism consisting of 3894 ncRNAs, 5783 proteins, and 92,998 interactions. The information was assembled into a bipartite network, which is characterized by two types of nodes (ncRNAs and proteins), and there are no interactions between nodes that belong to the same type. Using the proposed MDS-based controllability algorithm, we can identify driver nodes (a subset of ncRNAs) on this network (*see* Subheading 2 for details). Moreover, in order to evaluate an association between the identified MDS driver nodes (ncRNAs) and human diseases we used the HMDD database (version 2) [38] and the OMiR database [39].

2.2 Critical Control Node Computation in Bipartite Networks

2.2.1 MDS Algorithm in Bipartite Networks: The Case of Drug-Protein Target Network

Applications of MDS in bipartite networks are not limited to prediction of ncRNA-protein interactions. Here, we introduce MDS model by using an example of a drug-protein target network [28].

To investigate drug-protein interactions, we define a bipartite graph $G(V_T, V_B; E)$ that consists of a set of top nodes V_T (drugs) and a set of bottom nodes V_B (proteins). A set of edges connect both sets of nodes ($E \subseteq V_T \times V_B$). The edges have a direction defined from top nodes (drugs) V_T to bottom nodes V_B (proteins). The following set $S \subseteq V_T$ is called a dominating set of nodes in the graph G if for all nodes $v \in V_B$, there exists a node $u \in V_T$ such that $(u, v) \in E$. This dominating set is also called driver set of nodes and in our problem the set of controllers corresponds to a subset of V_T (drugs). The set with minimum cardinality is referred as minimum dominating set and is the focus of our study because we aim at identifying the minimum number of controllers to control the entire network. Figure 2a shows a solution for the given graph. There is, however, another MDS solution for this graph as shown in Fig. 2b. This modeling seems reasonable because many drugs control functions and/or activities of proteins and our purpose is to find a minimum set of drugs that can control all proteins.

An exact MDS solution in a bipartite network can be obtained by mathematically formalizing the above problem as an integer linear programming (ILP) expression as follows:

$$\text{minimize} \sum_{u \in V_T} z_u$$
$$\text{subject to} \sum_{\{u,v\} \in E} z_u \geq 1 \quad \forall v \in V_B \qquad (1)$$
$$z_u \in {0, 1} \quad \forall u \in V_T$$

By using network data, the optimal solution is calculated using this ILP instance. An MDS for a given bipartite network is obtained by the set $\{u | z_u = 1\}$. It is worth mentioning that in spite of the

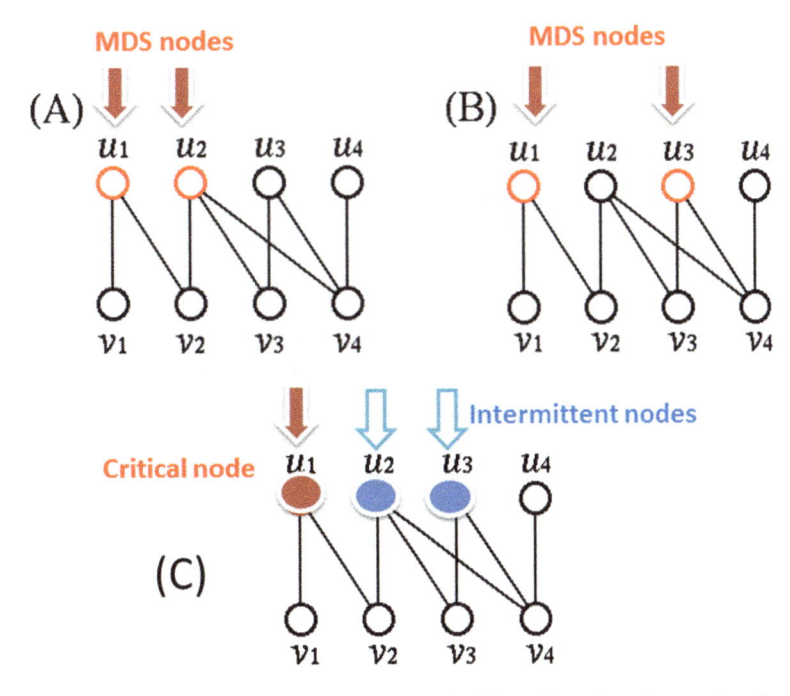

Fig. 2 For this bipartite graph, two possible MDS configurations (*see* red nodes denoted by arrows) can be obtained as solutions (**A**, **B**), where $V_T = \{u_1, u_2, u_3, u_4\}$, and $V_B = \{v_1, v_2, v_3, v_4\}$. These configurations indicate that node u_1 is present in both MDS solutions. Therefore, this node is regarded as critical control node (**C**) (filled red node). In contrast, the nodes u_2 and u_3 only appear in one of the MDS solutions (**A**) and (**B**), respectively. Therefore, these nodes are considered as intermitted control nodes (**C**) (filled blue nodes with empty arrows). Finally, u_4 is the redundant node

NP-hard nature of the MDS problem, the ILP can solve scale-free bipartite networks of up to approximately 110,000 nodes [28]. This model was used to examine the drug-target protein interaction system [28]. In the analysis, 888 approved drugs and 394 protein targets were considered. Moreover, when the protein was a disease-gene product, the drug targets were assigned the corresponding human disorder class using OMIM database. The analysis showed that by considering the giant component of the network, only 8% of the MDS drugs could control the entire known druggable proteome related to diseases. The MDS drugs, therefore, could interact with key disease-gene products corresponding to specific disorder classes, optimizing (minimizing) the total number of required drugs.

2.2.2 Control Categories Identification in Bipartite Networks

Since the MDS solution given by Eq. (1) is not unique, we need to define control categories for the driver nodes [27, 37]. For example, Fig. 2 shows that the given graph has two possible MDS solutions (a) and (b). The set of critical nodes is the most important as is defined as those nodes that belong to every MDS configuration. These nodes are always involved in network control. The set of redundant nodes consists of those nodes that never appear in any MDS configuration and therefore are never involved in network control. Next, intermittent nodes are those that appear in some MDS but do not belong to all MDS configurations (*see* Fig. 2c). The fraction of intermittent nodes f_i can be inferred from the above-calculated fractions of critical f_c and redundant f_r nodes as $f_i = 1 - f_c - f_r$.

In order to identify these control categories, we had to modify the bipartite network MDS algorithm shown in Eq. (1). The resulting algorithm combined the techniques presented in [27] to identify control categories in undirected and directed unipartite networks and extended them to be applied to bipartite networks.

The critical node detection algorithm for bipartite network is as follows:

1. Using an ILP as defined in Eq. (1), we identify an MDS of size M for a bipartite network $G(V_T, V_B; E)$.
2. Let B_{MDS} be an empty set.

Then, $\forall u \in M$ and we repeat **steps 3–6**.

3. By inserting a new constraint of $z_u \leq 0$ into the instance shown by Eq. (1), a new ILP instance I_u is defined.
4. Compute a solution of I_u and define $M_u = \{v|\ z_v = 1\}$.
5. If $|M_u| = 0$ or $|M_u| > |M|$, then we set $B_{MDS} \leftarrow B_{MDS} \cap \{u\}$.
6. Return B_{MDS}.

The identification of redundant nodes follows a conceptually similar algorithm. We refer to the original paper for details (*see* Ref. [40]). By combining critical and redundant algorithms with mathematically formalized ILP equations, we were able to investigate the controllability features of a real biological bipartite network system, such as the ncRNA-protein network.

3 Results

3.1 A Small Fraction of Nodes Control the Whole Network

The constructed network from the NPInter database included a total of 32 ncRNA classes, with lncRNA, lincRNA, mRNAlike lncRNA, and mirRNA being the most abundant. The application of the MDS algorithm for bipartite networks led to the identification of the minimum number of controllers needed to achieve full network control. Only 9.5% of all ncRNAs are needed to simultaneously control 5783 proteins using the MDS approach. In addition, the application of the control category algorithm for bipartite networks, as shown above, identified the number of critical (335) and redundant (3419) ncRNAs. The analysis of each ncRNA class

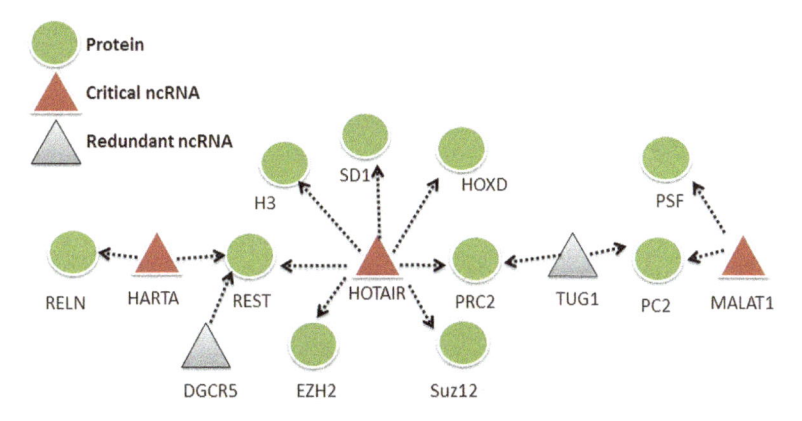

Fig. 3 A subset of the ncRNA-protein network that illustrates the application of the bipartite network MDS algorithm for identification of control categories. Only three critical ncRNAs (red colored triangles) are required to control the entire protein (green circles) subnetwork. The remaining two ncRNAs (grey-colored triangles) are redundant nodes and do not participate in network control

showed that most of the critical nodes belong to the miRNA class. Only 140 ncRNAs were identified as intermitted nodes, which is less than 3.5%. The total fraction of nodes involved in control (critical and intermittent classes) is less than 12.5%. *See* Fig. 3 for an illustration of a critical set of nodes in a protein-ncRNA subnetwork.

3.2 Enrichment of Critical Control in Human Disease-Related ncRNAs

Disease associations from HMDD database were mapped onto those ncRNAs collected from the NPInter database. This allowed us to classify ncRNAs intro two groups: associated and nonassociated to diseases. Similarly, we also classified the ncRNAs into two groups corresponding to critical and noncritical ncRNAs. A contingency table was prepared and the statistical significance of the association between critical control ncRNAs and disease-related ncRNAs was evaluated using Fisher's exact test. The results suggest a significant enrichment of the critical set of ncRNAs ($P = 9.8 \times 10^{-109}$). Similar statistical values were observed when the MDS set was used to group nodes into MDS and non-MDS features.

An enrichment analysis of the MDS and critical set of ncRNAs was also performed for each individual disease. Hepatocellular carcinoma and stomach, breast, and colorectal neoplasms had the largest number of ncRNAs involved in diseases and all of them exhibited a statistically significant association to MDS and critical control sets. In addition, we also investigated the ncRNA-disease associations available in the OMiR database. The statistical analysis also indicated a statistically significant association between the critical set of the ncRNAs and the set of "orphan" Mendelian diseases ($P = 3.3 \times 10^{-40}$).

Recently, the gene expression regulatory roles of long noncoding RNAs (lncRNAs) in mammals, from epigenetic to transcriptional stages, have been reported [43, 44]. Our analysis identified five lncRNAs as critical driver nodes. Interestingly, all of them were classified as regulatory class in NPInter v2.0 database. For example, the ncR-uPAR upregulated PAR-1 is especially active during embryogenesis and performs regulatory roles on the human protease-activated receptor-1 gene [45]. The PVT-1 has also been reported as a regulator of the c-Myc gene transcription [46].

On the other hand, the miR-17 family consists of several transcripts whose key roles in many biological processes, from cell cycle to embryonic development, have been investigated in depth. It is well known that all precursors give rise to miRNA with a characteristic seed sequence termed as "AAAGUG." Overexpression and downregulation of miRNA-17 family in leukemia and Alzheimer's, respectively, have been reported [47]. In our approach, a large number of miRNAs were also identified as critical nodes. However, not all members of miR-17 family were identified as network controllers. Combination of the MDS approach with gene expression data as well as making the model sensitive to network structure changes could be a further extension of this approach.

3.3 Integration of ncRNA Sequence Information with Multilayer Networks

The described controllability analysis for ncRNAs involved a bipartite network constructed using available ncRNA-protein interactions. Recently, in order to infer ncRNA-disease associations more complex topologies have been proposed [41]. Instead of a bipartite network, a tripartite network was considered. The first layer consists of available ncRNAs and they are connected to a second layer composed by target proteins. Because no links are allowed in the same layer, the ncRNA-protein target is a bipartite network. A third network layer is added by associating each target protein to a human disease if the protein encodes a disease-related gene. Again, associations in the same layer are not allowed, and the protein target-disease interaction system is considered as a bipartite network. These three layers (ncRNA-target-disease) and their corresponding adjacency matrices for the two defined bipartite networks were combined and used to predict ncRNA-disease associations using a multilayer resource-allocation methodology (see Fig. 4 for an illustration of the multilayer network architecture).

However, the sequence information was not included in the multilayer resource-allocation methodology. A recent work extended this approach, by adding sequence information for the ncRNAs and protein targets. The sequence similarity for each ncRNA and protein layers was evaluated independently by using a radial basis function (RBF) kernel which computed the similarity

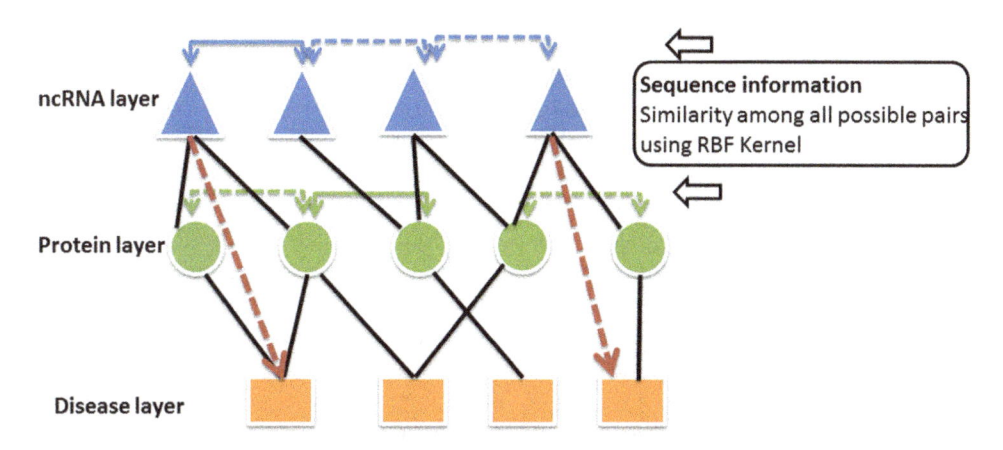

Fig. 4 Illustration of the tripartite network used to predict ncRNA-disease associations (*see* dashed red arrows) using a resource-allocation method. A new mathematical model integrates the sequence similarity with the resource-allocation method equations. For each ncRNA and protein layer, the sequence similarity is computed among all possible pairs using an RBF kernel function

between ncRNAs and proteins using l-gram string kernel features (*see* Fig. 4). The similarity information obtained from sequences was appropriately combined into the multilayer resource-allocation method equations leading to a new integrated mathematical model [42]. To evaluate the model performance, computational experiments were done by combining cross-validation techniques with deletion of existing links between ncRNAs and diseases so that the predictions were not sensitive to available information. The results suggested that the addition of sequence information improved the overall accuracy of the model. Indeed, the model led to identify 23 ncRNA-disease associations which were in agreement with other independent biological experimental works.

These findings suggest that the designed controllability approach for ncRNAs could benefit from integrating multilayer network information as well as sequence information.

4 Conclusion

In this book chapter, we have reviewed leading controllability methods that have been proposed to identify ncRNAs with critical roles in networks. The presented MDS controllability-based algorithm allowed us to identify control categories in large-scale ncRNA-protein bipartite networks for the first time. The MDS model can be applied to nonlinear systems, a feature shared by most of the natural systems. Among the three control categories, the analysis focused on the critical control nodes because they are always in any possible MDS solution. Therefore, they are engaged in control in any network configuration and considered the most important for control purposes.

By collecting data from polygenic diseases, such as cancer and cardiovascular disorders, and "orphan" Mendelian diseases, typically considered as single-gene diseases, we identified a statistically significant association between human diseases and critical control ncRNA molecules. This finding indicates that ncRNAs engaged in network control are also likely to play a role in human disease from a network regulation viewpoint.

As a future direction, it would be interesting to combine MDS modeling with dynamic information, such as gene expression data, in such a way that the ILP problem is also sensitive to network fluctuation changes. Similarly, the reviewed multilayer resource allocation-based methods for predicting ncRNAs-disease associations suggest that bipartite network controllability approach could benefit by adding new layer networks combined with sequence information in order to predict with higher accuracy critical ncRNAs associated to specific human diseases. Therefore, these studies are encouraged and left as future works.

Acknowledgments

J.C.N. was partially supported by JSPS KAKENHI Grant Number JP25330351, and T.A. was partially supported by JSPS KAKENHI Grant Number 26540125. This research was partially supported by the Collaborative Research Program of Institute for Chemical Research, Kyoto University.

References

1. Collins LJ (2011) The RNA infrastructure: an introduction to ncRNA networks. Landes Bioscience and Springer Science+Business Media
2. Collins LJ, Penny D (2009) RNA-infrastructure: dark matter of the eukaryotic cell? Trends Genet 25(3):120–128
3. Mattick JS, Makunin IV (2006) Non-coding RNA. Hum Mol Genet 1:R17–R29
4. Gesteland RF, Cech TR, Atkins JF (2006) The RNA World, 3rd edn. Cold Spring Harbor Laboratory Press, Cold Spring Harbor, NY
5. Pang KC, Stephen S, Engström PG, Tajul-Arifin K, Chen W, Wahlestedt C, Lenhard B, Hayashizaki Y, Mattick JS (2005) RNAdb—a comprehensive mammalian non-coding RNA database. Nucleic Acids Res 33(database issue):D125–D130
6. Esquela-Kerscher A, Slack FJ (2006) Oncomirs-microRNAs with a role in cancer. Nat Rev Cancer 6:259–269
7. Makeyev EV, Maniatis T (2008) Multilevel regulation of gene expression by MicroRNAs. Science 319:1789–1790
8. Royo H, Cavaille J (2008) Non-coding RNAs in imprinted gene clusters. Biol Cell 100(3):149–166
9. Costanzo M, Baryshnikova A, Bellay J et al (2010) The genetic landscape of a cell. Science 327(5964):425–431
10. Collins LJ, Chen XS, Schonfeld B (2010) The epigenetics of non-coding RNA. In: Tollefsbol T (ed) Handbook of epigenetics. Academic Press, Oxford, pp 49–61
11. Gangaraju VK, Lin H (2009) MicroRNAs: key regulators of stem cells. Nat Rev Mol Cell Biol 10(2):116–125
12. He L et al (2005) A microRNA plycistron as a potential human oncogene. Nature 435:828–833

13. Volinia S et al (2006) A microRNA expression signature of human solid tumors define gene targets. Proc Natl Acad Sci U S A 103:2257–2261
14. Medina PP, Slack FJ (2008) MicroRNAs and cancer: an overview. Cell Cycle 7:2485–2492
15. Drakaki A, Iliopoulos D (2009) MicroRNA gene networks in oncogenesis. Curr Genomics 10:35–41
16. Roberts APE, Lewis AP, Jopling CL (2011) The role of microRNAs in viral infection. Prog Mol Biol Transl Sci 102:101–139
17. Lin S, Gregory RI (2015) MicroRNA biogenesis pathways in cancer. Nat Rev Cancer 15:321–333
18. Alvarez-Garcia I, Miska EA (2005) MicroRNA functions in animal development and human disease. Development 132:4653–4662
19. van Rooij E, Olson EN (2007) MicroRNAs: powerful new regulators of heart disease and provocative therapeutic targets. J Clin Invest 117:2369–2376
20. Poy MN, Spranger M, Stoffel M (2007) MicroRNAs and the regulation of glucose and lipid metabolism. Diabetes Obes Metab 9:67–73
21. Lu M et al (2008) An analysis of human microRNA and disease associations. PLoS One 3(10):e3420
22. Van Rooij E, Olson EN (2012) MicroRNA therapeutics for cardiovascular disease: opportunities and obstacles. Nat Rev Drug Discov 11:860–872
23. Li Z, Rana TM (2014) Therapeutic targeting of microRNAs: current status and future challenges. Nat Rev Drug Discov 13:622–638
24. Yuan J et al (2014) NPInter v2.0: an updated database of ncRNA interactions. Nucleic Acids Res 42:D104–D108
25. Liu YY, Slotine JJ, Barabási AL (2011) Controllability of complex networks. Nature 473:167–173
26. Nacher JC, Akutsu T (2012) Dominating scale-free networks with variable scaling exponent: heterogeneous networks are not difficult to control. New J Phys 14:073005
27. Nacher JC, Akutsu T (2014) Analysis of critical and redundant nodes in controlling directed and undirected complex networks using dominating sets. J Complex Networks 2:394–412
28. Nacher JC, Akutsu T (2013) Structural controllability of unidirectional bipartite networks. Sci Rep 3:1647
29. Molnár F, Sreenivasan S, Szymanski BK, Korniss G (2013) Minimum dominating sets in scale-free network ensembles. Sci Rep 3:1736
30. Wuchty S (2014) Controllability in protein interaction networks. Proc Natl Acad Sci U S A 111:7156–7160
31. Sun PG (2015) Co-controllability of drug-disease-gene network. New J Phys 17:085009
32. Khuri S, Wuchty S (2015) Essentiality and centrality in protein interaction networks. BMC Bioinformatics 16:109
33. Zhang X-F, Ou-Yang L, Zhu Y, Wu M-Y, Dai D-Q (2015) Determining minimum set of driver nodes in protein-protein interaction networks. BMC Bioinformatics 16:146
34. Ishitsuka M, Akutsu T, Nacher JC (2016) Critical controllability in proteome-wide protein interaction network integrating transcriptome. Sci Rep 6:23541
35. Zhang X-F et al (2016) Comparative analysis of housekeeping and tissue-specific driver nodes in human protein interaction networks. BMC Bioinformatics 17:358
36. Nacher JC, Akutsu T (2016) Minimum dominating set-based methods for analyzing biological networks. Methods 102:57–63
37. Jia T et al (2013) Emergence of bimodality in controlling complex networks. Nat Commun 4:2002
38. Li Y et al (2014) HMDD v2.0: a database for experimentally supported human microRNA and disease associations. Nucleic Acids Res 42:D1070–D1074
39. Rossi S et al (2011) OMiR: identification of associations between OMIM diseases and microRNAs. Genomics 97:71–76
40. Kagami H, Akutsu T, Nacher JC (2015) Determining associations between human diseases and non-coding RNAs with critical roles in network control. Sci Rep 5:14577
41. Alaimo S, Giugno R, Pulvirenti A (2014) ncPred: ncRNA-disease association prediction through tripartite network-based inference. Front Bioeng Biotechnol 2:71
42. Mori T, Ngouv H, Hayashida M, Akutsu T, Nacher JC (2018) ncRNA-disease association prediction based on sequence information and tripartite network. BMC Syst Biol 12(Suppl 1):37
43. Zhu JJ, Fu HJ, Wu YG, Zheng XF (2013) Function of lncRNAs and approaches to lncRNA-protein interactions. Sci China Life Sci 56:876–885
44. Fatica A, Bozzoni I (2014) Long non-coding RNAs: new players in cell differentiation and development. Nat Rev Genet 15:7–21
45. Madamanchi NR et al (2002) A noncoding RNA regulates human protease activated

receptor-1 gene during embryogenesis. Biochim Biophys Acta 1576(3):237–245

46. Colombo T, Farina L, Macino G, Paci P (2015) PVT1: a rising star among oncogenic long noncoding RNAs. BioMed Res Int 2015:304208 10 pp

47. Mogilyansky E, Rigoutsos I (2013) The miR-17/92 cluster: a comprehensive update on its genomics, genetics, functions and increasingly important and numerous roles in health and disease. Cell Death Differ 20:1603–1614

Chapter 12

Network-Based Methods and Other Approaches for Predicting lncRNA Functions and Disease Associations

Rosario Michael Piro and Annalisa Marsico

Abstract

The discovery that a considerable portion of eukaryotic genomes is transcribed and gives rise to long noncoding RNAs (lncRNAs) provides an important new perspective on the transcriptome and raises questions about the centrality of these lncRNAs in gene-regulatory processes and diseases. The rapidly increasing number of mechanistically investigated lncRNAs has provided evidence for distinct functional classes, such as enhancer-like lncRNAs, which modulate gene expression via chromatin looping, and noncoding competing endogenous RNAs (ceRNAs), which act as microRNA decoys. Despite great progress in the last years, the majority of lncRNAs are functionally uncharacterized and their implication for disease biogenesis and progression is unknown. Here, we summarize recent developments in lncRNA function prediction in general and lncRNA–disease associations in particular, with emphasis on in silico methods based on network analysis and on ceRNA function prediction. We believe that such computational techniques provide a valuable aid to prioritize functional lncRNAs or disease-relevant lncRNAs for targeted, experimental follow-up studies.

Key words lncRNA, ceRNA, Function prediction, Disease-gene prediction, Network analysis, Chromatin interactions

1 Introduction

While protein-coding genes only account for approximately 2% of the human genome, noncoding RNAs (ncRNAs), i.e., transcripts that do not code for proteins, started receiving a lot of attention in the last decade as genome-wide sequencing efforts have shown that at least 75% of the human genome is actively transcribed into noncoding RNAs [1]. The size of ncRNAs varies extremely from small RNAs of a size of about 22 nucleotides, such as microRNAs, to about 100,000 nucleotides for some long noncoding RNAs (lncRNAs), a heterogeneous group of RNAs shown to play key roles in imprinting control, immune response, epigenetic regulation, and regulation of gene expression [2, 3]. Concomitantly with the increasing number of discovered lncRNAs, a number of resources collecting and curating functional information about lncRNAs have been built in recent years [4].

The most recent estimate by the Encyclopedia of DNA Elements (ENCODE) Project (GENCODE release 28) [5, 6] is that the human genome contains close to 16,000 genes which can produce more than 28,000 distinct lncRNA transcripts. The NONCODE database (version 4), an integrated collection of expressed lncRNAs, reported a total of 210,831 lncRNA transcripts with measured expression across 16 human tissues and six mouse tissues [7, 8]. Compared to protein-coding genes, most of which are highly conserved across vertebrates, lncRNA sequences evolve very rapidly: less than 6% of zebrafish lncRNAs are conserved on the primary sequence level in human or mouse; only 12% of human and mouse lncRNAs appear to be conserved, respectively [9]. Such poor sequence conservation and comparably low expression levels have initially hampered their functional characterization, and they have often been ignored in the study of biological systems. Despite extensive and continually growing catalogs of lncRNAs, only a small proportion have functionally characterized roles [10]. What is the fraction of transcribed lncRNAs which are functional in a certain tissue, as opposed to transcriptional noise, and which functions are encompassed by these lncRNAs, is an active area of research.

1.1 lncRNA Functions

Finding functional lncRNAs and understanding their molecular roles requires a combination of complementary approaches that will vary depending on their mode of action [11]. An emerging view of lncRNAs is that they are fundamental regulators of gene expression. Several modes of action have been proposed for lncRNAs, such as functioning as signal, decoy, scaffold, guide, enhancer RNAs, and short peptides [12]. Signal lncRNAs regulate transcription in response to various stimuli; decoy lncRNAs limit the availability of regulatory factors such as microRNAs by harboring "decoy" binding sites for these factors, therefore sequestrating them and reducing their abundance; scaffold lncRNAs play a structural role by providing platforms for assembly of multiple-component complexes, such as ribonucleoprotein (RNP) complexes; and guide lncRNAs interact with RNPs and direct them to specific target genes.

lncRNAs which regulate gene transcription may form physical interactions with DNA loci, direct DNA methylation, or interact either locally or distantly with chromatin remodelers [3]. A major category of well-studied regulatory lncRNAs are those implicated in coordinated gene silencing, either in *cis*, such as the lncRNA XIST, involved in X-chromosome inactivation, or in *trans* (e.g., HOTAIR). Both XIST and HOTAIR have been shown to mediate epigenetic mechanisms of gene silencing by recruiting polycomb repressive complexes PRC1 and PRC2, which mediate the gene silencing-associated marks, i.e., dimethylation of lysine

9 (H3K9me2) and trimethylation of lysine 27 on histone 3 (H3K27me3) [13–15].

The three-dimensional (3D) architectural landscape of the nucleus and the high density of chromatin long-range interactions have been shown to constitute an important layer in the regulation of gene expression [16]. Proximity ligation in combination with high-throughput sequencing has recently enabled the development of experimental chromosome conformation capture techniques, such as 3C, 4C, 5C, Hi-C, and ChIA-PET, in order to identify 3D interactions between distant genomic loci and explore the genome-wide cross-talk on the chromatin [17]. The regulatory potential of such contacts, especially when mediated by RNA polymerase II (PolII) or other factors, is an area of intense research. In particular, there is evidence that enhancer–promoter interactions might be induced by chromatin looping and mediated by enhancer-like lncRNA transcripts [18].

Structured lncRNAs transcribed from active enhancers, such as HOTTIP, A-ROD, and several others, have been shown to positively regulate their neighboring genes via recruitment of transcriptional activators and establishment of chromosomal looping between the lncRNA loci and the neighboring genes [14, 16, 18, 19]. In general, lncRNAs often engage with proteins, such as transcription factors (TFs) and RNA-binding proteins (RBPs) to perform their decoy, scaffold, and regulatory functions, exhibiting specific sequence–structure preferences for those proteins [3, 20–22].

1.1.1 lncRNAs as Competing Endogenous RNAs

Less than a decade ago, a novel layer of posttranscriptional gene regulation has been hypothesized, and partly confirmed by an increasing number of studies: indirect regulatory cross-talk of RNA transcripts that constitute so-called competing endogenous RNAs (ceRNAs) [23, 24]. These can act as natural microRNA sponges, or decoys, and de-repress other RNA transcripts by sequestration of shared microRNAs that otherwise would downregulate these other targets. Such an effect had thus far only been observed for artificial sponges designed for the specific task of inhibiting microRNA activity [25] or for endogenous transcripts in plants where it was named "target mimicry" [26]. This cross-talk can expand to form large-scale regulatory ceRNA networks based on the competition for the binding of a limited pool of shared microRNAs [24, 27] (see Fig. 2). Notably, this means that even messenger RNAs (mRNAs) can have a regulatory function which is not directly related to the protein they encode. Thus, the clear distinction between regulatory genes and structural genes postulated in the past may in some cases be difficult to justify [28].

Since the initial formulation of the ceRNA hypothesis, several experimental studies have identified ceRNAs and demonstrated that their indirect regulatory activity plays a role in both healthy cells [29, 30] and diseases including cancer [23, 30–34].

It is important to realize that all RNA species harboring microRNA response elements (MREs), including lncRNAs, can potentially show ceRNA activity [24, 27, 32, 34–37]. Mathematical models indicate that ceRNA interactions can be either mutually reciprocal or asymmetrical, where one ceRNA may influence another but not vice versa [38, 39].

The exact molecular requirements for microRNA-mediated cross-talk between RNA transcripts to occur remain unknown [38] but recent results have suggested that ceRNA competition depends on stoichiometry—that is, on the endogenous microRNA: target pool ratio [40–42]—and that optimal cross-talk is reached at a near-equimolar ratio of the interacting molecules [34, 38, 41, 43]. Beyond the relative concentration of ceRNA transcripts and microRNAs, additional factors determine the strength of cross-talk. These factors are: the number of microRNAs shared between two RNA transcripts, the number and binding affinities of MREs they target on these transcripts, the co-expression and subcellular localization of transcripts and microRNAs, interactions with RNA-binding proteins, microRNA-induced transcript degradation, and possibly even modifications due to RNA editing [24, 32, 34, 35, 38, 42, 44].

Among the lncRNAs that have already been demonstrated to act as ceRNAs, we find the muscle-specific linc-MD1 which has been implicated in the timing of muscle differentiation in human and mouse myoblasts because it sequesters miR-133 and miR-135 thus indirectly affecting the expression levels of MAML1 and MEF2C, respectively, two transcription factors that trigger muscle-specific gene expression [29].

1.2 lncRNAs in Disease

There is increasing evidence that mutations and dysregulation of lncRNAs are associated with a broad range of diseases, including cancer, cardiovascular diseases, and neurodegenerative disorders [45]. For example, based on its strong overexpression, the lncRNA HOTAIR is a potential biomarker for breast cancer and for the recurrence of hepatocellular carcinoma [46]. Many other lncRNAs are known to be associated with the formation and development of breast cancer, for example H19, whose downregulation reduces the anchorage-independent growth of breast cancer and lung cancer cells [47] and GAS5 and PVT1, whose transcript levels are significantly altered in breast cancer [48, 49]. Notably, GAS5 has also been shown to act as a ceRNA for the tumor suppressor PTEN through its interaction with miR-21 and miR-222 in different cancer types (reviewed in [34]).

The neural lncRNA BC200 is expressed in many kinds of cancers, such as breast, cervix, esophagus, lung, and ovary tumors but not in the respective normal tissues [50]. In addition, lncRNAs such as BC200 and BACE1-AS have been found to be related to Alzheimer's disease [51] and Fendrr, Trpm3, and Scarb2 to heart failure (reviewed in [4]). Recent studies also showed that some lncRNAs, such as MEG3, HOTAIR, lincRNA-p21, and MALAT-1 work as "tumor-suppressor ncRNAs" or "oncogenic ncRNAs" and play a major role in the development of various cancers [52].

A final and particularly striking example is lncRNA Activated in RCC with Sunitinib Resistance (lncARSR) which promotes sunitinib resistance in renal cancer by acting as a ceRNA for AXL and c-MET through sequestration of miR-34 and miR-449, and is intracellularly transferred by exosomes from resistant cells to sensitive cells, thus disseminating sunitinib resistance [33].

Identification of disease-related lncRNAs can therefore provide novel biomarkers for diagnosis, treatment, and prognosis. A few publicly available databases exist to study direct or indirect lncRNA–disease associations, for example LncRNADisease [53], which collects lncRNA–disease relationships from the scientific literature, interaction data of lncRNAs with mRNAs, proteins, microRNAs, and DNA, as well as predicted lncRNA–disease associations based on the computational methods available in the literature. Ning et al.'s Lnc2Cancer [54] is a manually curated database that provides a high-quality and integrated resource for exploring the mechanisms and functions of cancer-related lncRNAs. In addition, databases such as SNP@lincTFBS [55] and LncRNASNP [56] are resources to promote the study of lncRNA-associated variants and SNPs that affect their TF-binding sites or functional lncRNA--microRNA interactions and therefore link lncRNA expression to human diseases.

Despite these efforts, the number of lncRNA–disease associations reported in databases is still very limited compared to the number of annotated lncRNAs, and experimental methods to identify lncRNA functions and disease associations are expensive and time-consuming. Therefore, computational approaches to predict such associations based on multi-omics datasets are in high-demand.

1.3 lncRNA Function Prediction

Computational approaches for predicting lncRNA function and/or disease association are in high demand because they represent valuable means to prioritize candidates for further functional examination and guide the design of new experiments.

Conservation and sequence-alignment methods are powerful tools to elucidate the functional importance of particular sequences, especially for protein-coding genes. However, lncRNAs are in general much less conserved than protein-coding genes and evolve more quickly, rendering functional prediction by genomic comparison very difficult. Still, synteny is observed in several cases despite undetectable sequence conservation, meaning that some lncRNAs have conserved locations between orthologous genes

[57, 58]. It has also been proposed that lncRNAs mainly function via their secondary structure [59], in particular that they contain smaller structural domains or locally conserved structural motifs important for function [59–61]. The ability to accurately predict the structure of long RNAs as well as algorithms to identify shared sequence structure motifs are of high importance in order to understand which lncRNA functions are mediated by the RNA structure and are common to subgroups of functional lncRNAs.

In this chapter, we will discuss several methods for functional annotation of lncRNAs which make use of genomic, transcriptomic, as well as meta-data integration and network analysis techniques to infer lncRNA functions. In particular, we will first discuss the "guilt-by-association" concept to infer lncRNA functions based on co-expression of coding and noncoding transcripts; second, we will introduce principles and two specific examples of network analysis methods based on random walks and information diffusion to identify disease genes or genetic modules that drive diseases; third, we will introduce the construction of interaction networks from chromatin conformation data and the identification of lncRNA-mediated regulatory clusters within these networks; and fourth, we will have a look at the construction of ceRNA networks and how they can be used to predict lncRNA functions and their potential relevance for human diseases.

2 Methods for Function Prediction of lncRNAs

Expression data often constitute the first layer to gather clues about lncRNA functions. With a growing application of RNA sequencing comes a multitude of expression data that can be analyzed to generate hypotheses about functions and disease associations of lncRNAs. The most common method of studying lncRNA functions is by means of differential expression analysis. This helps prioritizing candidates for further examination, but differential expression alone does not provide any functional insights.

The immediate effect of expression of regulatory lncRNAs is to repress or induce target molecule expression, which can be measured by a positive or negative correlation of expression profiles. Co-expressed genes tend to share the same function [62], therefore clustering of expression profiles from both coding and noncoding RNAs can identify co-regulated transcripts which might be functionally related, and the functional enrichment of each cluster can help to understand functions of lncRNAs associated with the cluster.

Another useful approach is network analysis which has emerged as a powerful tool for obtaining novel insights into complex systems. The nonrandom topological properties of most real-world networks, including biological networks, are strongly associated with their functional organization [63]. Most cellular networks such as gene-regulatory, metabolic, protein–protein interaction, and signaling networks are being widely studied. The interplay between genes, regulatory elements in the genome, and lncRNAs can be analyzed in a network context as well. Given the complexity of lncRNA functions, diverse network-based approaches have been used to decipher them and will be described in more detail below. For example, a co-expression network of both coding and noncoding genes can be used to identify tightly linked modules involving both protein-coding genes and lncRNAs. Then, the unknown function of lncRNAs can be predicted by transferring functional annotation (e.g., GO terms) from protein-coding genes [64, 65]. Also, although not yet fully explored in the context of lncRNAs, a network can be built from chromatin conformation data, where nodes correspond to genomic regions and edges to 3D chromatin interactions between genomic loci. Then, interacting genomic locations can be annotated with protein-coding genes, lncRNAs, and other elements, and graph clustering and dimensionality reduction approaches can be used to identify regulatory modules.

Finally, promoter regions of lncRNAs may hold clues to their regulation and function. Transcription factor-binding sites (TFBSs) in lncRNA promoter regions—either experimentally determined by ChIP-seq or computationally predicted—might point at the regulatory network in which lncRNAs are involved [66] and might thus reveal sets of lncRNAs which are co-regulated and active in the same biological processes.

More sophisticated computational models could be effective ways for the prediction of potential lncRNA functions and in particular lncRNA–disease associations. Several powerful computational methods have been developed for this purpose in recent years [64, 67–76], and according to the strategies that they use they can be roughly divided into supervised machine learning methods and unsupervised or network-based methods. The former class involves training a model based on labeled data, i.e., known disease–lncRNA or known lncRNA–function associations and predicting unlabeled data, i.e., lncRNA without any known functional annotations or evidence of disease association [4, 77]. The latter class of computational methods often build heterogeneous "similarity" networks involving lncRNAs, protein-coding genes, and other entities to uncover potential associations. A common assumption of these methods is that functionally similar lncRNAs tend to be associated with similar diseases or share the same interaction partners. As we shall see below, for the identification of microRNA-mediated crosstalk between ceRNAs a third class of methods is based on the mathematical modeling of interaction dynamics.

2.1 Supervised Methods for lncRNA Function Prediction

One of the simplest supervised approaches is the one of Zhao et al. [69], a naïve Bayes classifier-based method to predict cancer-related lncRNAs. The authors use known cancer-related lncRNAs from the LncRNADisease database [53] and from the literature to build their labeled positive set, and define as negative set those lncRNAs without any evidence of tumorigenic mutations or cancer phenotype-related GWAS SNPs [78] at their genomic loci. Features used to represent both the "positive" and the "negative" set include: an exon sequence conservation score, the genomic location, the presence or absence of a microRNA gene within the lncRNA transcript, the number of repeat elements, the presence of AU-rich elements, the number of cancer-related microRNA target sites on the lncRNA transcript, a tissue-specificity score, and the number of co-expressed cancer-related protein-coding genes. These features are used to train the classifier which determines an lncRNA–cancer association by a probabilistic combination of distinct evidences.

Among the supervised machine learning methods, Chen and Yan [67] developed the Laplacian Regularized Least Squares for LncRNA–Disease Association (LRLSLDA) method, based on the framework of regularized least squares to predict potential disease-associated lncRNAs. They defined two nonparametric classifiers, one on the phenotypic space based on the assumption that similar diseases tend to be associated with functionally similar lncRNAs, and one on the molecular space based on expression similarity of lncRNAs computed by integrating several expression datasets.

2.2 Co-expression Networks and Guilt-by-Association

Co-expression relationships represent an extremely rich source of information for gene function annotation, as it has been shown that "functionally interacting" genes tend to show similar expression profiles. Accordingly, many potential functions of lncRNAs have been determined using methodologies which rely on expression patterns and the "guilt-by-association" principle: transcripts sharing common expression patterns tend to be co-regulated and are likely involved in common pathways.

The study of Liao et al. [64] performs a large-scale prediction of long noncoding RNA functions by building a co-expression network of both coding and noncoding genes from 34 microarray datasets collected from the Gene Expression Omnibus (GEO) database [79], each comprising several experimental conditions or cellular states. This multi-dataset approach allows to compute robust, significant correlations between genes, which are used to define edges of the network. To infer lncRNA functions, different network analysis methods are applied (Fig. 1a):

- Prediction via co-expression and genomic location: the function of an lncRNA is predicted based on the functional annotations of genes which are both co-expressed in the network and located nearby (within 100 kb) the lncRNA.

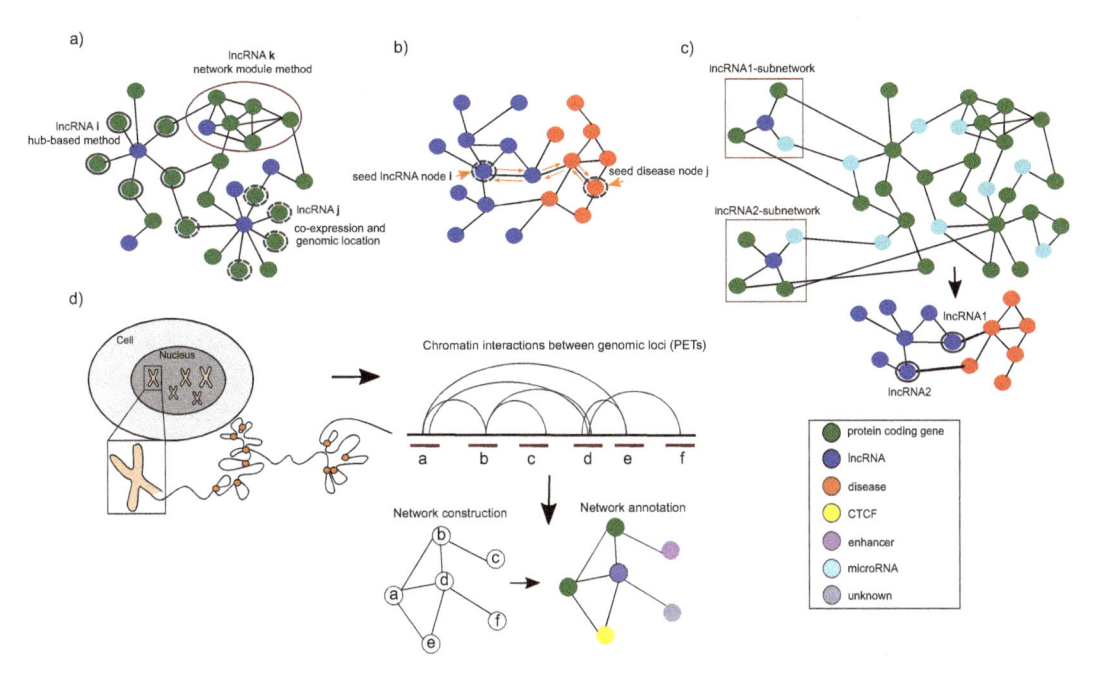

Fig. 1 Methods for lncRNA function prediction. (**a**) mRNA–lncRNA co-expression network and guilt-by-association inference via GO term enrichment; (**b**) random walk-based method on heterogeneous networks composed of lncRNAs and disease nodes; (**c**) random walk-based method based on similarities regarding mRNA or microRNA interaction partners of lncRNAs; (**d**) lncRNA function prediction using a chromatin interaction network constructed from ChIA-PET data

- Prediction via a hub-based method: functions are assigned to an uncharacterized lncRNA according to a GO term enrichment of its direct neighboring genes in the network. The accuracy of the method is assessed via cross-validation of coding genes with known GO terms.

- Prediction by network modules: network modules, which putatively correspond to similar functions, are detected with the Markov cluster algorithm (MCL), based on the simulation of random walks in a network (*see* below). Functional GO enrichment analysis is performed for each module, and enriched GO terms are transferred from protein-coding genes to lncRNAs of unknown function.

An extension of this framework has been developed by Guo et al. [80], who developed an lncRNA function predictor which relies on an integrated network from gene expression and protein interaction data. Since it has been shown in the past that the analysis of tissue-specific co-expression can better pinpoint functional relationships among genes which are highly tissue-specific, as most lncRNAs are, Perron et al. [76], extended the framework of co-expression networks for lncRNA characterization to include tissue-specificity, as well as co-expression in more than one species.

Both the tissue-specificity and the conservation of co-expression had already been successfully used for predicting the disease associations, and hence functions, of mRNAs [81, 82].

2.3 Random Walk-Based Prediction of lncRNA–Disease Associations

This class of computational methods predicts lncRNA–disease associations based on lncRNA similarities and disease similarities. An assumption of these methods is that functionally similar lncRNAs tend to be associated with phenotypically similar diseases, and vice versa. Thus many of them integrate an lncRNA–lncRNA network and a disease similarity network[1] and connect their nodes by known lncRNA–disease associations to form a heterogeneous, bipartite network which has the potential to uncover lncRNA–disease associations (Fig. 1b). Usually, a random walk on the heterogeneous network is implemented to predict novel lncRNA–disease associations [68, 70, 73, 84].

For example, Gu et al. [77] propose a random walk model for predicting potential human lncRNA–disease links, starting from a network which includes edges corresponding to known lncRNA–disease associations, disease–disease associations, and lncRNA–lncRNA associations. Disease similarities in the network are based on the semantic similarity of disease annotations from the MeSH database.[2] Pairwise functional similarities between lncRNAs are determined by computing the semantic similarity between the disease groups associated with the two lncRNAs using the LNCSIM method [68]. On this network, the authors use a random walk with restart (RWR) algorithm derived from graph theory which simulates a random walker's iterative transition from its current node to neighboring network nodes, starting at one of several given seed nodes. First, they run RWR using the lncRNAs associated with the query disease as seed nodes, and after several iterations the transition probability from lncRNA i to disease j is obtained. Then, they run RWR using disease nodes associated with a query lncRNA as seed nodes and again compute the transition probability from lncRNA i to disease j. Finally, the potential lncRNA–disease associations are predicted by integrating the results from these two steps.

Integrating several biological data (disease semantics, lncRNA expression profiles, and known lncRNA–disease associations) into a heterogeneous network, Chen et al. [85] proposed the method KATZLDA to predict disease-related lncRNAs, while Zhou et al. [84] integrated a microRNA-associated lncRNA–lncRNA crosstalk network, a disease-disease similarity network, and known lncRNA–disease associations into another heterogeneous network

---

[1] *See* [83] for a review on disease similarity networks, i.e., disease–disease networks in which nodes are diseases and edges represent phenotypic similarities.
[2] https://www.ncbi.nlm.nih.gov/mesh.

but also applied a random walk to prioritize candidate lncRNA–disease associations.

Similarly, Cheng et al. [75] developed the network analysis method IntNetLncSim to infer both lncRNA–disease associations and functional similarities between lncRNAs by modeling the information flow in a network which includes both lncRNA-related transcriptional and posttranscriptional information. In particular, the authors built an integrated network which comprises mRNA–lncRNA and microRNA–lncRNA interactions from star-Base [86], mRNA–mRNA interactions from the Human Protein Reference Database (HPRD) [87], microRNA–mRNA interactions from microRNA-target databases [88–90], and manually added lncRNA regulatory interactions from the literature. It is important to note that in this network, lncRNAs are not directly connected to each other but linked to the mRNAs or microRNAs they are associated with. IntNetLncSim is based on the expectation that functionally similar lncRNAs are associated with functionally similar mRNAs and microRNAs. A schematic representation is shown in Fig. 1c: the ITM Probe tool [91], based on a random walk with damping, is used to compute transition probabilities between lncRNAs and other network nodes, i.e., mRNAs and microRNAs. These probabilities, also called "weights" by the authors, are assigned to each lncRNA, which is then represented as a vector of weights whose dimension equals the number of mRNAs and microRNAs in the network. The similarity between lncRNAs is then defined as the cosine similarity between weight vectors. In a final step, once lncRNA–lncRNA associations have been determined, a heterogeneous lncRNA–disease network is built similarly to the one in Fig. 1b, and lncRNA–disease associations are predicted using an RWR analysis, similarly to Gu et al. [77].

2.4 3D Chromatin Interaction Networks to Predict Nuclear lncRNA Functions

To gain a better understanding of the role of chromatin interactions in gene regulation, several studies integrate high-confidence 3D chromatin interactions with other genomic datasets, such as expression data and epigenetic marks. The following two studies aim at linking chromatin structure and chromatin-related modules to gene regulation and gene function.

Siahpirani et al. [92] developed a novel, graph-based multitask clustering approach to find common and cell line-specific patterns of interacting chromosomal regions from Hi-C data. In particular, they apply spectral clustering to the Hi-C chromatin graph and identify biologically significant clusters which are statistically enriched in genomic regulatory signals, such as chromatin marks, basal transcription factors, cohesin components, repeat elements, and others. Unlike hierarchical and k-means clustering, which have been widely used to analyze functional genomic datasets, spectral clustering is a graph-based method that clusters the eigenvectors of the Laplacian operator on a graph and is therefore more suitable to incorporate both *cis* and *trans* interactions.

The ChIA-PET technology aims at detecting 3D chromatin interactions between distant genomic regions mediated by a transcription factor of interest, such as PolII or CTCF [93]. Sandhu et al. [16] construct the PolII-associated chromatin interaction network from ChIA-PET data by denoting the distinct genomic sites as vertices (nodes) and significant chromatin interactions among these sites as edges. Given that PolII is associated with active transcription, this network is expected to reflect chromatin interactions which are likely to be relevant for proper transcriptional gene regulation. Interestingly, in addition to the expected, widespread promoter–enhancer interactions, the authors identify also other types of chromatin cross-wiring, such as enhancer–enhancer, promoter–terminator, and promoter–promoter interactions. The authors perform module calling on the chromatin network and show that: (1) modules correlate significantly with gene expression, suggesting that long-range interactions are indeed important for transcriptional gene regulation and the global coordination of distant gene clusters; (2) distinct genomic elements and chromatin types contribute to different modules, and both promoter and enhancer elements show a greater module centrality than other categories, suggesting that the ChIA-PET network is shaped around these genomic elements; and (3) most of the called modules are enriched in at least one GO category: top hubs have a "rich-club" structure and are enriched in essential cellular functions such as chromatin assembly, cellular organization, and primary metabolic processes.

Although these two studies represent a big advancement in linking chromatin structure and chromatin-related modules to gene regulation and gene function, none of them focuses on the role of lncRNAs within the chromatin interaction networks or exploits network properties for the inference of lncRNA function.

To fill this gap, Thiel et al. [94] focus on lncRNA regulatory functions in the nucleus and build the chromatin interaction network involving lncRNAs, protein-coding genes, and other genomic elements (e.g., enhancers, promoters, and TF-binding sites) using PolII ChIA-PET data of the K562 cell line. In this network, nodes represent DNA segments, and edges are significant chromatin interactions between those segments (Fig. 1d). Given its scale-free degree distribution and modularity, network analysis can be performed to identify functionally essential elements and lncRNA-mediated *cis*-regulatory modules. The rationale for this analysis is that enhancer-like lncRNAs have been found to activate their neighboring genes using a *cis*-mediated mechanism and exploiting predefined chromatin looping [95].

To identify such lncRNA-mediated gene regulation, Thiel et al. take a hierarchical approach where they first compute several centrality measures and then focus on connected components of the chromatin graph, identifying lncRNA-containing modules with a variant of the Markov State Model (MSM) clustering algorithm [96]. The MSM algorithm uses properties of the random walk process to explore network topology and identify modules as regions of the network where the process is metastable, i.e., trapped for a longer period of time. Thiel et al. extend this approach to identify subgraphs of high connectivity while maximizing the expression correlation between protein-coding genes and lncRNAs in the same module, combining for the first time the topology of the 3D chromatin network with co-expression analysis in order to discover potential regulatory lncRNAs. The authors annotate interacting genomic regions with protein-coding genes, lncRNAs, and chromatin states and find that disease-associated, positionally conserved, and enhancer-like lncRNAs have a higher degree or betweenness centrality in the network. They validate well-known regulatory functions of lncRNAs and propose functional regulatory roles for so far uncharacterized lncRNAs.

2.5 Identification, Prediction, and Modeling of ceRNA Cross-talk

The identification or prediction of microRNA-mediated regulatory interactions between RNA transcripts is an emerging research subject [97]. Major challenges are the sheer number of candidate ceRNA–ceRNA interactions, if considering all types of sponges competing for shared microRNAs (mRNAs, pseudogenes, lncRNAs, circRNAs, ...), and the lack of datasets that constitute a gold standard or ground truth against which to evaluate the generated hypotheses [97].

Since the ceRNA cross-talk depends on microRNA-sequestering MREs located on the competing transcripts (*see* Fig. 2), the prediction of ceRNA cross-talk must necessarily start with the identification of said MREs on the RNA transcripts of interest [32, 35]. Although the computational prediction of microRNA binding remains challenging [98], several algorithms and databases have been developed for this purpose, including PicTar [99], TargetScan [100], RNA22 [101], and PITA [102], most of which, however, have concentrated on identifying MREs in the 3'UTRs of mRNAs (*see* [103] for a recent review). Only recently, databases for microRNA–lncRNA interactions have become available: starBase v2.0 [86], lncRNome [104], miRCode [105], and DIANA-lncBase [106] are prominent examples.

Apart from computational predictions, also high-throughput experiments for the genome-wide identification of microRNA binding have emerged [98]. Cross-linking immunoprecipitation followed by sequencing (CLIP-seq) of Argonaute-bound transcripts, for example, can be used to study microRNA activity [107, 108] because Argonaute protein family members are essential components of the RNA-induced silencing complex (RISC) [109, 110].

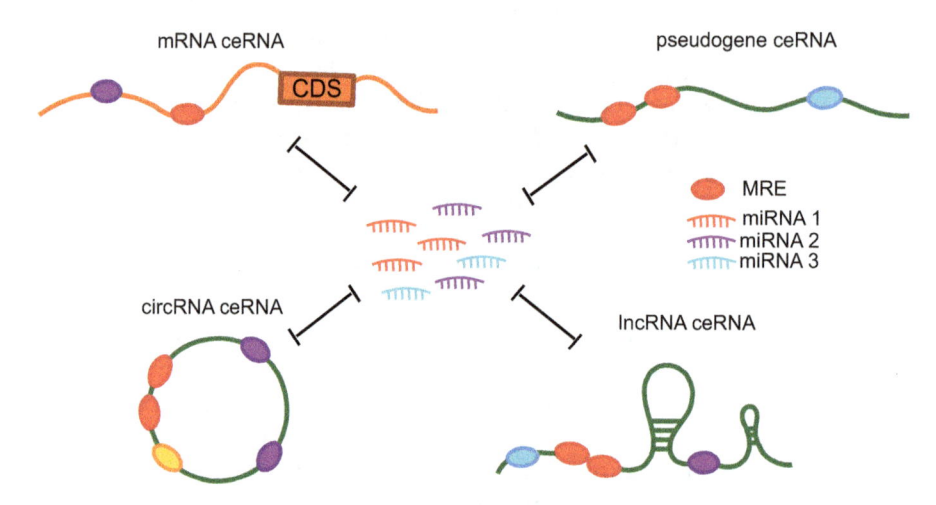

Fig. 2 Schematic representation of ceRNA-mediated gene regulation. Different RNA transcript classes can act as ceRNAs, including mRNAs, pseudogenes, lncRNAs, and circRNAs. The cross-talk between ceRNAs is mediated by shared microRNA response elements (MREs) and depends, among other factors, on the number and frequency of specific MREs and the presence of the microRNAs in the transcriptome of a given cell (figure adapted from [34])

Here, we will not go further into detail about the computational and experimental identification of microRNA–transcript interactions that underlie ceRNA–ceRNA cross-talk but will instead turn to the latter, taking given knowledge on the transcripts' MREs as a necessary input.

2.5.1 Example Methods and Studies

Generally, one can distinguish between condition-specific approaches, which try to identify transcripts that act as ceRNAs in a particular condition of interest (e.g., a particular tissue type), and condition-independent approaches, which try to identify pairs of transcripts that generally have the potential to interact as ceRNAs.

The maybe simplest method is cefinder [111] that was used to construct the "competing endogenous RNA database" (ceRDB), which unfortunately comprises only mRNAs. It uses a very simple, condition-independent scoring and ranking scheme: it counts the number of MREs of shared microRNAs in two transcripts, using microRNA–mRNA target predictions from TargetScan.

Tan et al. [42] analyzed transcriptome-wide expression changes caused by targeted knockdown of more than 100 lncRNAs in mouse embryonic stem cells (mESCs). They found mRNAs downregulated upon lncRNA knockdown to share a significantly higher number of predicted MREs with the knocked down lncRNAs than upregulated targets, especially when limiting the analysis to MREs overlapping Argonaute-bound regions in mESCs. Hence, they defined as potential protein-coding ceRNAs of lncRNAs those mRNAs that were downregulated upon lncRNA knockdown and contained predicted MREs for the same shared microRNAs.

Sumazin et al.'s HERMES algorithm [112] implements a multivariate analysis method that has been integrated into the Cupid software package [113]. The algorithm analyzes gene expression profiles and uses mutual information (MI) and conditional mutual information (CMI) to infer the dependency of the interaction between a microRNA and a targeted transcript upon another targeted transcript that acts as a ceRNA. In essence, this approach views a ceRNA as a modulator of the activity of microRNAs on other targeted RNA transcripts. Using their approach, the authors described an extensive ceRNA network which regulates oncogenic pathways in glioblastoma [112].

A very similar approach was published by Paci et al. [30] with the goal of identifying ceRNA interactions between lncRNAs and mRNAs in cancerous and normal breast tissues. They focused on lncRNA/mRNA pairs with a highly positive correlation and based their multivariate analysis on the notion of partial correlation rather than CMI. For each triplet of lncRNA, correlated mRNA, and microRNA, they computed the difference between the Pearson correlation coefficient and the partial correlation coefficient to define a sensitivity correlation S:

$$S = \text{corr}(\text{mRNA}, \text{lncRNA}) - \text{corr}(\text{mRNA}, \text{lncRNA}|\text{microRNA}) \quad (1)$$

They integrated the results of the multivariate analysis of the gene expression data with the results from an MRE seed match analysis to build microRNA-mediated interaction networks between lncRNAs and mRNAs [30].

MuTaME (mutually targeted MRE enrichment), the approach developed by Tay et al. [31], computes an interaction score between two potential ceRNAs which is based on: (1) the number of microRNAs shared between the two transcripts; (2) the number of MREs of each shared microRNA and their maximum sequence span on the transcripts, i.e., the maximum distance between the leftmost and the rightmost MRE of the same microRNA; (3) the uniformity of the distribution of the MREs within their maximum span; and (4) the relationship between the numbers of MREs to the number of microRNAs that bind to these MREs. These characteristics stem from the consideration that the expected number of spurious microRNA target predictions, i.e., false positive MREs, increases with the sequence length, and that true microRNA-binding events are more likely if for each shared microRNA many (predicted) MREs are evenly distributed in a relatively short span [31]. In a subsequent study, the same research group extended the MuTaME framework by developing analytical probabilistic measures for quantifying the statistical significance associated with abovementioned characteristics [114].

Finally, some groups have studied mathematical models to determine under which conditions ceRNA activity can be expected to occur in vivo. Ala et al. [38], for example, have described a kinetic mass-action model for the complex network environment in which ceRNA interaction would be optimal. For the same purpose, Figliuzzi et al. [39] have constructed a minimal rate equation-based model to characterize ceRNA interactions at steady-state condition. Both studies came to the conclusion that ceRNA cross-talk is most pronounced when the abundances of microRNA and their targeted ceRNAs are close to equimolarity.

Using these or similar methods to identify individual ceRNA–$32#ceRNA interactions or construct entire ceRNA networks, several publicly available ceRNA interaction databases have been set up, including starBase v2.0 [86], ceRDB [111], lnCeDB [115], miRSponge [116], and LncACTdb [117]. For a recent review, see [97].

2.5.2 Disease-Gene Prediction Based on the ceRNA Hypothesis

As a final consideration, we will have a brief look at how ceRNA networks can be used for disease-gene prioritization [118], more specifically the prioritization of potentially disease-associated lncRNAs.

For this purpose, Xu et al. [119] have developed LncNetP which exploits ceRNA networks to identify cancer-relevant lncRNAs. First, they constructed ceRNA interaction networks composed of lncRNAs for different cancer types, based on RNA-seq expression data and statistical enrichment of MREs shared between pairs of lncRNAs. On these networks, they applied the random walk with restart (RWR) algorithm, that had already been successfully used for prioritizing protein-coding genes [120], using known disease lncRNAs as seed genes, to obtain lists of candidate lncRNAs ranked according to their predicted microRNA-mediated ceRNA interaction strength with known disease-associated lncRNAs [119].

3 Conclusions and Future Prospects

Recently, lncRNAs have emerged as important key players in several biological processes and diseases. Indeed, RNA-sequencing experiments have shown that lncRNA transcripts exhibit tissue- and condition-specific expression patterns, indicating that most of them might have important functions in specialized gene-regulatory programs and participate in tissue- or cell type-specific gene networks. Although the functional roles of different types of lncRNAs are not yet fully understood, this does not mean that they do not contain information or lack relevant biological functions.

In this chapter, we have illustrated several approaches that may help to infer potential functional roles or disease associations of lncRNAs. Although differential expression analysis is useful to identify a set of candidate lncRNAs, it produces insufficient evidence to generate detailed hypotheses of lncRNA function. In fact, a multitude of lncRNAs can be differentially expressed between conditions but need not necessarily be functional.

The wide spectrum of recently developed, more sophistical computational methods can help to address key challenges of the functional assignment of lncRNAs and the identification of potential roles in human disease. In particular, network-based methods which integrate different types of data and link lncRNAs to protein-coding genes or diseases show great promise.

The data generated by chromatin conformation capture assays and ChIA-PET studies can be used to build physical chromatin interaction networks comprising loci of protein-coding genes, lncRNAs, and regulatory elements such as enhancers. The application of clustering techniques and module detection algorithms on such networks enables the discovery of putative lncRNA-mediated *cis*-regulatory modules and allows to predict functional annotations based on well-studied protein-coding genes that interact with lncRNAs within the identified chromatin modules.

Although the ceRNA hypothesis is still an emerging research topic and the extent to which it contributes to transcriptome-wide gene expression levels is still unknown, both experimental and computational studies have already demonstrated its potential to increase our understanding of disease mechanisms and have highlighted the existence of several intriguing microRNA-mediated regulatory relationships between coding and noncoding RNA transcripts.

Given the complexity of lncRNA functions and the experimental evidence confirming their heterogeneous mechanisms of action, in the future we will more and more need to develop integrative approaches—combining correlations between expression profiles, analysis of chromatin interaction networks, epigenetic data, DNA- and RNA-binding information, as well as RNA-binding protein (RBP) binding data—to classify lncRNAs into functional subclasses and uncover their intricate roles in human disease.

Acknowledgements

The authors kindly acknowledge Heike Siebert and Denise Thiel for insightful discussions. This study is supported by the DFG Grant MA 4454/3-1.

References

1. Djebali S et al (2012) Landscape of transcription in human cells. Nature 489(7414):101–108
2. Costa FF (2005) Non-coding RNAs: new players in eukaryotic biology. Gene 357(2):83–94
3. Bassett AR et al (2014) Considerations when investigating lncRNA function in vivo. eLife 3:e03058
4. Chen X et al (2017) Long non-coding RNAs and complex diseases: from experimental results to computational models. Brief Bioinform 18(4):558–576
5. The ENCODE Project Consortium (2012) An integrated encyclopedia of DNA elements in the human genome. Nature 489(7414):57–74
6. Harrow J et al (2012) GENCODE: the reference human genome annotation for The ENCODE Project. Genome Res 22(9):1760–1774
7. Xie C et al (2014) NONCODEv4: exploring the world of long non-coding RNA genes. Nucleic Acids Res 42(D1):D98–D103
8. Zhao Y et al (2016) NONCODEv4: annotation of noncoding RNAs with emphasis on long noncoding RNAs. Methods Mol Biol 1402:243–254
9. Ulitsky I, Bartel DP (2013) lincRNAs: genomics, evolution, and mechanisms. Cell 154(1):26–46
10. Quek XC et al (2015) lncRNAdb v2.0: expanding the reference database for functional long noncoding RNAs. Nucleic Acids Res 43(D1):D168–D173
11. Bedoya-Reina OC, Ponting CP (2017) Functional RNA classes: a matter of time? Nat Struct Mol Biol 24(1):7–8
12. Fang Y et al (2016) Mechanisms of long non-coding RNAs in cancer. Genomics Proteomics Bioinformatics 14(1):42–54
13. Mercer TR, Mattick JS (2013) Structure and function of long noncoding RNAs in epigenetic regulation. Nat Struct Mol Biol 20(3):300–307
14. Rinn JL, Chang HY (2012) Genome regulation by long noncoding RNAs. Annu Rev Biochem 81(1):145–166
15. Gendrel AV et al (2014) Noncoding RNAs and epigenetic mechanisms during X-chromosome inactivation. Annu Rev Cell Dev Biol 30:561–580
16. Sandhu KS et al (2012) Large-scale functional organization of long-range chromatin interaction networks. Cell Rep 2(5):1207–1219
17. Dekker J et al (2015) Long-range chromatin interactions. Cold Spring Harb Perspect Biol 7(10):a019356
18. Ørom UA et al (2010) Long noncoding RNAs with enhancer-like function in human cells. Cell 143(1):46–58
19. Ntini E et al (2018) Long ncRNA A-ROD activates its target gene DKK1 at its release from chromatin. Nat Commun 9(1):1636
20. Heller D et al (2017) ssHMM: extracting intuitive sequence structure motifs from high-throughput RNA-binding protein data. Nucleic Acids Res 45(19):11004–11018
21. Krakau S et al (2017) PureCLIP: capturing target-specific protein-RNA interaction footprints from single nucleotide CLIP-seq data. Genome Biol 18(1):240
22. Budach S, Marsico A (2018) Pysster: classification of biological sequences by learning sequence and structure motifs with convolutional neural networks. Bioinformatics 34(17):3035–3037
23. Poliseno L et al (2010) A coding-independent function of gene and pseudogene mRNAs regulates tumour biology. Nature 465:1033–1038
24. Salmena L et al (2011) A ceRNA hypothesis: the Rosetta Stone of a hidden RNA language? Cell 146(3):353–358
25. Ebert MS et al (2007) MicroRNA sponges: competitive inhibitors of small RNAs in mammalian cells. Nat Methods 4:721–726
26. Franco-Zorrilla JM et al (2007) Target mimicry provides a new mechanism for regulation of microRNA activity. Nat Genet 39(8):1033–1037
27. Karreth FA, Pandolfi PP (2013) ceRNA cross-talk in cancer: when ce-bling rivalries go awry. Cancer Discov 3(10):1113–1121
28. Piro RM (2011) Are all genes regulatory genes? Biol Philos 26(4):595–602
29. Cesana M et al (2011) A long noncoding RNA controls muscle differentiation by functioning as a competing endogenous RNA. Cell 147(2):358–369
30. Paci P, Colombo T, Farina L (2014) Computational analysis identifies a sponge interaction network between long non-coding RNAs and messenger RNAs in human breast cancer. BMC Syst Biol 8(1):83

31. Tay Y et al (2011) Coding-independent regulation of the tumor suppressor PTEN by competing endogenous mRNAs. Cell 147(2):344–357
32. Yang C et al (2016) Competing endogenous RNA networks in human cancer: hypothesis, validation, and perspectives. Oncotarget 7(12):13479–13490
33. Qu L et al (2016) Exosome-transmitted lncARSR promotes sunitinib resistance in renal cancer by acting as a competing endogenous RNA. Cancer Cell 29(5):653–668
34. Chan JJ, Tay Y (2018) Noncoding RNA:RNA regulatory networks in cancer. Int J Mol Sci 19(5):1310
35. Tay Y, Rinn J, Pandolfi PP (2014) The multilayered complexity of ceRNA crosstalk and competition. Nature 505(7483):344–352
36. Anastasiadou E, Jacob LS, Slack FJ (2018) Non-coding RNA networks in cancer. Nat Rev Cancer 18(1):5–18
37. Taulli R, Loretelli C, Pandolfi PP (2013) From pseudo-ceRNAs to circ-ceRNAs: a tale of cross-talk and competition. Nat Struct Mol Biol 20:541–543
38. Ala U et al (2013) Integrated transcriptional and competitive endogenous RNA networks are cross-regulated in permissive molecular environments. Proc Natl Acad Sci 110(18):7154–7159
39. Figliuzzi M, Marinari E, De Martino A (2013) MicroRNAs as a selective channel of communication between competing RNAs: a steady-state theory. Biophys J 104(5):1203–1213
40. Smillie CL, Sirey T, Ponting CP (2018) Complexities of post-transcriptional regulation and the modeling of ceRNA crosstalk. Crit Rev Biochem Mol Biol 53(3):231–245
41. Bosson AD, Zamudio JR, Sharp PA (2014) Endogenous miRNA and target concentrations determine susceptibility to potential ceRNA competition. Mol Cell 56(3):347–359
42. Tan JY et al (2015) Extensive microRNA-mediated crosstalk between lncRNAs and mRNAs in mouse embryonic stem cells. Genome Res 25(5):655–666
43. Denzler R et al (2014) Assessing the ceRNA hypothesis with quantitative measurements of miRNA and target abundance. Mol Cell 54(5):766–776
44. Kartha RV, Subramanian S (2014) Competing endogenous RNAs (ceRNAs): new entrants to the intricacies of gene regulation. Front Genet 5:8
45. Esteller M (2011) Non-coding RNAs in human disease. Nat Rev Genet 12(12):861–874
46. Yang F et al (2011) Long noncoding RNA high expression in hepatocellular carcinoma facilitates tumor growth through enhancer of zeste homolog 2 in humans. Hepatology 54(5):1679–1689
47. Tessier CR et al (2004) Mammary tumor induction in transgenic mice expressing an RNA-binding protein. Cancer Res 64(1):209–214
48. Mourtada-Maarabouni M et al (2009) GAS5, a non-protein-coding RNA, controls apoptosis and is downregulated in breast cancer. Oncogene 28(2):195–208
49. Guan Y et al (2007) Amplification of PVT1 contributes to the pathophysiology of ovarian and breast cancer. Clin Cancer Res 13(19):5745–5755
50. Chen W et al (1997) Expression of neural BC200 RNA in human tumours. J Pathol 183(3):345–351
51. Faghihi MA et al (2008) Expression of a non-coding RNA is elevated in Alzheimer's disease and drives rapid feed-forward regulation of beta-secretase. Nat Med 14(7):723–730
52. Zhang A et al (2014) Role of the lncRNA-p53 regulatory network in cancer. Mol Cell Biol 6(3):181–191
53. Cheng G et al (2013) LncRNADisease: a database for long-non-coding RNA-associated diseases. Nucleic Acids Res 41(Database issue):D983–D986
54. Ning S et al (2016) Lnc2Cancer: a manually curated database of experimentally supported lncRNAs associated with various human cancers. Nucleic Acids Res 44(D1):D980–D985
55. Ning S et al (2014) SNP@lincTFBS: an integrated database of polymorphisms in human lincRNA transcription factor binding sites. PLoS One 9(7):e103851
56. Gong J et al (2015) lncRNASNP: a database of SNPs in lncRNAs and their potential functions in human and mouse. Nucleic Acids Res 43(Database issue):D181–D186
57. Cabili MN et al (2011) Integrative annotation of human large intergenic noncoding RNAs reveals global properties and specific subclasses. Genes Dev 25(18):1915–1927
58. Iyer MK et al (2015) The landscape of long noncoding RNAs in the human transcriptome. Nat Genet 47(3):199–208
59. Guo X et al (2016) Advances in long noncoding RNAs: identification, structure prediction and function annotation. Brief Funct Genomics 15(1):38–46

60. Fiscon G, Paci P, Iannello G (2015) MONSTER v1.1: a tool to extract and search for RNA non-branching structures. BMC Genomics 16(6):S1
61. Wilusz JE, Freier SM, Spector DL (2008) 3' end processing of a long nuclear-retained noncoding RNA yields a tRNA-like cytoplasmic RNA. Cell 135(5):919–932
62. Stuart JM et al (2003) A gene-coexpression network for global discovery of conserved genetic modules. Science 302(5643):249–255
63. Barabási AL, Albert R (1999) Emergence of scaling in random networks. Science 286(5439):509–512
64. Liao Q et al (2011) Large-scale prediction of long non-coding RNA functions in a coding–non-coding gene co-expression network. Nucleic Acids Res 39(9):3864–3878
65. Spicuglia S et al (2013) An update on recent methods applied for deciphering the diversity of the noncoding RNA genome structure and function. Methods 63(1):3–17
66. Signal B, Gloss BS, Dinger ME (2016) Computational approaches for functional prediction and characterisation of long noncoding RNAs. Trends Genet 32(10):620–637
67. Chen X, Yan G-Y (2013) Novel human lncRNA–disease association inference based on lncRNA expression profiles. Bioinformatics 29(20):2617–2624
68. Chen X et al (2015) Constructing lncRNA functional similarity network based on lncRNA-disease associations and disease semantic similarity. Sci Rep 5:11338
69. Zhao T et al (2015) Identification of cancer-related lncRNAs through integrating genome, regulome and transcriptome features. Mol BioSyst 11:126–136
70. Sun J et al (2014) Inferring novel lncRNA-disease associations based on a random walk model of a lncRNA functional similarity network. Mol BioSyst 10:2074–2081
71. Chen X et al (2012) Prediction of disease-related interactions between microRNAs and environmental factors based on a semi-supervised classifier. PLoS One 7(8):e43425
72. Huang YA et al (2016) ILNCSIM: improved lncRNA functional similarity calculation model. Oncotarget 7(18):25902–25914
73. Ganegoda GU et al (2015) Heterogeneous network model to infer human disease-long intergenic non-coding RNA associations. IEEE Trans NanoBiosci 14(2):175–183
74. Liu M-X et al (2014) A computational framework to infer human disease-associated long noncoding RNAs. PLoS One 9(1):e84408
75. Cheng L et al (2016) IntNetLncSim: an integrative network analysis method to infer human lncRNA functional similarity. Oncotarget 7(30):47864–47874
76. Perron U, Provero P, Molineris I (2017) In silico prediction of lncRNA function using tissue specific and evolutionary conserved expression. BMC Bioinf 18(Suppl 5):144
77. Gu C et al (2017) Global network random walk for predicting potential human lncRNA-disease associations. Sci Rep 7(1):12442
78. Welter D et al (2014) The NHGRI GWAS Catalog, a curated resource of SNP-trait associations. Nucleic Acids Res 42(Database issue):D1001–D1006
79. Barrett T et al (2013) NCBI GEO: archive for functional genomics data sets–update. Nucleic Acids Res 41(D1):D991–D995
80. Guo X et al (2013) Long non-coding RNAs function annotation: a global prediction method based on bi-colored networks. Nucleic Acids Res 41(2):e35
81. Ala U et al (2008) Prediction of human disease genes by human-mouse conserved coexpression analysis. PLOS Comput Biol 4(3):e1000043
82. Piro RM et al (2011) An atlas of tissue-specific conserved coexpression for functional annotation and disease gene prediction. Eur J Hum Genet 19(11):1173–1180
83. Piro RM (2012) Network medicine: linking disorders. Hum Genet 131(12):1811–1820
84. Zhou M et al (2015) Prioritizing candidate disease-related long non-coding RNAs by walking on the heterogeneous lncRNA and disease network. Mol Biosyst 11(3):760–769
85. Chen X et al (2015) KATZLDA: KATZ measure for the lncRNA-disease association prediction. Sci Rep 5:16840
86. Li J-H et al (2014) starBase v2.0: decoding miRNA-ceRNA, miRNA-ncRNA and protein–RNA interaction networks from large-scale CLIP-Seq data. Nucleic Acids Res 42(D1):D92–D97
87. Keshava Prasad TS et al (2009) Human Protein Reference Database–2009 update. Nucleic Acids Res 37(Database issue):D767–D772
88. Vergoulis T et al (2012) TarBase 6.0: capturing the exponential growth of miRNA targets with experimental support. Nucleic Acids Res 40(D1):D222–D229
89. Hsu S-D et al (2014) miRTarBase update 2014: an information resource for experimentally validated miRNA-target interactions. Nucleic Acids Res 42(D1):D78–D85

90. Xiao F et al (2009) miRecords: an integrated resource for microRNA–target interactions. Nucleic Acids Res 37(Database issue): D105–D110
91. Stojmirovic A, Yu YK (2009) ITM Probe: analyzing information flow in protein networks. Bioinformatics 25(18):2447–2449
92. Siahpirani A et al (2016) A multi-task graph-clustering approach for chromosome conformation capture data sets identifies conserved modules of chromosomal interactions. Genome Biol 17(1):114
93. Li G et al (2010) ChIA-PET tool for comprehensive chromatin interaction analysis with paired-end tag sequencing. Genome Biol 11(2):R22
94. Thiel D et al (2018) Identifying lncRNA-mediated regulatory modules via ChIA-PET network analysis. bioRxiv
95. Lai F et al (2013) Activating RNAs associate with Mediator to enhance chromatin architecture and transcription. Nature 494(7438):497–501
96. Djurdjevac N et al (2011) Random walks on complex modular networks. J Numer Anal Ind Appl Math 6:29–50
97. Le TD et al (2017) Computational methods for identifying miRNA sponge interactions. Brief Bioinform 18(4):577–590
98. Thomas M, Lieberman J, Lal A (2010) Desperately seeking microRNA targets. Nat Struct Mol Biol 17(10):1169–1174
99. Krek A et al (2005) Combinatorial microRNA target predictions. Nat Genet 37(5):495–500
100. Agarwal V et al (2015) Predicting effective microRNA target sites in mammalian mRNAs. eLife 4:e05005
101. Miranda KC et al (2006) A pattern-based method for the identification of microRNA binding sites and their corresponding heteroduplexes. Cell 126(6):1203–1217
102. Kertesz M et al (2007) The role of site accessibility in microRNA target recognition. Nat Genet 39(10):1278–1284
103. Riffo-Campos ÁL, Riquelme I, Brebi-Mieville P (2016) Tools for sequence-based miRNA target prediction: what to choose? Int J Mol Sci 17(12):1987
104. Bhartiya D et al (2013) lncRNome: a comprehensive knowledgebase of human long noncoding RNAs. Database 2013:bat034
105. Jeggari A, Marks DS, Larsson E (2012) miRcode: a map of putative microRNA target sites in the long non-coding transcriptome. Bioinformatics 28(15):2062–2063
106. Paraskevopoulou MD et al (2013) DIANA-LncBase: experimentally verified and computationally predicted microRNA targets on long non-coding RNAs. Nucleic Acids Res 41(D1):D239–D245
107. Zisoulis DG et al (2010) Comprehensive discovery of endogenous Argonaute binding sites in Caenorhabditis elegans. Nat Struct Mol Biol 17(2):173–179
108. Clark PM et al (2014) Argonaute CLIP-Seq reveals miRNA targetome diversity across tissue types. Sci Rep 4:5947
109. Joshua-Tor L (2006) The Argonautes. Cold Spring Harb Symp Quant Biol 71:67–72
110. Kawamata T, Tomari Y (2010) Making RISC. Trends Biochem Sci 35(7):368–376
111. Sarver AL, Subramanian S (2012) Competing endogenous RNA database. Bioinformation 8(15):731–733
112. Sumazin P et al (2011) An extensive microRNA-mediated network of RNA-RNA interactions regulates established oncogenic pathways in glioblastoma. Cell 147(2):370–381
113. Chiu H-S et al (2015) Cupid: simultaneous reconstruction of microRNA-target and ceRNA networks. Genome Res 25(2):257–267
114. Zarringhalam K et al (2017) Identification of competing endogenous RNAs of the tumor suppressor gene PTEN: a probabilistic approach. Sci Rep 7(1):7755
115. Das S et al (2014) lnCeDB: database of human long noncoding RNA acting as competing endogenous RNA. PLoS One 9(6):e98965
116. Wang P et al (2015) miRSponge: a manually curated database for experimentally supported miRNA sponges and ceRNAs. Database 2015:bav098
117. Wang P et al (2015) Identification of lncRNA-associated competing triplets reveals global patterns and prognostic markers for cancer. Nucleic Acids Res 43(7):3478–3489
118. Piro RM, Di Cunto F (2012) Computational approaches to disease-gene prediction: rationale, classification and successes. FEBS J 279(5):678–696
119. Xu C et al (2017) LncNetP, a systematical lncRNA prioritization approach based on ceRNA and disease phenotype association assumptions. Oncotarget 8(70):114603–114612
120. Köhler S et al (2008) Walking the interactome for prioritization of candidate disease genes. Am J Hum Genet 82(4):949–958

Chapter 13

Integration of miRNA and mRNA Expression Data for Understanding Etiology of Gynecologic Cancers

Sushmita Paul

Abstract

Dysregulation of miRNA–mRNA regulatory networks is very common phenomenon in any diseases including cancer. Altered expression of biomarkers leads to these gynecologic cancers. Therefore, understanding the underlying biological mechanisms may help in developing a robust diagnostic as well as a prognostic tool. It has been demonstrated in various studies that the pathways associated with gynecologic cancer have dysregulated miRNA as well as mRNA expression. Identification of miRNA–mRNA regulatory modules may help in understanding the mechanism of altered gynecologic cancer pathways. In this regard, an existing robust mutual information-based Maximum-Relevance Maximum-Significance algorithm has been used for identification of miRNA–mRNA regulatory modules in gynecologic cancer. A set of miRNA–mRNA modules are identified first than their association with gynecologic cancer are studied exhaustively. The effectiveness of the proposed approach is compared with the existing methods. The proposed approach is found to generate more robust integrated networks of miRNA–mRNA in gynecologic cancer.

Key words Gynecologic cancer, miRNAs, Genes, Mutual information, MRMS, Ovarian cancer, Cervical cancer

1 Introduction

Gynecological malignancies, including cancers of endometrium, cervix, ovary, vulva, and vagina, account for 12.5% of all new cancers in women. According to the American Cancer Society, around 110,070 women will have been diagnosed with, and 32,120 women will have died of, cancer of the female genital tract in 2018 in the USA [2]. Hence, it is important to understand the mechanisms of carcinogenesis and progression in gynecological cancer. Among different gynecological cancers, cancer of cervix and ovarian cancer prevalence is higher.

Ovarian cancer has a distinctive biology and behavior at the clinical, cellular, and molecular levels. It is the most prevalent and lethal female reproductive cancer, accounting for 5% of female cancer deaths. According to the National Cancer Institute, around 22,440 women will get diagnosed by this disease and 14,080 cases will die due to the disease by 2017 [20]. The 5-year overall survival rate of this disease is 46.5% when untreated. Whereas, early detection of the disease with proper treatment can increase the overall survival rate of patients, that is, 92.5%. On the other hand, cervical cancers account for the second-most gynecological cancer death cases worldwide, and this situation is worse in developing countries due to the lack of adequate organized screening programs. It is believed that Human Papilloma Virus (HPV) infections are the major causes of invasive cervical cancer. Therefore, it is important to understand the role of biomarkers like miRNAs and mRNAs in various pathways of ovarian cancer and cervical cancer.

MicroRNAs (miRNAs) are small noncoding RNAs of size ~22-nucleotides. miRNA suppresses the expression of mRNA by binding to the 3′ untranslated region of the mRNA. When transcribed from DNA, they are of around 80 nucleotides size and are known as primary transcripts (pri-miRNA). Later, they are processed into hairpin precursors of ~70 nucleotides (pre-miRNA) that finally lead to mature miRNAs. They are found in many plants and animals. Extensive studies have been conducted to understand their role in different biological processes and diseases [1, 10, 15], including cell proliferation, apoptosis, differentiation, and development [5, 9, 10, 13], as well as in numerous human diseases, such as fragile X syndrome, chronic lymphocytic leukemia, and various types of cancers [1, 3, 15, 18].

Studies related to the role of miRNAs and their targets in ovarian cancer and cervical cancer is less studied. Only few papers related to this topic are available in PubMed. Therefore, there is dire need to conduct studies related to this topic, to come up with solutions for developing a diagnostic and prognostic tool against ovarian cancer and cervical cancer. Existing methods usually use sequence data for identification of miRNAs and their targets. However, there exists a higher possibility of false-positive rates. Therefore, few works have been done that used miRNA and mRNA expression data. However, few of them select miRNAs and mRNAs separately and then by using the correlation between the selected biomarkers reconstruction of regulatory modules takes place. Other methods use regression methods, they require more computational time. Hence, there is a need to develop a scalable approach for identification of miRNA–mRNA regulatory modules in ovarian cancer and cervical cancer.

In this chapter, a framework for selection of important miRNA–mRNA regulatory modules in ovarian cancer and cervical cancer is reported. For selection of regulatory modules, mutual information-based maximum-relevance maximum-significance (MIMRMS) [19] has been used. Here, a set of genes that are regulated by a particular miRNA is identified with MIMRMS. In the current study, the expression values of miRNAs are discretized and used as class labels. Whereas, the expression values of genes are considered features. Mutual information between two variables here miRNA and mRNA suggests about the interdependency between them. The MIMRMS algorithm selects a set of genes for a particular miRNA by maximizing both relevance and significance of the gene. In this manner, a set of gene is selected that is both relevant and significant with respect to that miRNA. The miRNA information is used as a class label, and mRNAs are later selected with the help of MIMRMS algorithm. For a particular miRNA, a set of 50 mRNAs is selected using the MIMRMS algorithm. The mRNAs of each module are evaluated further with the help of K-nearest neighbor classifier in order to reduce false positives. The effective mRNAs obtained represent a regulatory module, that is, an miRNA regulating a set of mRNAs. Next to avoid irrelevant modules, statistical significance of each module is computed using STRING database. Networks with P-value $<$ 0.05 are studied further. Pathway enrichment analysis, and disease ontology enrichment analysis revealed the importance of selected modules with respect to ovarian cancer. The modules generated by MIMRMS are compared with the modules generated by mRMR algorithm as well as MatrixEQTL. From the results, it is revealed that the MIMRMS-based approach generates more significant miRNA–mRNA regulatory modules for ovarian cancer data.

The structure of the rest of this chapter is as follows: Description of data set is provided in Subheading 2. Framework for automatic detection of regulatory modules is given in Subheading 3. Subheading 4 describes the importance of proposed approach for identification of important regulatory modules. The chapter finally concludes in Subheading 5.

2 Data Set Description

Both miRNA and gene expression data for serous ovarian cancer as well as cervical cancer were downloaded through the Cancer Genomics Browser of UC Santa Cruz [6]. Both expression data of ovarian cancer contain exactly the same samples, that is, 415. Whereas, the number of miRNAs and genes are 175 and 13,946, respectively. On the other hand, there are 308 samples in each gene expression data and miRNA expression data of cervical cancer. It contains 429 miRNAs and 15,607 genes. The platform used for generation of both the data sets is Illumina HiSeq.

3 Construction of miRNA–mRNA Modules

Automatic detection of miRNA–mRNA modules is very important to understand the underlying mechanism of the disease. This section describes the method that has been used for identification of miRNA–mRNA modules in ovarian cancer. In the present work, the MIMRMS [17] has been used to identify miRNA–mRNA regulatory modules.

Provided the matrices of miRNA expression and gene expression, a decision matrix is created first. A decision matrix contains a class label attribute and conditional attributes. Here, the expression values of each miRNA are discretized and later used as class labels. All the expression values of genes are considered as features or conditional attributes. The rows represent samples. Therefore, in total 175 decision matrices are created each having dimension of 415 rows and 13,946 columns and one class label. For each miRNA, a set of mRNAs is selected by implementing MIMRMS algorithm. Next, the K-nearest neighbor algorithm is applied on the genes for each module for selecting an effective set of genes that generates high classification accuracy. Biologically, it can be interpreted as those genes that are regulated by a particular miRNA. The aim of this study was to select a set of relevant as well as significant genes that can map on miRNA, thus generating a regulatory network that may potentially have some role in the onset and progression of ovarian cancer. Next, the existing MIMRMS algorithm and K-nearest neighbor algorithms are described.

3.1 The Gene Selection Algorithm

This section describes about the existing MIMRMS algorithm [17] that has been used in the current study. The MIMRMS generates a set of mRNAs by maximizing both relevance and significance. The MIMRMS algorithm is described next.

The MIMRMS algorithm selects a set of mRNAs Θ from a given microarray data set $\mathbb{C} = \{\mathscr{G}_1, \ldots, \mathscr{G}_i, \ldots, \mathscr{G}_j, \ldots, \mathscr{G}_m\}$ of m mRNAs. Relevance of an mRNA quantifies the correlation of the mRNA with respect to class label or miRNA. Also, it infers about the dependency of the class label $\mathbb{M}$ on an attribute. Here, the relevance of the mRNA $\mathscr{G}_i$ with respect to class labels/miRNAs $\mathbb{M}$ is defined as $\hat{f}(\mathscr{G}_i, \mathbb{M})$. Whereas, $\tilde{f}(\mathscr{G}_i, \mathscr{G}_j)$ is defined as the significance of the mRNA $\mathscr{G}_j$ with respect to the mRNA $\mathscr{G}_i$. In this study for calculation of both relevance and significance, mutual information [16] is used [17].

The relevance $\hat{f}(\mathscr{G}_i, \mathbb{M})$ of an mRNA $\mathscr{G}_i$ with respect to the class label or miRNA $\mathbb{M}$ using mutual information can be computed as follows:

$$\hat{f}(\mathscr{G}_i, \mathbb{M}) = I(\mathscr{G}_i, \mathbb{M}), \tag{1}$$

where $I(\mathscr{G}_i, \mathbb{M})$ represents the mutual information between attribute $\mathscr{G}_i$ and miRNA or class label $\mathbb{M}$ that is given by:

$$I(\mathscr{G}_i, \mathbb{M}) = H(\mathscr{G}_i) - H(\mathscr{G}_i \mid \mathbb{M}). \tag{2}$$

Here, $H(\mathscr{G}_i)$ and $H(\mathscr{G}_i \mid \mathbb{M})$ represent the entropy of mRNA $\mathscr{G}_i$ and the conditional entropy of $\mathscr{G}_i$ given class label $\mathbb{M}$, respectively. The entropy is a measure of uncertainty.

Provided a set of attributes, individual contribution of an attribute for calculation of dependency on decision attribute can be computed with the help of significance criterion. Hence, significance value of an attribute signifies its importance. Removal of an attribute from the set of condition attributes leads to change in dependency value. This change is the significance of the attribute. Its value ranges from 0 to 1. If its value is 0 (1), then the attribute is dispensable (indispensable).

Definition 1: Given $\mathbb{C}, \mathbb{M}$ and an attribute $\mathscr{G} \in \mathbb{C}$, the significance of the attribute $\mathscr{G}$ is defined as [17]:

$$\sigma_{\mathbb{C}}(\mathbb{M}, \mathscr{G}) = \hat{f}(\mathbb{C}, \mathbb{M}) - \hat{f}(\mathbb{C} - \{\mathscr{G}\}, \mathbb{M}) \qquad (3)$$

The total relevance of all selected mRNAs and total significance among the selected mRNAs are, therefore, given by:

$$\mathscr{I}_{\text{relev}} = \sum_{\mathscr{G}_i \in \Theta} \hat{f}(\mathscr{G}_i, \mathbb{M}) \qquad \mathscr{I}_{\text{signf}} = \sum_{\mathscr{G}_i \neq \mathscr{G}_j \in \Theta} \tilde{f}(\mathscr{G}_i, \mathscr{G}_j). \qquad (4)$$

For identification of miRNA–mRNA module, first of all, a decision table is created for each miRNA. The decision table contains gene or mRNA as conditional attributes and miRNA as class label. The rows are samples. The MIMRMS algorithm is implemented on each decision table for identification of genes or mRNAs that are associated with that particular miRNA. The MIMRMS process starts by initializing $\mathbb{C} \leftarrow \{\mathscr{G}_1, \ldots, \mathscr{G}_i, \ldots, \mathscr{G}_j, \ldots, \mathscr{G}_m\}, \Theta \leftarrow \emptyset$. Next, it calculates relevance $\hat{f}(\mathscr{G}_i, \mathbb{M})$ of each mRNA $\mathscr{G}_i \in \mathbb{C}$ with respect to class label or miRNA. Most relevant mRNA $\mathscr{G}_i$ is selected having the highest relevance value $\hat{f}(\mathscr{G}_i, \mathbb{M})$. In effect, $\mathscr{G}_i \in \Theta$ and $\mathbb{C} = \mathbb{C} \setminus \mathscr{G}_i$. The algorithm iteratively computes significance of each mRNA with respect to already selected mRNAs and selects the mRNA if it has maximum value for optimization function. As a result of that, $\mathscr{G}_j \in \Theta$ and $\mathbb{C} = \mathbb{C} \setminus \mathscr{G}_j$. This step occurs till the desired number of mRNAs is selected for corresponding miRNA or class label. The optimization function of the MIMRMS algorithm is

$$\hat{f}(\mathscr{G}_j, \mathbb{M}) + \frac{1}{|\Theta|} \sum_{\mathscr{G}_i \in \Theta} \tilde{f}(\mathscr{G}_i, \mathscr{G}_j). \qquad (5)$$

Mutual information is used to compute both relevance and significance of an mRNA. The relevance and significance of an mRNA are calculated using Eqs. 1 and 3, respectively.

The expression values of both miRNA and mRNA in a microarray data are continuous in nature. Continuous expression values of an miRNA and mRNA need to be discretized for calculation of relevance of an mRNA with respect to miRNA or clinical outcome using mutual information. The marginal probabilities and the joint probability are computed using discretized expression values of an mRNA and miRNA. These probabilities are later used to compute the mRNA-class/miRNA relevance. Therefore, discretization of continuous valued miRNAs and mRNAs is a very vital step in the current study. In the current study, discretization method mentioned in [7] is used. This method discretizes expression values of an miRNA and mRNA using mean μ and standard deviation σ that are computed over n expression values of that particular miRNAs or mRNA. Next, the values bigger than $(\mu + \sigma)$ are represented as 1, the values between $(\mu - \sigma)$ and $(\mu + \sigma)$ as 0, and the values smaller than $(\mu - \sigma)$ as -1. The overexpression, baseline, and underexpression of the miRNAs or mRNAs correspond to these three values.

3.2 K-Nearest Neighbor Rule

The K-nearest neighbor (K-NN) rule [8] is a classifier. It is used to evaluate the efficiency of a set of reduced mRNAs. It classifies an unknown sample by considering its nearest or closest training samples in the feature space. A sample is classified by a majority vote of its K-neighbors, with the sample being assigned to the class most common among its K-nearest neighbors. The value of K, chosen for the K-NN, is the square root of the number of samples in training set. In the current study, the mRNAs of each miRNA–mRNA module obtained using the MIMRMS algorithm are further processed. For each miRNA–mRNA module, K-NN is implemented for selecting best mRNAs for a particular miRNA. The mRNAs in a particular module generating the highest accuracy values are considered further. Biologically, it can be inferred that the mRNAs finally selected for a particular module are regulated by the miRNA of that module.

4 Experimental Results

In the current study, the existing MIMRMS algorithm is used to identify regulatory modules. Fifty top-ranked mRNAs are selected using the MIMRMS algorithm for further analysis. For Matrix-EQTL, top 50 mRNAs of each module are directly used for further analysis as the ranking of the mRNAs was not sure. In a module, filtering of mRNAs is further carried out to reduce false positives. Therefore, prediction accuracy of K-nearest neighbor (K-NN) rule along with leave-one-out cross-validation (LOOCV) is computed for the mRNAs of each module. Finally, the obtained modules are evaluated using STRING database [23], pathway enrichment analysis, and disease ontology.

The effectiveness of the proposed approach is demonstrated on ovarian cancer data as well as cervical cancer data, and the approach is also compared with the methods mentioned in Huang and Cai [11] and MatrixeQTL [21]. Huang and Cai used minimum-redundancy maximum-relevance criteria [7] for selection of modules.

4.1 Selection of Significant Regulatory Modules in Ovarian Cancer

In total, 175 modules are generated by implementing the MIMRMS and K-NN rule. The leave-one-out accuracy of each module varied from 0% to 48.43%. Next, STRING database [23] is used to generate connections between the genes of each module to check whether the genes of obtained modules are involved in the same biological function or not. The database uses information from experimentally validated connections, prediction, text mining, and so forth for creating a connection between two genes or proteins. STRING database stores the information of protein–protein interaction. It also provides the statistical significance for a particular protein–protein interaction network (PPIN). The statistical significance of networks is quantified using P-value.

Table 1 presents the total number of significant regulatory networks (P-value $<$ 0.05) generated by MatrixEQTL, mRMR, and MIMRMS algorithms. From the table, it is seen that MatrixEQTL, mRMR, and MIMRMS algorithms generate significant PPI network. However, only MatrixEQTL and MIMRMS algorithms generate network with very low P-value $=$ 0. The MatrixEQTL generates only one network with P-value $=$ 0, whereas the MIMRMS generates five highly significant P-value $=$ 0 networks. The details of all six (one MatrixEQTL and five MIMRMS) modules are presented in Table 2. The images of few networks are provided in Fig. 1. From the figure, it is seen that the networks

Table 1
Number of significant modules generated by MatrixEQTL, mRMR, and MIMRMS (ovarian cancer)

Algorithms/Methods	Number of Significant Modules
MatrixEQTL	56
mRMR	8
MIMRMS	42

Table 2
Description of most significant modules (ovarian cancer)

miRNA	No. of genes in module	Algorithm
hsa-mir-17	775	MatrixEQTL
hsa-mir-30e	40	MIMRMS
hsa-mir-100	50	MIMRMS
hsa-mir-181c	47	MIMRMS
hsa-let-7c	35	MIMRMS
hsa-mir-23a	36	MIMRMS

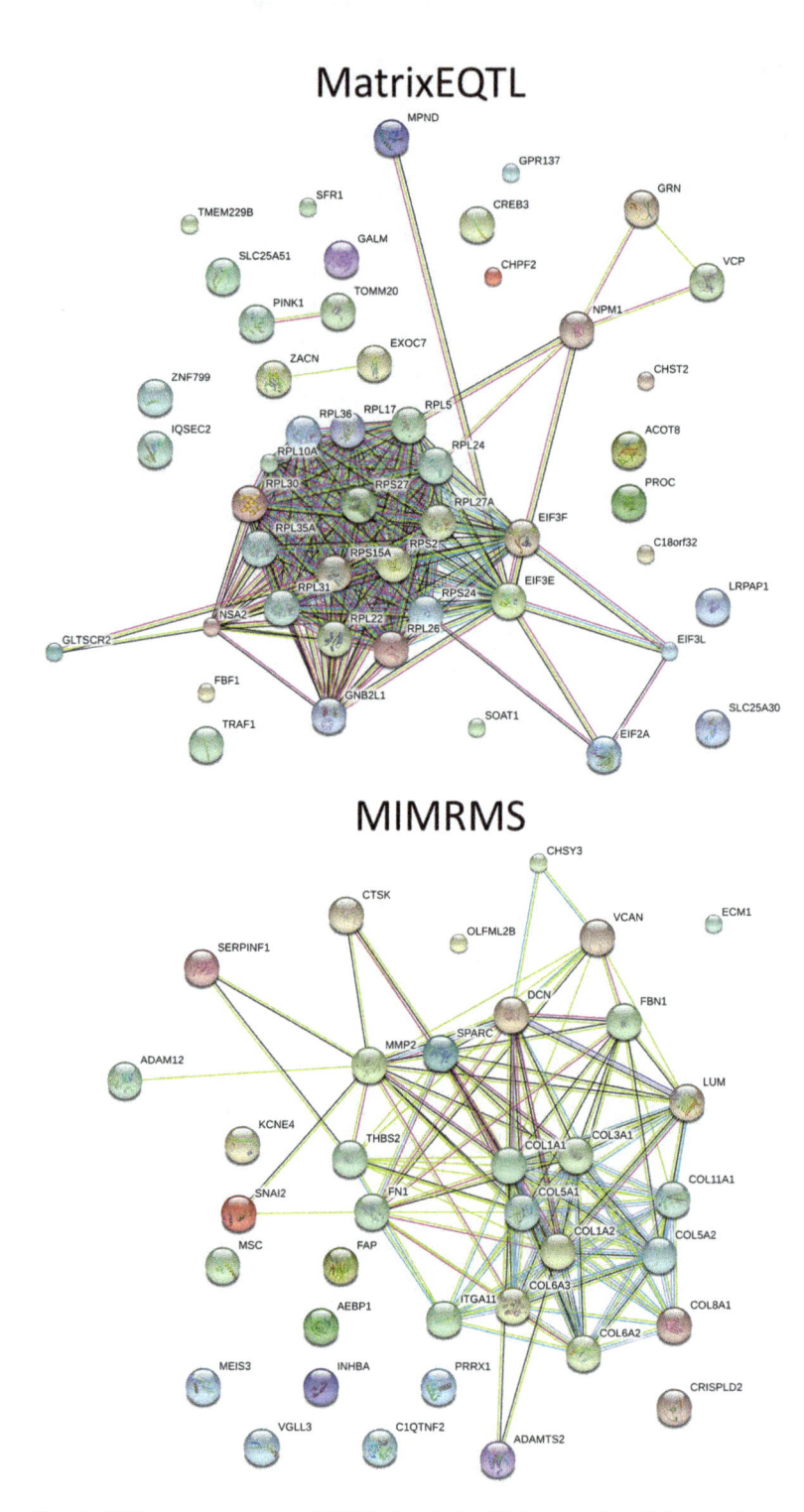

Fig. 1 PPINs generated by STRING for MatrixEQTL and MIMRMS algorithms (ovarian cancer)

Table 3
Number of significant modules generated by MatrixEQTL, mRMR, and MIMRMS (cervical cancer)

Algorithms/Methods	Number of significant modules
MatrixEQTL	90
mRMR	5
MIMRMS	83

generated are highly interconnected and compact. They also suggest that the MIMRMS-based approach selects significant regulatory modules.

4.2 Selection of Significant Regulatory Modules in Cervical Cancer

In total, 420 modules are generated by implementing the MIMRMS and K-NN rule. The leave-one-out accuracy of each module varied from 0% to 76.95%. Table 3 presents the total number of significant regulatory networks (P-value $<$ 0.05) generated by MatrixEQTL, mRMR, and MIMRMS algorithms. From the table, it is seen that MatrixEQTL, mRMR, and MIMRMS algorithms generate significant PPI network. For cervical cancer data, MatrixEQTL generates more number of networks having P-value $=$ 0 compared to mRMR and MIMRMS.

However, only MatrixEQTL and MIMRMS algorithms generate network with very low P-value $=$ 0. The MatrixEQTL generates four networks with P-value $=$ 0, whereas the MIMRMS generates eight highly significant P-value $=$ 0 networks. The details of all ten (four MatrixEQTL and eight MIMRMS) modules are presented in Table 4. The images of few networks are provided in Fig. 2. From the figure, it is seen that the networks generated are highly interconnected and compact. They also suggest that the MIMRMS-based approach selects significant regulatory modules.

4.3 Pathway Enrichment Analysis of Ovarian Cancer Modules and Cervical Cancer

For the biological interpretation of highly significant modules P-value $=$ 0, the Cytoscape [22] plug-in ClueGO [4] has been used to perform pathway enrichment analysis. Genes of significant modules are used for pathway enrichment analysis. ClueGO plugin has been used to perform pathway enrichment analysis. It identifies the pathway terms with which genes of a module are significantly associated. Moreover, it clusters similar pathway terms sharing common genes. ClueGO visualizes the nonredundant biological pathway terms for large clusters of genes in a functionally grouped network. The related terms that share similar associated genes can be fused to reduce redundancy, those terms have the same color code. The ClueGO network is created with kappa statistics and reflects the relationships between the terms based on the similarity of their associated genes. On the network, the node color can be

Table 4
Description of most significant modules (cervical cancer)

miRNA	No. of genes in module	Algorithm
hsa-mir-330	225	MatrixEQTL
hsa-mir-142	50	MatrixEQTL
hsa-mir-10a	160	MatrixEQTL
hsa-mir-27b	800	MatrixEQTL
hsa-mir-30d	34	MIMRMS
hsa-mir-205	20	MIMRMS
hsa-mir-25	50	MIMRMS
hsa-mir-101-2	25	MIMRMS
hsa-mir-93	42	MIMRMS
hsa-mir-29c	24	MIMRMS
hsa-let-7d	43	MIMRMS
hsa-mir-409	19	MIMRMS

switched between functional groups and clusters distribution. The association of genes with a term is calculated in terms of *P*-value. For correction of *P*-value, Bonferroni step down method is used. The lower the *P*-value, the more significant is the term. The size of the nodes in the network represents the significance level of that term. The largest node in the cluster denotes that the pathway term is the most significant one and it is also the representative of the cluster. For the current analysis, the threshold for *P*-value was set to 0.05 and the minimum number of genes associated with a term was set to 3. WikiPathways database [14] has been used as background database for the current study.

Figure 3 represents pathway terms obtained by MatrixEQTL and MIMRMS. From the figure, it is seen that the module selected by MatrixEQTL contains genes that are mainly associated with the process of protein synthesis. It indicates that they are housekeeping genes. On the other hand, modules generated by MIMRMS algorithm generate modules whose members are more associated with pathways in cancer. However, two modules from MIMRMS algorithm selected housekeeping genes. The terms generated for MIMRMS modules like *miRNA targets in ECM and membrane receptors*, *Senescence and Autophagy in Cancer*, and so forth are cancer associated pathways.

On the other hand, modules selected by MatrixEQTL contain genes that are mainly associated with the process of transcription

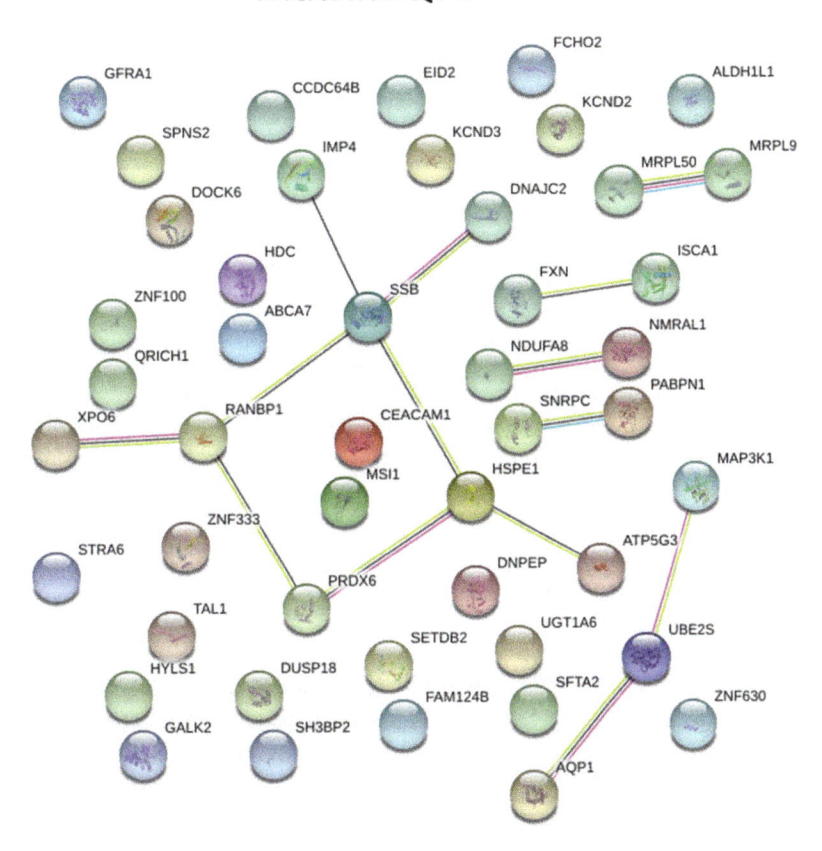

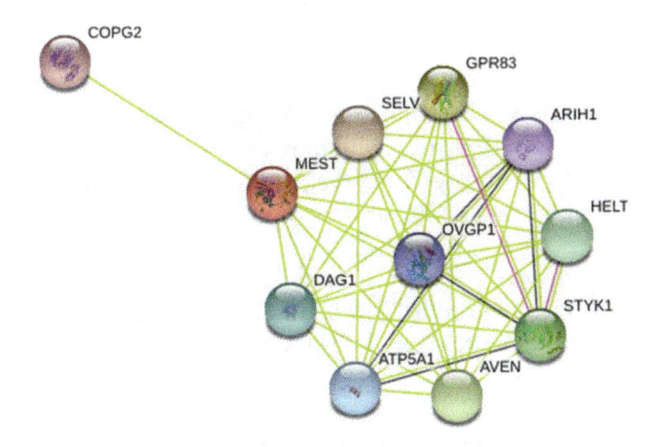

Fig. 2 PPINs generated by STRING for MatrixEQTL and MIMRMS algorithms (cervical cancer)

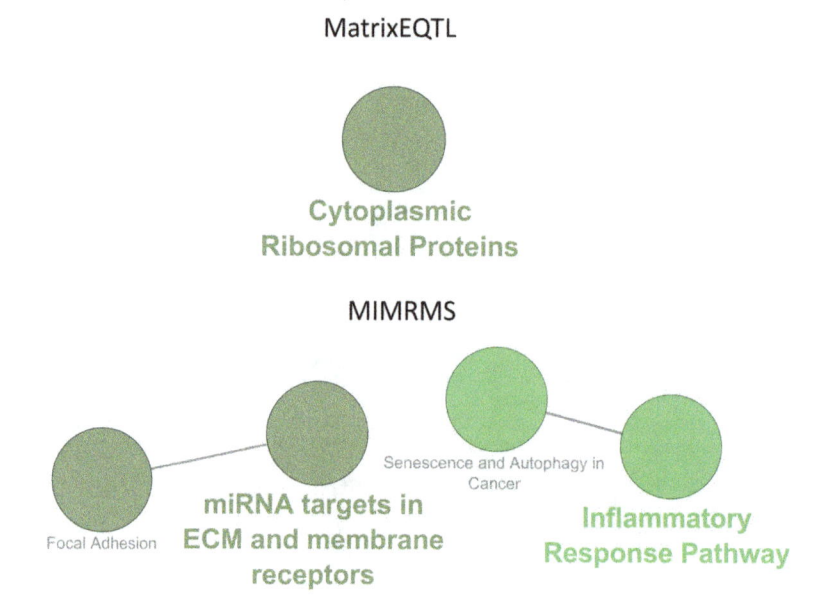

Fig. 3 Pathway enrichment analysis

like Spliceosome, Protein digestion, and absorption, so forth. It indicates that they are housekeeping genes. On the other hand, modules generated by MIMRMS algorithm generates modules that are found to be associated with terms like *Primary immunodeficiency, Th17 cell differentiation, IL-17 signaling pathway*, and so forth.

4.4 Disease Ontology Enrichment Analysis of Ovarian Cancer Modules

Further analysis of the most significant networks (one MatrixEQTL and five MIMRMS) was done using disease ontology (DO) enrichment analysis. The R package DOSE [24] was used. This package identifies a statistically significant disease ontology term that is associated with a set of genes. Here, DO id's with P-value < 0.05 are selected. Table 5 represents the DO terms and their respective P-values. From the table, it is seen that the MatrixEQTL does not generate any relevant DO term with respect to Ovarian cancer. Whereas, one of the modules of the MIMRMS generates DO term that is highly relevant to ovarian cancer (bold text). The result indicates that the MIMRMS algorithm efficiently selects regulatory networks compared to other existing methods. According to miR2Disease [12], all the miRNAs mentioned (both MatrixEQTL and MIMRMS) in Table 2 are associated with ovarian cancer.

4.5 Disease Ontology Enrichment Analysis of Cervical Cancer Modules

Table 6 represents the DO terms and their respective P-values. From the table, it is seen that the MatrixEQTL does not generate any relevant DO term with respect to cervical cancer. On the other hand, MIMRMS generates other cancer relevant DO terms.

Table 5
Comparative analysis of association of modules with diseases (ovarian cancer)

Algorithm	ID	Description	P-value	adj P-value
MatrixEQTL	DOID:1342	Congenital hypoplastic anemia	2.05E−03	1.89E−03
	DOID:9588	Encephalitis	2.04E−05	1.65E−05
MIMRMS-23	DOID:1883	Hepatitis C	5.07E−05	4.10E−05
	DOID:8469	Influenza	2.50E−04	2.02E−04
	DOID:2237	Hepatitis	1.47E−03	1.19E−03
MIMRMS-95	DOID:1342	Congenital hypoplastic anemia	1.37E−02	1.33E−02
MIMRMS-142	DOID:1342	Congenital hypoplastic anemia	1.69E−02	1.65E−02
MIMRMS-159	DOID:5683	Hereditary breast ovarian cancer	2.94E−02	2.77E−02
	DOID:0060095	Uterine benign neoplasm	4.69E−04	3.36E−04
	DOID:13223	Uterine fibroid	4.69E−04	3.36E−04
	DOID:0060086	Female reproductive organ benign neoplasm	5.01E−04	3.59E−04
MIMRMS-168	DOID:0050622	Reproductive organ benign neoplasm	5.37E−04	3.85E−04
	DOID:0060085	Organ system benign neoplasm	5.62E−03	4.04E−03
	DOID:3713	*Ovary adenocarcinoma*	4.37E−02	3.14E−02
	DOID:1790	Malignant mesothelioma	4.79E−02	3.44E−02

The result indicates that the MIMRMS algorithm efficiently selects regulatory networks compared to other existing methods. According to miR2Disease [12], all the miRNAs mentioned (both MatrixEQTL and MIMRMS) in Table 4 are associated with cervical cancer.

5 Conclusions

This chapter presents an integrative approach for automatic detection of regulatory network by applying the existing MIMRMS algorithm. The importance of MIMRMS algorithm over other existing algorithms is demonstrated in terms of identification of miRNA–mRNA regulatory modules. The MIMRMS algorithm generates more significant regulatory modules that are highly related to ovarian cancer and cervical cancer. The obtained regulatory modules may be helpful for understanding the underlying etiology of the disease.

Table 6
Comparative analysis of association of modules with diseases (cervical cancer)

Algorithm	ID	Description	P-value	adj P-value
	DOID:9352	Type 2 diabetes mellitus	2.25E−02	2.31E−02
MatrixEQTL-2	DOID:37	Skin disease	2.63E−02	2.60E−02
	DOID:1205	Hypersensitivity reaction type I disease	3.61E−02	3.39E−02
	DOID:1588	Thrombocytopenia	3.22E−02	3.10E−02
MatrixEQTL-50	DOID:2228	Thrombocytosis	4.46E−02	4.03E−02
	DOID:0060115	Nervous system benign neoplasm	4.46E−02	4.03E−02
	DOID:2001	Neuroma	3.30E−02	3.13E−02
MatrixEQTL-190	DOID:230	Lateral sclerosis	2.68E−02	2.63E−02
	DOID:4440	Seminoma	3.30E−02	3.13E−02
	DOID:272	Hepatic vascular disease	2.47E−02	2.46E−02
MatrixEQTL-287	DOID:2452	Thrombophilia	2.47E−02	2.46E−02
	DOID:8567	Hodgkin's lymphoma	1.51E−02	1.63E−02
	DOID:3376	Bone osteosarcoma	1.44E−02	1.55E−02
	DOID:120	Female reproductive organ cancer	7.59E−11	1.25E−09
	DOID:1380	Endometrial cancer	4.51E−04	1.01E−03
MIMRMS-50	DOID:231	Motor neuron disease	1.15E−04	3.10E−04
	DOID:1883	Hepatitis C	5.91E−04	1.24E−03
	DOID:10652	Alzheimer's disease	1.33E−03	2.31E−03
	DOID:14330	Parkinson's disease	2.94E−03	4.68E−03
	DOID:583	Hemolytic anemia	2.33E−02	2.37E−02
MIMRMS-68	DOID:720	Normocytic anemia	2.33E−02	2.37E−02
	DOID:169	Neuroendocrine tumor	2.47E−02	2.46E−02
MIMRMS-91	DOID:0050938	Breast lobular carcinoma	1.28E−02	1.42E−02
	DOID:8719	In situ carcinoma	1.43E−02	1.55E−02
	DOID:13223	Uterine fibroid	9.52E−03	1.13E−02
MIMRMS-108	DOID:2870	Endometrial adenocarcinoma	9.98E−03	1.16E−02
	DOID:0060086	Female reproductive organ benign neoplasm	1.08E−02	1.24E−02
	DOID:2615	Papilloma	1.13E−02	1.30E−02
	DOID:2615	Papilloma	1.13E−02	1.30E−02
MIMRMS-150	DOID:9973	Substance dependence	1.21E−02	1.38E−02
	DOID:0080014	Chromosomal disease	1.22E−02	1.39E−02

(continued)

Table 6
(continued)

Algorithm	ID	Description	P-value	adj P-value
	DOID:403	Mouth disease	1.24E−02	1.40E−02
	DOID:962	Neurofibroma	6.40E−03	8.41E−03
	DOID:0050737	Autosomal recessive disease	6.88E−03	8.95E−03
	DOID:288	Endometriosis of uterus	7.51E−03	9.59E−03
MIMRMS-222	DOID:1123	Spondyloarthropathy	7.85E−03	9.83E−03
	DOID:3149	Keratoacanthoma	8.70E−03	1.06E−02
	DOID:6204	Follicular adenoma	8.70E−03	1.06E−02
	DOID:0060095	Uterine benign neoplasm	9.52E−03	1.13E−02
MIMRMS-267	DOID:303	Substance-related disorder	4.07E−03	6.11E−03
	DOID:2706	Synovium cancer	4.15E−03	6.15E−03
	DOID:10747	Lymphoid leukemia	8.15E−04	1.59E−03
MIMRMS-301	DOID:5409	Lung small cell carcinoma	8.15E−04	1.59E−03
	DOID:2355	Anemia	8.24E−04	1.60E−03
	DOID:1319	Brain cancer	8.51E−04	1.64E−03

Acknowledgements

This work is partially supported by the seed grant program of the Indian Institute of Technology Jodhpur, India (grant no. I/SEED/SPU/20160010). The author wants to acknowledge Mr. Shubham Talbar, Indian Institute of Technology Jodhpur, India for his contribution in implementing certain bioinformatics tools.

References

1. Alvarez-Garcia I, Miska EA (2005) MicroRNA functions in animal development and human disease. Development 132(21):4653–4662
2. American Cancer Society (n.d.) https://www.cancer.org/. Accessed 13 June 2018
3. Beezhold K, Castranova V, Chen F (2010) Microprocessor of MicroRNAs: regulation and potential for therapeutic intervention. Mol Cancer 9(134). https://doi.org/10.1186/1476-4598-9-134
4. Bindea G, Mlecnik B, Hackl H et al (2009) ClueGO: a cytoscape plug-in to decipher functionally grouped gene ontology and pathway annotation networks. Bioinformatics 25(8):1091
5. Bushati N, Cohen SM (2007) microRNA functions. Annu Rev Cell Dev Biol 23(1):175–205
6. Cline MS, Craft B, Swatloski T et al (2013) Exploring TCGA Pan-Cancer data at the UCSC cancer genomics browser. Sci Rep 3 (2652):1–6
7. Ding C, Peng H (2005) Minimum redundancy feature selection from microarray gene

expression data. J Bioinform Comput Biol 3 (02):185–205
8. Duda RO, Hart PE (1973) Pattern classification and scene analysis. Wiley, Hoboken
9. Harfe BD (2005) MicroRNAs in vertebrate development. Curr Opin Genes Dev 15 (4):410–415
10. He L, Hannon GJ (2004) MicroRNAs: small RNAs with a big role in gene regulation. Nat Rev Genet 5:522–531
11. Huang T, Cai YD (2013) An information-theoretic machine learning approach to expression QTL analysis. PLoS One 8(6):1–9
12. Jiang Q, Wang Y, Hao Y et al (2009) miR2Disease: a manually curated database for microRNA deregulation in human disease. Nucleic Acids Res 37:D98
13. Krol J, Loedige I, Filipowicz W (2010) The widespread regulation of MicroRNA biogenesis, function and decay. Nat Rev Genet 11:597–610
14. Kutmon M, Riutta A, Nunes N et al (2016) WikiPathways: capturing the full diversity of pathway knowledge. Nucleic Acids Res 44 (D1):D488
15. Li J, Liu Y, Wang C et al (2015) Serum miRNA expression profile as a prognostic biomarker of stage II/III colorectal adenocarcinoma. Sci Rep 5(12921). https://doi.org/10.1038/srep12921
16. Maji P (2012) Mutual information-based supervised attribute clustering for microarray sample classification. IEEE Trans Knowl Data Eng 24(1):127–140
17. Maji P, Paul S (2011) Rough set based maximum relevance-maximum significance criterion and gene selection from microarray data. Int J Approx Reason 52(3):408–426
18. Mitchell PS, Parkin RK, Kroh EM et al (2008) Circulating MicroRNAs as stable blood-based markers for cancer detection. Proc Natl Acad Sci 105(30):10513–10518
19. Paul S, Maji P (2016) Gene expression and protein-protein interaction data for identification of colon cancer related genes using f-information measures. Nat Comput 15 (3):449–463
20. Quitadamo A, Tian L, Hall B et al (2015) An integrated network of MicroRNA and gene expression in ovarian cancer. BMC Bioinf 16 (5):S5
21. Shabalin AA (2012) Matrix eQTL: ultra fast eQTL analysis via large matrix operations. Bioinformatics 28(10):1353
22. Shannon P, Markiel A, Ozier O et al (2003) Cytoscape: a software environment for integrated models of biomolecular interaction networks. Genome Res 13(11):2498–2504
23. Szklarczyk D, Franceschini A, Wyder S et al (2015) STRING v10: protein-protein interaction networks, integrated over the tree of life. Nucleic Acids Res 43(D1):D447
24. Yu G, Wang LG, Yan GR et al (2015) DOSE: an R/Bioconductor package for disease ontology semantic and enrichment analysis. Bioinformatics 31(4):608

Part V

Kinetic Modeling of ncRNA-Mediated Gene Regulation

Chapter 14

Quantitative Characteristic of ncRNA Regulation in Gene Regulatory Networks

Federico Bocci, Mohit Kumar Jolly, Herbert Levine, and José Nelson Onuchic

Abstract

RNA is mostly known for its role in protein synthesis, where it encodes information for protein sequence in its messenger RNA (mRNA) form (translation). Yet, RNA molecules regulate several cellular processes other than translation. Here, we present an overview of several mathematical models that help understanding and characterizing the role of noncoding RNA molecules (ncRNAs) in regulating gene expression and protein synthesis. First, we discuss relatively simple models where ncRNAs can modulate protein synthesis via targeting a mRNA. Then, we consider the case of feedback interactions between ncRNAs and their target proteins, and discuss several biological applications where these feedback architectures modulate a cellular phenotype and control the levels of intrinsic and extrinsic noise. Building from these simple circuit motifs, we examine feed-forward circuit motifs involving ncRNAs that generate precise spatial and temporal patterns of protein expression. Further, we investigate the competition between ncRNAs and other endogenous RNA molecules and show that the cross talk between coding and noncoding RNAs can form large genetic circuits that involve up to hundreds of chemical species. Finally, we discuss the role of ncRNAs in modulating cell-cell signaling pathways and therefore the dynamics of spatiotemporal pattern formation in a tissue.

Key words Noncoding RNA, Gene network, Mathematical model, Network motif, Feedback loops

1 Introduction

The central dogma of molecular biology is regarded as a golden rule to understand how genetic information is transferred from DNA and ultimately used to synthetize proteins [1]. It states that genetic information encoded in the DNA which is present in the cell nucleus is transferred as messenger RNA (mRNA) that is then used to synthetize proteins [1]. This simple, deterministic model of information transfer in the cell does not consider several layers of regulation such as molecular interactions between different components or fluctuations in protein levels due to stochastic effects at different steps of the protein production chain.

Here, we specifically investigate the different functions of RNA in shaping the cell's phenotype besides encoding information for protein sequences and carrying it from DNA to ribosomes [2, 3]. Although RNA's primary and perhaps central role is to provide information transfer from cell nucleus via its messenger RNA (mRNA) form, several RNA molecules do not encode for proteins and are instead designed to fulfil different roles in the modulation of gene and protein expression levels. This set of non-coding RNA molecules (ncRNAs) provides the cell with additional control mechanisms over several cell fate-level processes including adjusting gene expression in the presence of changes of environmental conditions and controlling noise along the various steps of protein synthesis [2–5].

Several different types of ncRNAs have been identified. For example, bacterial small RNAs (sRNAs) are found in bacteria that are 50–500 nucleotides long. Similarly, microRNAs (often abbreviated as miRs) are RNA molecules that are found in animals, plants, and some viruses. They are about 22 nucleotides long and play a major role in targeting mRNA at a posttranslational level, therefore inhibiting mRNA translation and protein synthesis [6, 7]. Similarly, small interfering RNAs (siRNAs) or silencing RNAs are 20–25 nucleotides long RNA molecules that prevent mRNA translation via rapid degradation after transcription [8]. Piwi-interacting RNAs (piRNAs) and small nucleolar RNAs (snRNAs), are other relatively shorter ncRNAs, while long noncoding RNAs (lnc-RNAs) are the transcripts over 200 nucleotides in length that are not translated into protein.

Understanding the role of ncRNAs in cellular decision-making requires a quantitative characterization of the mechanisms underlying ncRNA action. In this chapter, we review some of the mathematical models formulated in recent years that provide a quantitative characterization of several processes mediated by ncRNAs, and elucidate emergent dynamics of various cross-talking networks formed by interactions of ncRNAs with other cellular components. In doing so, we mostly focus on short ncRNA molecules, such as microRNAs, small bacteria RNAs, or siRNAs. Long noncoding RNAs (Lnc-RNAs) represent an additional class of RNA molecules that do not contribute to cell dynamics via translation of sequence into protein. These molecules are significantly larger than short ncRNAs and can reach a length of up to 200 nucleotides [9]. Although Lnc-RNAs play a paramount role in diverse cellular processes in physiological cell dynamics [10] and tumorigenesis [11, 12], the larger body of work about quantitative characteristic of ncRNA action relates to short RNAs, which is therefore the focus of this chapter.

To investigate mathematical modeling of ncRNAs, we adopt a bottom-up approach and start by considering the simplest scenario where a single ncRNA modulates the expression of a mRNA. From this elementary system, we add layers of complexity such as regulation of multiple mRNA targets by multiple ncRNAs, feedback interactions that regulate the expression of the ncRNA as well, and regulatory networks with complex functions. Further, we review the competition of ncRNA with competing endogenous RNAs and finally look beyond a cell-autonomous role by investigating the role of ncRNA in modulating cell-cell signaling, and thus modulating tissue patterning.

2 Targeting mRNA: The Control on Protein Expression

The simplest conceptual model of ncRNA-mediated regulation considers the ability of ncRNA to bind to a target mRNA and inhibit the translation of a protein via rapid degradation of the mRNA-ncRNA complex (Fig. 1a). Such class of models [4, 5] considers the kinetic steps in the transcription of mRNA and ncRNA, and the mRNA-ncRNA complex formation. The temporal dynamics of the mRNA (m), the ncRNA (n), and the mRNA-ncRNA complex ($[mn]$) can be generally described via a system of ordinary differential equations (ODEs):

$$\frac{dm}{dt} = g_m - k_+ m\, n + k_- [mn] - \gamma_m m$$

$$\frac{dn}{dt} = g_n - k_+ m\, n + k_- [mn] - \gamma_n n$$

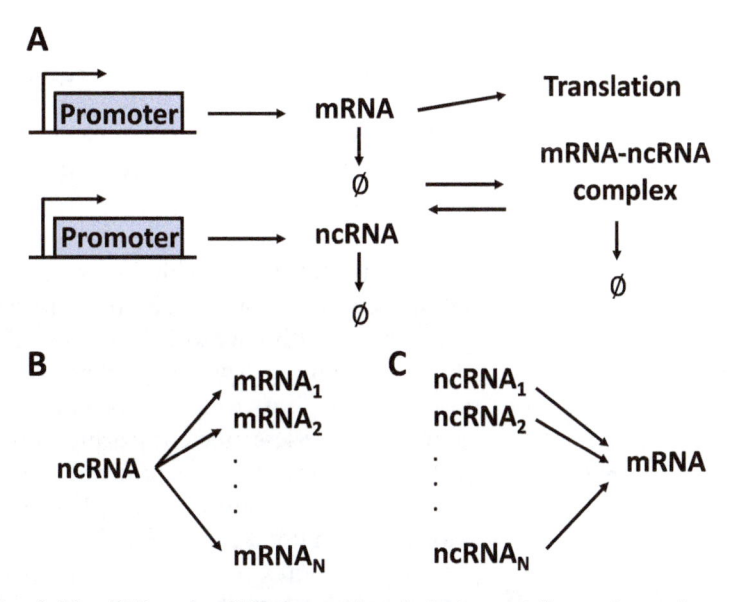

Fig. 1 (a) mRNA and ncRNA are produced via transcription and can degrade independently. mRNA can encode for its target protein (translation) or can bind to the ncRNA and form the mRNA-ncRNA complex, which can also degrade. (b) A single ncRNA can regulate up to hundreds of target mRNAs. (c) A single mRNA can be targeted by several ncRNAs

$$\frac{d[mn]}{dt} = k_+ m\, n - k_-[mn] - \gamma_{[mn]}[mn]$$

where the terms g_m and g_n describe the production rates of mRNA (i.e., transcription) and ncRNA in number of molecules per unit time. Further, both m and n can degrade independent of their interactions, represented by the terms $\gamma_m m$ and $\gamma_n n$, where γ_m and γ_n are the degradation rates of mRNA and ncRNA. Assuming first-order kinetics, the degradation rate of a molecule (γ) can be calculated from a molecule's half-life (τ) via $\gamma = \ln(2)/\tau$. In addition, the ncRNA can reversibly bind to the mRNA with binding rate constant k_+ and unbinding rate constant k_-, hence forming the mRNA-ncRNA complex $[mn]$ that degrades at rate $\gamma_{[mn]}[mn]$. Usually, the degradation rate of the mRNA-ncRNA complex is larger than that of mRNA or ncRNA [13–15]. For this reason, several models treat the mRNA-ncRNA binding as an irreversible process [4]. Such simplified class of models can, nonetheless, shed light into several key aspects of ncRNA-mediated regulation, including the regulation of protein levels in response to external cues, and controlling intrinsic and extrinsic noise that could affect the gene expression profile in a cell. Bacterial small RNA (sRNA) perhaps represents the simplest and most well-studied example of ncRNA regulation on target mRNA. These molecules are usually 50 to 500 nucleotides long, and are involved in several key processes of bacterial life including metabolism, cell cycle, and stress response [16, 17].

Mitarai et al. [18] developed a mathematical model to investigate the dynamical properties of regulation by RyhB and Spot42, two well-known bacterial sRNA involved in iron uptake [19, 20] and bacterial metabolism [21]. They observed that the efficiency of sRNA-mediated regulation of mRNA is dictated by two parameters—first, the bacterium's capacity to overexpress the sRNA relatively to its target mRNA. Second, the rate of inactivation of the target mRNA due to binding with sRNA. Applying experimental and computational tools, the authors further measured these parameters for the two sRNAs and concluded that they operated in different regimes: RyhB is weakly overexpressed compared to its target mRNA but compensates with an efficient mRNA deactivation, while the less efficient pairing between Spot42 and its target mRNA is balanced by a stronger sRNA overexpression.

Additionally, Semsei et al. [22] implemented a mathematical model of iron uptake in *Escherichia coli* and showed that the sRNA RyhB contributes to an optimal redistribution of iron and thus avoids possible toxicity generated by a high intracellular iron concentration. Moreover, the authors showed that sRNA-based regulation guarantees a smaller turnover of target mRNAs compared to transcriptional regulation [23]. Legewie et al. [24] modeled the iron stress response in cyanobacteria and showed that the sRNA

IsrR generated is responsible for a delay in the accumulation of the stress protein IsiA, therefore preventing the activation of the upstream response pathway in the presence of a stress signal that is too short. Indeed, Shimoni et al. [25] developed a mathematical model of sRNA posttranslational regulation and showed that such inhibitory strategy is preferable when a fast response is needed, as in the case of external stresses that require a quick fine-tuning of a cell's gene expression profile. It should be pointed out that direct mRNA targeting is not the only strategy for signal sensing: more elaborate network motifs, such as a feed-forward loop including ncRNAs [26, 27], are discussed in the following sections. Furthermore, Levine et al. [28] studied the stress response mechanism of *Escherichia coli* and showed that regulation by small RNAs can tightly repress target genes as soon as the small RNA synthesis rate becomes higher than a certain threshold, hence allowing a precise and global response to stress. Such thresholds can efficiently filter out stochastic fluctuations arising from the environment and can be adjusted dynamically to adapt to changing conditions [29, 30]. In short, one important role of sRNA-mediated regulation can be to help the cell distinguish between a transient signal and a sustained stimulus.

Besides extrinsic noise sources, the relatively low copy number of ncRNA and mRNA, as compared to typical number of protein molecules in a cell [31], makes the role of random fluctuations of copy number, or intrinsic noise, very important [32–34]. In fact, mRNA copy number usually oscillates between 1000 and 10,000 units per cell, roughly about a 1000-fold smaller than the usual protein copy number [35]. Jia et al. [36] investigated the effect of noise in posttranslational gene regulation due to ncRNA and found that intrinsic noise in gene expression does not propagate to mRNA copy number if the cellular production rates of ncRNA and mRNA are very different, but can result in strong fluctuation in mRNA level when the cellular production rates of ncRNA and mRNA are comparable. Further, Elgart et al. [37, 38] developed a model of stochastic expression bursts of sRNA and derived mathematical expressions for the propagation of such bursts on the mean steady-state protein levels using different types of stochastic laws for the waiting time distribution of the bursting events. The authors showed that the extent of sRNA transcriptional bursting can be derived from the steady-state levels of mRNA and translated protein, hence building a bridge between dynamics and equilibrium properties of the model.

The associations among various mRNAs and ncRNAs are rarely one to one. Multiple mRNAs can be targeted by a single ncRNA species [13]. In fact, most miRNAs have up to hundreds of target mRNAs (Fig. 1b). Zhdanov [39] modeled the case of one ncRNA regulating 100 target mRNAs and showed that such regulation is possible only if the ncRNA population is much larger than the population size of any of its targets. In other words, the synthesis rate of ncRNA must guarantee enough influx to balance the turnover due to ncRNA loss in mRNA-ncRNA complex degradation. Interestingly, Jost et al. [40] observed that most target mRNAs respond weakly to ncRNA regulation, and the cell phenotype is usually determined by only few targets. Therefore, they proposed that the weak inhibition of the ncRNA on most targets confers robustness to the regulation of the few strongly regulated targets.

Conversely, certain genes can transcribe for mRNA which is in turn targeted by several ncRNAs. In such case, the gene is a ncRNA target hub [41, 42] (Fig. 1c). Lai et al. developed a kinetic model for the miRNA target hub CDKN1A which encodes for the protein p21, a regulator of cell cycle and apoptosis [43, 44]. The authors showed that the expression of p21 can be fine-tuned by selectively expressing CDKN1A-targeting miRNAs due to the cooperative nature of their regulation. In other words, the spatial proximity of different miRNA-binding sites on the same target can drastically increase their inhibitory effect [45, 46]. Starting from this modeling scheme, the authors performed a systematic search for human genes that are regulated by multiple cooperative ncRNAs and organized the information in an online dataset [47]. Such cooperativity is a hallmark of both ncRNA-mediated interactions and transcriptional regulation, hence motivating the frequent use of Hill functions in the mathematical modeling of genetic circuits involving ncRNAs, mRNAs, and transcription factors [4, 5]. Nonetheless, transcriptional and posttranslational interactions differ because ncRNA-mediated posttranslational regulation involves degradation of the ncRNA in addition to the inhibition of the target mRNA. Moreover, the number of ncRNAs that are bound to the mRNA changes the degradation rate of the mRNA-ncRNA complex as well as the loss in ncRNA copy number. For these reasons, Lu et al. [48] developed a mathematical description of ncRNA-mediated posttranslational inhibition that takes into account the cooperative binding of ncRNA molecules to the mRNA as well as the increasing mRNA-ncRNA complex degradation rate as a function of the number of bound ncRNAs [48]. Thus, unlike many previous models discussed above, this model does not explicitly consider the formation of a ncRNA-mRNA complex, but rather model ncRNA action in a more implicit way as an additional degradation of the ncRNA and mRNA species. In this case, a fast equilibration for the ncRNA-mRNA complex is assumed. Therefore, effective degradation terms for ncRNA and mRNA can be derived to replace the loss due to complex formation and degradation [48].

In contrast to the canonical roles of inhibitory regulation of mRNA by ncRNA, there have been a few reports of positive modulation of mRNA by ncRNA. For instance, miR369-3 can upregulate the translation of TNF-alpha [49], miR-34 can upregulate its target protein expression [50], and inhibiting miR4661 degradation leads to upregulation of IL10 [51]. To elucidate this effect, Gokhale and Gadgil [52] developed a model that describes the different kinetic steps of the ncRNA-mRNA interaction and showed that the dynamics can be described by a restricted set of parameters, further deriving the parameter range allowing for increase in protein level due to ncRNA.

3 Shaping the Phenotype of the Cell: Genetic Loops with ncRNAs

The previous section dealt with models where the ncRNAs regulate the expression of their target mRNA, but the protein encoded by the mRNA does not affect the transcription of the ncRNA directly. Several situations in biological and synthetic systems, however, are characterized by feedback interactions and the production of ncRNA in the cell is modulated by its own target mRNA/protein. In the simplest scenario, the target mRNA translates for a protein which, in turn, activates or inhibits the transcription of the ncRNA (Fig. 2a). This schematic can be described by generalizing the equation presented in the previous section to the following set of ODEs:

$$\frac{dm}{dt} = g_m f^{+/-}(p) - k_+ m\, n + k_-[mn] - \gamma_m m$$

$$\frac{dn}{dt} = g_n - k_+ m\, n + k_-[mn] - \gamma_n n$$

$$\frac{d[mn]}{dt} = k_+ m\, n - k_-[mn] - \gamma_{[mn]}[mn]$$

$$\frac{dp}{dt} = k_p m - \gamma_p p$$

where an equation for the dynamics of intracellular protein p has been introduced. p is translated from the target mRNA with translation rate constant k_p and degrades with a rate constant γ_p. In this description, k_p represents the number of proteins that are translated per mRNA molecule. Furthermore, the produced protein modulates the transcription of the ncRNA m via a term $f^{+/-}(p)$, where "+" stands for a transcriptional activation, i.e., the protein increases the level of ncRNA, and "−" stands for a transcriptional inhibition, i.e., the protein decreases the level of ncRNA. Different mathematical forms have been used to model $f^{+/-}(p)$, with the common requirement that the strength of the modulation (positive or negative) increases with the level of protein p.

If the protein inhibits the production of ncRNA, the system can be represented at a more coarse-grained level as two nodes (the ncRNA and the protein) that mutually repress each other (Fig. 2b). This architecture is referred to as a double-negative feedback loop,

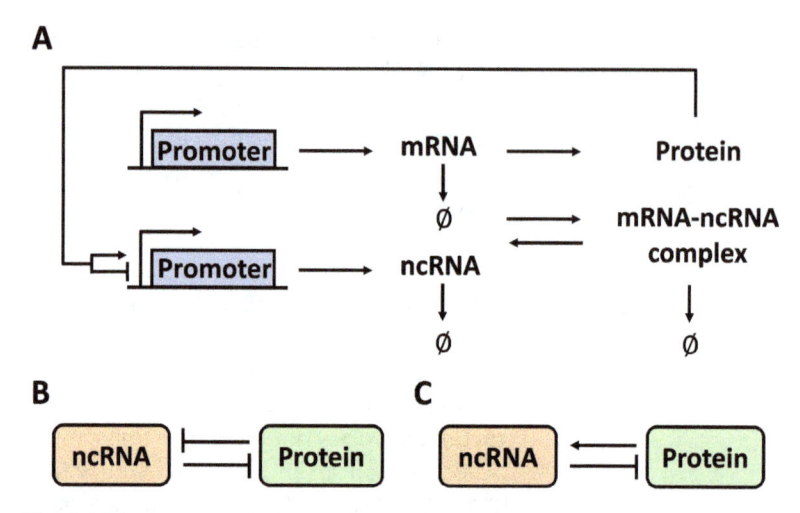

Fig. 2 (**a**) A schematic for feedback interaction between a ncRNA and a protein includes the additional steps of (1) protein translation from target mRNA and (2) protein modulation of ncRNA transcription. The regulation of ncRNA can be positive (normal arrow) or negative (T-shaped arrow). (**b**) Coarse-grained representation of a double-negative feedback loop between a ncRNA and a protein. (**c**) Coarse-grained representation of a feedback loop where the protein activates the ncRNA

or toggle switch, and has been observed in multiple biological systems and engineered in synthetic systems. The toggle switch is a recurrent biological motif because, with a strong enough repression, it admits two stable solutions: the former has a high level of ncRNA and a low level of target protein, while the latter has a low level of ncRNA and a high level of target protein. Therefore, this motif offers a basic framework to model a system that can assume one of the two alternative states, thus achieving bistability.

Zhdanov [53, 54] examined a simple, generic kinetic model of ncRNA that is transcriptionally inhibited by its target protein, hence giving rise to a bistable regime. Going further, the author considered the case of a ncRNA modulating multiple (100) mRNAs (such as shown in Fig. 1b) and showed the existence of a regime that enables bistable switches in the levels of all the mRNAs. Finally, the bistable regime could still be observed in the presence of stochastic bursts of RNA transcription [55].

A biological application of the bistable toggle switch motif between a ncRNA and a protein is the model of epithelial-mesenchymal transition (EMT) developed by Tian et al. [56]. EMT is the biophysical process where epithelial cells, which are adhesive and exhibit apicobasal polarity, lose their cell-cell adhesion and become motile cells [57]. This process is crucial in both physiological biological processes such as morphogenesis or wound healing and in pathological contexts such as cancer metastases [58]. Tian et al. considered two negative feedback motifs: the first motif is formed by the epithelial state-promoting microRNA, miR-34, and the mesenchymal state-inducing transcription factor (TF) SNAIL. MicroRNAs are a specific class of ncRNA molecules usually composed of around 22 nucleotides [6, 7]. Tian et al. showed that such system can reversibly switch from an epithelial phenotype (high miR-34, low SNAIL) to a partially EMT phenotype (high SNAIL, low miR-34), when exposed to an EMT-inducing signal such as TGF-β [56]. Further, they described a second toggle switch motif between the epithelial state-promoting micro-RNA miR-200 and the mesenchymal state-inducing TF ZEB, and showed that this latter motif is responsible for a second, irreversible switch to establish a fully mesenchymal phenotype characterized by high levels of ZEB and SNAIL, and low levels of miR-200 and miR-34 [56]. Thus, their mathematical model proposed that EMT is a cascade of two-step process, each step being regulated by a mutually inhibitory loop between a microRNA and a TF.

Although the architecture of the toggle switch motif is usually associated with bistability, several examples in ncRNA-TF interplay show that additional factors or modulation of the interactions can lead to different scenarios. A core genetic network that regulates stemness properties in embryonic stem cells as well as cancer cells is formed by a toggle switch between the microRNA family of let-7 and the reprogramming factor LIN-28 [59]. Jolly and collaborators [60, 61] showed that this circuit can exhibit three different stable states due to the self-activation of both let-7 and LIN28; while let-7 can autoregulate its maturation, LIN28, a RNA-binding factor, can accelerate its translation [62–64]. Thus, compared to a canonical bistable toggle switch, this architecture can exhibit a third solution with intermediate levels of both let-7 and LIN-28 that is further shown to be biologically relevant as it can connect to a stem-like phenotype [59].

Another toggle switch motif that can give rise to tristability is the EMT genetic circuit modeled by Lu et al. [48]. Similar to Tian et al. [56], this motif considers the interaction between miR-200 and ZEB (Fig. 3a). Lu et al. showed that the nonlinear response of the feedback loop between miR-200 and ZEB can result in a tristable regime (Fig. 3b). In this model, the third solution is characterized by intermediate levels of both miR-200 and ZEB, hence being associated with a partial EMT or a hybrid epithelial/mesenchymal (E/M) phenotype capable of exhibiting both cell-cell adhesion and motility [48, 58]. The other two solutions are associated with an epithelial phenotype (high miR-200, low ZEB) and mesenchymal phenotype (low miR-200, high ZEB), respectively. Additionally, the authors showed that the toggle switch between miR-34 and SNAIL (described by Tian et al. [56] as well) need not even be bistable, but can exhibit monostability due to the differences in the strength of interactions among miR-34 and SNAIL, as

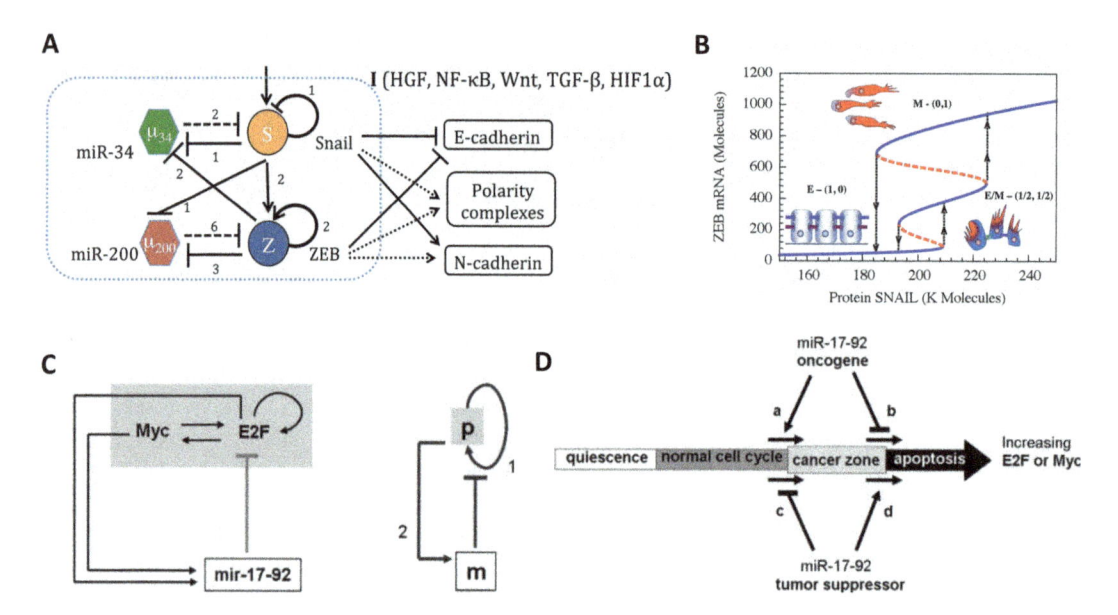

Fig. 3 (**a**) The core genetic circuit that regulates epithelial-mesenchymal transition (EMT) as proposed by Lu et al. [48]. The circuit is composed by two double-negative feedback motifs. The first loop is between miR-200 and ZEB (bottom) while the second considers miR-34 and SNAIL (top). Further cross talk between the motifs includes inhibition of miRNAs by ZEB and SNAIL and activation of ZEB by SNAIL. (**b**) Bifurcation diagram of ZEB mRNA as a function of the level of protein SNAIL computed by Lu et al. [48]. ZEB mRNA is low (lower branch, epithelial phenotype) in response to a low EMT-inducing signal and becomes first intermediate (second branch, hybrid E/M phenotype) and finally high (upper branch, mesenchymal phenotype) when the signal is increased. (**c**) The E2F-Myc-miR-17-92 regulatory circuit as described by Aguda et al. [67]. Left: miR-17-92 targets E2F, which in turn activates miR-17-92 and itself. Further, E2F and Myc form a double-positive feedback loop. Right: A reduced version of the circuit that considers solely the feedback loop between miR-17-92 and E2F. (**d**) Cell phenotype as a function of E2F/Myc levels as predicted by Aguda et al. [67]. The cell can attain four phenotypes in order of increasing E2F/Myc levels: (i) quiescent state; (ii) normal cell cycle; (iii) cancer zone; and (iv) apoptosis. miR-17-92 can act as an oncogene via stabilizing the cancer phenotype or as tumor suppressor via destabilizing the cancer phenotype

compared to that between miR-200 and ZEB (Fig. 3a). When coupled to the miR-200/ZEB loop, this additional circuit can act as a modulator for external noise, therefore stabilizing the output of the core EMT circuit and filtering out transient fluctuations [48]. Thus, these authors also proposed that EMT is a two-step process but suggested a different role for both mutually inhibitory feedback loops—miR-34/SNAIL and miR-200/ZEB—in EMT dynamics. These differences largely relate to the model formulation used to capture miRNA-mediated regulation—while Tian et al. [56] used Hill functions typically used for transcriptional regulation, Lu et al. [48] modeled miRNA action by incorporating their effects in promoting the degradation of the ncRNA and mRNA species, implicitly representing the mRNA-miRNA complex formation.

Next, let us consider the case when the target protein activates the transcription of the ncRNA (Fig. 2c). A well-studied biological example of such an architecture is the TF family E2F, which is related to several hallmarks of cancer, including angiogenesis, chemoresistance, and proliferation [65]. E2F is targeted by a cluster of microRNAs miR-17-92 (a family of six microRNAs) and can in turn activate the transcription of miR-17-92 family, hence forming a feedback loop [66]. Aguda et al. [67] developed a mathematical model for the temporal dynamics of the interplay between E2F and miR-17-92 (Fig. 3c). First, the authors showed that E2F can independently assume two steady states (i.e., bistability) via positive feedback interactions with Myc, another protein participating in proliferation and apoptosis (Fig. 3c). Simulating the cell dynamics under various parameter conditions unraveled a rich ensemble of phenotypes that the cell can assume depending on how strongly E2F, miR-17-92, and Myc mutually interact. As the cellular level of E2F or Myc increases, the cell transits through four different phenotypes: (1) a resting or quiescent state, (2) a "normal" state corresponding to a normal cell cycle, (3) an "abnormal" state corresponding to cancer cell proliferation, and (4) a "death" state corresponding to cell apoptosis (Fig. 3d). Specifically, in the "abnormal" cancer regime, the fast cell growth is not optimally balanced by apoptosis, hence resulting in hyperproliferation. Due to the feedbacks among miR-17-92, E2F, and Myc, miR-17-92 can play two possible roles: (1) maintaining E2F/Myc levels in the cancer window via activation from the "cell cycle" region or inhibition of the transition to the "apoptosis" phase, hence acting as an oncogene, or (2) inhibiting the cancer phenotype via either stabilizing the normal "cell cycle" state or activating the apoptosis program from the "cancer" state, therefore acting as a tumor suppressor (Fig. 3d). Conversely, the quiescent state is characterized by a low expression of E2F and it is hypothesized to be inactive. Further, introducing intrinsic noise due to low RNA copy number revealed that the combination of negative and positive feedback loops on the E2F-Myc-miR-17-92 circuit can act as a noise buffer [68, 69]. Specifically, the negative feedback motif E2F and miR-17-92 suppresses the noise, but the positive feedback motif between E2F and Myc has the opposite effect [68, 69]. Overall, it can be concluded from this theoretical study that the regulation of miR-17-92 improves the robustness of the E2F dynamics. Further, the stochastic behavior of the E2F-miR-17-92 interaction was investigated by Giampieri et al. [70] with chemical master equations. Notably, the stochastic version of the E2F-Myc-miR-17-92 model predicts bistability between a resting state with low levels of E2F and a proliferating state with high levels of E2F over a larger range of model's parameters compared to the deterministic (i.e., without considering intrinsic noise) model.

4 From Simple to Complex Networks: Feed-Forward Loops with ncRNAs

The simple ncRNA-protein feedbacks presented in Fig. 2 can be considered as the building blocks to construct larger gene regulatory networks able to perform higher level operations and modulate the expression of large repertoires of genes in a cell. The simpler circuit architecture with more than two species is the feed-forward loop (FFL) [27, 71]. FFLs including ncRNAs are often structured as follows: a mRNA is a common target of a transcription factor (TF) and a ncRNA. Additionally, one of the two following scenarios happens: (1) the TF transcriptionally regulates the ncRNA or (2) the TF is targeted by the ncRNA [27]. Detailed investigation of gene network motifs highlighted that such architecture is very common in the mammalian genome [71, 72]. In general, FFLs are classified into two main categories based on the sign of the regulation on the common target node. If the signs of the direct and indirect target regulation are consistent (i.e., both are either activation or inhibition signals), the FFL is said to be coherent (Fig. 4a). Conversely, if the signs are opposite the FFL is said to be incoherent [27, 73, 74] (Fig. 4b).

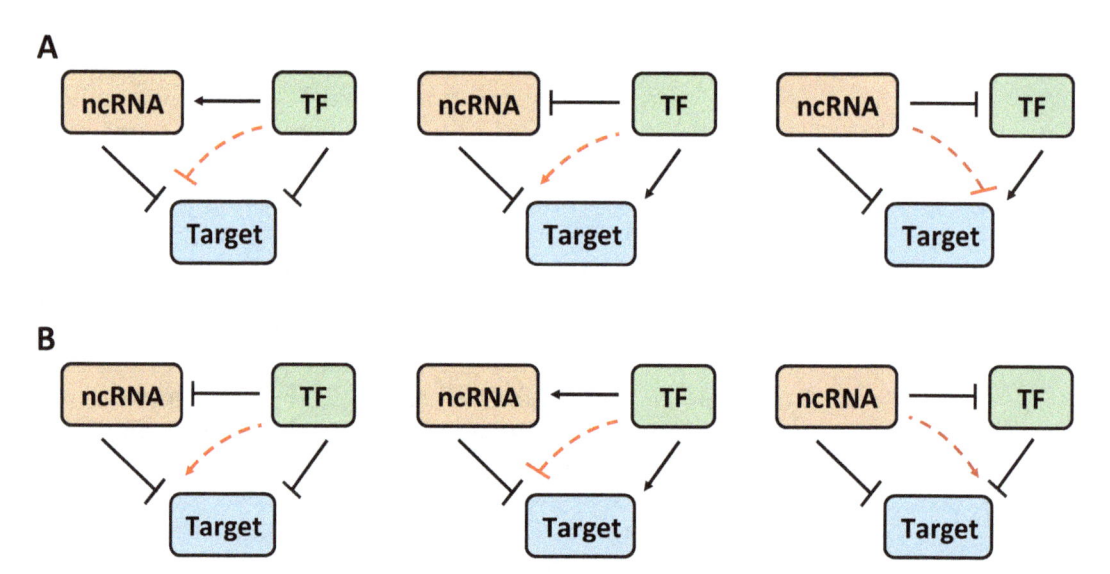

Fig. 4 (**a**) In coherent FFLs, the direct and indirect (red arrows) actions of the "master regulator" on the target have the same sign. The "master regulator" is the species that directly regulates the two other species in the network (the TF in left and center panels, the ncRNA in the right panel). The red dashed arrow exemplifies the sign of the "effective" regulatory action that the master regulator has on the target via the second intermediate regulator (the ncRNA in left and center panels, the TF in right panel). Pointing arrows represent activation and T-shaped arrows represent inhibition (interactions can be either transcriptional, translational, etc.). (**b**) In incoherent FFLs, the direct and indirect (red arrows) actions of the "master regulator" on the target have opposite signs

The structure of the coherent FFL can avoid the spatial co-expression of a ncRNA and its target gene [75]. If the TF targets both the ncRNA and the common target (left and center panels of Fig. 4a), the signs of the two regulations must be opposite to comply with the definition of coherent FFL. Therefore, the only two possible states are (high ncRNA, low target) and (low ncRNA, high target). If, otherwise, the ncRNA regulates both TF and the common target (right panel of Fig. 4a), the inhibition of the common target by the ncRNA directly avoids high co-expression of both of them simultaneously. For example, Kabayashi et al. [76] studied the miR-199/Brm/EGR1 regulatory axis in epithelial tumor cells. Due to a negative feedback interaction between miR-199 and Brm mediated by EGR1, these cells tend to fall into either one of the two following phenotypes: a (low miR-199, low Brm, high EGR1) state that is capable of forming colonies in soft agar, and a second, opposite state with (high miR-199, low Brm, high EGR1) that is unable to form colonies. The authors showed that the architecture of the network can be understood in terms of two interacting coherent FFLs. The first FFL considers miR-199, Brm, and a group of miR-199 targets that are activated by Brm. The second FFL is formed by EGR1, miR-199, and another set of miR-199 targets that are inhibited by EGR1. Therefore, cells with different phenotypes, and therefore either low or high expression of EGR1, have either low or high expression of miR-199 and a resulting opposite expression profile of the shared target genes of EGR1 and miR-199 [76].

A second function played by coherent FFLs is to avoid leaky transcription of the common gene targeted by the TF and the ncRNA. For instance, Shalgi et al. [75] discussed how the E2F/miR-17 regulatory axis (already discussed in the previous section) modulates RB1. RB1 is transcriptionally activated by E2F, which is in turn targeted by miR-17. Furthermore, miR-17 targets RB1 mRNA, hence creating a coherent FFL (as shown in the right panel of Fig. 4a). Therefore, when miR-17 is upregulated, the levels of E2F are diminished and thus BR1 is silenced at the transcriptional level. Without a direct action of miR-17 on BR1, the already transcribed BR1 mRNA, or leaky transcripts, would remain in the cell for as long as allowed by the natural half-life of BR1 mRNA. However, miR-17 accelerates the degradation of BR1 mRNA molecules via miRNA-mRNA binding, hence resulting in a very effective silencing of BR1 [75].

While coherent FFLs can modulate spatial expression of ncRNAs and target genes, incoherent FFLs have a dynamical effect on the temporal dynamics of the common target of a ncRNA and a TF. When the common master regulator of the incoherent FFL (the TF in the left and center panels and the ncRNA in the right panel of Fig. 4b) is upregulated by some triggering signal, the common target of the FFL is activated by a branch of the FFL but inhibited by the other branch. Yet, the interaction mediated by second regulator of the common target (the ncRNA in the left and center panels and the TF in the right panel of Fig. 4b) requires more time to become active because the expression level must first exceed some concentration threshold. Therefore, the incoherent FFL generates a time window when the common target is modulated by only one regulator. Therefore, this delay mechanism can avoid high temporal co-expression of the ncRNA and the target.

An example of this function is the network formed by Myc, E2F, and miR-17 already discussed in the previous section. The connections between the three players can be represented as follows: Myc directly activates E2F and miR-17 while miR-17, in turn, targets and inhibits E2F, hence generating the incoherent FFL. Shalgi et al. [75] discussed that the upregulation of Myc results in increased levels of both E2F and its miRNA antagonist miR-17. Nonetheless, the upregulation of miR-17 is slower compared to the one of E2F due to delays in biogenesis. Therefore, this network generates a temporal window when E2F is high because it is not targeted by miR-17. Later on, miR-17 is successfully upregulated and can target E2F, hence decreasing its cellular level [75].

5 Modulating ncRNA Regulation: The Competition of Endogenous RNAs

Beyond targeting mRNA, a ncRNA (specifically, mostly miRNAs) can target another ncRNA that possesses a specific miRNA recognition element (MRE), or binding site, and therefore attenuate the regulatory effect on the primary target mRNA [77–79]. In this case, the second target ncRNA is referred to as a competing endogenous RNA (ceRNA). The higher the number of MREs on the ceRNA, the stronger the co-regulation that attenuates the repression on the target mRNA [80]. Moreover, it has been hypothesized that the cross talk between coding and noncoding RNAs forms a large-scale regulatory network where the expression of ceRNAs is subjected to modulation as well, the so-called ceRNA hypothesis [81].

Several mathematical models have been proposed in recent years to achieve a mechanism-based understanding of how ncRNA-ceRNA interactions help precisely shaping the genetic landscape of a cell. Ala et al. [82] developed a mathematical mass-action model for the competition of ceRNAs over shared miRNAs and investigated how that modulates their cross-regulation (Fig. 5a). Using mechanism-based models and bioinformatics, they found several factors contributing to an effective ceRNA regulation: first, the relative abundance of the ceRNA and miRNA is crucial (Fig. 5b). Further, the stoichiometry, or number of available binding sites on the ceRNA, plays an important role as well [82]. Finally, the authors discussed the connection between

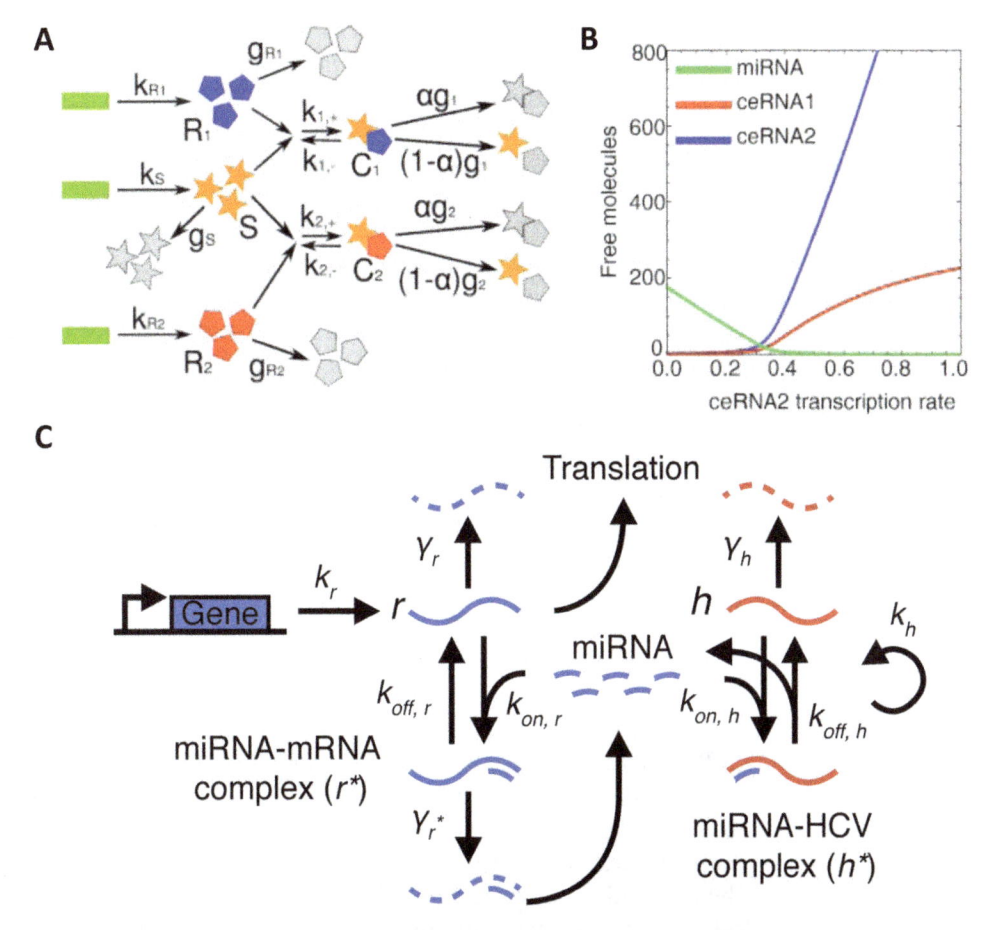

Fig. 5 (**a**) Schematic of the mass-action model of ceRNA-ncRNA interaction as proposed by Ala et al. [82]. Two species of mRNAs (R_1 and R_2, red and blue pentagons) and one species of microRNA molecules (S, orange stars) are transcribed from DNA (green). microRNAs can bind to both mRNAs and form the complexes C_1 and C_2. When the complexes degrade, the microRNA is lost with probability α or recycled with probability $1 - \alpha$. (**b**) Levels of ceRNAs and microRNA as a function of the transcription rate of one of the two ceRNAs (ceRNA$_2$) as predicted by the model of Ala et al. [82]. When the transcription rate of ceRNA$_2$ is small, the level of miRNA is higher due to the reduced turnover of the miRNA-ceRNA$_2$ complex. Conversely, the miRNA is shut down for a large ceRNA$_2$ transcription rate. (**c**) The dynamical model of miR-122 sequestration of Luna et al. [91] considers the transcription and translation of mRNA (r), the binding of mRNA with miR-122 (miRNA), and the degradation of the miRNA-mRNA complex. Additionally, HCV (h) can bind and form a complex with miR-122 (miRNA-HCV complex), therefore sequestrating it

ceRNA and transcription factors, suggesting that the potency of a TF has even broader implications in establishing a cell's phenotype due to their cross talk with RNA transcripts [82]. In a follow-up study, Bosia et al. [83] generalized the model to a case with multiple miRNAs and ceRNAs and also introduced stochastic effects. The model shows that not only a group of ceRNAs competes for a common set of miRNAs but, symmetrically, a group of miRNAs cross talk through a common set of ceRNAs, hence resonating well with the ceRNA hypothesis mentioned above [81, 83]. Another model by Martirosyan et al. [84] showed that the regulation of a gene by the miRNA-ceRNA network can be more efficient as compared to that of a transcription factor, hence proposing that ceRNAs not only participate in fine-tuning but rather play a major role in establishing a cell's gene expression.

Gerard and Novak [85] proposed a model where the ncRNA can bind to multiple different RNAs to simulate the presence of a ceRNA network. They show that a TF that only directly regulates the transcription of the ncRNA can still modulate the level of ceRNAs due to ncRNA-ceRNA binding interactions. Interestingly, if the expression of the ncRNA is controlled by a sharp threshold in TF level, such threshold propagates to the cellular production of the other ceRNAs due to the ncRNA-ceRNA cross talk, hence offering a possible mechanism to limit the propagation of noise in the genetic network [80, 86].

Intriguingly, it has been shown experimentally that miRNA activity can be modulated not only by endogenous RNA but also by exogenous RNA, such as in the case of viral RNA [87–89]. McCaskill et al. [90] proposed that viral ncRNA can promote miRNA degradation with three main mechanisms. First, viral RNA can recruit proteins to the RNA that are required for turnover. Second, viral RNA can mediate the localization of the miRNA to sites where it can be then degraded. Last, it can induce conformational changes facilitating miRNA exposure and degradation by proteins [90]. As an example, Luna et al. [91] developed a mathematical model for the sequestration of miRNA miR-122 in the liver induced by the hepatitis C virus (HCV) (Fig. 5c). Specifically, it was shown that inhibition by HCV results in a global decrease in miR-122 levels and therefore promotes a fertile oncogenic environment for HCV [91].

6 Beyond the Single-Cell Picture: ncRNAs Mediate Cell-Cell Signaling

In addition to intracellular signaling, phenotypic decisions can often be driven by neighboring cells. A cell in close spatial proximity can modulate the phenotypic state of another "receiver" cell in multiple ways. Juxtacrine signaling, i.e., contact-based ligand-receptor interactions, such as in the case of Notch signaling, allows a direct transfer of information between cells in close proximity via binding of an endogenous membrane receptor with an exogenous ligand [92–94]. Further, cells can secrete ligands as soluble signals that diffuse in the extracellular environment and bind to a much distant target cell's receptor, hence not requiring cell-cell contact (i.e., paracrine signaling). Tunneling nanotubes (TNTs) are a further example of cell-cell signaling that can elevate information transfer beyond nearest neighbors [95, 96]. TNTs are membranous bridges of variable diameter able to connect cells along large distances (in this context, "large" implies greater than cell diameter and hence beyond nearest neighbor) and allowing transfer of diverse cellular components [95, 96]. Another mechanism for cell-to-cell transfer is the secretion of exosomes, small vesicles that are released by a sender cell into the extracellular environment [97, 98].

A biological case of interest where intracellular miRNA-mediated posttranslational regulations modulates cell-cell communication is the case of the Notch signaling pathway. Notch signaling is an evolutionarily conserved pathway that operates via binding between the Notch transmembrane receptor and a transmembrane ligand of a neighboring cell. Upon binding, the Notch intracellular domain (NICD) is released and translocated to cell nucleus where it starts a regulatory cascade by acting as a transcriptional cofactor (Fig. 6a, left). This pathway is well studied in the context of embryonic development for its ability to generate fine-grained phenotypic patterning. On the one hand, NICD inhibits the transcription of a family of Notch ligands, called Delta or Delta-like (Dll) (Fig. 5a). Thus, the cell with high levels of NICD expresses low levels of Delta, and consequently Notch signaling is not activated in the neighboring cell (i.e., low levels of NICD). In other words, one cell behaves as a "receiver" (high Notch, low Delta), while the other behaves as a "sender" (low Notch, high Delta) [99, 100] (Fig. 6a, right). Such a double-negative feedback loop drives the neighboring cells communicating via Notch-Delta pathway to acquire opposite phenotypes, leading to so-called checkerboard-like or salt-and-pepper patterns. On the other hand, NICD transcriptionally activates a second family of ligands, called Jagged (Fig. 6a, left), hence generating a double-positive feedback between neighboring cell that will be both high in Notch receptor and Jagged ligand, or both behave as hybrid sender/receiver cells [101–103] (Fig. 6a, right).

The tissue-level outcomes of Notch signaling can also be influenced by multiple ncRNAs. In the context of embryonic development, Chen et al. [104] modeled the downregulation of Notch by the microRNA miR-124 to explain why the spatial cell patterning deviates from the expected chessboard-like distribution of Notch-Delta signaling in neuronal cells of *Ciona intestinalis*, a species of sea squirt. The authors could explain the variable spacing between phenotypes in terms of a single model parameter that depends on the cellular concentration of miR-124 [104].

Similarly, Goodfellow et al. [105] investigated temporal dynamics of gene expression during neuronal cell differentiation and discovered that Hes1, an effector of Notch, and its target genes undergo an oscillatory dynamic due to a double-negative feedback with the microRNA miR-9 (Fig. 6b). The authors observed that, while mother cells exhibit such oscillations, differentiated cells are either stably low or high in Hes1 levels. Therefore, such oscillations

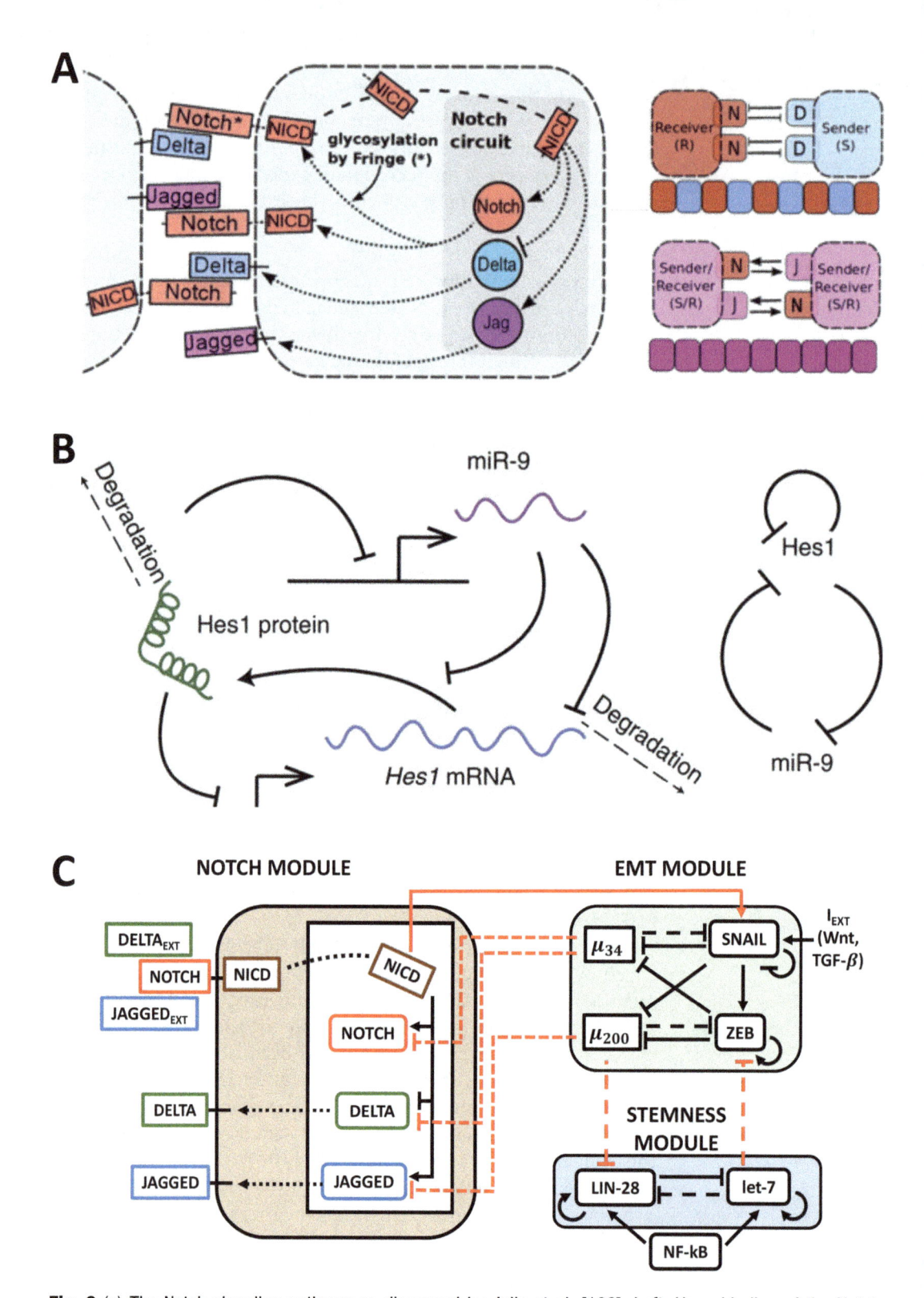

Fig. 6 (a) The Notch signaling pathway as discussed by Jolly et al. [102]. Left: Upon binding of the Notch transmembrane receptor with a ligand (Delta or Jagged) from a neighboring cell, the Notch intracellular domain (NICD) translocates to cell nucleus where it activates Notch and Jagged but inhibits Delta. Further,

are crucial in triggering stem cell differentiation with the correct timing [105]. Further, in a follow-up study on neuronal oscillations during development, Roese-Koerner et al. [106] elucidated how an additional, direct interaction between miR-9 and Notch2 (a member of the Notch receptor family) combined with the miR-9/Hes1 loop provides an additional layer of control to avoid errors in differentiation timing.

Phenotype patterning via the Notch pathway also plays a role in the context of cancer development. The microRNAs miR-34 and miR-200 regulated during EMT (*see* the previous section) are implicated in modulating Notch signaling. Receptor and ligands that are not attached to the cell's membrane can bind with microRNAs (miR-34 targets Notch and Delta while miR-200 targets miR-200), hence preferentially inhibiting Notch signaling when the cell is in an epithelial state (Fig. 6c, Notch-EMT connections). Boareto et al. [107] developed a mathematical model to elucidate the connection between Notch signaling and the EMT core regulatory circuit and found that a partial or complete EMT (i.e., a hybrid E/M or mesenchymal phenotype) is accompanied by a switch to a hybrid sender/receiver Notch state. Further, multicellular simulations showed that Notch-Jagged signaling can act as a pro-metastatic factor because it facilitates spatial co-localization of cells undergoing a partial EMT [107]. Such cells can therefore form a motile cluster of circulating tumor cells that can enter the bloodstream, the first step toward metastasis [58, 108]. In a follow-up work on the Notch-EMT interaction, Bocci et al. [109] showed via mathematical modeling and experiments that the inhibition of Notch signaling by the protein Numb has an indirect effect in regulating the expression of miR-34 and miR-200. In fact, the activation of Notch signaling promotes EMT by activating the mesenchymal TF SNAIL (Fig. 6c, Notch-EMT connections). Therefore, inhibitors of Notch such as Numb can play a role as "phenotypic stability factors" (PSF) that stabilize a hybrid E/M

Fig. 6 (continued) glycosylation by Fringe, a glycosyltransferase, increases Notch receptor's binding affinity for Delta while decreasing binding affinity to Jagged. Right: Notch-Delta signaling generates alternate cell fate between neighboring cells (sender-receiver dynamics). Notch-Jagged signaling generates convergent cell fate among neighboring cells (sender/receiver-sender/received dynamics). (**b**) The Hes1/miR-9 interaction network as modeled by Goodfellow et al. [105]. Left: Hes1 protein (green) inhibits the transcription of Hes1 mRNA (blue) and miR-9 (magenta) while miR-9 targets Hes1 mRNA. Right: A simplified scheme includes the double-negative feedback loop between Hes1 and miR-9 and Hes1 self-inhibition. (**c**) Connections between Notch pathway, EMT circuit, and stemness circuit as modeled by Bocci et al. [111]. Notch and EMT modules are connected as follows: NICD transcriptionally activates the mesenchymal TF SNAIL while the epithelial microRNAs miR-34 and miR-200 target Notch, Delta, and Jagged. The EMT and stemness modules are connected as follows: the epithelial microRNA miR-200 targets the reprogramming factor LIN-28 while the microRNA let-7 targets the mesenchymal TF ZEB

state by stopping cellular transformation midway *en route* to a complete EMT [110].

Generalizing further the model, Bocci et al. [111] developed a mechanism-based mathematical framework to elucidate the interconnections between Notch signaling, EMT, and the let-7/LIN-28 stemness circuit discussed in Subheading 3 (Fig. 6c). They showed that, due to the connections between the three circuits, the sender/receiver Notch-Jagged state not only correlates with a partial EMT but further relates with the stem-like phenotype as predicted in terms of let-7/LIN-28 levels and observed experimentally [111, 112]. Therefore, this model provides a general framework to understand the connection between several hallmarks of cancer including EMT, stemness, and therapeutic resistance related to increased Notch-Jagged signaling pathway [113].

7 Conclusion

ncRNAs constitute a crucial component of gene regulation with implications in both physiological and pathological contexts. Here, we discuss various mathematical models that have helped identify various functions of ncRNAs and their networks, including (1) fast variation of the gene expression profile in response to change in the extracellular environment, such as in the case of iron uptake in bacteria [18, 22]; (2) conferring robustness to protein synthesis and limiting the effect of noise [36]; (3) mediating cell fate decision via forming feedback loops with other proteins that result in genetic switches associated with distinct cellular phenotypes, such as in the case of epithelial-mesenchymal transition (EMT) [48, 56]; (4) providing mechanisms for precise spatial and temporal patterns of expressions via coherent and incoherent feed-forward loops [75, 76]; (5) creating complex architectures by interacting with other RNA species, the competing endogenous RNAs, that can potentially regulate up to hundreds of genes [82, 83]; and (6) affecting neighboring cell's phenotype by modulating cell-cell signaling pathways such as in the case of Notch signaling [107, 109].

8 Copyright Note

All figure panels were taken from published articles and comply with the journal copyright policies. Figure 3c, d has been taken from reference [67] (Copyright (2008) National Academy of Sciences, USA); Fig. 5c has been taken from reference [91]. This chapter was published in Cell, Vol. 160, Luna et al., "Hepatitis C Virus RNA Functionally Sequesters miR-122", 1099–1110, Copyright Elsevier (2015). Figure 6b has been originally published in reference [105] under a Creative Commons Attribution 3.0 Unported License (CC-BY 3.0).

References

1. Crick F (1970) Central dogma of molecular biology. Nature 227:561–563
2. Mattick JS, Makunin IV (2006) Non-coding RNA. Hum Mol Genet 15(Spec 1):17–29. https://doi.org/10.1093/hmg/ddl046
3. Eddy SR (2001) Non-coding RNA genes and the modern RNA world. Nat Rev Genet 2:919–929. https://doi.org/10.1038/35103511
4. Zhdanov VP (2011) Kinetic models of gene expression including non-coding RNAs. Phys Rep 500:1–42
5. Lai X, Wolkenhauer O, Vera J (2016) Understanding microRNA-mediated gene regulatory networks through mathematical modelling. Nucleic Acids Res 44:6019–6035
6. Ambros V (2004) The functions of animal microRNAs. Nature 431:350–355. https://doi.org/10.1038/nature02871
7. Bartel DP (2004) MicroRNAs: genomics, biogenesis, mechanism, and function. Cell 116:281–297. https://doi.org/10.1016/S0092-8674(04)00045-5
8. Neema A, Dasaradhi PVN, Asif Mohmmed PM, Bhatnagar RK, SKM (2004) RNA interference: biology, mechanism, and applications. Microbiol Mol Biol Rev 38:285–294. https://doi.org/10.1128/MMBR.67.4.657
9. Ma L, Bajic VB, Zhang Z (2013) On the classification of long non-coding RNAs. RNA Biol 10:924–933. https://doi.org/10.4161/rna.24604
10. Geisler S, Coller J (2013) RNA in unexpected places: long non-coding RNA functions in diverse cellular contexts. Nat Rev Mol Cell Biol 14:699–712. https://doi.org/10.1038/nrm3679
11. Gibb EA, Brown CJ, Lam WL (2011) The functional role of long non-coding RNA in human carcinomas. Mol Cancer 10:38. https://doi.org/10.1186/1476-4598-10-38
12. Gutschner T, Diederichs S (2012) The hallmarks of cancer: a long non-coding RNA point of view. RNA Biol 9:703–709. https://doi.org/10.4161/rna.20481
13. Bartel DP (2009) MicroRNAs: target recognition and regulatory functions. Cell 136:215–233. https://doi.org/10.1016/j.cell.2009.01.002
14. Ghildiyal M, Zamore PD (2009) Small silencing RNAs: an expanding universe. Nat Rev Genet 10:94–108. https://doi.org/10.1038/nrg2504
15. Yang E, van Nimwegen E, Zavolan M, Rajewsky N, Schroeder M, Magnasco M, Darnell JE (2003) Decay rates of human mRNAs: correlation with functional characteristics and sequence attributes. Genome Res 13:1863–1872. https://doi.org/10.1101/gr.1272403
16. Gottesman S (2005) Micros for microbes: non-coding regulatory RNAs in bacteria. Trends Genet 21:399–404. https://doi.org/10.1016/j.tig.2005.05.008
17. Storz G, Vogel J, Wassarman KM (2011) Regulation by small RNAs in bacteria: expanding frontiers. Mol Cell 43:880–891. https://doi.org/10.1016/j.molcel.2011.08.022
18. Mitarai N, Benjamin J-AM, Krishna S, Semsey S, Csiszovszki Z, Massé E, Sneppen K (2009) Dynamic features of gene expression control by small regulatory RNAs. Proc Natl Acad Sci U S A 106:10655–10659. https://doi.org/10.1073/pnas.0901466106
19. Massé E, Vanderpool CK, Gottesman S (2005) Effect of RyhB small RNA on global iron use in Escherichia coli. J Bacteriol 187:6962–6971. https://doi.org/10.1128/JB.187.20.6962-6971.2005
20. Večerek B, Moll I, Bläsi U (2007) Control of Fur synthesis by the non-coding RNA RyhB and iron-responsive decoding. EMBO J 26:965–975. https://doi.org/10.1038/sj.emboj.7601553
21. Hansen GT, Ahmad R, Hjerde E, Fenton CG, Willassen NP, Haugen P (2012) Expression profiling reveals Spot 42 small RNA as a key regulator in the central metabolism of Aliivibrio salmonicida. BMC Genomics 13:37. https://doi.org/10.1186/1471-2164-13-37
22. Semsey S, Andersson AMC, Krishna S, Jensen MH, Massé E, Sneppen K (2006) Genetic regulation of fluxes: iron homeostasis of Escherichia coli. Nucleic Acids Res 34:4960–4967. https://doi.org/10.1093/nar/gkl627
23. Mitarai N, Andersson AMC, Krishna S, Semsey S, Sneppen K (2007) Efficient degradation and expression prioritization with small RNAs. Phys Biol 4:164–171. https://doi.org/10.1088/1478-3975/4/3/003
24. Legewie S, Dienst D, Wilde A, Herzel H, Axmann IM (2008) Small RNAs establish delays and temporal thresholds in gene expression. Biophys J 95:3232–3238. https://doi.org/10.1529/biophysj.108.133819

25. Shimoni Y, Friedlander G, Hetzroni G, Niv G, Altuvia S, Biham O, Margalit H (2007) Regulation of gene expression by small non-coding RNAs: a quantitative view. Mol Syst Biol 3:1–9. https://doi.org/10.1038/msb4100181

26. Alon U (2007) Network motifs: theory and experimental approaches. Nat Rev Genet 8:450–461. https://doi.org/10.1038/nrg2102

27. Mangan S, Alon U (2003) Structure and function of the feed-forward loop network motif. Proc Natl Acad Sci U S A 100:11980–11985. https://doi.org/10.1073/pnas.2133841100

28. Levine E, Zhang Z, Kuhlman T, Hwa T (2007) Quantitative characteristics of gene regulation by small RNA. PLoS Biol 5:1998–2010. https://doi.org/10.1371/journal.pbio.0050229

29. Levine E, Hwa T (2008) Small RNAs establish gene expression thresholds. Curr Opin Microbiol 11:574–579. https://doi.org/10.1016/j.mib.2008.09.016

30. Levine E, Ben JE, Levine H (2007) Target-specific and global effectors in gene regulation by microRNA. Biophys J 93:L52–L54. https://doi.org/10.1529/biophysj.107.118448

31. Milo R, Jorgensen P, Moran U, Weber G, Springer M (2009) BioNumbers The database of key numbers in molecular and cell biology. Nucleic Acids Res 38:750–753. https://doi.org/10.1093/nar/gkp889

32. Swain PS, Elowitz MB, Siggia ED (2002) Intrinsic and extrinsic contributions to stochasticity in gene expression. Proc Natl Acad Sci 99:12795–12800. https://doi.org/10.1073/pnas.162041399

33. Elowitz MB, Levine AJ, Siggia ED, Swain PS (2002) Stochastic gene expression in a single cell. Science 297:1183–1186

34. Raser JM, O'Shea EK (2010) Noise in gene expression. Science 309:2010–2014. https://doi.org/10.1126/science.1105891

35. Schwanhüusser B, Busse D, Li N, Dittmar G, Schuchhardt J, Wolf J, Chen W, Selbach M (2011) Global quantification of mammalian gene expression control. Nature 473:337–342. https://doi.org/10.1038/nature10098

36. Jia Y, Liu W, Li A, Yang L, Zhan X (2009) Intrinsic noise in post-transcriptional gene regulation by small non-coding RNA. Biophys Chem 143:60–69. https://doi.org/10.1016/j.bpc.2009.04.001

37. Elgart V, Jia T, Kulkarni RV (2010) Applications of Little's Law to stochastic models of gene expression. Phys Rev E Stat Nonlinear Soft Matter Phys 82:1–6. https://doi.org/10.1103/PhysRevE.82.021901

38. Elgart V, Jia T, Kulkarni R (2010) Quantifying mRNA synthesis and decay rates using small RNAs. Biophys J 98:2780–2784. https://doi.org/10.1016/j.bpj.2010.03.022

39. Zhdanov VP (2009) Conditions of appreciable influence of microRNA on a large number of target mRNAs. Mol BioSyst 5:638. https://doi.org/10.1039/b808095j

40. Jost D, Nowojewski A, Levine E (2013) Regulating the many to benefit the few: role of weak small RNA targets. Biophys J 104:1773–1782. https://doi.org/10.1016/j.bpj.2013.02.020

41. Shalgi R, Lieber D, Oren M, Pilpel Y (2007) Global and local architecture of the mammalian microRNA-transcription factor regulatory network. PLoS Comput Biol 3:1291–1304. https://doi.org/10.1371/journal.pcbi.0030131

42. Wu S, Huang S, Ding J, Zhao Y, Liang L, Liu T, Zhan R, He X (2010) Multiple microRNAs modulate p21Cip1/Waf1 expression by directly targeting its 3′ untranslated region. Oncogene 29:2302–2308. https://doi.org/10.1038/onc.2010.34

43. Lai X, Schmitz U, Gupta SK, Bhattacharya A, Kunz M, Wolkenhauer O, Vera J (2012) Computational analysis of target hub gene repression regulated by multiple and cooperative miRNAs. Nucleic Acids Res 40:8818–8834. https://doi.org/10.1093/nar/gks657

44. Lai X, Bhattacharya A, Schmitz U, Kunz M, Vera J, Wolkenhauer O (2013) A systems' biology approach to study microrna-mediated gene regulatory networks. Biomed Res Int 2013:703849. https://doi.org/10.1155/2013/703849

45. Doench JG, Sharp PA (2004) Specificity of microRNA target selection in translational repression. Genes (Basel) 504:504–511. https://doi.org/10.1101/gad.1184404.species

46. Sætrom P, Heale BSE, Snøve O, Aagaard L, Alluin J, Rossi JJ (2007) Distance constraints between microRNA target sites dictate efficacy and cooperativity. Nucleic Acids Res 35:2333–2342. https://doi.org/10.1093/nar/gkm133

47. Schmitz U, Lai X, Winter F, Wolkenhauer O, Vera J, Gupta SK (2014) Cooperative gene

47. (continued) regulation by microRNA pairs and their identification using a computational workflow. Nucleic Acids Res 42:7539–7552. https://doi.org/10.1093/nar/gku465
48. Lu M, Jolly MK, Levine H, Onuchic JN, Ben-Jacob E (2013) MicroRNA-based regulation of epithelial-hybrid-mesenchymal fate determination. Proc Natl Acad Sci U S A 110:18174–18179. https://doi.org/10.1073/pnas.1318192110
49. Vasudevan S, Tong Y, Steitz JA (2007) Switching from repression to activation: MicroRNAs can up-regulate translation. Science 318:1931–1934
50. Ghosh T, Soni K, Scaria V, Halimani M, Bhattacharjee C, Pillai B (2008) MicroRNA-mediated up-regulation of an alternatively polyadenylated variant of the mouse cytoplasmic β-actin gene. Nucleic Acids Res 36:6318–6332. https://doi.org/10.1093/nar/gkn624
51. Ma F, Liu X, Li D, Wang P, Li N, Lu L, Cao X (2010) MicroRNA-466l upregulates IL-10 expression in TLR-triggered macrophages by antagonizing RNA-binding protein tristetraprolin-mediated IL-10 mRNA degradation. J Immunol 184:6053–6059. https://doi.org/10.4049/jimmunol.0902308
52. Gokhale SA, Gadgil CJ (2012) Analysis of miRNA regulation suggests an explanation for 'unexpected' increase in target protein levels. Mol BioSyst 8:760–765. https://doi.org/10.1039/C1MB05368J
53. Zhdanov VP (2010) ncRNA-mediated bistability in the synthesis of hundreds of distinct mRNAs and proteins. Phys A Stat Mech Appl 389:887–890. https://doi.org/10.1016/j.physa.2009.11.028
54. Zhdanov VP (2010) Effect of non-coding RNA on bistability and oscillations in the mRNA-protein interplay. Biophys Rev Lett 05:89–107
55. Zhdanov VP (2006) Transient stochastic bistable kinetics of gene transcription during the cellular growth. Chem Phys Lett 424:394–398. https://doi.org/10.1016/j.cplett.2006.05.024
56. Tian XJ, Zhang H, Xing J (2013) Coupled reversible and irreversible bistable switches underlying TGFβ-induced epithelial to mesenchymal transition. Biophys J 105:1079–1089. https://doi.org/10.1016/j.bpj.2013.07.011
57. Nieto MA, Huang RY, Jackson RA, Thiery JP (2016) EMT: 2016. Cell 166:21–45
58. Jolly MK, Boareto M, Huang B, Jia D, Lu M, Ben-Jacob E, Onuchic JN, Levine H (2015) Implications of the hybrid epithelial/mesenchymal phenotype in metastasis. Front Oncol 5:155. https://doi.org/10.3389/fonc.2015.00155
59. Yang X, Lin X, Zhong X, Kaur S, Li N, Liang S, Lassus H, Wang L, Katsaros D, Montone K, Zhao X, Zhang Y, Bützow R, Coukos G, Zhang L (2010) Double-negative feedback loop between reprogramming factor LIN28 and microRNA let-7 regulates aldehyde dehydrogenase 1-positive cancer stem cells. Cancer Res 70:9463–9472. https://doi.org/10.1158/0008-5472.CAN-10-2388
60. Jolly MK, Huang B, Lu M, Mani SA, Levine H, Ben-Jacob E (2014) Towards elucidating the connection between epithelial-mesenchymal transitions and stemness. J R Soc Interface 11:20140962
61. Jolly MK, Jia D, Boareto M, Mani SA, Pienta KJ, Ben-Jacob E, Levine H (2015) Coupling the modules of EMT and stemness: a tunable 'stemness window' model. Oncotarget 6:25161–25174
62. Hafner M, Max KEA, Bandaru P, Morozov P, Gerstberger S, Brown M, Molina H, Tuschl T (2013) Identification of mRNAs bound and regulated by human LIN28 proteins and molecular requirements for RNA recognition. RNA 19:613–626. https://doi.org/10.1261/rna.036491.112
63. Wilbert ML, Huelga SC, Kapeli K, Stark TJ, Liang TY, Chen SX, Yan BY, Nathanson JL, Hutt KR, Lovci MT, Kazan H, Vu AQ, Massirer KB, Morris Q, Hoon S, Yeo GW (2012) LIN28 binds messenger RNAs at GGAGA motifs and regulates splicing factor abundance. Mol Cell 48:195–206. https://doi.org/10.1016/j.molcel.2012.08.004
64. Zisoulis DG, Kai ZS, Chang RK, Pasquinelli AE (2012) Autoregulation of microRNA biogenesis by let-7 and Argonaute. Nature 486:541–544. https://doi.org/10.1038/nature11134
65. Emmrich S, Pützer BM (2010) Checks and balances: E2F – MicroRNA crosstalk in cancer control. Cell Cycle 9:2555–2567. https://doi.org/10.4161/cc.9.13.12061
66. Concepcion CP, Bonetti C, Ventura A (2012) The MicroRNA-17-92 family of MicroRNA clusters in development and disease. Cancer J (United States) 18:262–267. https://doi.org/10.1097/PPO.0b013e318258b60a

67. Aguda BD, Kim Y, Piper-Hunter MG, Friedman A, Marsh CB (2008) MicroRNA regulation of a cancer network: consequences of the feedback loops involving miR-17-92, E2F, and Myc. Proc Natl Acad Sci 105:19678–19683. https://doi.org/10.1073/pnas.0811166106
68. Zhang H, Chen Y, Chen Y (2012) Noise propagation in gene regulation networks involving interlinked positive and negative feedback loops. PLoS One 7:1–8. https://doi.org/10.1371/journal.pone.0051840
69. Li Y, Li Y, Zhang H, Chen Y (2011) Microrna-mediated positive feedback loop and optimized bistable switch in a cancer network involving miR-17-92. PLoS One 6:2–10. https://doi.org/10.1371/journal.pone.0026302
70. Giampieri E, Remondini D, de Oliveira L, Castellani G, Lió P (2011) Stochastic analysis of a miRNA-protein toggle switch. Mol BioSyst 7:2796–2803. https://doi.org/10.1039/c1mb05086a
71. Tsang J, Zhu J, van Oudenaarden A (2007) MicroRNA-mediated feedback and feedforward loops are recurrent network motifs in mammals. Mol Cell 26:753–767. https://doi.org/10.1016/j.molcel.2007.05.018
72. Re A, Cora D, Taverna D, Caselle M (2009) Genome-wide survey of MicroRNA – transcription factor feed-forward regulatory circuits in human. BMC Bioinformatics 5:51. https://doi.org/10.1039/b900177h
73. Wall ME, Dunlop MJ, Hlavacek WS (2005) Multiple functions of a feed-forward-loop gene circuit. J Mol Biol 349:501–514. https://doi.org/10.1016/j.jmb.2005.04.022
74. Herranz H, Cohen SM (2010) MicroRNAs and gene regulatory networks: managing the impact of noise in biological systems. Genes Dev 24:1339–1344. https://doi.org/10.1101/gad.1937010
75. Shalgi R, Brosh R, Oren M, Pilpel Y, Rotter V (2009) Coupling transcriptional and post-transcriptional miRNA regulation in the control of cell fate. Aging (Albany NY) 1:762–770
76. Kobayashi K, Sakurai K, Hiramatsu H, Inada KI, Shiogama K, Nakamura S, Suemasa F, Kobayashi K, Imoto S, Haraguchi T, Ito H, Ishizaka A, Tsutsumi Y, Iba H (2015) The miR-199a/Brm/EGR1 axis is a determinant of anchorage-independent growth in epithelial tumor cell lines. Sci Rep 5:8428. https://doi.org/10.1038/srep08428
77. Taulli R, Loretelli C, Pandolfi PP (2013) From pseudo-ceRNAs to circ-ceRNAs: a tale of cross-talk and competition. Nat Struct Mol Biol 20:541–543. https://doi.org/10.1038/nsmb2580
78. Tay Y, Rinn J, Pandolfi PP (2014) The multilayered complexity of ceRNA crosstalk and competition. Nature 505:344–352. https://doi.org/10.1038/nature12986
79. Kartha RV, Subramanian S (2014) Competing endogenous RNAs (ceRNAs): new entrants to the intricacies of gene regulation. Front Genet 5:1–9. https://doi.org/10.3389/fgene.2014.00008
80. Ebert MS, Sharp PA (2012) Roles for MicroRNAs in conferring robustness to biological processes. Cell 149:505–524. https://doi.org/10.1016/j.cell.2012.04.005
81. Salmena L, Poliseno L, Tay Y, Kats L, Pandolfi PP (2011) A ceRNA hypothesis: the rosetta stone of a hidden RNA language? Cell 146:353–358. https://doi.org/10.1016/j.cell.2011.07.014
82. Ala U, Karreth FA, Bosia C, Pagnani A, Taulli R, Leopold V, Tay Y, Provero P, Zecchina R, Pandolfi PP (2013) Integrated transcriptional and competitive endogenous RNA networks are cross-regulated in permissive molecular environments. Proc Natl Acad Sci U S A 110:7154–7159. https://doi.org/10.1073/pnas.1222509110
83. Bosia C, Pagnani A, Zecchina R (2013) Modelling competing endogenous RNA networks. PLoS One 8:e66609. https://doi.org/10.1371/journal.pone.0066609
84. Martirosyan A, Figliuzzi M, Marinari E, De Martino A (2016) Probing the limits to MicroRNA-mediated control of gene expression. PLoS Comput Biol 12:1–23. https://doi.org/10.1371/journal.pcbi.1004715
85. Gérard C, Novák B (2013) microRNA as a potential vector for the propagation of robustness in protein expression and oscillatory dynamics within a ceRNA network. PLoS One 8:e83372. https://doi.org/10.1371/journal.pone.0083372
86. Pedraza JM, van Oudenaarden A (2005) Noise propagation in gene networks. Science 307:1965–1970
87. Skalsky RL, Cullen BR (2010) Viruses, microRNAs, and host interactions. Annu Rev

Microbiol 64:123–141. https://doi.org/10.1146/annurev.micro.112408.134243
88. Grundhoff A, Sullivan CS (2011) Virus-encoded microRNAs. Virology 411:325–343. https://doi.org/10.1016/j.virol.2011.01.002
89. Cullen BR (2009) Viral and cellular messenger RNA targets of viral microRNAs. Nature 457:421–425. https://doi.org/10.1038/nature07757
90. McCaskill J, Praihirunkit P, Sharp PM, Buck AH (2015) RNA-mediated degradation of microRNAs: a widespread viral strategy? RNA Biol 12:579–585. https://doi.org/10.1080/15476286.2015.1034912
91. Luna JM, Scheel TKH, Danino T, Shaw KS, Mele A, Fak JJ, Nishiuchi E, Takacs CN, Catanese MT, De Jong YP, Jacobson IM, Rice CM, Darnell RB (2015) Hepatitis C virus RNA functionally sequesters miR-122. Cell 160:1099–1110. https://doi.org/10.1016/j.cell.2015.02.025
92. Bray SJ (2006) Notch signalling: a simple pathway becomes complex. Nat Rev Mol Cell Biol 7:678–689. https://doi.org/10.1038/nrm2009
93. Andersson ER, Sandberg R, Lendahl U (2011) Notch signaling: simplicity in design, versatility in function. Development 138:3593–3612
94. Bolos V, Grego-Bessa J, De La Pompa JL (2007) Notch signaling in development and cancer. Endocr Rev 28:339–363. https://doi.org/10.1210/er.2006-0046
95. Gerdes HH, Rustom A, Wang X (2013) Tunneling nanotubes, an emerging intercellular communication route in development. Mech Dev 130:381–387. https://doi.org/10.1016/j.mod.2012.11.006
96. Gerdes HH, Bukoreshtliev NV, Barroso JFV (2007) Tunneling nanotubes: a new route for the exchange of components between animal cells. FEBS Lett 581:2194–2201. https://doi.org/10.1016/j.febslet.2007.03.071
97. Théry C, Zitvogel L, Amigorena S (2002) Exosomes: composition, biogenesis and function. Nat Rev Immunol 2:569–579. https://doi.org/10.1038/nri855
98. Raposo G, Stoorvogel W (2013) Extracellular vesicles: exosomes, microvesicles, and friends. J Cell Biol 200:373–383. https://doi.org/10.1083/jcb.201211138
99. Shaya O, Sprinzak D (2011) From Notch signaling to fine-grained patterning: modeling meets experiments. Curr Opin Genet Dev 21:732–739
100. Beatus P, Lendahl U (1998) Notch and neurogenesis. J Neurosci Res 54:125–136. https://doi.org/10.1002/(SICI)1097-4547(19981015)54:2<125::AID-JNR1>3.0.CO;2-G
101. Boareto M, Jolly MK, Lu M, Onuchic JN, Clementi C, Ben-Jacob E (2015) Jagged-Delta asymmetry in Notch signaling can give rise to a Sender/Receiver hybrid phenotype. Proc Natl Acad Sci U S A 112:402–409
102. Jolly MK, Boareto M, Lu M, Onuchic JN, Clementi C, Ben-Jacob E, Jose'N O, Clementi C, Ben-Jacob E (2015) Operating principles of Notch-Delta-Jagged module of cell-cell communication. New J Phys 17:55021
103. Hartman BH, Reh TA, Bermingham-McDonogh O (2010) Notch signaling specifies prosensory domains via lateral induction in the developing mammalian inner ear. Proc Natl Acad Sci U S A 107:15792–15797
104. Chen JS, Gumbayan AM, Zeller RW, Mahaffy JM (2014) An expanded Notch-Delta model exhibiting long-range patterning and incorporating MicroRNA regulation. PLoS Comput Biol 10:e1003655. https://doi.org/10.1371/journal.pcbi.1003655
105. Goodfellow M, Phillips NE, Manning C, Galla T, Papalopulu N (2014) MicroRNA input into a neural ultradian oscillator controls emergence and timing of alternative cell states. Nat Commun 5:1–10. https://doi.org/10.1038/ncomms4399
106. Roese-Koerner B, Stappert L, Brüstle O (2017) Notch/Hes signaling and miR-9 engage in complex feedback interactions controlling neural progenitor cell proliferation and differentiation. Neurogenesis 4:e1313647. https://doi.org/10.1080/23262133.2017.1313647
107. Boareto M, Jolly MK, Goldman A, Pietilä M, Mani SA, Sengupta S, Ben-Jacob E, Levine H, Jose'N O (2016) Notch-Jagged signalling can give rise to clusters of cells exhibiting a hybrid epithelial/mesenchymal phenotype. J R Soc Interface 13:20151106
108. Fabisiewicz A, Grzybowska E, Grybowska E (2017) CTC clusters in cancer progression and metastasis. Med Oncol 34:12
109. Bocci F, Jolly MK, Tripathi SC, Aguilar M, Onuchic N, Hanash SM, Levine H, Levine H (2017) Numb prevents a complete epithelial – mesenchymal transition by modulating Notch signalling. J R Soc Interface 14:20170512
110. Jolly MK, Tripathi SC, Jia D, Mooney SM, Celiktas M, Hanash SM, Mani SA, Pienta KJ,

Ben-Jacob E, Levine H (2016) Stability of the hybrid epithelial/mesenchymal phenotype. Oncotarget 7:27067–27084. https://doi.org/10.18632/oncotarget.8166

111. Bocci F, Jolly MK, George J, Levine H, Onuchic JN (2018) A mechanism-based computational model to capture the interconnections among epithelial-mesenchymal transition, cancer stem cells and Notch-Jagged signaling. Oncotarget 9:29906–29920. https://doi.org/10.1101/314187

112. Bocci F, Gearhart-Serna L, Boareto M, Ribeiro M, Ben-Jacob E, Devi GR, Levine H, Onuchic JN, Jolly MK (2018) Towards understanding cancer stem cell heterogeneity in the tumor microenvironment. BiorXiv https://doi.org/10.1101/408823

113. Bocci F, Levine H, Onuchic JN, Jolly MK (2018) Deciphering the dynamics of Epithelial-Mesenchymal Transition and Cancer Stem Cells in tumor progression. Arxiv https://arxiv.org/pdf/1808.09113

Chapter 15

Kinetic Modelling of Competition and Depletion of Shared miRNAs by Competing Endogenous RNAs

Araks Martirosyan, Marco Del Giudice, Chiara Enrico Bena, Andrea Pagnani, Carla Bosia, and Andrea De Martino

Abstract

Non-coding RNAs play a key role in the post-transcriptional regulation of mRNA translation and turnover in eukaryotes. miRNAs, in particular, interact with their target RNAs through protein-mediated, sequence-specific binding, giving rise to extended and highly heterogeneous miRNA–RNA interaction networks. Within such networks, competition to bind miRNAs can generate an effective positive coupling between their targets. Competing endogenous RNAs (ceRNAs) can in turn regulate each other through miRNA-mediated crosstalk. Albeit potentially weak, ceRNA interactions can occur both dynamically, affecting, e.g., the regulatory clock, and at stationarity, in which case ceRNA networks as a whole can be implicated in the composition of the cell's proteome. Many features of ceRNA interactions, including the conditions under which they become significant, can be unraveled by mathematical and in silico models. We review the understanding of the ceRNA effect obtained within such frameworks, focusing on the methods employed to quantify it, its role in the processing of gene expression noise, and how network topology can determine its reach.

Key words miRNA, ceRNA, Competition, Sponging, Mathematical modeling

1 Introduction

microRNAs (miRNAs)—short, endogenous, non-coding RNAs that operate post-transcriptionally via sequence-specific binding to target RNAs—are increasingly recognized as key actors in the regulation of eukaryotic gene expression [1–5]. Following transcription (either from introns of protein-coding genes or from miRNA-specific genes) and maturation, miRNAs get incorporated into specialized, multiprotein complexes known as RISCs (short for RNA-induced silencing complexes) [6]. Once within a RISC, the miRNA provides the pattern to bind specific sites called miRNA response elements (MREs) found on their target RNAs [7, 8]. Effective base pairing typically requires 6- to 9-nucleotide complementarity, and leads to negative gene expression control through either mRNA destabilization or translational repression [9–11]. The fact that miRNA expression is significantly tissue-specific places miRNAs at the centre of the regulatory layer that controls the composition of the protein repertoire and cell type specificity [12–15]. Still, many aspects of miRNA biology suggest that this role might be exerted through a broader and more complex, yet possibly more subtle, class of mechanisms.

In the first place, miRNAs appear to be highly conserved in vertebrates and invertebrates, and their mRNA target structure displays a significant degree of conservation in higher organisms [16, 17]. For instance, more than half of human genes are conserved miRNA targets, including a large number of weak-interacting sites that appear to be under selective pressure to be maintained [18]. Such a strong degree of conservation suggests that protein levels may need to be fine-tuned within extremely precise ranges [19]. Quantitative studies together with the statistical overrepresentation of noise-buffering motifs within the miRNA-RNA network indeed support this idea [20–22], and recent experiments have confirmed miRNA's ability to stabilize output levels for lowly expressed proteins [23]. Yet, the amount of noise reduction that can be achieved even in optimal conditions does not seem to justify a view of noise suppression as the key evolutionary driver for a significantly conserved miRNA targeting pattern [24–27].

Secondly, miRNA targets are known to include, together with messenger RNAs, a host of ncRNA species like lncRNAs as well as pseudogenes [28–30]. On the one hand, miRNA sponging by ncRNAs can clearly be critical in determining both miRNA levels and their potential for translational repression. On the other, it substantially increases the complexity of the network of miRNA–RNA interactions. It is now clear that each long RNA molecule can typically be targeted by multiple miRNAs, while every miRNA can interact with a very large number of distinct RNAs, generating an extended interaction network stretching across the entire transcriptome [31–34]. Now the ability of miRNAs to regulate gene expression is ultimately linked to the overall target availability, and tends to get weaker as the number of targets (more precisely, of potential binding sites) increases, the so-called dilution effect [35]. This leaves room to search for alternative mechanisms through which miRNAs could exert a regulatory function, even at the non-local (up to system-scale) level.

The heterogeneity of the miRNA-RNA network and the fact that repression potential depends tightly on molecular levels suggest that competition to bind miRNAs might be a contributing factor in the establishment of robust protein profiles [36, 37]. In rough terms, the essence of the so-called ceRNA hypothesis (whereby 'ceRNA' stands for 'competing endogenous RNA') is that, due to a cross-correlation of molecular levels, competition can induce an effective positive coupling between miRNA targets, such that a perturbation affecting the level of one target could be broadcast to its competitor via the subsequent shift in miRNA availability [38]. In this respect, one might say that RNAs form a sort of 'molecular ecosystem', where mutual dependencies can be established post-transcriptionally via miRNA-mediated interactions driven by competition. The ceRNA scenario has received much attention since its formulation, both ex vivo and in synthetic systems (*see*, e.g., [39–44]). Effective interactions coupling RNAs targeted by the same miRNAs (which can be probed, e.g., by over-expressing miRNAs or targets) are now known to be implicated in a variety of processes, from development and differentiation [45], to stress response [46] and disease [47, 48], and have been investigated in connection to their perspective therapeutic usefulness [49].

Still, it has also become clear that the theoretical appeal of the ceRNA effect is not easily translated into quantitative understanding. A key issue is that of fine-tuning. Several conditions clearly factor in the emergence of the ceRNA scenario. The possibility to turn competition between miRNA targets into an effective positive coupling between them presupposes, for instance, a cross-coordination of molecular levels, as a large excess (resp. scarcity) of miRNAs with respect to targets or binding sites will necessarily result into a completely repressed (resp. unrepressed) profile [50, 51]. The ceRNA scenario would naturally become less realistic if kinetic parameters had to be tightly tuned in order to allow for ceRNA crosstalk conditions to arise. In addition, experiments suggest that a relatively small number of targets are usually sensitive to modulation in miRNA availability. Moreover, which targets are responsive depends on miRNA levels [52–54]. The emergent selectivity and adaptability of ceRNA interactions should be reconciled with the heterogeneity observed in the miRNA–RNA interaction network in which each miRNA can regulate up to hundreds of targets.

Mathematical and in silico models developed in recent years have shed light on several of these issues and revealed many unexpected traits [55, 56]. This chapter aims at reviewing the methods employed and the key features of the ceRNA scenario that such studies suggest.

Our starting point is a generic, minimal deterministic mathematical model of post-transcriptional regulation whose steady states can be fully characterized analytically and numerically. Despite its roughness, it allows to precisely quantify the sensitivity of a ceRNA to alterations in the level of one of its competitors, sufficing to capture many of the central characteristics of miRNA-based regulation from basic assumptions about the underlying processes. In particular, miRNA–ceRNA interaction strengths and silencing/sequestration mechanisms emerge, together with the relative abundance of regulators and targets, as key factors for the onset and character of ceRNA crosstalk, including its selectivity. Moreover, heterogeneities in kinetic parameters as well as in miRNA–ceRNA interaction topology are major drivers of ceRNA crosstalk in a broad range of parameter values. The picture obtained at stationarity can be extended to out-of-equilibrium regimes. In particular, one can characterize a 'dynamical' ceRNA effect, which can be stronger than the equilibrium one, as well as the typical timescales required to reach stationary crosstalk.

Passing from a deterministic to a stochastic description, one can address the behaviour of fluctuations in molecular levels and evaluate the ability of miRNA-based regulatory elements to process noise. We will show in particular that the ceRNA mechanism can provide a generic pathway to the reduction of intrinsic noise both for individual proteins and for complexes formed by sub-units sharing a miRNA regulator (which might explain why interacting proteins are frequently regulated by miRNA clusters). The processing of extrinsic (transcriptional) noise is more involved. While ceRNA crosstalk is generically hampered by it, specific patterns of transcriptional correlations can actually result in enhanced noise buffering and in the emergence of complex (e.g. bistable) expression patterns. On the other hand, one can quantify the physical limits to crosstalk intensity by considering how different sources of noise affect it. It turns out that the size of target derepression upon the activation of its competitor is a crucial determinant of the strength of miRNA-mediated ceRNA regulation. When it is sufficiently large, post-transcriptional crosstalk can be as effective as direct transcriptional regulation in controlling expression levels. In specific cases, ceRNA crosstalk may even represent the most effective mechanism to tune gene expression.

An especially important question (and a difficult one, in view of the fact that the effect can be rather modest) concerns the quantification of ceRNA crosstalk intensity, and specifically the identification of unambiguous crosstalk markers that can be validated both experimentally and through the analysis of transcriptional data. We shall examine a few alternatives that have been employed, highlighting the different motivations underlying their use, their physical meaning and their respective limitations.

2 Models and Methods

2.1 Deterministic Model

The simplest mathematical representation of the dynamics of N ceRNA species and M miRNA species interacting in a miRNA-ceRNA network is based on deterministic mass-action kinetics. We shall denote by m_i the level of ceRNA species i (with i ranging from 1 to N), by μ_a the level of miRNA species a (ranging from 1 to M), and by c_{ia} the levels of miRNA-ceRNA complexes. Based on the experimental evidence, one can assume that all miRNA molecules

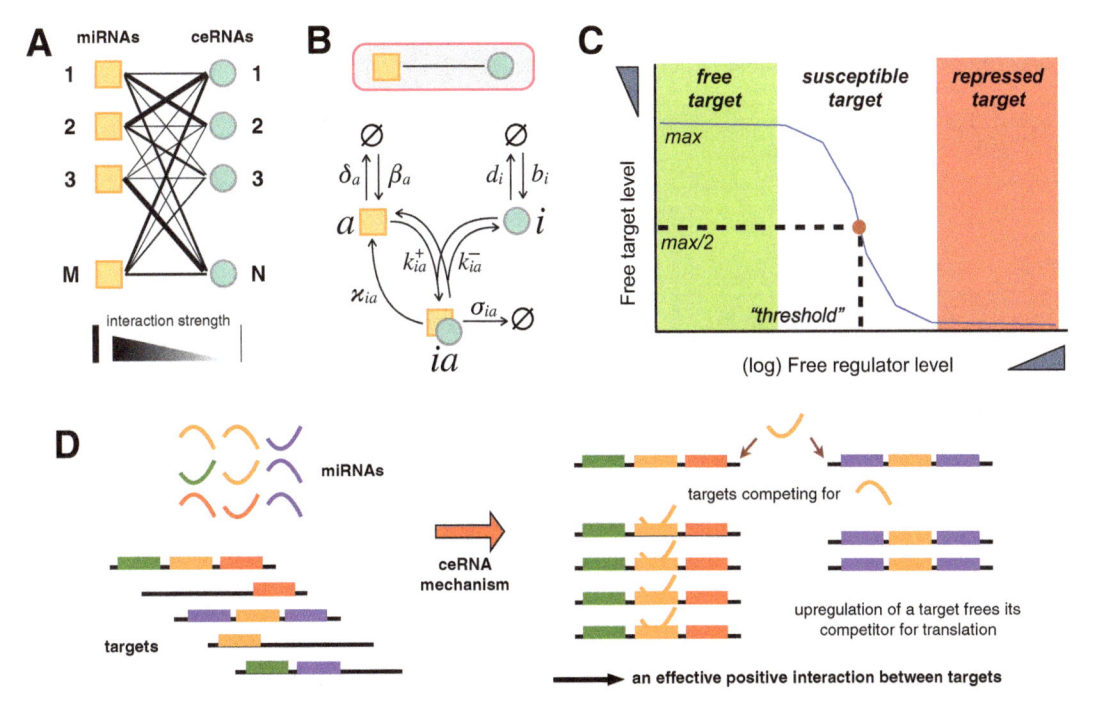

Fig. 1 (**a**) Sketch of an interaction network formed by miRNAs and their targets (ceRNAs). The network is a weighted bipartite graph. Line thickness is proportional to the coupling strength (i.e. to the miRNA-ceRNA binding affinity). (**b**) Sketch of the individual processes lumped in each interaction represented in (**a**). Details of reactions and rates are given in Eq. 1. (**c**) Sketch of the behaviour of the level of free targets (ceRNA or miRNA) as a function of the level of free regulators (miRNA or ceRNA, respectively). (**d**) Sketch of the ceRNA mechanism: competition to bind a miRNA can induce an effective positive coupling between its targets

are 'active', i.e. bound to an Argonaute protein and ready to attach to a target ceRNA. This allows to discard the kinetic steps leading to the formation of the RNA-induced silencing complex (RISC). In such conditions, concentration variables evolve in time due to

1. synthesis and degradation events,
2. complex binding and unbinding events,
3. the processing of complexes.

The latter in turn can follow two distinct pathways: a catalytic one, leading to the degradation of the ceRNA with the re-cycling of the miRNA; and a stoichiometric one, where both molecules are degraded, possibly after sequestration into P-bodies [57, 58]. The relevant processes (*see* Fig. 1a and b for a sketch) are therefore

$$\emptyset \underset{d_i}{\overset{b_i}{\rightleftharpoons}} m_i \qquad \emptyset \underset{\delta_a}{\overset{\beta_a}{\rightleftharpoons}} \mu_a \qquad \mu_a + m_i \underset{k_{ia}^-}{\overset{k_{ia}^+}{\rightleftharpoons}} c_{ia}$$

$$c_{ia} \xrightarrow{\sigma_{ia}} \emptyset \qquad c_{ia} \xrightarrow{\kappa_{ia}} \mu_a$$

(1)

Table 1
Variables and parameters appearing in the basic model, Eq. 2

Variable	Units	Description
m_i	molecules	Number of free copies of ceRNA species i
μ_a	molecules	Number of free copies of miRNA species a
c_{ia}	molecules	Number of copies of $i - a$ complex

Parameter	Units	Description
b_i	molecule min^{-1}	Transcription rate of ceRNA species i
d_i	min^{-1}	Degradation rate of ceRNA species i
β_a	molecule min^{-1}	Transcription rate of miRNA species a
δ_a	min^{-1}	Degradation rate of miRNA species i
k_{ia}^+	molecule^{-1} min^{-1}	$i - a$ complex association rate
k_{ia}^-	min^{-1}	$i - a$ complex dissociation rate
κ_{ia}	min^{-1}	Catalytic decay rate (with miRNA re-cycling) of $i - a$ complex
σ_{ia}	min^{-1}	Stoichiometric decay rate (without miRNA re-cycling) of $i - a$ complex

Note that the levels of molecular species can be specified by copy numbers (as indicated below) as well as by (continuous) concentrations, depending on whether the modelling framework is stochastic (*see* Subheading 2.3) or deterministic (as in Eq. 2), respectively

Correspondingly, the mass action kinetic equations take the form (*see*, e.g., [59–61])

$$\frac{dm_i}{dt} = b_i - d_i m_i - \sum_a k_{ia}^+ m_i \mu_a + \sum_a k_{ia}^- c_{ia},$$

$$\frac{d\mu_a}{dt} = \beta_a - \delta_a \mu_a - \sum_i k_{ia}^+ m_i \mu_a + \sum_i (k_{ia}^- + \kappa_{ia}) c_{ia}, \quad (2)$$

$$\frac{dc_{ia}}{dt} = k_{ia}^+ m_i \mu_a - (\sigma_{ia} + \kappa_{ia} + k_{ia}^-) c_{ia},$$

where the physical meaning of parameters is summarized in Table 1 and where the indices i and a range from 1 to N and from 1 to M, respectively. For several purposes it is useful to introduce the 'stoichiometricity ratio'

$$\alpha_{ia} = \frac{\sigma_{ia}}{\sigma_{ia} + \kappa_{ia}} \quad (3)$$

quantifying the probability that the $i - a$ complex is processed without miRNA re-cycling.

Note: *The model just described, that is the one on which we will mostly focus, is limited to miRNAs and ceRNAs and excludes, for instance, upstream regulators (e.g. transcription factors, TFs) and downstream products (e.g. proteins). Integrating some of these ingredients is however straightforward and it has been done in the literature. For instance, upstream TFs independently regulating the synthesis of ceRNAs and miRNA can be accounted for by assuming that transcription requires the cooperative binding of H TF molecules for each of the RNA species involved (labeled ℓ, including both miRNAs and ceRNAs). Denoting by k_{on} and k_{off} the binding and unbinding rates of TFs to DNA, respectively, the fractional occupancies of TF binding sites on the DNA evolve as*

$$\frac{dn_\ell}{dt} = k_{\rm on}(1 - n_\ell)f_\ell^H - k_{\rm off} n_\ell, \quad (4)$$

where n_ℓ ($0 \le n_\ell \le 1$) stands for the probability that the binding site for the TF controlling the transcription of species ℓ is occupied and f_ℓ stands for the level of the TF controlling species ℓ. In most cases, the variables n_ℓ will equilibrate on timescales much shorter than those characterizing the dynamics of molecular levels [62]. In such conditions, each n_ℓ can be thought to take on its stationary value, i.e.

$$\langle n_\ell \rangle = \frac{f_\ell^H}{f_\ell^H + K^H}, \quad K = \left(\frac{k_{\rm off}}{k_{\rm on}}\right)^{1/H}. \quad (5)$$

Such occupancies in turn modulate the transcription rates appearing in Eq. 2. In particular, the effective transcription rate of ceRNA (resp. miRNA) species i (resp. a) becomes $b_{i,\rm eff} = b_i \langle n_i \rangle$ (resp. $\beta_{a,\rm eff} = \beta_a \langle n_a \rangle$) [63].

An extension of Eq. 2 including downstream species (proteins) is briefly discussed in Subheading 2.6.4.

2.2 Analysis of the Steady State: Threshold Behaviour and Competition-Induced Responses

At steady state, molecular populations evolving according to Eq. 2 are given by the solutions of

$$\langle m_i \rangle = \frac{b_i + \sum_a k_{ia}^- \langle c_{ia} \rangle}{d_i + \sum_a k_{ia}^+ \langle \mu_a \rangle}$$

$$\langle \mu_a \rangle = \frac{\beta_a + \sum_i (k_{ia}^- + \kappa_{ia})\langle c_{ia} \rangle}{\delta_a + \sum_i k_{ia}^+ \langle m_i \rangle} \quad (6)$$

$$\langle c_{ia} \rangle = \frac{k_{ia}^+ \langle \mu_a \rangle \langle m_i \rangle}{\sigma_{ia} + \kappa_{ia} + k_{ia}^-}$$

(We shall henceforth represent the steady-state level of species x by angular brackets, i.e. $\langle x \rangle$.) These conditions have been rigorously shown to describe the unique, asymptotically stable steady state of Eq. 2 [64]. Equation 6 provides a full description of the molecular network in terms of the populations of all species at sufficiently long times, given all kinetic parameters, and are easily solved numerically for any N and M. It is however possible to get a mathematical intuition about how miRNAs affect ceRNA levels at stationarity by eliminating complexes (i.e. $\langle c_{ia}\rangle$) from Eq. 6. This allows to re-cast the steady state in terms of miRNA and ceRNA levels only. Specifically, one gets

$$\langle m_i \rangle = \frac{m_i^\star}{1 + \sum_a \mu_a/\mu_{0,ia}}, \qquad (7)$$
$$\langle \mu_a \rangle = \frac{\mu_a^\star}{1 + \sum_i m_i/m_{0,ia}},$$

where $m_i^\star \equiv b_i/d_i$ and $\mu_a^\star = \beta_a/\delta_a$ stand for the maximum values achievable by ceRNA and miRNA levels at stationarity, while

$$m_{0,ia} = \frac{\delta_a}{k_{ia}^+}\left(1 + \frac{k_{ia}^- + \kappa_{ia}}{\sigma_{ia}}\right), \qquad (8)$$
$$\mu_{0,ia} = \frac{d_i}{k_{ia}^+}\left(1 + \frac{k_{ia}^-}{\sigma_{ia} + \kappa_{ia}}\right)$$

represent 'reference' concentrations that depend on the specific miRNA-ceRNA pair. For the sake of simplicity, we shall refer to these values as 'thresholds'. The gist of Eq. 7 is the following (*see* Fig. 1c) [59]:

Free or unrepressed regime: If the levels of all miRNA species interacting with ceRNA i are sufficiently low (specifically, much lower than the respective thresholds $\mu_{0,ia}$, so that $\Sigma_a \mu_a/\mu_{0,ia} \ll 1$), then the steady-state level of ceRNA i will be very close to the maximum possible, $m_i^\star$. In such conditions, ceRNA species i will be roughly insensitive to changes in miRNA levels. We call this the 'unrepressed' or 'free' regime for ceRNA i.

Susceptible regime: As the quantity $\Sigma_a \mu_a/\mu_{0,ia}$ increases, e.g. following an increase in the level of one or more miRNA species, $\langle m_i \rangle$ deceases in a sigmoidal fashion. This occurs most notably when $\Sigma_a \mu_a/\mu_{0,ia} \simeq 1$ (corresponding, for $M=1$, to a miRNA level close to the threshold value $\mu_{0,ia}$). Here ceRNA i is very sensitive to a change in miRNA levels. We shall therefore term this the 'susceptible' regime for ceRNA i.

Repressed regime: When miRNA levels become sufficiently large, ceRNA i will eventually become fully repressed. In order for this to occur, it suffices that $\Sigma_a \mu_a/\mu_{0,ia} \gg 1$ (which occurs, e.g., when the level of at least one of the miRNA species targeting i significantly exceeds its corresponding threshold $\mu_{0,ia}$). We shall call this the 'repressed' regime for ceRNA i.

(Notice that, because the role of miRNAs and ceRNAs is fully interchangeable, similar regimes can be defined for miRNAs, with the reference concentrations $m_{0,ia}$ playing the role of the threshold

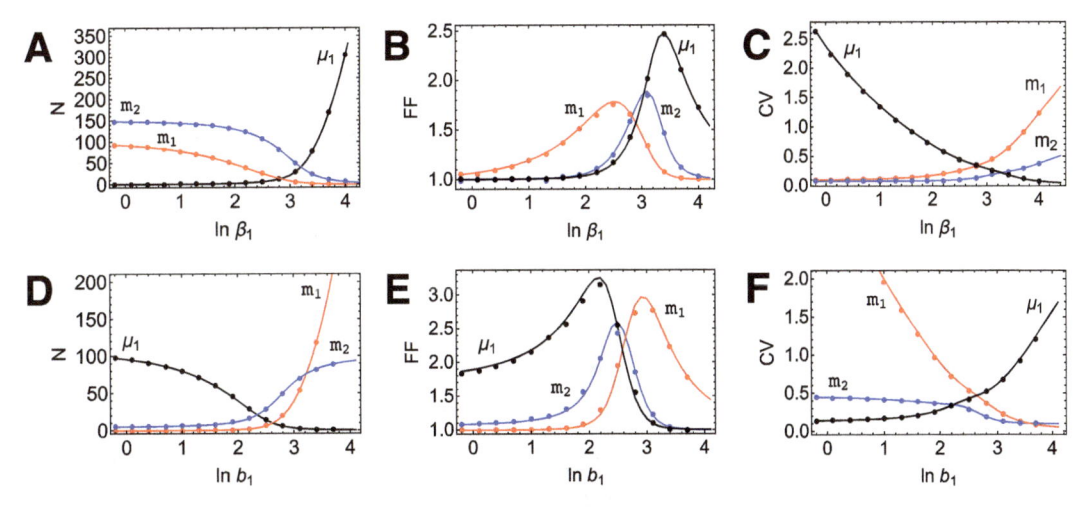

Fig. 2 Characterization of the steady state for a system with 2 ceRNA species competing for one miRNA species. (**a**) Steady-state molecular levels as a function of the miRNA transcription rate β_1. (**b**) Fano Factor (FF) of each molecular species versus β_1. (**c**) Coefficient of variation (CV) of each molecular species versus β_1. (**d**) Steady-state molecular levels as a function of the transcription rate of ceRNA 1, b_1. (**e**) Fano Factor of each molecular species versus b_1. (**f**) Coefficient of variation of each molecular species versus b_1. In panels (**a**) and (**d**), continuous lines describe analytical results (from Eq. 7) while markers denote mean values obtained from stochastic simulations performed using the Gillespie algorithm (*see* Subheading 2.3.5). In panels (**b**), (**c**), (**e**) and (**f**), continuous lines describe analytical results obtained by the linear noise approximation (*see* Subheading 2.3.4) while markers represent numerical results derived from stochastic simulations. Parameter values are reported in Table 2

ceRNA levels characterizing the distinct regimes.) Figure 2a and d reports results obtained for the case $N = 2$, $M = 1$ (two ceRNA species competing for a single miRNA regulator). One sees that ceRNA levels get increasingly repressed as the miRNA transcription rate increases while all other parameters remain fixed (Fig. 2a). The range of values of β_1 where ceRNA levels change most strongly corresponds to the susceptible regime. One also sees that ceRNAs 1 and 2 have slightly different thresholds ($\mu_{0,11} \simeq 2$ and $\mu_{0,21} \simeq 15$), as ceRNA 1 is clearly sensitive to variations in miRNA availability for smaller values of β_1 compared to ceRNA 2. Figure 2d shows instead how molecular levels change upon modulating the transcription rate of ceRNA species 1. As b_1 increases, m_1 grows as expected while concentration of free miRNAs decreases as they increasingly engage targets. This in turn derepresses the other ceRNA species, whose level also increases as the transcription rate of ceRNA 1 is upregulated. That the level of ceRNA 2 can increase upon changing b_1 is the key signature of the miRNA-mediated crosstalk that can be established between competing RNAs.

Note: *The reference levels (Eq. 8) ultimately represent the combinations of parameters that are most relevant in order to elucidate many of the network's features. As one would expect, the leading behaviour for $\mu_{0,ia}$ is determined by the ratio d_i/k_{ia}^+: the threshold gets smaller as the*

Table 2
Values of kinetic parameters used in the different figures

Parameter	Figure 2A–C	2D–F	3A	3B	3C	3D	5	6A, B	6C, D
b_1 [molecule min^{-1}]	10	–	10	20	2	1	1 (mean)	–	10
b_2 [molecule min^{-1}]	15	10	15	10	10	10	1 (mean)	0	0
β_1 [molecule min^{-1}]	–	20	15	–	15	–	–	15	15
d_1 [min^{-1}]	0.1	0.1	0.1	0.1	0.1	0.1	0.005	0.1	0.1
d_2 [min^{-1}]	0.1	0.1	0.1	0.1	0.1	0.1	0.005	0	0.1
δ_1 [min^{-1}]	0.1	0.1	0.1	0.1	0.1	0.1	0.01	0.1	0.1
k_{11}^+ [molecule^{-1} min^{-1}]	e^{-3}	e^{-2}	1	e^{-2}	e^{15}	e^{15}	e^{-2}	Shown	Caption
k_{21}^+ [molecule^{-1} min^{-1}]	e^{-5}	e^{-4}	e^{-3}	e^{-3}	e^{-4}	e^{-4}	e^{-3}	0	Caption
k_{11}^- [min^{-1}]	0.001	0.001	0.001	0.001	0.001	0.001	0.1	0.001	0.001
k_{21}^- [min^{-1}]	0.001	0.001	0.001	0.001	0.001	0.001	0.1	0	0.001
κ_{11} [min^{-1}]	0.001	0.001	0.001	0.001	0.1	0.1	0.05	0.001	0.001
κ_{21} [min^{-1}]	0.001	0.001	0.001	0.001	0.001	0.001	0.05	0	0.001
σ_{11} [min^{-1}]	1	1	1	1	1	1	0.001	1	1
σ_{21} [min^{-1}]	1	1	1	1	1	1	0.001	0	1

miRNA–ceRNA interaction gets stronger (i.e. lower miRNA levels suffice to repress a target in presence of stronger coupling), whereas larger intrinsic ceRNA decay rates impose larger repression thresholds. Expectedly, catalytic decay rate affects the thresholds $\mu_{0,ia}$ and $m_{0,ia}$ differentially: while the former decreases as catalytic processing gets more efficient (i.e. miRNA recycling strengthens repression by effectively increasing miRNA availability), $m_{0,ia}$ increases as κ_{ia} gets larger (i.e. higher ceRNA levels are required to repress miRNAs at high catalytic processing rates). Note however that $m_{0,ia}$ diverges as $\sigma_{ia} \to 0$, i.e. when all miRNAs are recycled after complex degradation. In other words, in absence of stoichiometric processing of the i − a complex, miRNA a can never be repressed by ceRNA i. This implies that, in order for the ceRNA scenario described above to take place, it is necessary that the stoichiometricity ratio α_{ia}, Eq. 3, is strictly positive.

2.3 Stochastic Model

Like all regulatory processes [65], the individual reactions reported in Eq. 1, i.e. transcription, degradation and titration events due to miRNA–ceRNA interactions, are intrinsically stochastic. This means in practice that molecular levels evolving in time according to Eq. 1 are bound to be subject to random fluctuations, with the strength of the noise affecting each molecular species roughly proportional to the square root of its mean. After a transient, concentrations will stabilize and fluctuate around the steady state of the deterministic model (Eq. 2), described by Eq. 6. The deterministic model thereby yields a description of the miRNA-ceRNA network that is all the more accurate when the system is well mixed and concentrations are sufficiently large, making noise negligible. Besides giving a more realistic description of the dynamics of molecular populations, accounting for randomness is however crucial to characterize ceRNA crosstalk in detail, and particularly to disentangle competition-induced effects from fluctuation-induced ones. We shall now therefore briefly review some of the frameworks that have been employed to analyse the stochastic dynamics of Eq. 1.

2.3.1 The Master Equation

The direct mathematical route to account for stochasticity is based on the chemical Master Equation (ME) [66], which describes the time evolution of the probability $P(\mu, m, c, t)$ to find the system with prescribed molecular levels $\mathbf{m} = \{m_i\}_{i \in \{1,...,N\}}$ for ceRNAs, $\mu = \{\mu_a\}_{a \in \{1,...,M\}}$ for miRNAs and $\mathbf{c} = \{c_\ell\}_{\ell \in 1,...,M\cdot N}$ for the $N \cdot M$ species of miRNA-ceRNA complexes at time t. The ME reads

$$\frac{\partial P}{\partial t} = \sum_{a=1}^{M} \beta_a (P_{\mu_a - 1} - P) \qquad \emptyset \xrightarrow{\beta_a} \mu_a$$

$$+ \sum_{i=1}^{N} b_i (P_{m_i - 1} - P) \qquad \emptyset \xrightarrow{b_i} m_i$$

$$+ \sum_{a=1}^{M} \delta_a [(\mu_a + 1) P_{\mu_a + 1} - \mu_a P] \qquad \mu_a \xrightarrow{\delta_a} \emptyset$$

$$+ \sum_{i=1}^{N} d_i [(m_i + 1) P_{m_i + 1} - m_i P] \qquad m_i \xrightarrow{d_i} \emptyset$$

$$+ \sum_{i=1}^{N} \sum_{a=1}^{M} k_{ia}^+ [(\mu_a + 1)(m_i + 1)$$
$$\times P_{\mu_a + 1, m_i + 1, c_{ia} - 1} - \mu_a m_i P] \qquad \mu_a + m_i \xrightarrow{k_{ia}^+} c_{ia}$$

$$+ \sum_{i=1}^{N} \sum_{a=1}^{M} k_{ia}^- [(c_{ia} + 1)$$
$$\times P_{\mu_a - 1, m_i - 1, c_{ia} + 1} - c_{ia} P] \qquad c_{ia} \xrightarrow{k_{ia}^-} \mu_a + m_i$$

$$+ \sum_{i=1}^{N} \sum_{a=1}^{M} \sigma_{ia} [(c_{ia} + 1) P_{c_{ia} + 1} - c_{ia} P] \qquad c_{ia} \xrightarrow{\sigma_{ia}} \emptyset$$

$$+ \sum_{i=1}^{N} \sum_{a=1}^{M} \sum_{i=M+1}^{M+N} \kappa_{ia} [(c_{ia} + 1) P_{\mu_a - 1, c_{ia} + 1} - c_{ia} P] \qquad c_{ia} \xrightarrow{\kappa_{ia}} \mu_a$$

(9)

where we adopted for simplicity the compact notation $P_{x_i \pm 1} := P(x_1, \ldots, x_i \pm 1, \ldots, x_{N+M+NM})$. Equation 9 relies on the (unrealistic) hypothesis that chemical species live in a well-mixed environment without compartments, so that they are all in principle capable of interacting. An interesting and fundamental connection between the mass action kinetics in Eq. 2 and the ME is provided by the so-called *mean field approximation*, which assumes a simplified factorized form for the joint probability distribution P:

$$P(\{\mu_a\}, \{m_i\}, \{c_{ia}\}, t) = \prod_{i=1}^{N} P_i(m_i) \prod_{a=1}^{M} P_a(\mu_a) \prod_{\ell=1}^{N \cdot M} P_\ell(c_\ell) \qquad (10)$$

Plugging Eq. 10 into Eq. 9 and computing the mean value of all chemical species, one can see that the differential equation governing their time evolution coincides with Eq. 2. This point of view casts in a new perspective the deterministic mass action kinetics: as long as the correlations between the different variables can be neglected, the deterministic scheme is expected to provide an accurate description of the dynamics of the model. On the other hand, by construction, the deterministic mass action kinetic is blind to statistical correlations between variables. If one is interested in this aspect, Eq. 9 provides the correct theoretical framework.

Unfortunately, the ME is notoriously hard to handle analytically. Therefore, in the following, we will outline different approximation schemes that have been used to obtain useful indications about fluctuations and correlations between molecular levels.

2.3.2 Gaussian Approximation

The Gaussian approximation is probably the simplest one going beyond mean-field. The rationale of the method is rooted in Van Kampen's expansion [66], and specifically in the fact that, if molecules are assumed to be enclosed in a sufficiently large volume, the solution of the ME is Gaussian except for small corrections. Adopting the following vector notation already implicitly used in Eq. 10, i.e.

$$\mathbf{x} := \{x_1, \ldots x_M, x_{M+1}, \ldots, x_{M+N}, x_{M+N+1}, \ldots, x_{M+N+MN}\}$$
$$= \{\mu_1, \ldots, \mu_M, m_1, \ldots, m_N, c_{11}, \ldots, c_{NM}\}, \qquad (11)$$

the Gaussian approximation assumes that $\mathbf{x}$ is distributed as a multivariate Gaussian, namely

$$P(\mathbf{x}) \simeq G(\mathbf{x}|\mathbf{a}, \Sigma^{-1}) = \frac{\exp\left[-\frac{1}{2}(\mathbf{x}-\mathbf{a})^T \Sigma^{-1}(\mathbf{x}-\mathbf{a})\right]}{\sqrt{(2\pi)^{M+N+MN} \det(\Sigma)}} \qquad (12)$$

where the covariance matrix Σ has element $\Sigma_{ij} = E(x_i x_j) - E(x_i) E(x_j)$, the vector $\mathbf{a}$ has coordinates $a_i = E(x_i)$, and the expectation value $E(\cdot)$ is with respect to the Gaussian measure G defined in Eq. 12. One of the characteristics that make Gaussian distributions useful in this context lies in the property that all moments of a Gaussian measure can be expressed in terms of the mean $\mathbf{a}$ and the covariance matrix Σ, so that, for instance, the generic third and fourth order moments read $E(x_i x_j x_k) = \Sigma_{ij} a_k + \Sigma_{ik} a_j + \Sigma_{jk} a_i$ and $E(x_i x_j x_k x_l) = \Sigma_{ij}\Sigma_{kl} + \Sigma_{ik}\Sigma_{jl} + \Sigma_{il}\Sigma_{jk}$, respectively. In analogy with the closure of the system of equations in the first moments that the factorization hypothesis in Eq. 10 induces, a shrewd use of the moment generating function produces a closed system of equations for $\mathbf{a}$ and Σ. The natural formalism to impose this moment closure is that of the *moment-generating function*, defined as

$$F(\mathbf{z},t) = \sum_{\mathbf{x}} \prod_{i=1}^{N+M+N\cdot M} z_i^{x_i} P(\mathbf{x},t). \qquad (13)$$

It is simple to show that the time evolution of $F(\mathbf{z}, t)$ is ruled the second-order partial differential equation

$$\partial_t F(\mathbf{z},t) = \mathscr{H}(\mathbf{z}) F(\mathbf{z},t), \qquad (14)$$

where, for the miRNA-ceRNA network, the operator $\mathscr{H}$ is defined as

$$\begin{aligned}
\mathscr{H}(\mathbf{z}) =& \sum_{a=1}^{M} \beta_a (z_a - 1) + \sum_{i=M+1}^{M+N} b_i (z_i - 1) \\
&+ \sum_{a=1}^{M} \delta_a (\partial_{z_a} - z_a \partial_{z_a}) + \sum_{i=M+1}^{M+N} d_i (\partial_{z_i} - z_i \partial_{z_i}) \\
&+ \sum_{l=N+M+1}^{N+M+N\cdot M} \sigma_l (\partial_{z_l} - z_l \partial_{z_l}) \\
&+ \sum_{a=1}^{M} \sum_{i=M+1}^{M+N} k^{+}_{ia} (z_{ia} \partial^2_{z_i z_a} - z_i z_a \partial^2_{z_i z_a}) \\
&+ \sum_{a=1}^{M} \sum_{i=M+1}^{M+N} k^{-}_{ia} (z_i z_a \partial_{z_{ia}} - z_{ia} \partial_{z_{ia}}) \\
&+ \sum_{a=1}^{M} \sum_{i=M+1}^{M+N} \kappa_{ia} (z_i \partial_{z_{ia}} - z_{ia} \partial_{z_{ia}}).
\end{aligned} \qquad (15)$$

The moment-generating function F owes its name to the following constitutive property:

$$\partial^{l_1 + l_2 + \cdots + l_k}_{z_{i_1}^{l_1}, z_{i_2}^{l_2}, \ldots, z_{i_k}^{l_k}} F(\mathbf{z},t)|_{\mathbf{z}=1} = \left\langle x_{i_1}^{l_1} x_{i_2}^{l_2} \cdots x_{i_k}^{l_k} \right\rangle_{P(\mathbf{x},t)}. \qquad (16)$$

In other terms, consecutive derivatives of F generate all moments of the distribution P. The ME (Eq. 9) allows us to write a hierarchy of equations for the moments. However, it turns out that moments of order k are usually expressed in terms of moments of order $k + 1$, not allowing to close the system of equations for the moments. The Gaussian approximation truncates the hierarchy of moment dependencies by expressing third-order cumulants in terms of second-order ones (an approximation that turns out to be correct for Gaussian distributions). Thanks to this moment-closure approximation one ends up with a complete system of $N + M + N \cdot M + \binom{N+M+N\cdot M}{2}$ equations for the mean molecular levels and all covariances.

2.3.3 The Langevin Approach

A possibly more intuitive description of the stochastic dynamics is obtained by noting that, under broad conditions [66], one can effectively represent molecular fluctuations by adding specific noise terms to each of the factors appearing in the kinetic equations 2. This leads to a Langevin dynamics given by

$$\frac{dm_i}{dt} = b_i - d_i m_i + \underline{\xi_i} - \sum_a k_{ia}^+ m_i \mu_a + \sum_a \underline{\xi_{ia}^+} + \sum_a k_{ia}^- c_{ia} + \sum_a \underline{\xi_{ia}^-},$$

$$\frac{d\mu_a}{dt} = \beta_a - \delta_a \mu_a + \underline{\xi_a} - \sum_i k_{ia}^+ m_i \mu_a + \sum_i \underline{\xi_{ia}^+} + \sum_i (k_{ia}^- + \kappa_{ia}) c_{ia} + \sum_i (\underline{\xi_{ia}^- + \xi_{ia}^{\text{cat}}}),$$

$$\frac{dc_{ia}}{dt} = k_{ia}^+ m_i \mu_a + \underline{\xi_{ia}^+} - (\sigma_{ia} + \kappa_{ia} + k_{ia}^-) c_{ia} + (\underline{\xi_{ia}^{\text{st}} + \xi_{ia}^{\text{cat}} + \xi_{ia}^-}),$$

(17)

where the mutually independent stochastic 'forces' associated with each process have been inserted after the corresponding term and underlined. In specific,

- ξ_i and ξ_a represent the intrinsic noise due to random synthesis and degradation events that affect m_i and μ_a, respectively;
- ξ_{ia}^+ and ξ_{ia}^- model the noise affecting the random association and dissociation of complexes, respectively;
- ξ_{ia}^{cat} and ξ_{ia}^{st} represent the noise of catalytic and stoichiometric complex processing events, respectively.

Each of these noise terms has zero mean. Correlations are instead given by

$$\begin{aligned}
\langle \xi_i(t)\xi_i(t')\rangle &= (b_i + d_i \langle m_i\rangle)\,\delta(t-t'),\\
\langle \xi_a(t)\xi_a(t')\rangle &= (\beta_a + \delta_a \langle \mu_a\rangle)\,\delta(t-t'),\\
\langle \xi_{ia}^+(t)\xi_{ia}^+(t')\rangle &= k_{ia}^+ \langle m_i\rangle\langle \mu_a\rangle\,\delta(t-t'),\\
\langle \xi_{ia}^-(t)\xi_{ia}^-(t')\rangle &= k_{ia}^- \langle c_{ia}\rangle\,\delta(t-t'),\\
\langle \xi_i^{\text{cat}}(t)\xi_i^{\text{cat}}(t')\rangle &= \kappa_{ia}\langle c_{ia}\rangle\,\delta(t-t'),\\
\langle \xi_i^{\text{st}}(t)\xi_i^{\text{st}}(t')\rangle &= \sigma_{ia}\langle c_{ia}\rangle\,\delta(t-t'),
\end{aligned}$$

(18)

where steady-state abundances (in angular brackets) are given by the solutions of Eq. 6. The specific form (Eq. 18), involving steady-state values, can be derived within the so-called linear noise approximation (LNA, [66]), assuming that stationary molecular levels are sufficiently large [67]. As we show next, the LNA also provides direct access to the covariances of molecular levels.

2.3.4 Linear Noise Approximation

Denoting by **x** the vector of molecular levels of all species involved, i.e. $\mathbf{x} = (\{m_i\}, \{\mu_a\}, \{c_{ia}\})$, the stochastic dynamics (Eq. 17) can be written in vector notation as

$$\frac{d\mathbf{x}}{dt} = \mathbf{f}(\mathbf{x}) + \boldsymbol{\xi}, \qquad (19)$$

where the vector function **f** accounts for the deterministic terms in Eq. 17 while the vector noise $\boldsymbol{\xi}$ contains the overall noise affecting each component. The LNA is based on the assumption that, at stationarity, random fluctuations cause **x** to deviate from its steady-state value $\langle \mathbf{x} \rangle$ by a quantity $\delta \mathbf{x} = \mathbf{x} - \langle \mathbf{x} \rangle$ that is small enough to allow for the linearization of Eq. 19 around $\langle \mathbf{x} \rangle$. In such conditions, $\delta \mathbf{x}$ changes in time as [66]

$$\frac{d}{dt}\delta \mathbf{x} = \mathbf{S}\delta \mathbf{x} + \boldsymbol{\xi}, \qquad \mathbf{S} = \frac{d\mathbf{f}}{d\mathbf{x}}\bigg|_{\mathbf{x}=\langle \mathbf{x} \rangle}, \qquad (20)$$

where **S** is the stability matrix of first-order derivatives evaluated at the steady state. Assuming that $\boldsymbol{\xi}$ is a Gaussian noise with zero mean and cross-correlations described by a matrix $\boldsymbol{\Gamma}$, i.e. $\langle \xi_s(t)\xi_{s'}(t') \rangle = \Gamma_{ss'}\delta(t - t')$ (where the indices s and s' range over the components of **x**), one can show that the covariances of molecular levels at steady state obey [67]

$$\langle \delta x_a \delta x_b \rangle = -\sum_{i,l,s,r} B_{as}B_{br}\frac{\Gamma_{il}}{\lambda_s + \lambda_r}(B^{-1})_{si}(B^{-1})_{rl}, \qquad (21)$$

where λ denotes the vector of eigenvalues of the stability matrix, while **B** stands for its eigenvectors (i.e. $\Sigma_b S_{ab}B_{br} = \lambda_r B_{ar}$).

The above formula provides a way to estimate correlations (and hence Pearson coefficients) of all molecular species involved in the system. The continuous lines in Fig. 2b, c, e and f have indeed been obtained by the LNA.

2.3.5 The Gillespie Algorithm

The standard numerical route to simulate systems like Eq. 17 relies on the Gillespie algorithm (GA), a classical stochastic simulation method that computes the dynamics of a well-mixed system of molecular species interacting through a set of possible processes [68]. The GA allows to simulate the dynamics of systems like Eq. 9 without solving the ME, i.e. without the full knowledge of the probability $P(\mathbf{x}, t)$ of the system being in state vector **x** (encoding for the population of each molecular species) at time t. In short (see however [69] for a more detailed presentation), one can say that the GA essentially relies on two assumptions: (1) each process occurs with a specific rate constant; and (2) the current state of the system (in terms of the number of molecules of each species) determines which process is going to occur next, independently of the previous history. Under these conditions, one can simulate trajectories of a system described by a set of processes such as Eq. 9 simply from the knowledge of the probability density $P(k, \tau|\mathbf{x}, t)$ that process k takes place between time points $t + \tau$ and $t + \tau + d\tau$ given that the state of the system at time t is $\mathbf{x}$ (with no other processes occurring between time t and time $t + \tau$). Because the dynamics is memoryless, $P(k, \tau|\mathbf{x}, t)$ factorizes as

$$P(k,\tau|\mathbf{x},t)d\tau = \text{Prob}\{\text{no process between time } t \text{ and time } t+\tau\}$$
$$\times \text{Prob}\{\text{process } k \text{ between time } t+\tau \text{ and time } t+\tau+d\tau\}$$
$$\equiv P_0 \times P_k. \qquad (22)$$

The probability P_k is given by the intrinsic rate of process k (c_k) times a function of $\mathbf{x}$ ($g_k(\mathbf{x})$) that quantifies the number of different ways in which process k might occur and which basically encodes for the law of mass action. We shall use the shorthand $c_k g_k(\mathbf{x}) = f_k(\mathbf{x})$. Hence $P_k = f_k(\mathbf{x}(t + \tau))d\tau$.

P_0 can instead be evaluated by sub-dividing the interval $[t, t + \tau]$ in K parts ($K \gg 1$), each of duration τ/K. If f_k denotes the rate of process k, then P_0 is just the probability that no process occurs in any of the K sub-intervals, i.e.

$$P_0 = \left(1 - \sum_{k'} f_{k'} \frac{\tau}{K}\right)^K \simeq e^{-\tau \sum_{k'} f_{k'}} \quad (K \gg 1). \qquad (23)$$

Hence

$$P(k, \tau|\mathbf{x}, t) \simeq f_k \, e^{-\tau \sum_{k'} f_{k'}}, \qquad (24)$$

which can also be re-cast as

$$P(k, \tau|\mathbf{x}, t) \simeq \underbrace{\left(\sum_{k'} f_{k'}\right) e^{-\tau \sum_{k'} f_{k'}}}_{\text{prob. of waiting time } \tau} \times \underbrace{\frac{f_k}{\sum_{k'} f_{k'}}}_{\text{prob. of process } k}. \qquad (25)$$

A value of τ sampled from the above distribution of waiting times is easily obtained by noting that, if u denotes a random variable uniformly distributed in $[0, 1]$, then

$$\tau = -\frac{\ln(u)}{\sum_{k'} f_{k'}} \qquad (26)$$

is actually distributed according to the exponential function given in Eq. 25. This allows to formulate the GA in the following scheme:

> **Gillespie Algorithm**
> Step 1: Initialization: set initial populations for all molecular species (vector $\mathbf{x}(0)$) together with the rate c_k of each process k and an end-time T
> Step 2: Evaluate reaction probabilities f_k for each k as well as $\sum_k f_k \equiv Z$
> Step 3: Generate a pair (k, τ) from Eq. 25
> Step 4: Update molecular populations according to the selected process k and advance time by τ
> Step 5: Iterate from Step 2 or stop if the end-time T has been reached

Figure 2b, c, e and f shows how mean molecular levels obtained by the GA (markers) compare against analytic results (lines). One sees that the Fano Factor (FF) markedly peaks when molecular levels become roughly equimolar, i.e. close to the threshold where the system becomes susceptible to changes in the modulated parameter (in this case, the miRNA transcription rate or the transcription rate of ceRNA 1). The coefficient of variation (CV) also modifies its qualitative behaviour in the same range, although this feature generically appears to be less drastic (*see* however [60]). This shows that when ceRNAs become susceptible and cross-talk is established, fluctuations in molecular levels become strongly correlated.

The fluctuation scenario just described is clearly connected to the establishment of miRNA-mediated crosstalk. How exactly, and how it relates to other signatures of cross-talk, is the subject of the following section.

2.4 Quantifying miRNA-Mediated Crosstalk at Steady State

The competing endogenous RNA scenario concerns the possibility that, as a result of competition to bind miRNAs, ceRNAs could cross-regulate each other. We have so far identified two signatures that accompany the establishment of miRNA-mediated crosstalk at stationarity:

(a) a change in the steady-state level of a ceRNA following a change of the level of a competitor (i.e. a response following a perturbation);

(b) an increase of connected ceRNA-ceRNA correlations.

Both are clearly defined and testable in experiments and from data (at least in principle). Yet, despite the apparent simplicity, the reliable detection of the ceRNA mechanism in experiments or data is far from simple. The key issue lies in the fact that several mechanisms, both involving miRNAs and involving other molecular actors, potentially bear similar effects on transcripts and, as the cause differs, so do their consequences. Disentangling the competition-driven ceRNA effect from other processes is in many ways essential to be able to predict how a miRNA-ceRNA network will react to perturbations. We shall recap below how the ceRNA crosstalk scenario looks when seen through different glasses. While each allows to capture certain aspects of the ceRNA mechanism, different quantities employed to quantify crosstalk intensity focus on slightly different physical features and therefore can be useful in different situations. Understanding such differences is however crucial both for applications and for the unambiguous identification of biological drivers.

2.4.1 Pearson Correlation Coefficient

Since an increase of correlations between molecular levels accompanies the establishment of crosstalk, it is reasonable to view the Pearson correlation coefficient between two ceRNAs as a basic proxy for crosstalk intensity [52, 60, 61]. For ceRNAs i and j, it is defined as

$$\rho_{ij} = \frac{\langle m_i\, m_j \rangle - \langle m_i \rangle \langle m_j \rangle}{\sqrt{\langle m_i^2 \rangle - \langle m_i \rangle^2} \sqrt{\langle m_j^2 \rangle - \langle m_j \rangle^2}} \equiv \frac{\mathrm{cov}(m_i, m_j)}{\sqrt{\langle \delta m_i^2 \rangle} \sqrt{\langle \delta m_j^2 \rangle}}, \quad (27)$$

where averages are taken over random fluctuations in the steady state of a stochastic dynamics. (When the interaction network is conserved across different cellular samples and single snapshots of molecular levels are available for each sample, the $\langle \cdots \rangle$ average can also be taken over different samples, as long as each sample can be considered to be stationary.) Note that $-1 \leq \rho_{ij} \leq 1$.

The rationale for using Eq. 27 as a measure of crosstalk intensity is roughly the following. In a network of N ceRNA species interacting with M miRNA species, both ceRNA and miRNA levels will fluctuate stochastically over time at stationarity. A large positive value of ρ_{ij} points to the existence of a positive (linear) correlation between m_i and m_j, i.e. to the fact that $m_i \simeq cm_j + d$ + noise, with constants $c > 0$ and d. In such conditions, it is reasonable to expect that an increase in the level of ceRNA i, whichever its origin, will divert part of the miRNA population currently targeting ceRNA j to bind to i, thereby freeing up molecules of j for translation. In practice, with a large ρ_{ij}, perturbations affecting ceRNA i could be 'broadcast' to ceRNA j because of the miRNA-mediated statistical correlation existing between their respective levels.

The Pearson correlation coefficient between competing ceRNAs indeed attains a maximum in a specific range of values for the

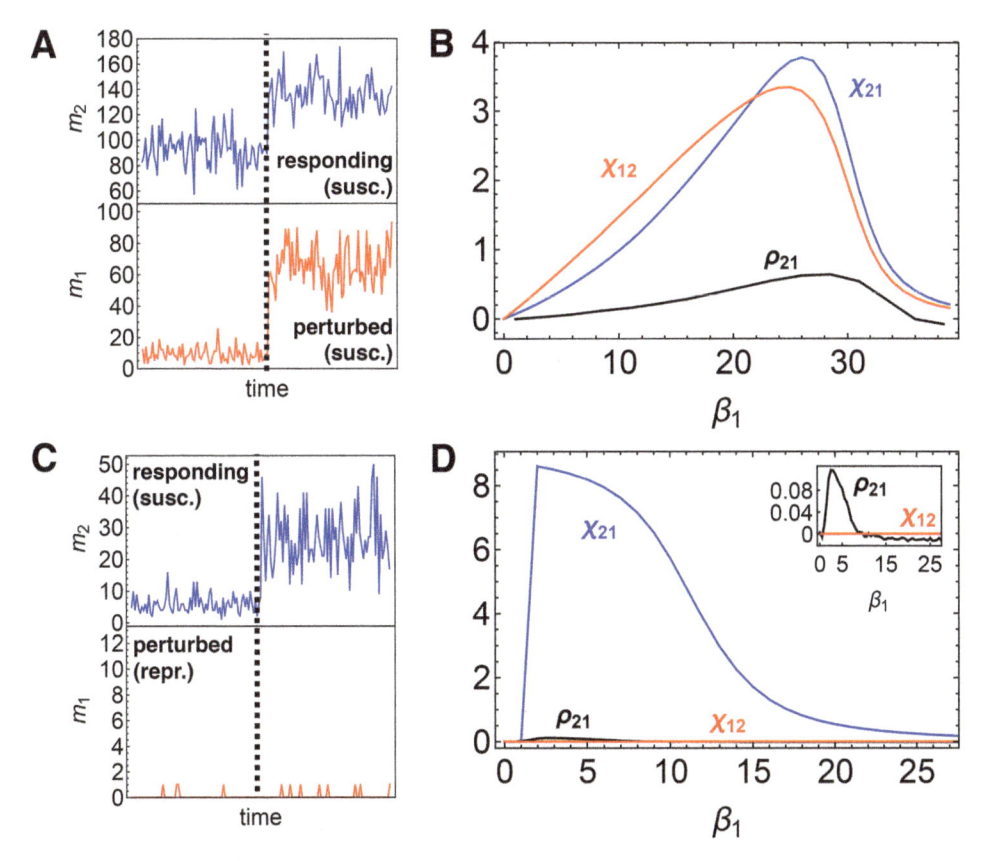

Fig. 3 (**a**) Stochastic simulation showing the free levels of two ceRNA species co-regulated by a miRNA species (not shown). Both ceRNAs are susceptible with respect to changes in the miRNA level. The transcription rate of ceRNA 1 is perturbed at the time indicated by the dashed line. ceRNA 2 responds by increasing its amount. (**b**) Susceptibilities and Pearson coefficients for two ceRNAs co-regulated by a miRNA species for moderate miRNA repression strength. All three quantifiers of ceRNA crosstalk are significantly different from zero and $\chi_{12} \simeq \chi_{21}$. (**c**) Same as (**a**) but now ceRNA 1 is fully repressed by the miRNA. Still, an increase of its transcription rate yields an upregulation of m_2. (**d**) Same as (**b**) but for strong miRNA repression on ceRNA 1. Both the Pearson coefficient ρ_{21} and χ_{12} (quantifying the response of ceRNA 1 to a perturbation affecting ceRNA 2) are effectively zero, whereas χ_{21} is not. Parameter values are given in Table 2

transcription rates, *see*, e.g., Fig. 3b. Expectedly, this happens when the levels of the different molecular species become comparable (or, more precisely, when the number of miRNA binding sites becomes similar to that of miRNA molecules) [60]. Here, ceRNA fluctuations become strongly correlated and one might expect ceRNA crosstalk to be active, so that a perturbation affecting one ceRNA will result in a shift in the level of its competitor. In other words, this regime is characterized by significant crosstalk effects.

2.4.2 Susceptibility

A mechanistic (as opposed to statistical) quantification of the magnitude of the ceRNA effect can be obtained by computing derivatives of steady-state ceRNA levels like [59]

$$\chi_{ij} = \frac{\partial \langle m_i \rangle}{\partial b_j} \geq 0$$
$$\chi_{ia} = \frac{\partial \langle m_i \rangle}{\partial \beta_a} \leq 0 \qquad (28)$$

where b_j (resp. β_a) stands for the transcription rate of ceRNA j (resp. miRNA a). We shall term quantities like Eq. 28 *susceptibilities*. In short, χ_{ij} measures the variation in the mean level of ceRNA i caused by a (small) change in b_j. As an increase of b_j leads to an increase of the level of ceRNA j by titration of miRNAs away from it, χ_{ij} is bound to be non-negative. A similar straightforward interpretation applies to χ_{ia}, which is non-positive since an increase of β_a is bound to cause a decrease of $\langle m_i \rangle$. The central hypothesis behind Eq. 28 is that small perturbations cause small changes in molecular levels, or, more precisely, that the latter will be proportional to the former if the perturbation is sufficiently small (linear response scenario).

Assuming no direct control of ceRNA i by ceRNA j, a large value of χ_{ij} directly points to the existence of miRNA-mediated crosstalk in terms of a change in the level of a target upon perturbing the level of a competitor. Hence χ_{ij} focuses on the response part of the ceRNA effect rather than on the fluctuation-related aspects.

Quantities like χ_{ij} can be directly computed from the steady-state conditions and in numerical simulations upon probing the system with the desired perturbation. A susceptibility-based theory of ceRNA crosstalk at steady state has indeed been presented in [59]. When quantified through χ_{ij}, ceRNA crosstalk displays the following key features:

Selectivity: When a miRNA targets multiple ceRNA species, crosstalk may occur only among a subset of them. This effect is related to the fact that different ceRNAs can have different thresholds for repression by the miRNA and is enhanced by heterogeneities in the thresholds;

Directionality (asymmetry): In general, $\chi_{ij} \neq \chi_{ji}$, i.e. ceRNA i may respond to a perturbation affecting ceRNA j but not the reverse;

Plasticity: The pattern of miRNA-mediated ceRNA crosstalk, whereby ceRNA j is linked to ceRNA i when χ_{ij} is sufficiently large, is modulated by kinetic parameters, and particularly by miRNA levels (in other words, changes in miRNA availability modify the ceRNA crosstalk network);

Dependency on stoichiometric processing: If all miRNA-ceRNA complexes formed by ceRNA j are degraded in a purely catalytic way, then $\chi_{ij} = 0$ (i.e. stoichiometric processing is necessary for ceRNA crosstalk at stationarity).

Like the Pearson coefficient ρ_{ij}, the ceRNA-ceRNA susceptibility χ_{ij} also peaks when ceRNA crosstalk is strongest (*see* Fig. 3b). However, the fact that susceptibilities are perturbation-specific makes their usefulness for data analysis and the interpretation of experiments less immediate compared to Pearson coefficients. Ideally, one would like to connect susceptibilities like Eq. 28 to simpler quantities like correlation functions. A more refined mathematical analysis of the stochastic dynamics shows that this is indeed possible.

2.4.3 Fluctuations Versus Response

It is important to understand the physical meaning and therefore the crosstalk scenarios underlined by ρ_{ij} and χ_{ij} are rather different. The fact that χ_{ij} is asymmetric under exchange of its indices (i.e. $\chi_{ij} \neq \chi_{ji}$ in general) whereas ρ_{ij} is necessarily symmetric already pointed in this direction. Other subtle differences however emerge when the two quantities are compared in greater detail.

In first place, χ_{ij} can be non-zero (and possibly large) even for a completely deterministic system like Eq. 2, as it simply measures how a target's steady-state level is modulated by changes affecting the transcription rate of one of its competitors, independently of the presence of stochastic fluctuations around the steady state. In this sense, χ_{ij} focuses exclusively on the effects induced by competition. On the other hand, in absence of fluctuations ρ_{ij} is identically zero.

Second, and related to this, is the fact that a large value of ρ_{ij} can occur when both ceRNAs respond to fluctuations in miRNA levels ('indirect correlation'). This however does not imply that m_i and m_j are directly correlated. (If variables X and Y are both correlated with Z, they will be correlated too. However, in absence of a direct correlation between X and Y, upon conditioning over the value of Z one will observe that X and Y are uncorrelated.) The same holds in presence of extrinsic noise, in which case averages are performed over different samples rather than over time in a single sample. To see this directly, one can consider a system formed by N ceRNA species (labeled $i, j, k, \ldots$) and M miRNA species (labeled $a, b, c, \ldots$) [59]. If transcription rates fluctuate across cells and if fluctuations are sufficiently small, ceRNA levels at steady state will be approximately given by

$$\langle m_i \rangle \simeq \overline{\langle m_i \rangle} + \sum_j \frac{\partial \langle m_i \rangle}{\partial b_j}(b_j - \overline{b_j}) + \sum_a \frac{\partial \langle m_i \rangle}{\partial \beta_a}(\beta_a - \overline{\beta_a})$$
$$\equiv \overline{\langle m_i \rangle} + \sum_j \chi_{ij}\delta b_j + \sum_a \chi_{ia}\delta\beta_a, \quad (29)$$

the over-bar denoting an average over transcription rates. Assuming that transcription rates of different species are mutually independent, the Pearson correlation coefficient ρ_{ij} can be seen to be given by

$$\rho_{ij} = A \left(\sum_k \chi_{ik}\chi_{jk}\overline{\delta b_k^2} + \sum_a \chi_{ia}\chi_{ja}\overline{\delta \beta_a^2} \right) \quad (30)$$

where $A > 0$ is a constant, the index k runs over ceRNAs, the index a runs over miRNAs and $\overline{\delta b_k^2}$ (resp. $\overline{\delta \beta_a^2}$) is the variance of the transcription rate of ceRNA species k (resp. miRNA species a). Now one sees that, if all ceRNA-ceRNA susceptibilities are zero (i.e. in absence of competition-induced crosstalk),

$$\rho_{ij} \propto \sum_a \chi_{ia}\chi_{ja}\overline{\delta \beta_a^2} . \quad (31)$$

Because ceRNAs always respond to fluctuations in miRNA levels, susceptibilities on the right-hand side are not zero. In particular, both χ_{ia} and χ_{ja} are negative, as an increase in miRNA levels causes a decrease in the level of free ceRNAs. One therefore concludes that $\rho_{ij} > 0$ even though all ceRNA-ceRNA susceptibilities are nil. This explicitly shows that χ_{ij} and ρ_{ij} describe a priori different crosstalk mechanisms.

A mathematical analysis of susceptibilities and fluctuations shows that crosstalk intensity ultimately depends on whether the involved ceRNAs are unrepressed, susceptible or repressed by miRNAs. In particular, it turns out that the ceRNA-ceRNA susceptibility χ_{ij} is qualitatively described by a matrix whose entries depend only on the state of repression of i (the responding ceRNA) and j (the perturbed one), given by [59]

		\multicolumn{3}{c}{j (perturbed)}
$\chi_{ij} =$	i (resp.)	Unrepr.	Susc.	Repr.
	Unrepr.	$\simeq 0$	$\simeq 0$	$\simeq 0$
	Susc.	$\simeq 0$	> 0	> 0
	Repr.	$\simeq 0$	$\simeq 0$	$\simeq 0$

(32)

Besides showing explicitly that $\chi_{ij} \neq \chi_{ji}$, the above matrix clarifies that a non-zero χ_{ij} (and therefore competition-driven response of i to a change in the transcription rate of j) occurs (1) symmetrically, when both ceRNAs are susceptible to the miRNA (as in Fig. 3a), and (2) asymmetrically, when the perturbed ceRNA is repressed while the responding one is susceptible (as in Fig. 3c). Along the same lines, one finds that [70]

$$\rho_{ij} = \begin{array}{c} \\ i \text{ (resp.)} \end{array} \begin{array}{|c|c|c|c|} \hline & \multicolumn{3}{c|}{j \text{ (perturbed)}} \\ \hline & \text{Unrepr.} & \text{Susc.} & \text{Repr.} \\ \hline \text{Unrepr.} & \simeq 0 & \simeq 0 & \simeq 0 \\ \hline \text{Susc.} & \simeq 0 & > 0 & \simeq 0 \\ \hline \text{Repr.} & \simeq 0 & \simeq 0 & \simeq 0 \\ \hline \end{array} \qquad (33)$$

i.e. the Pearson coefficient is expected to be significantly different from zero only when both ceRNAs are susceptible to changes in miRNA levels, as is clear by comparing Fig. 3b and d.

The quantitative relationship linking susceptibilities to fluctuations emerges through a more careful mathematical analysis of Eq. 17 based on approximating the stochastic variability affecting molecular levels with a thermal-like noise. This leads to a set of results closely related to the Fluctuation-Dissipation Relations that characterize the linear-response regime of multi-particle systems in statistical physics. Specifically, one finds that, under broad conditions, susceptibilities can be expressed in terms of covariances of molecular levels or functions thereof. In particular, in Ref. [70] it is shown that

$$\begin{aligned} \chi_{ij} &\equiv \frac{\partial \langle m_i \rangle}{\partial b_j} = \gamma \operatorname{cov}(m_i, \log m_j) \geq 0, \\ \omega_{ij} &\equiv \frac{\partial \langle m_i \rangle}{\partial d_j} = -\gamma \operatorname{cov}(m_i, m_j) \leq 0, \\ \chi_{ia} &\equiv \frac{\partial \langle m_i \rangle}{\partial \beta_a} = \gamma \operatorname{cov}(m_i, \log \mu_a) \leq 0, \\ \omega_{ia} &\equiv \frac{\partial \langle m_i \rangle}{\partial \delta_a} = -\gamma \operatorname{cov}(m_i, \mu_a) \geq 0, \end{aligned} \qquad (34)$$

where $\gamma > 0$ is a constant. In other terms, the response χ_{ij} of $\langle m_i \rangle$ to a perturbation affecting the transcription rate of ceRNA j is proportional to the covariance function $\operatorname{cov}(m_i, \log m_j)$ which incidentally, like χ_{ij}, is not symmetric under the exchange of i and j. Similarly, the bare ceRNA-ceRNA covariance $\operatorname{cov}(m_i, m_j)$ describes the response of $\langle m_i \rangle$ to (small) change of the intrinsic *degradation rate* d_j of ceRNA j. Importantly, by comparing Eq. 27 with ω_{ij}, Eq. 34, one sees that, perhaps unexpectedly, the Pearson coefficient ρ_{ij} is related to ω_{ij} (rather than χ_{ij}). (Likewise, one could calculate ceRNA-miRNA susceptibilities like χ_{ia} and ω_{ia} by evaluating bare covariances of ceRNA and miRNA levels as shown in Eq. 34.)

Generically, covariances are as easy to estimate from transcriptional data as Pearson coefficients, from which they only differ by the (crucial) normalization factor corresponding to the magnitude of fluctuations of individual variables. Relationships (Eq. 34) have been used to infer different features of ceRNA crosstalk network generated by the tumour suppressor gene PTEN from transcriptional data, in particular directionality [70]. The large-scale use of such quantities might provide detailed transcriptome-wide crosstalk patterns, open for analysis and further validation.

2.5 The Role of Network Topology

The topology of the miRNA-ceRNA provides an additional degree of freedom through which the effectiveness of ceRNA crosstalk can be influenced. To understand how, we assume that the miRNA-ceRNA network is sufficiently sparse and that connectivity correlations are absent. In such conditions, one can reasonably neglect ceRNA-ceRNA couplings involving more than one miRNA species and express the ceRNA-ceRNA susceptibility as

$$\chi_{ij} \simeq \sum_a \underbrace{\frac{\partial m_i}{\partial \mu_a} \frac{\partial \mu_a}{\partial b_j}}_{\chi_{ij,a}}. \qquad (35)$$

One sees that if $\chi_{ij,a}$, i.e. the ceRNA-ceRNA susceptibility mediated by miRNA species a, is roughly the same for all miRNA regulators shared by i and j, i.e. if $\chi_{ij,a} \simeq \chi_{ij}^{(0)}$ for all a, then $\chi_{ij} \simeq n_{ij}\chi_{ij}^{(0)}$, with n_{ij} the number of miRNA species that target both ceRNAs i and j. In other words, χ_{ij} increases with the number n_{ij} of miRNA species shared by i and j. This dependence can become especially significant in presence of strong degree correlations in the miRNA-ceRNA network, explaining why clustered networks such as those addressed in [60] generically lead to more intense crosstalk patterns than random networks.

The role of topology is however most clearly isolated when ingredients other than strictly topological ones are as homogeneous as possible. We therefore assume that

(a) all kinetic parameters are homogeneous (i.e. independent of the molecular species); in particular, $\mu_{0,ia} \equiv \mu_0$ for all miRNA-ceRNA pairs;

(b) miRNA levels are homogeneous, that is $\mu_a = \mu$ for each a.

Based on these, one can show that, when the number n_i (resp. n_j) of miRNAs targeting ceRNA i (resp. ceRNA j) is sufficiently large, each shared miRNA contributes a quantity [59]

$$\chi_{ij,a} \simeq \frac{1}{d} \frac{\widetilde{\mu}}{A + \sum_{k \in a} \frac{1}{1 + n_k \widetilde{\mu}}} \frac{1}{(1 + n_i \widetilde{\mu})^2 (1 + n_j \widetilde{\mu})} \qquad (36)$$

to the overall susceptibility Eq. 35, where $\widetilde{\mu} \equiv \mu/\mu_0$ is the miRNA level expressed in units of μ_0, $A > 0$ is a constant while $k \in a$ denotes the set of ceRNAs that interact with miRNA a. Hence $\chi_{ij,a}$

decreases (i.e. crosstalk intensity is diluted) as n_i increases, as n_j increases, and/or as the number of targets of miRNA a increases.

This suggests that a particularly intriguing scenario arises when a large number of miRNA species target i and j and when $\mu \ll \mu_0$, i.e. when all ceRNA species are unrepressed by miRNAs. For simplicity, we assume the miRNA–ceRNA interaction network to be a regular bipartite graph where each ceRNA interacts with $n_i = n$ miRNAs while each miRNA interacts with $\nu_\alpha = \nu$ ceRNAs. In this case, Eq. 36 takes the form

$$\chi_{ij,a} \simeq \frac{1}{d} \frac{\widetilde{\mu}}{\nu + A(1 + n\widetilde{\mu})} \frac{1}{(1 + n\widetilde{\mu})^2}. \qquad (37)$$

Now the value of $\chi_{ij,a}$ clearly depends on $\widetilde{\mu}$. In particular, one sees that

$$\chi_{ij,a} \begin{cases} \ll \dfrac{1}{dn} & \text{for } \mu \ll \mu_0/n \\ \simeq \dfrac{1}{dn} & \text{for } \mu \simeq \mu_0/n \\ \ll \dfrac{1}{dn} & \text{for } \mu \gg \mu_0/n \end{cases} \qquad (38)$$

In other terms, $\chi_{ij,a}$ is maximum when miRNA levels are close to μ_0/n, i.e. (for sufficiently large n) when each is well below the susceptibility threshold.

Formula 38 essentially reproduces the standard 3-regime scenario (unrepressed, susceptible, repressed) in a network context, albeit starting from the assumption that ceRNAs are unrepressed by each individual miRNA species. In this sense, it describes a 'distributed' effect: many weakly interacting miRNA species can collectively mediate efficient ceRNA crosstalk. Recalling Eq. 35, we see that when $\mu \simeq \mu_0/n$ the overall susceptibility is given by

$$\chi_{ij} \propto \frac{n_{ij}}{dn}, \qquad (39)$$

which becomes comparable to the self-susceptibility χ_{ii} for $n_{ij} \simeq n$. A sketch summarizing the results just described is shown in Fig. 4.

When connectivity correlations are not negligible and the approximation (Eq. 35) fails, χ_{ij} can in principle be expressed as

$$\chi_{ij} = \sum_{n \geq 0} \chi_{ij}^{(n)}, \qquad (40)$$

where $\chi_{ij}^{(n)}$ stands for the contribution to the $i - j$ susceptibility given by crosstalk interactions mediated by chains formed by n miRNA species. Starting from the steady-state conditions (Eq. 6), one can compute $\chi_{ij}^{(n)}$ exactly in the limit where the stoichiometricity ratio α_{ia} is the same for all pairs, i.e. $\alpha_{ia} = \alpha$ for each i and a, finding

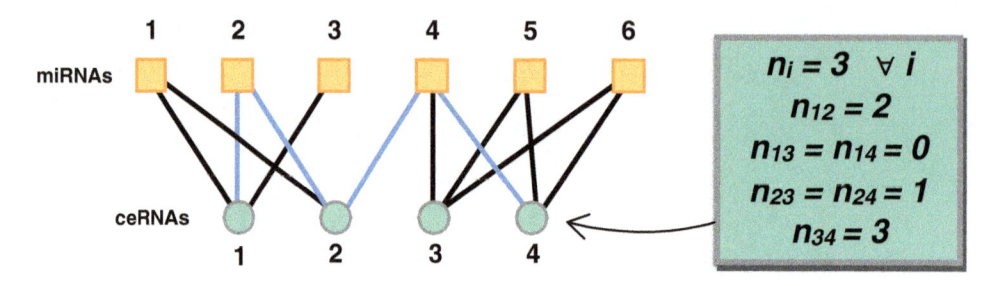

Fig. 4 Sketch of a miRNA-ceRNA network with $N = 4$ and $M = 6$. Each ceRNA species is regulated by 3 miRNA species, but ceRNA pairs (1, 2) and (3, 4) share more regulators than other pairs. Crosstalk between 1 and 2 and between 3 and 4 should therefore generically be stronger than for other ceRNA pairs. On the other hand, ceRNAs 1 and 4 don't have any regulator in common. Still, they may be able to crosstalk through the chain of miRNA-mediated interactions shown in light blue

$$\chi_{ij}^{(n)} = \frac{1}{d_i} \frac{\alpha^n}{1 + \sum_a \frac{\mu_a}{\mu_{0,ia}}} (\mathbf{X}^n)_{ij},$$

$$X_{ij} = \frac{m_i^\star}{\alpha \left(1 + \sum_a \frac{\mu_a}{\mu_{0,ia}}\right)^2} \sum_a A_{ai} A_{aj} \frac{\mu_a^\star}{\mu_{0,ia} m_{0,ja}} \left(1 + \sum_\ell \frac{m_\ell}{m_{0,\ell a}}\right)^2,$$

(41)

where $\mathbf{X}$ is the matrix with elements X_{ij} and $A_{ai} = 1$ if miRNA a targets ceRNA i and zero otherwise. Because $\alpha < 1$, one sees that the contribution coming from chains of n miRNA-mediated couplings becomes smaller and smaller (exponentially fast) as n increases. Equation 41 shows explicitly that ceRNAs i and j can crosstalk even when they have no miRNA regulator in common (in which case $X_{ij} = 0$), provided there is a path of miRNA-mediated interactions connecting them (as suggested, e.g., in [59, 71]; *see* also Fig. 4). Hence, clearly, the topological structure of the miRNA-ceRNA network can strongly influence the emergent crosstalk scenario. The discussion presented here does virtually nothing to address the ensuing complexity. A deeper understanding of the interplay between topological and kinetic heterogeneities might shed light on the evolutionary drivers of miRNA targeting patterns and of the ceRNA mechanism.

2.6 Noise Processing

2.6.1 Noise Buffering in Small Regulatory Motifs

Together with transcription factors (TFs), miRNAs form a highly interconnected network whose structure can be decomposed in small regulatory patterns or circuits. Few of them, hereafter call *motifs*, are overrepresented and thus expected to perform regulatory functions. In particular, it has been proven that all these miRNA-mediated motifs play some role in stabilizing the expression of the miRNA-target against fluctuations [72–76]. Amongst others, a special role is performed by feedforward loops involving one miRNA, one TF and one target. Both the miRNA and the TF can play the role of the master regulator, while the target is down-regulated by the miRNA and activated or inhibited by the TF. The incoherent version of this motif, where the TF activates the expression of miRNA and target, can couple fine-tuning of the target together with an efficient noise control [72, 76]. Intuitively, this can be understood by noting that fluctuations that propagate from TF to target and miRNA are correlated, so that an increase or decrease in the amount of miRNA will coincide with a decrease in the amount of target. The theoretical framework for the analysis of these effects is that of the ME, which in this case takes into account five different variables, one for each of the involved molecular species (mRNA and protein for the TF, mRNA and protein for the target, and the miRNA). The transcriptional activation of miRNA and target is modelled via a non-linear increasing Hill function of the number of TF, i.e.

$$b_m(f) = \frac{b_m f^c}{h_m^c + f^c}, \quad \beta_\mu(f) = \frac{\beta_\mu f^c}{h_\mu^c + f^c}, \qquad (42)$$

where b_m and β_μ are the transcription rates of target m and miRNA μ, respectively, c is the Hill coefficient setting the steepness of the sigmoidal function and h_m and h_μ are the dissociation constants, that specify the amount of TF proteins f at which the transcription rate is half of its maximal value (b_m and β_μ, respectively). The miRNA interaction can be either modelled via a repressive Hill function of the number of miRNA molecules, i.e. as $b_f(\mu) = \frac{b_p h^c}{h^c + \mu^c}$, or via a titration-based mechanism. In the Hill function, c is again the Hill coefficient and h set the amount of miRNAs necessary to halve the maximum target translation rate b_f. In the first case, one implicitly assumes that the miRNA action is catalytic (that is, the miRNA is never affected by the interaction with the target) and directs translational repression. In the second case, instead, one assumes that the miRNA action is stoichiometric, via binding and unbinding reactions (with rates $k_{m\mu}^+$ and $k_{m\mu}^-$, respectively). As long as miRNA and target mRNA are bound, the target cannot be translated. The miRNA might be affected by the interaction with the target (with recycling rate α) and the target itself has an effective degradation rate that depends on the binding and unbinding rates and that is bigger than its intrinsic value d_i. In this case the miRNA actively promotes the degradation of the target. It is possible to show analytically and by numerical simulations that the maximal noise attenuation for the target is obtained for a moderate miRNA repression, independently of the way the miRNA interaction is modelled [72]. This prediction, besides being in agreement with experimental observations of the impact of a wide class of microRNAs on their target proteins, also suggests that an optimal noise reduction might be achieved even when the miRNA repression is diluted over multiple targets, provided these ceRNAs are not too noisy.

The analysis of data from the *Encyclopedia of DNA Elements* (ENCODE) [77] revealed that two other classes of miRNA-mediated circuits are enriched over the mixed network of miRNAs and TFs. One of them has a miRNAs that regulates two different genes that can eventually dimerize; the second has a miRNA that interacts with two TFs which in turn regulate the same gene. In both cases, the miRNA seems to have a role in stabilizing the relative concentration of their targets. The interesting fact is that a further enrichment appears when looking for those circuits in which there is a transcriptional connection between the two miRNA targets, i.e. one of them is a TF of the other. This TF, together with the miRNA, can in turn regulate multiple targets. This motif is again a feedforward loop where the miRNA plays the role of the master regulator and the TF and targets are ceRNAs. When modelling the motif with a titrative interaction for miRNA and target, in line with Eq. 6, and with an activatory Hill function from the TF to the target, it becomes clear that the topology of the circuit, together with the ceRNA interaction, enhances the coordination of the targets [74]. This aspect is useful when TF and target have to maintain a fixed concentration ratio, which might be the case when they interact under a given stoichiometry.

2.6.2 Transcriptional Noise and the Role of Transcriptional Correlations

miRNA-mediated crosstalk can also provide a pathway to processing extrinsic noise, specifically cell-to-cell variability in transcription rates. Generalizing the lines that brought us to Eq. 29, one can say that if such a noise is sufficiently small, each component $\langle x_k \rangle$ of the steady-state concentration vector $\langle \mathbf{x} \rangle = (\{\langle m_i \rangle\}_{i=1}^{N}, \{\langle \mu_a \rangle\}_{a=1}^{M})$ can be written as

$$\langle x_k \rangle \simeq \overline{\langle x_k \rangle} + \sum_s \chi_{ks}(r_s - \overline{r}_s), \quad \chi_{ks} = \frac{\partial \langle x_k \rangle}{\partial r_s}, \qquad (43)$$

where $\overline{\langle \mathbf{x} \rangle}$ stands for the mean steady-state vector (averaged over transcriptional noise), r_s denotes the components of the vector $\mathbf{r} = (\{b_i\}_{i=1}^{N}, \{\beta_a\}_{a=1}^{M})$ of transcription rates (including both those relative to ceRNAs and miRNAs), and the sum runs over all ceRNA and miRNA species. In turn, transcriptional noise induces fluctuations in the level of molecular species k described by [59]

$$\sigma_k^2 \equiv \overline{(\langle x_k \rangle - \overline{\langle x_k \rangle})^2} = \sum_{s,s'} \chi_{ks} \chi_{ks'} \Sigma_{ss'}, \qquad (44)$$

where Σ denotes the covariance matrix of transcription rates. If Σ is diagonal, i.e. if transcription rates are mutually independent, the above expression reduces to

$$\sigma_k^2 = \sum_s \chi_{ks}^2 \, \Sigma_{ss}. \tag{45}$$

This means that, in absence of transcriptional correlations, each molecular species in the network (both ceRNAs and miRNAs) contributes a positive quantity to the overall level of noise affecting species k. In such conditions, the latter clearly exceeds the intrinsic noise level Σ_{kk}. In particular, large competition-driven susceptibilities (both to perturbations affecting ceRNAs and to perturbations affecting miRNAs) may cause σ_k^2 to be much larger than Σ_{kk}, eventually leading to a loss of resolution in molecular levels that will necessarily limit crosstalk effectiveness.

Interestingly, though, Eq. 44 suggests that the presence of transcriptional correlations (i.e. of off-diagonal terms in Σ) can compensate for this effect [59]. For instance, negative correlations between ceRNA transcription rates tend to reduce the overall noise level affecting ceRNA k with respect to the fully uncorrelated case (since both χ_{ks} and $\chi_{ks'}$ are non-negative if k, s and s' are ceRNAs). The same holds for positive correlations between the transcription rates of miRNAs and ceRNAs. In both cases, specific patterns of transcriptional correlations coupled with competition may confer a miRNA-ceRNA network the ability to buffer extrinsic noise. On the contrary, anti-correlated miRNA-ceRNA transcription rates or positively correlated ceRNA transcription rates tend to amplify extrinsic noise. These effects are displayed in Fig. 5, where we

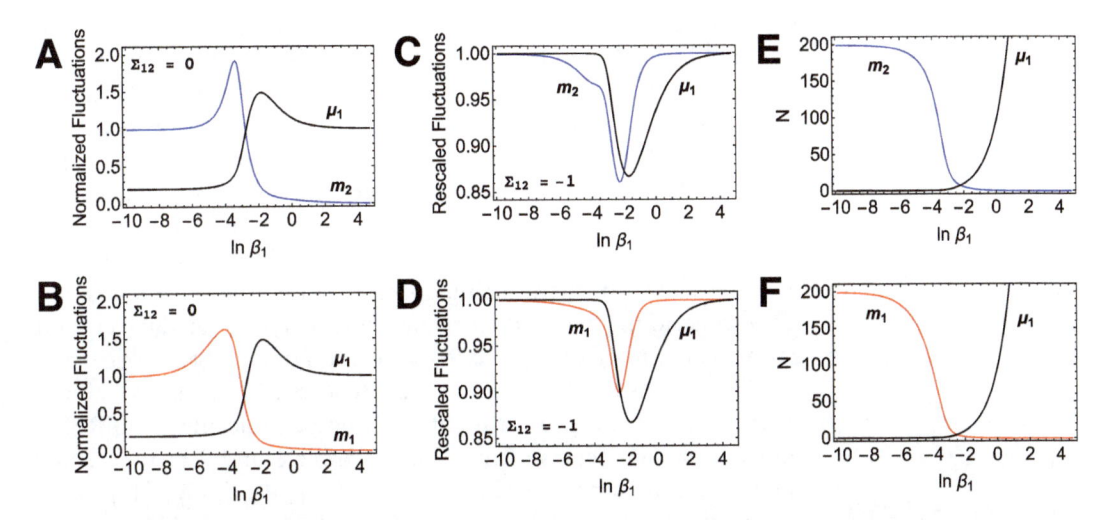

Fig. 5 (**a**, **b**) Ratio between the magnitude of fluctuations for each molecular species for an interacting ($k_{11}^+, k_{21}^+ > 0$) and a non-interacting ($k_{11}^+, k_{21}^+ = 0$) system with 2 ceRNAs and one miRNA ('normalized fluctuations') for uncorrelated ceRNA transcription rates ($\Sigma_{12} = 0$) as a function of the miRNA transcription rate. (**c**, **d**) Ratio between the normalized fluctuations obtained for maximally anti-correlated transcription rates ($\Sigma_{12} = -1$) and for the fully uncorrelated case ($\Sigma_{12} = 0$) as a function of the miRNA transcription rate. (**e**, **f**) Mean steady-state molecular levels as a function of the miRNA transcription rate. All results were obtained by averaging steady-state solutions over transcriptional noise. Parameter values are reported in Table 2

show the fluctuation picture arising when all transcription rates are Gaussian distributed, with a fixed ratio between the average and the width. Uncorrelated ceRNA transcription rates lead to an enhancement of fluctuations with respect to the case in which the miRNA is absent, while anti-correlated ceRNA transcription rates can attenuate this effect.

The noise-processing capacity of crosstalk patterns, and hence ultimately their effectiveness, is therefore strongly linked to the statistics of transcription rates. We shall see below that such correlations can indeed be exploited for the stabilization of the expression levels of protein complexes via the ceRNA mechanism.

2.6.3 Emergence of Bimodal Gene Expression

As shown above, one of the main properties of molecular sequestration is the possibility to obtain threshold responses and ultrasensitivity in absence of molecular cooperativity (a property found also when one or more genes are regulated by miRNAs). We also recalled that the system Eq. 2 possesses a unique, asymptotically stable steady state [64]. However, both theoretical and experimental studies have shown that miRNAs, in peculiar conditions of stoichiometry, induce bimodal distributions in the expression level of their targets [44, 60, 78]. As reviewed in [79] and shown in [80], some biological systems may present bimodality just as a consequence of stochasticity and despite being monostable at the deterministic level. The titrative interaction between miRNA and targets places targets, and ceRNAs in general, into this class of systems. Indeed, when the target expression level is around the threshold established by the amounts of miRNA, if the interaction is sufficiently strong, a small fluctuation in the amount of miRNA or target molecules makes the system jump from the repressed to the unrepressed regime and vice versa. The direct outcome is a bimodal distribution of the targets around the threshold, whose modes are related to the repressed and unrepressed regimes.

The constraint of strong miRNA–target interaction can however be relaxed by introducing some extrinsic noise in the system. This scenario has been exhaustively addressed, both analytically and numerically, in [81]. Let us focus on a simple system with two ceRNAs and one miRNA. The system is described by the probability distribution $P(\mu, m_1, m_2, t|\mathbf{K})$ of observing μ molecules of miRNAs and m_1, m_2 molecules of mRNAs of target 1 and 2 at time t, for a given set of parameters $\mathbf{K} = \{b_1, b_2, \beta, d_1, d_2, \delta, k_{1\mu}^+, k_{2\mu}^+\}$. Such a probability distribution evolves according to the ME (Eq. 9) with $N = 2$ and $M = 1$. Fluctuations in $\mathbf{K}$ should be taken into account in order to obtain the full distribution at the steady-state $P(\mu, m_1, m_2)$. For the sake of simplicity, now assume that β is the only fluctuating rate, drawn from a Gaussian distribution centred around $\langle \beta \rangle$, with variance σ_β^2 and defined for $\beta > 0$. We can obtain the steady-state probability distribution $P(\mu, m_1, m_2|\beta)$ conditional on a specific β by applying, e.g., the LNA or the Gaussian approximation to the ME. Once this is done, the joint distribution $P(\mu, m_1, m_2)$ is found by performing a weighted average over all possible values of β, i.e. by applying the law of total probability: $P(\mu, m_1, m_2) = \int P(\beta) P(\mu, m_1, m_2|\beta) d\beta$.

The presence of extrinsic noise in terms of fluctuating parameters is such that the miRNA transcription rate β is not the same for every cell as for the pure intrinsic noise case (indeed, we are extracting β from a Gaussian distribution). This implies that picking values of β above or below the threshold has the consequence of placing the system in the repressed or unrepressed regime, respectively. Again, the outcome is a bimodal distribution, which is this time at the population level. Then, the larger the variance σ_β^2 (i.e. the extrinsic noise), the broader the ranges of expressed target explored by the left-tails of the Gaussian distribution that will superimpose in the unrepressed mode. The right-tail instead will accumulate cells in the repressed mode. This makes the threshold/noise coupling an efficient tool to filter the variability introduced by extrinsic noise.

2.6.4 Impact on Protein Expression

The ability of generic regulatory systems to process noise is most crucial for the fine-tuning of protein levels [82]. Interestingly, the control exerted by miRNAs on a single target has been found to be capable of buffering its expression noise [83], especially for sufficiently low expression levels [27]. Given this scenario, one can ask whether the presence of a competitor would improve noise processing, especially at high expression, with the rationale that fluctuations affecting the target mRNA will be smaller (at fixed average) if a competitor titrates regulatory miRNAs away from it. This idea has been tested in simulations after modifying the basic model, Eq. 17, to account for protein production [84]. This is done by simply including the extra equation

$$\frac{dp_i}{dt} = g_i m_i - q_i p_i, \qquad (46)$$

which, for each mRNA species i, describes the time evolution of the level p_i of proteins of type i due to synthesis (occurring at rate g_i per substrate molecule) and degradation (occurring at rate q_i per protein). Fluctuations affecting p_i depend on the strength of the interaction between the miRNA and the target's competitor. A weak coupling is insufficient to draw miRNAs away from the target, leading (expectedly) to the same qualitative picture found in absence of the competitor. Likewise, very strong miRNA-competitor coupling leaves the target free from miRNAs, in which case its noise level is comparable to that attained in absence of miRNAs. However, for an intermediate value of the miRNA-competitor binding rate, titration by the competitor appears to be optimally tuned to reduce target fluctuations even at high

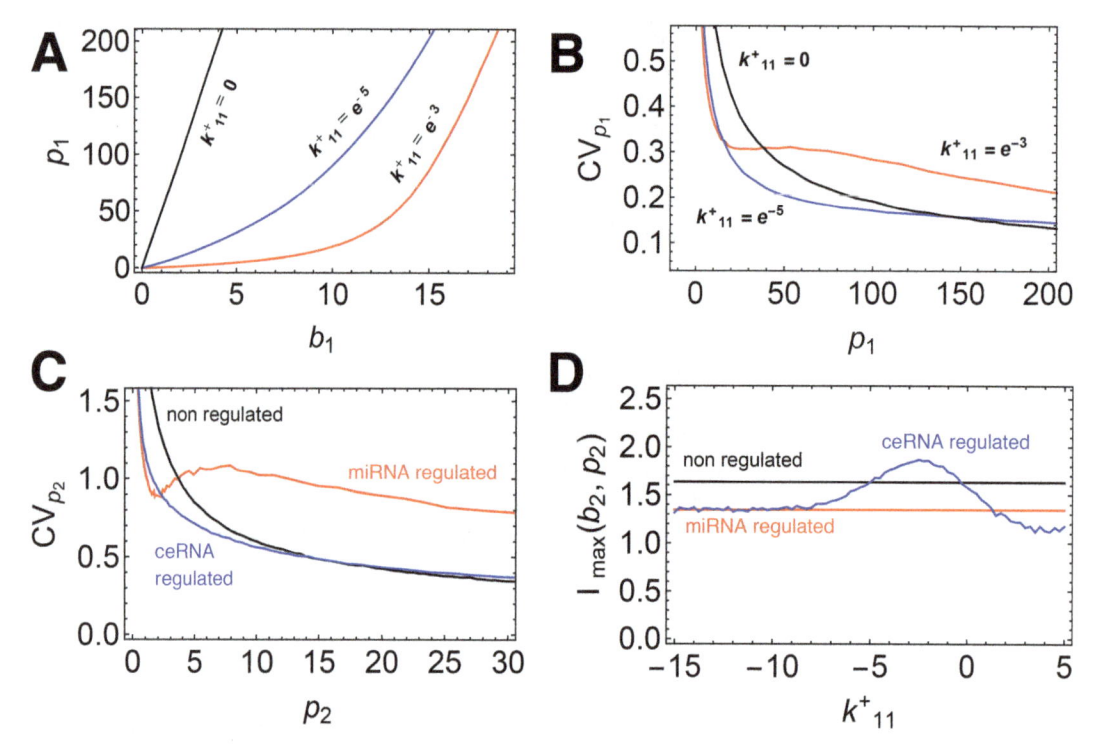

Fig. 6 (a) Mean level of a protein (p_1) interacting with a miRNA versus the transcription rate of its mRNA (b_1) for different values of the miRNA–mRNA interaction strength. No competitor is present. Expression of p_1 gets a stronger threshold-linear behaviour as the miRNA–mRNA interaction strength increases. (b) Relative fluctuations (CV) of p_1 versus b_1, again in absence of competition. (c) CV of a target protein (p_2) as a function of the mean protein level for the case in which the protein is not interacting with a miRNA (black line, $k_{11}^+ = k_{21}^- = 0$), is miRNA-regulated but has no competitor (red line, $k_{11}^+ = 0$, $k_{21}^+ = 1$) and is miRNA-regulated and has a competitor (blue line, $k_{11}^+ = e^{-2}$, $k_{21}^+ = 1$). (d) Maximal mutual information between p_2 and its transcription rate b_2 for the three regulatory modes presented in (c), plotted as a function of the miRNA–competitor interaction strength. The ceRNA-effect provides the most efficient fine-tuning pathway for intermediate strengths. We used $g_1 = 0.5/\text{min}$ and $q_1 = 0.1/\text{min}$ for panels (a) and (b); $g_1 = g_2 = 0.5/\text{min}$, $q_1 = q_2 = 0.1/\text{min}$ for panels (c) and (d). Remaining parameter values are reported in Table 2

expression levels (*see* Fig. 6). In this regime, the competitor is maximally derepressed. Remarkably, the overall behaviour of relative fluctuations is close to the Poissonian scenario obtained for an unregulated protein, implying that target derepression plays the main role in reducing fluctuations. Moreover, when crosstalk is most efficient, noise at low expression levels is still efficiently buffered with respect to the case in which miRNAs are absent. A more refined analysis shows that miRNA recycling generically provides enhanced fine-tuning by increasing the effective miRNA level.

The fact that, in the human PPI network, the functional products of mRNAs targeted by the same miRNAs are more strongly connected than would be expected by chance strongly suggests that miRNA-mediated regulation, and by extension the ceRNA mechanisms, might play a role in the regulation of protein complex levels [85–87]. In particular, protein forming the subunits of larger complexes tend to be regulated by miRNA clusters, i.e. by groups of miRNA species that are co-expressed [88]. When competing RNAs are the substrate for the synthesis of interacting proteins, the onset of the ceRNA mechanism modifies the correlation pattern of the two sub-units, specifically changing the sign of correlations from negative (corresponding to sub-units that are not co-regulated) to positive (reflective the positive correlation that is established between ceRNAs in crosstalk conditions). Such a modification has been observed experimentally [89–91], suggesting that it might provide a biological (albeit non-universal) signature of the ceRNA effect in action.

2.6.5 Limits to Crosstalk Effectiveness

From the previous discussion it is clear that the effectiveness of the ceRNA mechanism is dictated in large part by the relative levels of the molecular species involved and is ultimately limited by noise. An important question in this respect is whether one can characterize the optimal performance that miRNA-mediated regulation can achieve in controlling gene expression. In general, the optimal properties achievable by a regulatory circuit describe fundamental physical limits to its performance, which cannot be overcome independently of kinetic details, and point to the individual processes constituting, in some sense, the bottlenecks for regulatory effectiveness. It is clear that this requires, on the one hand, a quantitative definition of 'regulatory effectiveness' and, on the other, a benchmark. To fix ideas, one can focus on the system formed by a single miRNA connecting two competing RNAs. Following [92], a natural definition for the effectiveness of ceRNA crosstalk is represented by the degree to which one can control the level of one of the ceRNAs, say ceRNA i, by modulating the level of its competitor (ceRNA j). In a stochastic setting, the miRNA-mediated interaction linking i and j can be seen as a 'communication channel' that probabilistically translates the transcription rate of j into a value of m_i. This channel is fully described by the conditional probability density $p(m_i|b_j)$, returning a random value of m_i (the contributing noise coming from all involved processes) upon presenting input b_j. In turn, miRNA-mediated regulation consists in processing, via $p(m_i|b_j)$ a distribution of values for b_j (denoted by $p(b_j)$) into a distribution of values of m_i. For any given $p(m_i|b_j)$ and $p(b_j)$, the strength of the mutual dependence between these variables is quantified by the mutual information [63]

$$I(b_j, m_i) = \int_{b_j^{\min}}^{b_j^{\max}} db_j\, p(b_j) \int_{m_i^{\min}}^{m_i^{\max}} dm_i\, p(m_i|b_j) \log_2 \frac{p(m_i|b_j)}{p(m_i)}, \tag{47}$$

with $p(m_i) = \int_{b_j^{\min}}^{b_j^{\max}} db_j p(m_i|b_j) p(b_j)$ the output distribution of m_i. Assuming that the channel is fixed, i.e. that $p(m_i|b_j)$ is given, the optimal regulatory effectiveness is obtained when the input distribution $p(b_j)$ is such that I is maximized:

$$\max_{p(b_j)} I(b_j, m_i) \equiv I_{\max}. \tag{48}$$

$I_{\max}$ is called the *capacity* in information-theoretic terms and ultimately measures how much information (in bits) can be conveyed at most from input (b_j) to output (m_i) by a given input–output relationship $p(m_i|b_j)$. In loose but intuitive terms, $I_{\max}$ describes the number N of different values m_i that can be distinguished in a reliable way given the noise, which is roughly given by $N \sim 2^{I_{\max}}$. If $I_{\max} \simeq 0$ (note that $I \geq 0$ by definition), the noise only allows to distinguish at most one level of m_i; for $I_{\max} \simeq 1$ two levels (high/low) can be separated; and so on.

The effectiveness of miRNA-mediated crosstalk has been characterized within the above setup starting from numerical simulations of the stochastic dynamics and using direct transcriptional regulation of m_i (i.e. the capacity of the corresponding miRNA-independent regulatory channel) as the benchmark against which miRNA-mediated information flow was evaluated. In particular, the dependence of $I_{\max}$ on kinetic parameters was analysed to identify optimal parameter regions and limiting processes. The emergent scenario can be summarized as follows [63]:

1. As might have been expected, the capacity of miRNA-mediated regulation is optimal in a specific range of values for the target's repression strength. Intuitively, a tight control of m_i based on b_j requires ceRNA i to be sensitive to changes in miRNA levels. Too weak (resp. too strong) repression causes ceRNA i to become fully unrepressed (resp. fully repressed), so that the optimal range lies between these extremes. Quite remarkably, though, optimal ceRNA crosstalk can be more effective than direct transcriptional control.

2. In presence of significantly different catalytic degradation rates (faster for m_i, slower for m_j) ceRNA crosstalk outperforms direct transcriptional regulation. Intuitively, the above situation makes transcriptional control especially inefficient since m_i is going to be strongly repressed by miRNAs. miRNA-mediated control, instead, benefits from the fact that m_j can de-repress ceRNA i by lifting miRNAs away from it.

3. When miRNA populations are sufficiently large and miRNA-ceRNA couplings are weak, miRNA-mediated regulation is as effective as a direct transcriptional control. This is intuitively due to the fact that, in this limit, the relative noise affecting miRNA levels becomes negligible. This removes the additional source of noise affecting the post-transcriptional channel compared to the transcriptional one, effectively making the two regulatory modes comparable.

The outlook is that, besides generically contributing to noise buffering, ceRNA crosstalk can control gene expression to a degree that is tightly connected to the ability of the competitor (m_j) to de-repress the target (m_i). When the controller's kinetics does not suffice to titrate miRNAs away from ceRNA i, miRNA-mediated regulation is ineffective. Otherwise, it provides a high (and, possibly, the highest achievable) degree of control over expression levels, especially when kinetic parameters are sufficiently heterogeneous.

2.7 ceRNA Crosstalk Away from Stationarity

2.7.1 Equilibration Times

The titrative miRNA–target interaction entails both susceptibility and statistical correlation between the competing chemical species. We have seen before how all these effects become maximal at quasi-equimolar ratio. One can however also study how fast the system responds to an external perturbation. To fix ideas, we will focus as usual on the case of a single miRNA targeting 2 ceRNAs. In particular, we want to quantify the time needed for a particular ceRNA (here ceRNA$_1$) to reach the new stationary state after

- A sudden increase of the transcriptional activity of ceRNA$_2$ at time $t = 0$, i.e.

$$b_2(t = 0^-) = 0 \quad \text{and} \quad b_2(t = 0^+) = b^*$$

- A sudden decrease of the transcriptional activity of ceRNA$_2$ at time $t = 0$, i.e.

$$b_2(t = 0^-) = b^* \quad \text{and} \quad b_2(t = 0^+) = 0$$

We define the response time as the time needed for ceRNA$_1$ to reach half the way between the initial (before perturbation) and final (after perturbation) steady-state levels. In particular one can evaluate the response times T_{ON} and T_{OFF} for both the switch-on and switch-off scenarios (i.e. for ceRNA$_2$ OFF $\rightarrow$ ON and ON $\rightarrow$ OFF, respectively) by numerically integrating Eq. 2 to estimate $T_{\text{ON/OFF}}$ as the times at which the following relations hold:

$$m_1(T_{\text{ON}}) = m_1(0) + \frac{1}{2}\left(\lim_{t\to\infty} m_1(t) - m_1(0)\right) \quad (49)$$
$$m_1(0) = m_1^{\text{ss}}, \quad m_2(0) = 0$$

$$m_1(T_{\text{OFF}}) = m_1(0) - \frac{1}{2}\left(m_1(0) - \lim_{t\to\infty} m_1(t)\right) \quad (50)$$
$$m_1(0) = m_1^{\text{ss}}, \quad m_2(0) = m_2^{\text{ss}}$$

In this framework we can easily study the dependence of the response times $T_{\text{ON/OFF}}$ on the basal miRNA concentration (i.e. on β_1 in this case). Results (*see* Fig. 7) show a non-monotonous dependence of $T_{\text{ON/OFF}}$ on the transcriptional activity of the miRNA. In particular, T_{ON} (resp. T_{OFF}) displays a maximum (resp. a minimum) in correspondence with the threshold between the repressed and unrepressed phase.

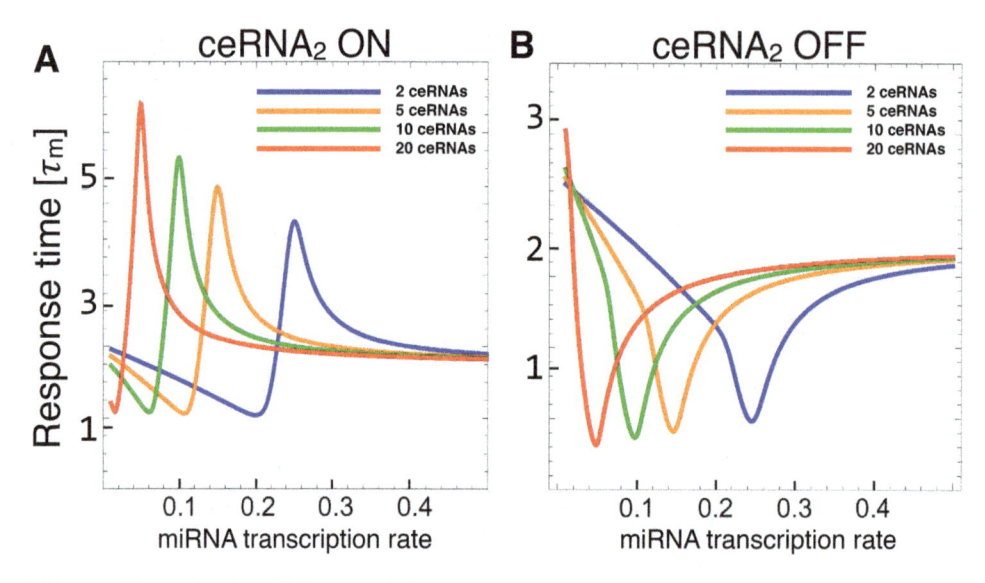

Fig. 7 We consider a simple ceRNA network for an increasing number of targets ranging from 2 to 20 and a single microRNA. (**a**) Equilibration time of ceRNA$_1$ when ceRNA$_2$ is induced, as a function of the miRNA transcription rate. Around threshold, we observe a critical slowing down in the response time. (**b**) Same as (**a**), but now the response time is measured after a knock-down of one of the competitors. In this case, we observe a speed-up of the response time at threshold

A natural question is how the presence of more ceRNAs changes the scenario we just described for the simple one miRNA two ceRNAs network. The same in silico experiment can be generalized to an arbitrary number of ceRNAs where all but one (say ceRNA$_2$) is either knock-out or induced. Perhaps unsurprisingly (*see* Fig. 7), one again sees a dilution effect: upon increasing the number of ceRNAs from 2 to 20 the relevance of the effect—measured in terms of the distance between the initial and final state of the system—becomes quantitatively less relevant.

2.7.2 Out-of-Equilibrium Dynamics

The out-of-equilibrium dynamics of the miRNA-ceRNA system has been studied in [93]. The emergent crosstalk scenario is substantially richer than the stationary one. For simplicity, we shall limit ourselves to describing results obtained for a system with N ceRNAs interacting with and a single miRNA species. From a physical viewpoint, the quantities

$$\tau_0 = \delta^{-1}, \quad \tau_{1,i} = d_i^{-1}, \quad \tau_{2,i} = (\sigma_i + \kappa_i + k_i^-)^{-1}$$
$$\tau_{3,i} = (\sigma_i + \kappa_i)^{-1}, \quad \tau_{4,i} = \sigma_i^{-1}, \quad \tau_{5,i} = \kappa_i^{-1} \tag{51}$$

represent the relevant characteristic intrinsic time scales of this system. Based on Eq. 2, they represent, respectively, the mean lifetime of miRNA species a (τ_0) and of ceRNA species i ($\tau_{1,i}$), the mean lifetime of the complex formed by ceRNA i ($\tau_{2,i}$), and the mean time required for complex degradation ($\tau_{3,i}$), stoichiometric complex degradation ($\tau_{4,i}$) and catalytic complex degradation ($\tau_{5,i}$). The features characterizing dynamical crosstalk can change depending on how these time scales are related. To get some insight, one can focus on how the system relaxes back to the steady state following a small perturbation away from it. Upon linearizing the system (Eq. 2), one can derive equations for the deviations of each molecular species from the steady state, i.e. for the quantities

$$x_i(t) \equiv m_i(t) - \langle m_i \rangle$$
$$y(t) \equiv \mu(t) - \langle \mu \rangle. \quad (52)$$
$$z_i(t) \equiv c_i(t) - \langle c_i \rangle$$

(We have suppressed the miRNA index for the sake of simplicity.) Introducing (small) time-dependent additive perturbations of the transcription rates of the form $b_i^o(t)$ and $\beta_o(t)$, the above variables can be seen to evolve in time according to

$$\frac{d}{dt}x_i = -d_i x_i + b_i^o - k_i^+(\mu\, x_i + m_i\, y) + k_i^- z_i$$
$$\frac{d}{dt}y = -\delta y + \beta_o - \sum_i k_i^+(\mu\, x_i + m_i\, y) + \sum_i (k_i^- + \kappa_i) z_i, \quad (53)$$
$$\frac{d}{dt}z_i = -(\sigma_i + k_i^- + \kappa_i) z_i + k_i^+(\mu\, x_i + m_i\, y)$$

This system can be analysed in the frequency domain (ω) by Fourier-transforming (Eq. 53). This allows to define the *dynamical susceptibility*

$$\chi_{ij}(\omega) = \frac{\partial \hat{x}_i}{\partial \hat{b}_j^o}, \quad (54)$$

where $\hat{f}$ is the Fourier transform of f. The general study of this quantity is possible while not straightforward [94]. However, $\chi_{ij}(\omega)$ can be estimated in a relatively simple way in few instructive limiting cases in which timescales are sufficiently separated. For instance, when $\tau_{3,j} \ll 1/k_j^-$ and $\tau_{1,j} < \tau_{5,j}$, complexes formed by ceRNA j will typically keep miRNAs blocked for times longer than the intrinsic ceRNA degradation timescale. This may allow for ceRNA i to get de-repressed and hence for the establishment of crosstalk, independently of whether stoichiometric processing takes place. Indeed one finds that, when $\kappa_j \ll \omega \ll d_j$ (i.e. for timescales intermediate between $\tau_{1,j}$ and $\tau_{5,j}$),

$$\chi_{ij}(\omega) \simeq \frac{\sigma_j + \kappa_j}{\sigma_j} \chi_{ij} \quad (55)$$

where $\chi_{ij} = \frac{\partial \langle m_i \rangle}{\partial b_j}$ stands for the steady-state susceptibility [93]. Remarkably, the quantity on the left-hand-side of Eq. 55 can be shown to remain finite for $\sigma_j \to 0$, providing quantitative support to the observation that ceRNA crosstalk can be active dynamically even in purely catalytic systems (where no crosstalk occurs at stationarity and χ_{ij} vanishes). In other words, then, in this limiting case and in an intermediate frequency window, the dynamical susceptibility is comparable to the steady-state value and occurs even for $\sigma_j = 0$. Away from this window, instead, crosstalk in this limit is weaker than it is at stationarity.

A more careful analysis shows that, in certain regimes, the dynamical response can even exceed the stationary one. This happens, for instance, when complex dissociation is much faster than other processing pathways and ceRNAs are fully repressed, implying that dynamical crosstalk can occur even between pairs of ceRNAs that could not interact at steady state [93]. In this sense, the ceRNA mechanism out of equilibrium is substantially more complex and richer than its stationary counterpart. In addition, the possibility to modulate the time scales of different interactions allows to construct systems in which static and dynamic responses are tuned so as to ensure the correct transient activation of a specific gene and the long-term stabilization of expression levels. An example of such a coordination, based on findings related to skeletal muscle cell differentiation [95], has been studied in [96].

3 Outlook

Mathematical models developed to elucidate the emergent features of ceRNA crosstalk have so far mainly relied on computational schemes for stochastic simulations (Gillespie algorithm) and on analytical approximations of the master equation associated with the system of interacting molecules (LNA, Gaussian, Langevin). On the other hand, a full understanding of competition-driven coupling requires, as we have seen, disentangling it from concurrent effects. Indeed, the identification of crosstalk from transcriptional data is in our view especially hard since statistical correlations between RNAs sharing a common miRNA regulator can arise just due to the fact that they both respond to fluctuating miRNA levels. Once the relationship between competition- and fluctuations-related features is clarified, ceRNA crosstalk patterns display strong intrinsic specificities like

1. selectivity,
2. asymmetry,
3. plasticity (i.e. sensitivity to kinetic parameters),
4. sensitivity to the degree of parameter heterogeneity, and
5. the possibility to aggregate a large number of weak interactions to significantly impact molecular levels.

These features in turn allow for the establishment of complex noise-processing properties. Note that, unsurprisingly, some of these features characterize other competition scenarios in regulatory system (e.g. competition to bind transcription factors, σ-factors, ribosomes, etc. [97–99]).

We have reviewed these aspects together with the methods that can be employed to quantify them. Several important points might however deserve equal consideration. In first place, miRNAs can also crosstalk through ceRNAs, generating a very similar phenomenology whose impact has been, to the best of our knowledge, far less clarified [100]. Secondly, the modelling framework we discussed ignores some kinetic steps assuming essentially that they are non rate-limiting. Still, it is known that in some cases binding to Argonaute (Ago), the catalytic component of the RNA-induced silencing complex, represents a kinetic bottleneck [101]. Likewise, crosstalk can be affected by the competition to bind Ago [100]. Third, a rich trafficking of miRNAs and their targets is known to occur between the cell nucleus and the cytoplasm, leading to remarkable localization effects whose biological significance is largely unexplored [102]. Well-mixed models like those discussed here are clearly unable to deal with such effects; spatial generalizations are mandatory [103, 104]. Finally, the phenomenology derived from small modules can integrate in highly non-trivial ways at the scale of the transcriptome, where topology provides additional degrees of freedom to modulate crosstalk patterns. While, as shown here, some (basic) things about the role of network structure can be understood with simple calculations, a more thorough data-based analysis of these aspects would be greatly welcome.

Acknowledgements

Work supported by the European Union's Horizon 2020 research and innovation programme MSCA-RISE-2016 under grant agreement No 734439 INFERNET. We are indebted with Matteo Figliuzzi, Enzo Marinari, Matteo Marsili and Riccardo Zecchina for our fruitful and enjoyable collaboration.

References

1. Bartel DP (2009) MicroRNAs: target recognition and regulatory functions. Cell 136:215–233
2. Flynt AS, Lai EC (2008) Biological principles of microRNA-mediated regulation: shared themes amid diversity. Nat Rev Genet 9:831
3. Cech TR, Steitz JA (2014) The noncoding RNA revolution–trashing old rules to forge new ones. Cell 157:77–94
4. Gurtan AM, Sharp PA (2013) The role of miRNAs in regulating gene expression networks. J Mol Biol 425:3582–3600

5. Bartel DP (2018) Metazoan microRNAs. Cell 173:20–51
6. Gregory RI, Chendrimada TP, Cooch N, Shiekhattar R (2005) Human RISC couples microRNA biogenesis and posttranscriptional gene silencing. Cell 123:631–640
7. Chandradoss SD, Schirle NT, Szczepaniak M, MacRae IJ, Joo C (2015) A dynamic search process underlies microRNA targeting. Cell 162:96–107
8. Klein M, Chandradoss SD, Depken M, Joo C (2017) Why Argonaute is needed to make microRNA target search fast and reliable. In: Seminars in cell & developmental biology, vol. 65. Academic, New York, pp 20–28
9. Chekulaeva M, Filipowicz W (2009) Mechanisms of miRNA-mediated post-transcriptional regulation in animal cells. Curr Opin Cell Biol 21:452–60
10. Jonas S, Izaurralde E (2015) Towards a molecular understanding of microRNA-mediated gene silencing. Nat Rev Genet 16:421
11. Djuranovic S, Nahvi A, Green R (2012) miRNA-mediated gene silencing by translational repression followed by mRNA deadenylation and decay. Science 336:237–240
12. Bartel DP (2004) MicroRNAs: genomics, biogenesis, mechanism, and function. Cell 116:281–297
13. Liang Y, Ridzon D, Wong L, Chen C (2007) Characterization of microRNA expression profiles in normal human tissues. BMC Genomics 8:166
14. Franks A, Airoldi E, Slavov N (2017) Post-transcriptional regulation across human tissues. PLoS Comput Biol 13:e1005535
15. Ebert MS, Sharp PA (2012) Roles for microRNAs in conferring robustness to biological processes. Cell 149:515–524
16. Berezikov E (2011) Evolution of microRNA diversity and regulation in animals. Nat Rev Genet 12:846
17. Joshi A, Beck Y, Michoel T (2012) Post-transcriptional regulatory networks play a key role in noise reduction that is conserved from micro-organisms to mammals. FEBS J 279:3501–3512
18. Friedman RC, Farh KK, Burge CB, Bartel DP (2009) Most mammalian mRNAs are conserved targets of microRNAs. Genome Res 19:92–105
19. Baek D, Villén J, Shin C, Camargo FD, Gygi SP, Bartel DP (2008) The impact of microRNAs on protein output. Nature 455:64

20. Shimoni Y, Friedlander G, Hetzroni G, Niv G, Altuvia S, Biham O, Margalit H (2007) Regulation of gene expression by small non-coding RNAs: a quantitative view. Mol Syst Biol 3:138
21. Tsang J, Zhu J, van Oudenaarden A (2007) MicroRNA-mediated feedback and feedforward loops are recurrent network motifs in mammals. Mol Cell 26:753–767
22. Re A, Corá D, Taverna D, Caselle M (2009) Genome-wide survey of microRNA-transcription factor feed-forward regulatory circuits in human. Mol BioSyst 5:854–867
23. Siciliano V, Garzilli I, Fracassi C, Criscuolo S, Ventre S, Di Bernardo D (2013) MiRNAs confer phenotypic robustness to gene networks by suppressing biological noise. Nat Commun 30:2364
24. Wang S, Raghavachari S (2011) Quantifying negative feedback regulation by micro-RNAs. Phys Biol 8:055002
25. Das J, Chakraborty S, Podder S, Ghosh TC (2013) Complex-forming proteins escape the robust regulations of miRNA in human. FEBS Lett 587:2284–2287
26. Obermayer B, Levine E (2014) Exploring the miRNA regulatory network using evolutionary correlations. PLoS Comput Biol 10: e1003860
27. Schmiedel JM, Klemm SL, Zheng Y, Sahay A, Blüthgen N, Marks DS, van Oudenaarden A (2015) MicroRNA control of protein expression noise. Science 348:128–132
28. Guil S, Esteller M (2015) RNA-RNA interactions in gene regulation: the coding and noncoding players. Trends Biochem Sci 40:248–256
29. Hansen TB, Jensen TI, Clausen BH, Bramsen JB, Finsen B, Damgaard CK, Kjems J (2013) Natural RNA circles function as efficient microRNA sponges. Nature 495:384
30. Ebert MS, Neilson JR, Sharp PA (2007) MicroRNA sponges: competitive inhibitors of small RNAs in mammalian cells. Nat Methods 4:721
31. Sumazin P, Yang X, Chiu HS, Chung WJ, Iyer A, Llobet-Navas D, Rajbhandari P, Bansal M, Guarnieri P, Silva J, Califano A (2011) An extensive microRNA-mediated network of RNA-RNA interactions regulates established oncogenic pathways in glioblastoma. Cell 147:370–381
32. Helwak A, Kudla G, Dudnakova T, Tollervey D (2013) Mapping the human miRNA interactome by CLASH reveals frequent noncanonical binding. Cell 153:654–665

33. Kim D, Sung YM, Park J, Kim S, Kim J, Park J, Ha H, Bae JY, Kim S, Baek D (2016) General rules for functional microRNA targeting. Nat Genet 48:1517
34. Breda J, Rzepiela AJ, Gumienny R, van Nimwegen E, Zavolan M (2015) Quantifying the strength of miRNA-target interactions. Methods 85:90–99
35. Arvey A, Larsson E, Sander C, Leslie CS, Marks DS (2010) Target mRNA abundance dilutes microRNA and siRNA activity. Mol Syst Biol 6:363
36. Levine E, Zhang Z, Kuhlman T, Hwa T (2007) Quantitative characteristics of gene regulation by small RNA. PLoS Biol 5:e229
37. Franco-Zorrilla JM, Valli A, Todesco M, Mateos I, Puga MI, Rubio-Somoza I, Leyva A, Weigel D, García JA, Paz-Ares J (2007) Target mimicry provides a new mechanism for regulation of microRNA activity. Nat Genet 39:1033
38. Salmena L, Poliseno L, Tay Y, Kats L, Pandolfi PP (2011) A ceRNA hypothesis: the Rosetta Stone of a hidden RNA language? Cell 146:353–358
39. Tay Y, Kats L, Salmena L, Weiss D, Tan SM, Ala U, Karreth F, Poliseno L, Provero P, Di Cunto F, Lieberman J (2011) Coding-independent regulation of the tumor suppressor PTEN by competing endogenous mRNAs. Cell 147:344–357
40. Mukherji S, Ebert MS, Zheng GX, Tsang JS, Sharp PA, van Oudenaarden A (2011) MicroRNAs can generate thresholds in target gene expression. Nat Genet 43:854
41. Karreth FA, Tay Y, Perna D, Ala U, Tan SM, Rust AG, DeNicola G, Webster KA, Weiss D, Perez-Mancera PA, Krauthammer M (2011) In vivo identification of tumor-suppressive PTEN ceRNAs in an oncogenic BRAF-induced mouse model of melanoma. Cell 147:382–395
42. Tay Y, Rinn J, Pandolfi PP (2014) The multilayered complexity of ceRNA crosstalk and competition. Nature 505:344
43. Yuan Y, Liu B, Xie P, Zhang MQ, Li Y, Xie Z, Wang X (2015) Model-guided quantitative analysis of microRNA-mediated regulation on competing endogenous RNAs using a synthetic gene circuit. Proc Natl Acad Sci 112:3158–3163
44. Bosia C, Sgrò F, Conti L, Baldassi C, Brusa D, Cavallo F, Di Cunto F, Turco E, Pagnani A, Zecchina R (2017) RNAs competing for microRNAs mutually influence their fluctuations in a highly non-linear microRNA-dependent manner in single cells. Genome Biol 18:37
45. Fatica A, Bozzoni I (2014) Long non-coding RNAs: new players in cell differentiation and development. Nat Rev Genet 15:7
46. Leung AK, Sharp PA (2010) MicroRNA functions in stress responses. Mol Cell 22:205–215
47. Alvarez-Garcia I, Miska EA (2005) MicroRNA functions in animal development and human disease. Development 132:4653–4662
48. Anastasiadou E, Jacob LS, Slack FJ (2018) Non-coding RNA networks in cancer. Nat Rev Cancer 18:5
49. Sanchez-Mejias A, Tay Y (2015) Competing endogenous RNA networks: tying the essential knots for cancer biology and therapeutics. J Hematol Oncol 8:30
50. Jens M, Rajewsky N (2015) Competition between target sites of regulators shapes post-transcriptional gene regulation. Nat Rev Genet 16:113
51. Denzler R, Agarwal V, Stefano J, Bartel DP, Stoffel M (2014) Assessing the ceRNA hypothesis with quantitative measurements of miRNA and target abundance. Mol Cell 54:766–776
52. Ala U, Karreth FA, Bosia C, Pagnani A, Taulli R, Léopold V, Tay Y, Provero P, Zecchina R, Pandolfi PP (2013) Integrated transcriptional and competitive endogenous RNA networks are cross-regulated in permissive molecular environments. Proc Natl Acad Sci 110:7154–7159
53. Bosson AD, Zamudio JR, Sharp PA (2014) Endogenous miRNA and target concentrations determine susceptibility to potential ceRNA competition. Mol Cell 56:347–359
54. Denzler R, McGeary SE, Agarwal V, Bartel DP, Stoffel M (2016) Impact of microRNA levels, target-site complementarity, and cooperativity on competing endogenous RNA-regulated gene expression. Mol Cell 64:565–579
55. Wang X, Li Y, Xu X, Wang YH (2010) Toward a system-level understanding of microRNA pathway via mathematical modeling. Biosystems 100:31–38
56. Lai X, Wolkenhauer O, Vera J (2016) Understanding microRNA-mediated gene regulatory networks through mathematical modelling. Nucleic Acids Res 44:6019–6035
57. Valencia-Sanchez MA, Liu J, Hannon GJ, Parker R (2006) Control of translation and mRNA degradation by miRNAs and siRNAs. Genes Dev 20:515–524

58. Baccarini A, Chauhan H, Gardner TJ, Jayaprakash AD, Sachidanandam R, Brown BD (2011) Kinetic analysis reveals the fate of a microRNA following target regulation in mammalian cells. Curr Biol 21:369–376
59. Figliuzzi M, Marinari E, De Martino A (2013) MicroRNAs as a selective channel of communication between competing RNAs: a steady-state theory. Biophys J 104:1203–1213
60. Bosia C, Pagnani A, Zecchina R (2013) Modelling competing endogenous RNA networks. PLoS One 8:e66609
61. Noorbakhsh J, Lang AH, Mehta P (2013) Intrinsic noise of microRNA-regulated genes and the ceRNA hypothesis. PLoS One 8:e72676
62. Alon U (2006) An introduction to systems biology: design principles of biological circuits. CRC Press, Boca Raton
63. Martirosyan A, Figliuzzi M, Marinari E, De Martino A (2016) Probing the limits to microRNA-mediated control of gene expression. PLoS Comput Biol 12:e1004715
64. Flondor P, Olteanu M, Stefan R (2018) Qualitative analysis of an ODE model of a class of enzymatic reactions. Bull Math Biol 80:32–45
65. Sanchez A, Choubey S, Kondev J (2013) Regulation of noise in gene expression. Annu Rev Biophys 42:469–491
66. Van Kampen NG (1992) Stochastic processes in physics and chemistry. Elsevier, Amsterdam
67. Swain PS (2004) Efficient attenuation of stochasticity in gene expression through post-transcriptional control. J Mol Biol 344:965–976
68. Gillespie DT (1977) Exact stochastic simulation of coupled chemical reactions. J Phys Chem 81:2340–2361
69. Gibson MA, Bruck J (2000) Efficient exact stochastic simulation of chemical systems with many species and many channels. J Phys Chem A 104:1876–1889
70. Martirosyan A, Marsili M, De Martino A (2017) Translating ceRNA susceptibilities into correlation functions. Biophys J 113:206–213
71. Nitzan M, Steiman-Shimony A, Altuvia Y, Biham O, Margalit H (2014) Interactions between distant ceRNAs in regulatory networks. Biophys J 106:2254–2266
72. Osella M, Bosia C, Corá D, Caselle M (2011) The role of incoherent microRNA-mediated feedforward loops in noise buffering. PLoS Comput Biol 7(3):e1001101
73. Bosia C, Osella M, El Baroudi M, Corá D, Caselle M (2012) Gene autoregulation via intronic microRNAs and its functions. BMC Syst Biol 6:131
74. Riba A, Bosia C, El Baroudi M, Ollino L, Caselle M (2014) A combination of transcriptional and MicroRNA regulation improves the stability of the relative concentrations of target genes. PLoS Comput Biol 10(2):e1003490
75. Osella M, Riba A, Testori A, Corá D, Caselle M (2014) Interplay of microRNA and epigenetic regulation in the human regulatory network. Front Genet 5:345
76. Grigolon S, Di Patti F, De Martino A, Marinari E (2016) Noise processing by microRNA-mediated circuits: the incoherent feed-forward loop, revisited. Heliyon 2(4):e00095
77. Gerstein M, Kundaje A, Hariharan M, Landt S, Yan K et al (2012) Architecture of the human regulatory network derived from ENCODE data. Nature 489:91–100
78. Bose I, Ghosh S (2012) Origins of binary gene expression in post-transcriptional regulation by microRNAs. Eur Phys J E 35:102
79. Tsimring L (2014) Noise in biology. Rep Prog Phys 77:026601
80. Samoilov M, Plyasunov S, Arkin A (2005) Stochastic amplification and signaling in enzymatic futile cycles through noise-induced bistability with oscillations. Proc Natl Acad Sci USA 102(7):2310–2315
81. Del Giudice M, Bo S, Grigolon S, Bosia C (2018, in press) On the role of microRNA-mediated bimodal gene expression. PLoS Comput Biol
82. López-Maury L, Marguerat S, Bähler J (2008) Tuning gene expression to changing environments: from rapid responses to evolutionary adaptation. Nat Rev Genet 9:583
83. Mehta P, Goyal S, Wingreen NS (2008) A quantitative comparison of sRNA-based and protein-based gene regulation. Mol Syst Biol 4:221
84. Martirosyan A, De Martino A, Pagnani A, Marinari E (2017) ceRNA crosstalk stabilizes protein expression and affects the correlation pattern of interacting proteins. Sci Rep 7:43673
85. Liang H, Li WH (2007) MicroRNA regulation of human protein-protein interaction network. RNA 13:1402–1408
86. Yuan X, Liu C, Yang P, He S, Liao Q, Kang S, Zhao Y (2009) Clustered microRNAs' coordination in regulating protein-protein interaction network. BMC Syst Biol 3:65

87. Sass S, Dietmann S, Burk UC, Brabletz S, Lutter D, Kowarsch A, Mayer KF, Brabletz T, Ruepp A, Theis FJ, Wang Y (2011) MicroRNAs coordinately regulate protein complexes. BMC Syst Biol 5:136
88. Hsu CW, Juan HF, Huang HC (2008) Characterization of microRNA-regulated protein-protein interaction network. Proteomics 8:1975–1979
89. Du B, Wang Z, Zhang X, Feng S, Wang G, He J, Zhang B (2014) MicroRNA-545 suppresses cell proliferation by targeting cyclin D1 and CDK4 in lung cancer cells. PLoS One 9:e88022
90. Nadal A, Jares P, Pinyol M, Conde L, Romeu C, Fernández PL, Campo E, Cardesa A (2007) Association of CDK4 and CCND1 mRNA overexpression in laryngeal squamous cell carcinomas occurs without CDK4 amplification. Virchows Arch 450:161–167
91. Kwon J, Lee TS, Lee HW, Kang MC, Yoon HJ, Kim JH, Park JH (2013) Integrin alpha 6: a novel therapeutic target in esophageal squamous cell carcinoma. Int J Oncol 43:1523–1530
92. Tkacik G, Callan Jr CG, Bialek W (2008) Information capacity of genetic regulatory elements. Phys Rev E 78:011910
93. Figliuzzi M, De Martino A, Marinari E (2014) RNA-based regulation: dynamics and response to perturbations of competing RNAs. Biophys J 107:1011–1022
94. Detwiler PB, Ramanathan S, Sengupta A, Shraiman BI (2000) Engineering aspects of enzymatic signal transduction: photoreceptors in the retina. Biophys J 79:2801–2817
95. Legnini I, Morlando M, Mangiavacchi A, Fatica A, Bozzoni I (2014) A feedforward regulatory loop between HuR and the long noncoding RNA linc-MD1 controls early phases of myogenesis. Mol Cell 53:506–514
96. Fiorentino J, De Martino A (2017) Independent channels for miRNA biosynthesis ensure efficient static and dynamic control in the regulation of the early stages of myogenesis. J Theor Biol 430:53–63
97. Mauri M, Klumpp S (2014) A model for sigma factor competition in bacterial cells. PLoS Comput Biol 10:e1003845
98. Brewster RC, Weinert FM, Garcia HG, Song D, Rydenfelt M, Phillips R (2014) The transcription factor titration effect dictates level of gene expression. Cell 156:1312–1323
99. Raveh A, Margaliot M, Sontag ED, Tuller T (2016) A model for competition for ribosomes in the cell. J R Soc Interface 13:20151062
100. Loinger A, Shemla Y, Simon I, Margalit H, Biham O (2012) Competition between small RNAs: a quantitative view. Biophys J 102:1712–1721
101. Koller E, Propp S, Murray H, Lima W, Bhat B, Prakash TP, Allerson CR, Swayze EE, Marcusson EG, Dean NM (2006) Competition for RISC binding predicts in vitro potency of siRNA. Nucleic Acids Res 34:4467–4476
102. Pitchiaya S, Heinicke LA, Park JI, Cameron EL, Walter NG (2017) Resolving subcellular miRNA trafficking and turnover at single-molecule resolution. Cell Rep 19:630–642
103. Levine E, McHale P, Levine H (2007) Small regulatory RNAs may sharpen spatial expression patterns. PLoS Comput Biol 3:e233
104. Teimouri H, Korkmazhan E, Stavans J, Levine E (2017) Sub-cellular mRNA localization modulates the regulation of gene expression by small RNAs in bacteria. Phys Biol 14:056001

Chapter 16

Modeling ncRNA-Mediated Circuits in Cell Fate Decision

Xiao-Jun Tian, Manuela Vanegas Ferro, and Hanah Goetz

Abstract

Noncoding RNAs (ncRNAs) play critical roles in essential cell fate decisions. However, the exact molecular mechanisms underlying ncRNA-mediated bistable switches remain elusive and controversial. In recent years, systematic mathematical and quantitative experimental analyses have made significant contributions on elucidating the molecular mechanisms of controlling ncRNA-mediated cell fate decision processes. In this chapter, we review and summarize the general framework of mathematical modeling of ncRNA in a pedagogical way and the application of this general framework on real biological processes. We discuss the emerging properties resulting from the reciprocal regulation between mRNA, miRNA, and competing endogenous mRNA (ceRNA), as well as the role of mathematical modeling of ncRNA in synthetic biology. Both the positive feedback loops between ncRNAs and transcription factors and the emerging properties from the miRNA-mRNA reciprocal regulation enable bistable switches to direct cell fate decision.

Key words Ultrasensitivity, Competing endogenous mRNA, Posttranscriptional, Mathematical modeling, Bistability, Cell fate decision

1 Introduction

Although 90% of the eukaryotic genome is transcribed, only 1–2% of these transcripts are translated into proteins while most of them remain as noncoding RNAs (ncRNAs) [1, 2]. Decades of study have shown the importance of ncRNAs and the crucial roles they play in the signal transduction of various biological processes [3], especially in processes that determine cell fates, such as embryogenesis [4], neuronal development [5], epithelial-mesenchymal transition (EMT) [6–8], cell death [9], cell cycle [10], and cell reprogramming [11–13]. Tight orchestration of miRNA regulation is of vital importance, as shown by the association of miRNA dysregulation with complex diseases including cancer [14, 15] and neurodegenerative diseases [16].

Although the details about how these miRNAs are involved in regulating various cell fates are becoming clear, due to the complexity and nonlinearity of these regulatory networks, an intuitive linear understanding is not enough to explain some interesting phenomena. Many research efforts have been devoted to dissecting the role of miRNAs in cell fate by integrating mathematical modeling and quantitative experiments. The predictive power of mathematical modeling has been proven by several successful examples for modeling ncRNA-mediated circuits in a wide range of cell fate decisions, including EMT [17–19], restriction-point switch of the cell cycle [20, 21], apoptosis mediated by p53 [22, 23], and cell transformation [24, 25]. In the following sections, we introduce the general regulatory motifs found between transcription factors and miRNAs, the general modeling framework of miRNA-mRNA regulation, and the application of this framework in specific biological systems. We have the objective of demonstrating how mathematical modeling can be used to study the nonlinearity of miRNA regulation and how this nonlinearity contributes to cell fate decision.

2 miRNA-Mediated Positive Feedback Loops

Even if biological regulatory networks are very complex, we can still find the recurrence of some specific regulatory or interaction patterns. They are known as network motifs and function as the building blocks for the large regulatory networks [26, 27]. Different network motifs can achieve different functions. For instance, positive feedback loops can generate bistability and are used frequently in cell fate decision [28, 29], while negative feedback loops are used in oscillatory systems [30]. To achieve specific functions, we can only choose some network motifs. For example, only negative feedback loops or feed-forward loops are able to achieve adaptation, i.e., the capability of a system to return to the initial state following response to a constant stimulus [31]. In some cases, these basic motifs are coupled together to either make the system more robust or achieve new emerging properties. Coupling a positive feedback loop to an oscillating negative feedback loop not only makes the oscillation more robust [32] but can also make the system excitable [33, 34], whereas coupling multiple positive feedback loops produces more robust and reliable cell decisions [35, 36].

While bistability mediated by positive feedback loops is a paradigm for cell fate decision, a special kind of positive feedback loop is particularly enriched in the complex biological regulatory networks. As shown in Fig. 1, this particular motif consists of a miRNA that is negatively regulated by a transcription factor (TF), which is in turn negatively regulated by the miRNA. In this TF-miRNA motif, TF inhibits the transcription of miRNA by binding to the promoter of the miRNA gene. Meanwhile, miRNA binds to the TF mRNA through complementary sequences and thus inhibits its translation or promotes its degradation.

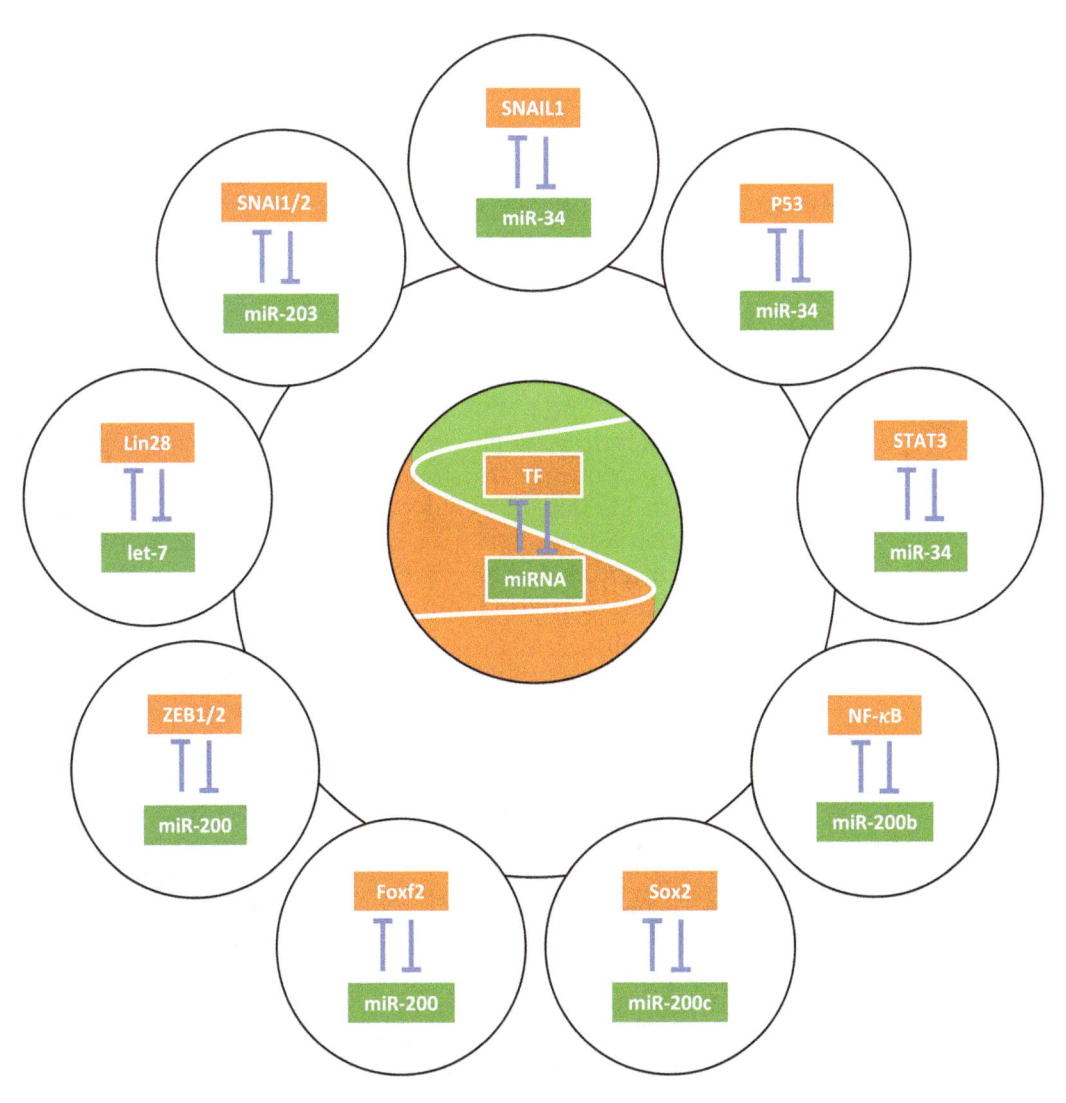

Fig. 1 The general TF-miRNA motif. The regulatory network motif between transcription factor (TF; orange) and miRNA (green) in cell fate determination processes. The center circle depicts an idealized bifurcation diagram for the general TF-miRNA motif, where the TF is prevalent for certain conditions and the miRNA is prevalent for other conditions. Biological examples of TF miRNA motif couples are shown in the external circumference (*see* detail in the text)

Therefore, this positive feedback loop has one transcriptional arm and one posttranscriptional arm. A few examples in real biological systems, as shown in Fig. 1, are SNAIL1-*miR-34* [37], ZEB1/2-*miR-200* [38], and p53-*miR-34* [39], STAT3-*miR-34a* [40], NF-κB-*miR-200b* [41], Sox2-*miR-200c* [42], Foxf2-*miR-200* [43], SNAI1/2-*miR-203* [44], and Lin28-*let-7* [45, 46]. Although the specific details of the two arms, such as the binding affinity or complementary sequence of miRNA and TF mRNA, vary on a case-by-case basis, the repetition of this topology suggests that these motifs play a similar role in the regulation of cell fates. In the following sections, we will introduce a general modeling framework for this motif and its application to real systems.

3 Mathematical Model of miRNA-mRNA Mutual Regulation

The general modeling framework for miRNA-mRNA regulation is introduced here. In the general model, mRNA is produced with a rate k_{mR} and degraded with a rate constant d_{mR}, while miRNA is produced with a rate constant k_{miR} and degraded with a rate constant d_{miR}. The translation rate of mRNA is k_{s0}. mRNA and miRNA can form a complex mRNA-miRNA with a binding affinity K, which depends on the complementary sequence and binding free energy. The degradation rate of the mRNA-miRNA complex is d_{R1}, which is usually much larger than d_{mR}. The translation rate of the mRNA-miRNA complex is k_{s1}, which is much smaller than k_{s0}. Upon the degradation of the mRNA-miRNA complex, miRNA can be recycled with a ratio λ_1, enabling each single miRNA to be reused and regulate several rounds of mRNA. miRNA inhibits mRNA in two ways, either by promoting its degradation or by inhibiting its translation. It is noted that under some conditions, miRNA is able to promote the translation of mRNA [47]. The inhibition of miRNA transcription by TF is described with a Hill function $[TF]^n/([TF]^n + J^n)$ with Hill coefficient n which depends on the number of the TF-binding sites on the promoter of the miNRA gene. If there are multiple ($=N$) miRNA-binding sites on the mRNA sequence, the number of complex mRNA-miRNA$_n$ with n copies of miRNA is $N!/[n!(N-n)!]$. The total number of complexes increases combinatorially with the number of binding sites. Therefore, a quasi-equilibrium approximation is used for binding and unbinding between miRNAs and mRNAs by assuming that these processes are much faster than transcription, translation, and degradation.

3.1 Model for mRNA-miRNA Mutual Regulation with One miRNA-Binding Site

For the system with only one miRNA-binding site, the reaction diagram is shown in Fig. 2a. We have the following ordinary differential equations (ODEs):

$$\frac{d[miR]_t}{dt} = k_{miR} - kd_{miR} \cdot [miR] - (1-\lambda_1) \cdot kd_{R1} \cdot [R_1],$$

$$\frac{d[mR]_t}{dt} = k_{mR} - kd_{mR} \cdot [mR] - kd_{R1} \cdot [R_1],$$

$$\frac{d[TF]}{dt} = k_{s0} \cdot [mR] + k_{s1} \cdot [R_1] - k_{dp} \cdot [TF],$$

and the relationships derived from the quasi-equilibrium approximation:

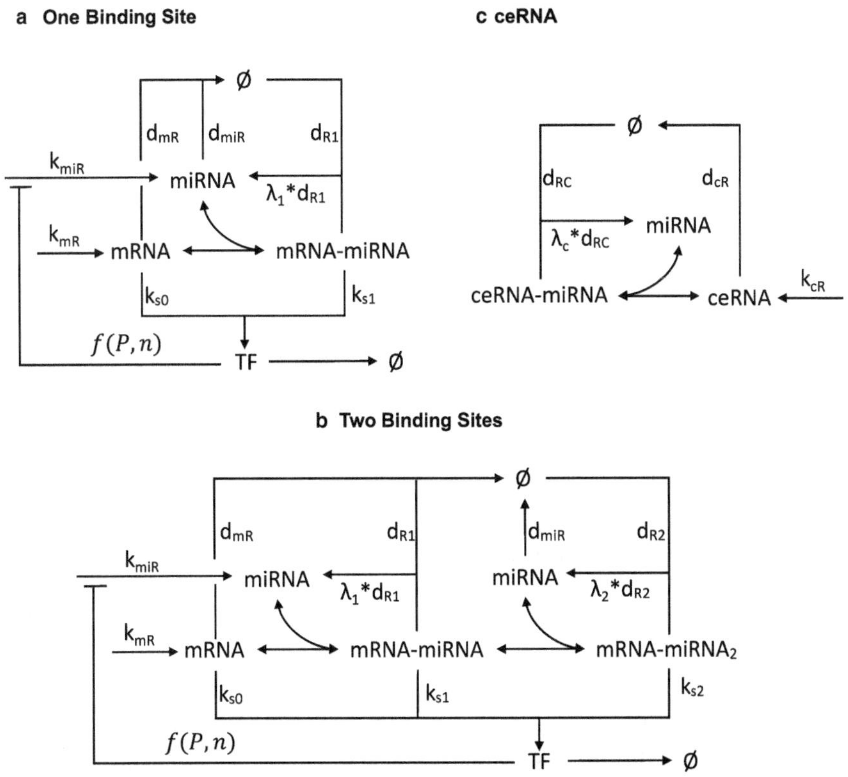

a One Binding Site

c ceRNA

b Two Binding Sites

Fig. 2 The general modeling framework for the miRNA-TF motif. The reaction diagrams for one mRNA with one miRNA-binding site (**a**) or with two miRNA-binding sites (**b**). (**c**) The reaction diagram for one additional ceRNA which competes for miRNA with mRNA by binding to the same miRNA with one binding site

$$[mR] = [mR]_t - [R_1],$$
$$[miR] = [miR]_t - [R_1],$$
$$[R_1] = K_1 \cdot [mR] \cdot [miR].$$

Here $[miR]_t$, $[mR]_t$, and $[TF]$ are the total level of miRNA, mRNA, and protein, respectively, while $[miR]$ and $[mR]$ are the cellular level of the free miRNA and mRNA, respectively, and $[R_1]$ is the concentration of the miRNA-mRNA complex. K_1 is the binding constant between miRNA and mRNA. d_{R1} is the degradation rate constant of complex mRNA-miRNA. k_{s0} and k_{s1} are the translation rates of mRNA and mRNA-miRNA R_1. miRNA is recycled with a ratio λ_1 ($0 \leq \lambda_1 \leq 1$) upon degradation of the complex mRNA-miRNA R_1.

3.2 Model for mRNA-miRNA Mutual Regulation with Two miRNA-Binding Sites

For the system with two miRNA-binding sites, the reaction diagram is shown in Fig. 2b and we use the following ODEs:

$$\frac{d[miR]_t}{dt} = k_{miR} - kd_{miR} \cdot [miR] - (1-\lambda_1) \cdot kd_{R1} \cdot 2 \cdot [R_1]$$
$$- (1-\lambda_2) \cdot kd_{R2} \cdot 2 \cdot [R_2],$$
$$\frac{d[mR]_t}{dt} = k_{mR} - kd_{mR} \cdot [mR] - kd_{R1} \cdot 2 \cdot [R_1] - kd_{R2} \cdot [R_2],$$
$$\frac{d[TF]}{dt} = k_{s0} \cdot [mR] + k_{s1} \cdot [R_1] + k_{s2} \cdot [R_2] - k_{dp} \cdot [TF],$$

and the relationships derived from the quasi-equilibrium approximation:

$$[mR] = [mR]_t - 2 \cdot [R_1] - [R_2],$$
$$[miR] = [miR]_t - 2 \cdot [R_1] - 2 \cdot [R_2],$$
$$[R_1] = K_1 \cdot [mR] \cdot [miR],$$
$$[R_2] = K_2 \cdot [R1] \cdot [miR].$$

Here $[R_2]$ is the level of complex mRNA-miRNA$_2$, while K_2 is the binding constant between miRNA and complex mRNA-miRNA R_1. d_{R2} is the degradation rate constant of $[R_2]$ and k_{s2} is the translation rate of $[R_2]$. miRNA is recycled with a ratio λ_2 ($0 \leq \lambda_2 \leq 1$) upon degradation of the complex mRNA-miRNA R_2. The other parameters are the same as the system with one miRNA-binding site.

3.3 Model for mRNA-miRNA Mutual Regulation with One miRNA-Binding Site and One ceRNA

If we also have a ceRNA which competes with mRNA for miRNA, the additional reaction diagram for the case with one miRNA-binding site on ceRNA is shown in Fig. 2c. We need to revise the ODE as follows:

$$\frac{d[miR]_t}{dt} = k_{miR} - kd_{miR} \cdot [miR] - (1-\lambda_1) \cdot kd_{R1} \cdot [R_1]$$
$$- (1-\lambda_c) \cdot kd_{Rc} \cdot [R_c],$$
$$\frac{d[mR]_t}{dt} = k_{mR} - kd_{mR} \cdot [mR] - kd_{R1} \cdot [R_1],$$
$$\frac{d[mRc]_t}{dt} = k_{mRc} - kd_{mRc} \cdot [mRc] - kd_{Rc} \cdot [R_c],$$
$$\frac{d[TF]}{dt} = k_{s0} \cdot [mR] + k_{s1} \cdot [R_1] - k_{dp} \cdot [TF],$$

and the relationships derived from the quasi-equilibrium approximation:

$$[mR] = [mR]_t - [R_1],$$
$$[mRc] = [mRc]_t - [R_c],$$
$$[miR] = [miR]_t - [R_1] - [R_c],$$
$$[R_1] = K_1 \cdot [mR] \cdot [miR],$$
$$[Rc] = K_c \cdot [mRc] \cdot [miR].$$

Here $[mRc]_t$ is the total level of ceRNA, $[mRc]$ is the cellular level of free ceRNA, and K_c is the binding constant between miRNA and ceRNA. d_{Rc} is the degradation rate constant of complex ceRNA-miRNA. miRNA is recycled with a ratio λ_c ($0 \leq \lambda_c \leq 1$) upon degradation of the complex ceRNA-miRNA. The other parameters are the same as the system with one binding site.

3.4 Model for mRNA-miRNA Mutual Regulation with Two miRNA-Binding Sites and One ceRNA

If we also have a ceRNA which competes with mRNA for miRNA in the two miRNA-binding sites system, we need to revise the ODE as follows:

$$\frac{d[miR]_t}{dt} = k_{miR} - kd_{miR} \cdot [miR] - (1-\lambda_1) \cdot kd_{R1} \cdot 2 \cdot [R_1]$$
$$- (1-\lambda_2) \cdot kd_{R2} \cdot 2 \cdot [R_2] - (1-\lambda_c) \cdot kd_{Rc} \cdot [R_c],$$

$$\frac{d[mR]_t}{dt} = k_{mR} - kd_{mR} \cdot [mR] - kd_{R1} \cdot 2 \cdot [R_1] - kd_{R2} \cdot [R_2],$$

$$\frac{d[mRc]_t}{dt} = k_{mRc} - kd_{mRc} \cdot [mRc] - kd_{Rc} \cdot [R_c],$$

$$\frac{d[TF]}{dt} = k_{s0} \cdot [mR] + k_{s1} \cdot [R_1] + k_{s2} \cdot [R_2] - k_{dp} \cdot [TF],$$

and the relationship derived from the quasi-equilibrium approximation:

$$[mR] = [mR]_t - 2 \cdot [R_1] - [R_2],$$
$$[mRc] = [mRc]_t - [R_c],$$
$$[miR] = [miR]_t - 2 \cdot [R_1] - 2 \cdot [R_2] - [R_c],$$
$$[R_1] = K_1 \cdot [mR] \cdot [miR],$$
$$[R_2] = K_2 \cdot [R1] \cdot [miR],$$
$$[Rc] = K_c \cdot [mRc] \cdot [miR].$$

The parameters are the same as the above systems.

When a TF-mediated positive feedback loop is considered, we revise the synthesis rate of miRNA from k_{miR} to $\frac{k_{miR}}{[TF]^n/J^n + 1}$, where a repressive Hill function with a Hill coefficient n is used to represent the inhibition of TF to miRNA transcription. It is straightforward to extend these models into other systems with multiple binding sites of miRNA or other ncRNA on mRNA and ceRNA.

4 Bistability from Reciprocal Regulation Between mRNA and microRNA

It has been demonstrated that mRNA-miRNA regulation can generate ultrasensitivity [48]. In our previous work, we have shown that this ultrasensitivity helps the system to be bistable and direct the cell fate decisions [17]. The system becomes bistable when one positive feedback loop is added to the system with miRNA-binding site. As shown in Fig. 3a, three steady states can be found from the nullclines of [TF] and $[miR]_t$, two of which are stable while the other is unstable. The system can be in either the "ON" state with high level of [TF] and low level of $[miR]_t$ or the "OFF" state with low level of [TF] and high level of $[miR]_t$. From the one-parameter bifurcation of [TF] on the production rate of miRNA k_{miR} (Fig. 3b), one can see that there is a large range between SN1 and SN2 where the system is bistable. In order to clearly see the contribution of the ultrasensitivity from miRNA-mRNA mutual regulation on the bistability, the bistable region in the parameter space of Hill coefficient n and k_{miR} is analyzed with the two-parameter bifurcation diagram (Fig. 3c). It is noted that bistability can be found even when $n < 1$, where there is no ultrasensitivity from the Hill function. That is, in this region, the bistability solely relies on the ultrasensitivity resulting from the mRNA-miRNA mutual regulation. We also demonstrated that the sensitivity from the mRNA-miRNA regulation decreases with the recycle ratio of miRNA λ_1. The minimal Hill coefficient n for bistability increases with the recycle ratio λ_1, as shown in Fig. 3d. That is, when some systems do not have cooperativity for the TF-meditated transcription, bistability can still be generated with a lower recycle ratio of miRNA. Taken together, both the ultrasensitivity from the mutual regulation between mRNA and miRNA and the one from protein-regulated miRNA synthesis with a Hill function contribute to bistability. While the former depends on the recycle ratio λ_1, the latter depends on the Hill coefficient n.

Interestingly, we found that bistability can be generated without a positive feedback loop if multiple miRNA-binding sites exist on its target mRNA (Fig. 3e, f) [49]. Furthermore, ceRNA with a strong binding affinity can expand the bistable region. As shown in Fig. 3f, the bistable region in the space of Hill coefficient n and k_{miR} is increased if one ceRNA is considered in the system (blue curve). Other mechanisms of bistability without positive feedback have been reported, such as multisite phosphorylation [50, 51]. One can imagine that the bistability becomes more robust when several of these mechanisms are combined together. For instance, one system has a double-negative feedback loop between TF and miRNA, multiple binding sites on the mRNA of TF, and multiple phosphorylation sites on TF.

5 Examples of miRNA Regulation in EMT

Two different mathematical models are proposed for EMT and they predict different partial (intermediate or hybrid) EMT states in addition to the epithelial and mesenchymal states [17–19]. In the cascaded bistable switch (CBS) model [17], we predicted that the

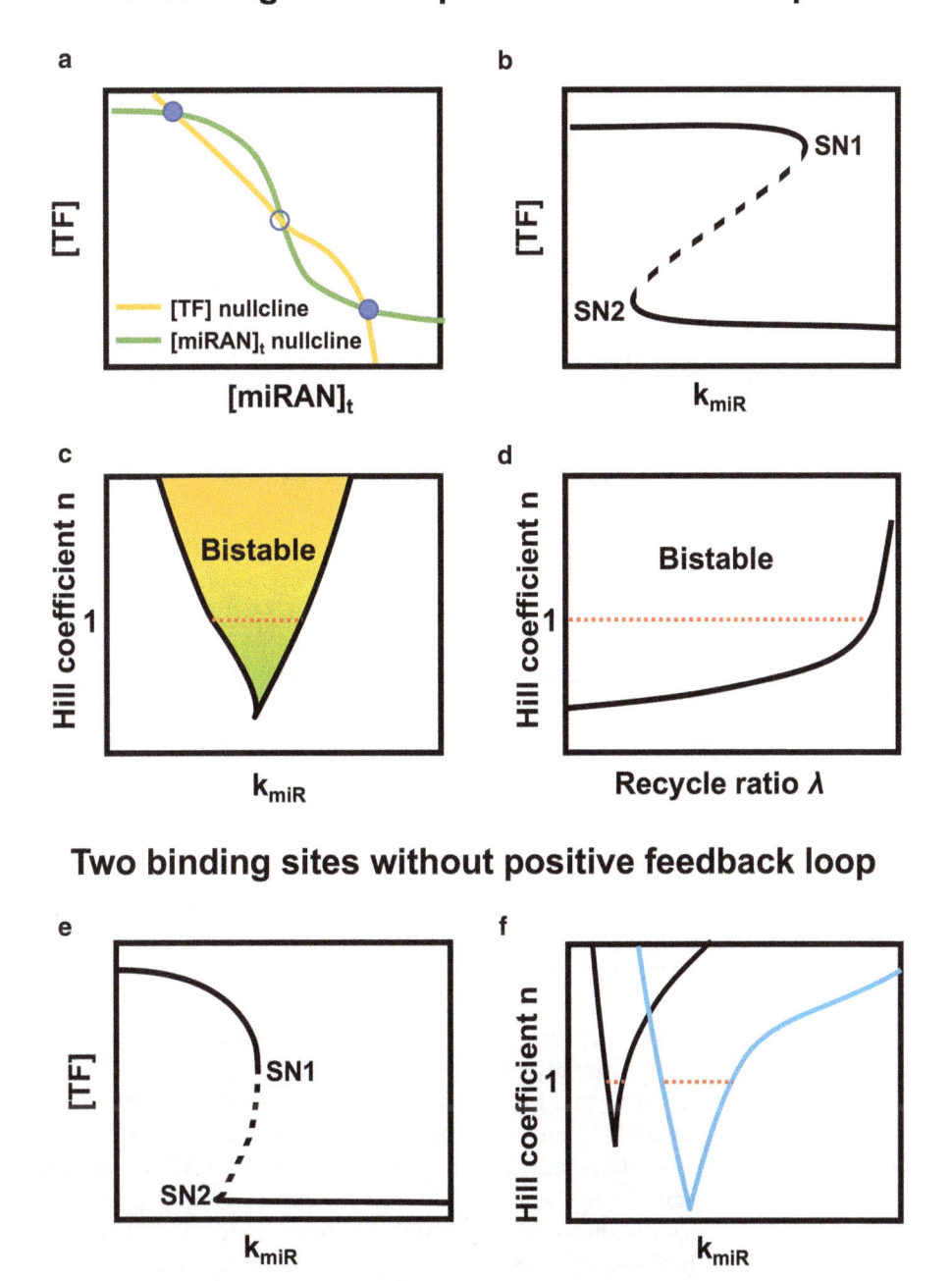

Fig. 3 The contribution of mRNA-miRNA mutual regulation to bistability. (**a–d**) In the system with only one binding site of miRNA on mRNA, ultrasensitivity from mRNA-miRNA mutual regulation helps to generate bistability with one positive feedback. (**a**) Nullclines of TF and miRNA. (**b**) The dependence of [TF] on the transcription rate of miRNA k_{miR}. (**c**) The bistable region in the space of k_{miR} and Hill coefficient n. (**d**) The minimal Hill coefficient for bistability as a function of recycle ratio. (**e, f**) In the system with two binding sites of miRNA on mRNA, bistability can be generated without any positive feedback loop. (**e**) The dependence of [TF] on k_{miR}. (**f**) The bistable region in the space of Hill coefficient *n* and k_{miR} without ceRNA (black curve) and with ceRNA (blue curve) (adapted from Ref. [49])

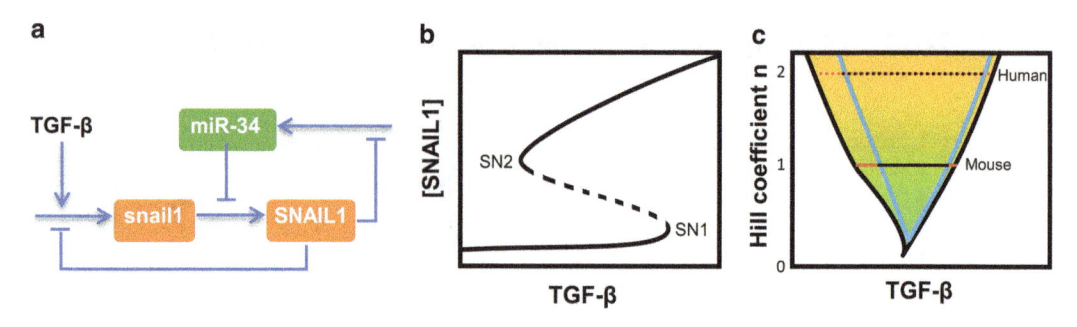

Fig. 4 The bistability of SNAIL1-*miR-34* module directs the transition of epithelial to one partial EMT state. (**a**) The regulatory network for SNAIL1-*miR-34* module in EMT. (**b**) The dependence of [SNAIL1] on TGF-β. (**c**) The bistable region without SNAIL1 self-inhibition (black curve) or with SNAIL1 self-inhibition (blue curve) in the space of Hill coefficient *n* and TGF-β. The dotted horizontal lines indicate the human cell system with $n = 2$ and the solid horizontal lines indicate the mouse cell system with $n = 1$

SNAIL1-*miR-34* double-negative feedback loop is able to generate a bistable switch and regulates the transition of one partial EMT state from the epithelial state. Here, we will take SNAIL1-*miR-34* module as an example to show how the mutual regulation between miRNA and mRNA contributes to its bistability.

The regulatory network is shown in Fig. 4a. TGF-β promotes the transcription of *SNAIL1*, while *miR-34* promotes its degradation or inhibits its translation to the protein SNAIL1. SNAIL1 inhibits the transcription of *miR-34* as a transcription factor. In addition, SNAIL1 weakly inhibits its own transcription and thus has an additional self-inhibition. There is only one binding site for *miR-34* on the sequence of *Snail1* mRNA [37]. There are two SNAIL1-binding sites on the *miR-34* promoter in human cells, but only one in mouse cells [37]. Thus, for human cells we have two sources of ultrasensitivity for bistability: one from the SNAIL1-mediated inhibition of *miR-34* transcription with Hill coefficient $n = 2$ and the other from the mutual regulation between *Snail1* mRNA and *miR-34*. While the Hill coefficient in murine cells is $n = 1$ instead, the ultrasensitivity arising from miRNA-mRNA mutual regulation still compensates for the low Hill coefficient. The bistable range might be smaller than in human cells (Fig. 4b, c). In addition, the weak self-inhibition of SNAIL1 may shrink the bistable range as shown by the blue curve, compared with the black curve without SNAIL1 self-inhibition (Fig. 4c). Thus, the self-inhibition of SNAIL1 has a function for fine-tuning the bistable region and the dynamics of the transition between the partial EMT state and epithelial state.

The bistability of the SNAIL1-*miR-34* module was experimentally verified in the MCF10A cell line through a flow cytometry analysis [18]. It is also reported that SNAIL1-mediated partial EMT in tubular epithelial cells drives renal fibrosis by secreting growth factors [52, 53]. Although we have not found the coexistence of both partial EMT states in the same cell, it has been suggested that there is a spectrum of EMT states among different cancer cell lines [54–56]. Nevertheless, the existence of the multiple partial EMT states is already well accepted and plays an important role in different EMT-regulated processes. Future studies will reveal the mechanism of these partial EMT states and their functions.

6 Synthetic Biology of miRNA

While the real regulatory network in living cells is too complex to study isolated miRNA-mRNA interactions systematically, synthetic biology has been used to quantitatively study small miRNA-mRNA circuits. Modeling ncRNA-mediated circuits enables a better understanding of these systems by providing predictions that can be experimentally verified. In addition, the quantitative feature of the predictive power of models can be harnessed to rationally design ncRNA circuits with specific functions that manipulate cell fates. Indeed, synthetic biology of ncRNA is guided by model-derived predictions and the models can be informed by the experimental results related to these synthetic molecules.

Following this idea, fundamental features of miRNA regulation have been proven by studies that combine mathematical modeling and quantitative synthetic biology. A seminal study which combined experimental data from a synthetic reporter system in mammalian cells with a mathematical model has shown that miRNA-mediated regulation generates mRNA expression thresholds, where production of target protein responds sensitively only after the mRNA level is above a threshold [48]. The effect of ceRNA has also been studied from a theoretical perspective [57, 58]. Through a combination of mathematical modeling and synthetic biological experiments, these studies [59, 60] yields the understanding that ceRNA regulation is mostly controlled by the relative expression level and binding affinities of miRNA and ceRNAs.

A synthetic translational negative feedback control system was built as a proportional controller and maintains target intracellular protein levels [61]. It will be very interesting to couple this synthetic negative feedback control system with one additional positive feedback loop to control the cell fate in the future. Interestingly, a cell classifier circuit was built as a miRNA sensor to specifically induce apoptosis of Hela cells in a mixed-cell population [62]. Synthetic miRNA switches are designed to purify hPSC-derived cardiomyocyte populations without the need for cell sorting and other significant side effects [63]. These RNAi-based control systems have a great application in the specific cell phenotype diagnosis and manipulation. For example, one target circuit can be designed to detect the circulating tumor cells (CTCs) in the blood given that (partial) EMT is associated with the CTC clusters [64, 65]. However, an inherent sensitivity-specificity trade-off limits its application due to cellular variability and stochasticity. Thus, we need to optimize the targeting circuits to increase the efficiency and develop a better practical implementation [66]. We may need to increase the complexity of the circuit by adding more positive and negative feedback control to break this trade-off and the guidance from the sophisticated modeling analysis.

7 Discussion

In this chapter, we reviewed the roles of miRNA in cell fate decision from a mathematical modeling perspective. Many emerging properties, such as ultrasensitivity and bistability, were found with the help of mathematical modeling. All these findings suggest that the regulation of mRNA by miRNA is not just a mechanism for fine-tuning the gene expression at the posttranscriptional level. Instead, it can play indispensable roles [5]. These nontrivial important roles of miRNA are emerging and becoming evident with the aid of mathematical modeling.

An additional consideration supporting efforts on miRNA modeling and experimentation is the exceptional stability of these molecules in plasma and serum. This suggests that miRNA could be a very good marker of cancer, especially circulating cancer. Indeed, it is found that the serum level of *miR-141* is a marker with a very good specificity and sensitivity that can be used to distinguish patients with prostate cancer from the healthy control group [67]. *miR-141* is a member of the *miR-200* family, which together with ZEB1/2 regulates the partial EMT transition to the mesenchymal cell according to the CBS model of EMT [18–17]. Therefore, according to the CBS model, the marker might be produced by circulating prostate cancer cells that are in the partial EMT state, which is most likely regulated by the SNAIL1-*miR-34* motif. The contribution of EMT or partial EMT on cancer metastasis is still under debate. Two recent reports argue that EMT is dispensable for metastasis but induces chemoresistance for both lung and pancreatic cancer metastasis [68, 69]. The challenge is to find good miRNA markers to detect various cancers in circulation at early stages. However, the mechanism underlying the decision of cancer cells to enter the blood circulation is not clear. We need to integrate computational and experimental approaches that are necessary for synthetic biology and systems biology guided by computational mathematical modeling to tackle the molecular and cellular regulation mechanisms and design more efficient and effective diagnosis and treatment strategies.

In addition to miRNA, other types of ncRNA, such as circular RNA (circRNA), Piwi-interacting RNAs (piRNAs), and long noncoding RNAs (lncRNAs), have also emerged as key regulators in cell fate decision processes [13]. Different functional interactions between miRNAs and lncRNAs have been summarized [70], including the expected miRNA-triggered lncRNA degradation, and the function of lncRNAs as sponges that reduce the effective concentration of miRNA. Given that lncRNAs have multiple miRNA-binding sites, they can also function as a very good ceRNAs. Additionally, it was noted that circRNA could also serve as a particularly stable ceRNA [71], since it is resistant to miRNA-mediated degradation [72]. However, it is necessary to develop additional work involving experimental characterization and mathematical modeling to investigate the mutual regulation between these ncRNAs and mRNA. Future studies with combined mathematical modeling and quantitative experiments will provide novel emerging properties from these new types of ncRNA-mRNA regulation.

8 Acknowledgments

H.G. was supported by the Arizona State University Dean's Fellowship and M.V.F. was supported by Arizona State University Biological Design Program. This study was financially supported by the ASU School of Biological and Health Systems Engineering (X-J.T.). We apologize to the authors whose work could not be cited due to space constraints.

References

1. Carninci P et al (2005) The transcriptional landscape of the mammalian genome. Science 309:1559–1563
2. Birney E et al (2007) Identification and analysis of functional elements in 1% of the human genome by the ENCODE pilot project. Nature 447:799–816
3. Inui M, Martello G, Piccolo S (2010) MicroRNA control of signal transduction. Nat Rev Mol Cell Biol 11:252–263
4. Pauli A, Rinn JL, Schier AF (2011) Non-coding RNAs as regulators of embryogenesis. Nat Rev Genet 12:136–149
5. Davis GM, Haas MA, Pocock R (2015) MicroRNAs: not "fine-tuners" but key regulators of neuronal development and function. Front Neurol 6:245. https://doi.org/10.3389/fneur.2015.00245
6. Zhang J, Ma L (2012) MicroRNA control of epithelial-mesenchymal transition and metastasis. Cancer Metastasis Rev 31:653–662
7. Gregory PA, Bracken CP, Bert AG, Goodall GJ (2008) MicroRNAs as regulators of epithelial-mesenchymal transition. Cell Cycle 7:3112–3117
8. Guo F, Kerrigan BCP, Yang D, Hu L, Shmulevich I, Sood AK, Xue F, Zhang W (2014) Post-transcriptional regulatory network of epithelial-to-mesenchymal and mesenchymal-to-epithelial transitions. J Hematol Oncol 7:19
9. Jovanovic M, Hengartner MO (2006) miRNAs and apoptosis: RNAs to die for. Oncogene 25:6176–6187
10. Shurin MR (2010) MicroRNAs are invading the tumor microenvironment: fibroblast microRNAs regulate tumor cell motility and invasiveness. Cell Cycle 9:4430–4430
11. Bao X, Zhu X, Liao B, Benda C, Zhuang Q, Pei D, Qin B, Esteban MA (2013) MicroRNAs in somatic cell reprogramming. Curr Opin Cell Biol 25:208–214
12. Lüningschrör P, Hauser S, Kaltschmidt B, Kaltschmidt C (2013) MicroRNAs in pluripotency reprogramming and cell fate induction.

Biochimica et Biophysica Acta (BBA) - Molecular Cell Research 1833:1894–1903
13. Flynn RA, Chang HY (2014) Long noncoding RNAs in cell-fate programming and reprogramming. Cell Stem Cell 14:752–761
14. Iorio MV, Croce CM (2012) MicroRNA dysregulation in cancer: diagnostics monitoring and therapeutics. A comprehensive review. EMBO Mol Med 4:143–159
15. Bracken CP, Scott HS, Goodall GJ (2016) A network-biology perspective of microRNA function and dysfunction in cancer. Nat Rev Genet 17:719–732
16. Tan L, Yu J-T, Tan L (2014) Causes and consequences of MicroRNA dysregulation in neurodegenerative diseases. Mol Neurobiol 51:1249–1262
17. Tian X-J, Zhang H, Xing J (2013) Coupled reversible and irreversible bistable switches underlying TGFβ-induced epithelial to mesenchymal transition. Biophys J 105:1079–1089
18. Zhang J, Tian X-J, Zhang H, Teng Y, Li R, Bai F, Elankumaran S, Xing J (2014) TGF-β-induced epithelial-to-mesenchymal transition proceeds through stepwise activation of multiple feedback loops. Sci Signal 7:ra91
19. Lu M, Jolly MK, Levine H, Onuchic JN, Ben-Jacob E (2013) MicroRNA-based regulation of epithelial-hybrid-mesenchymal fate determination. Proc Natl Acad Sci U S A 110:18144–18149
20. Aguda BD, Kim Y, Piper-Hunter MG, Friedman A, Marsh CB (2008) MicroRNA regulation of a cancer network: consequences of the feedback loops involving miR-17-92 E2F, and Myc. Proc Natl Acad Sci U S A 105:19678–19683
21. Sengupta D, Govindaraj V, Kar S (2017) Subtle alteration in microRNA dynamics accounts for differential nature of cellular proliferation. https://doi.org/10.1101/214429
22. Zhou C-H, Zhang X-P, Liu F, Wang W (2014) Involvement of miR-605 and miR-34a in the DNA damage response promotes apoptosis induction. Biophys J 106:1792–1800
23. Lai X, Wolkenhauer O, Vera J (2012) Modeling miRNA regulation in cancer signaling systems: miR-34a regulation of the p53/Sirt1 signaling module. Methods Mol Biol 880:87–108
24. Gérard C, Gonze D, Lemaigre F, Novák B (2014) A model for the epigenetic switch linking inflammation to cell transformation: deterministic and stochastic approaches. PLoS Comput Biol 10:e1003455
25. Lee J, Lee J, Farquhar KS, Yun J, Frankenberger CA, Bevilacqua E, Yeung K, Kim E-J, Balazsi G, Rosner MR (2014) Network of mutually repressive metastasis regulators can promote cell heterogeneity and metastatic transitions. Proc Natl Acad Sci U S A 111: E364–E373
26. Milo R (2002) Network motifs: simple building blocks of complex networks. Science 298:824–827
27. Alon U (2007) Network motifs: theory and experimental approaches. Nat Rev Genet 8:450–461
28. Ferrell JE, Xiong W (2001) Bistability in cell signaling: how to make continuous processes discontinuous and reversible processes irreversible. Chaos 11:227
29. Tyson JJ, Chen KC, Novak B (2003) Sniffers buzzers, toggles and blinkers: dynamics of regulatory and signaling pathways in the cell. Curr Opin Cell Biol 15:221–231
30. Novák B, Tyson JJ (2008) Design principles of biochemical oscillators. Nat Rev Mol Cell Biol 9:981–991
31. Ma W, Trusina A, El-Samad H, Lim WA, Tang C (2009) Defining network topologies that can achieve biochemical adaptation. Cell 138:760–773
32. Tsai TY-C, Choi YS, Ma W, Pomerening JR, Tang C, Ferrell JE (2008) Robust tunable biological oscillations from interlinked positive and negative feedback loops. Science 321:126–129
33. Tian X-J, Zhang X-P, Liu F, Wang W (2009) Interlinking positive and negative feedback loops creates a tunable motif in gene regulatory networks. Phys Rev E Stat Nonlin Soft Matter Phys 80(1 Pt 1):011926. https://doi.org/10.1103/physreve.80.011926
34. Suel GM, Kulkarni RP, Dworkin J, Garcia-Ojalvo J, Elowitz MB (2007) Tunability and noise dependence in differentiation dynamics. Science 315:1716–1719
35. Brandman O (2005) Interlinked fast and slow positive feedback loops drive reliable cell decisions. Science 310:496–498
36. Zhang X-P, Cheng Z, Liu F, Wang W (2007) Linking fast and slow positive feedback loops creates an optimal bistable switch in cell signaling. Phys Rev E Stat Nonlin Soft Matter PhysPhys Rev E 76(3 Pt 1):031924. https://doi.org/10.1103/physreve.76.031924
37. Siemens H, Jackstadt R, Hünten S, Kaller M, Menssen A, Götz U, Hermeking H (2011) miR-34 and SNAIL form a double-negative feedback loop to regulate epithelial-mesenchymal transitions. Cell Cycle 10:4256–4271
38. Brabletz S, Brabletz T (2010) The ZEB/miR-200 feedback loopa motor of cellular plasticity

39. Yamakuchi M, Lowenstein CJ (2009) MiR-34 SIRT1, and p53: The feedback loop. Cell Cycle 8:712–715
40. Rokavec M, Ö-ner MG, Li H et al (2014) IL-6R/STAT3/miR-34a feedback loop promotes EMT-mediated colorectal cancer invasion and metastasis. J Clin Investig 124:1853–1867
41. Wu H, Wang G, Wang Z, An S, Ye P, Luo S (2016) A negative feedback loop between miR-200b and the nuclear factor-κB pathway via IKBKB/IKK-β in breast cancer cells. FEBS J 283:2259–2271
42. Lu Y-X, Yuan L, Xue X-L, Zhou M, Liu Y, Zhang C, Li J-P, Zheng L, Hong M, Li X-N (2014) Regulation of colorectal carcinoma stemness growth, and metastasis by an miR-200c-Sox2-negative feedback loop mechanism. Clin Cancer Res 20:2631–2642
43. Kundu ST, Byers LA, Peng DH, Roybal JD, Diao L, Wang J, Tong P, Creighton CJ, Gibbons DL (2015) The miR-200 family and the miR-183~96~182 cluster target Foxf2 to inhibit invasion and metastasis in lung cancers. Oncogene 35:173–186
44. Ding X, Park SI, McCauley LK, Wang C-Y (2013) Signaling between Transforming Growth Factor β (TGF-β) and Transcription Factor SNAI2 Represses Expression of MicroRNA miR-203 to Promote Epithelial-Mesenchymal Transition and Tumor Metastasis. J Biol Chem 288:10241–10253
45. Yang X, Lin X, Zhong X et al (2010) Double-negative feedback loop between reprogramming factor LIN28 and microRNA let-7 regulates aldehyde dehydrogenase 1-positive cancer stem cells. Cancer Res 70:9463–9472
46. Iliopoulos D, Hirsch HA, Struhl K (2009) An epigenetic switch involving NF-κB Lin28, Let-7 MicroRNA and IL6 links inflammation to cell transformation. Cell 139:693–706
47. Pasquinelli AE (2012) MicroRNAs and their targets: recognition regulation and an emerging reciprocal relationship. Nat Rev Genet 13:271–282
48. Mukherji S, Ebert MS, Zheng GXY, Tsang JS, Sharp PA, van Oudenaarden A (2011) MicroRNAs can generate thresholds in target gene expression. Nat Genet 43:854–859
49. Tian X-J, Zhang H, Zhang J, Xing J (2016) Reciprocal regulation between mRNA and microRNA enables a bistable switch that directs cell fate decisions. FEBS Lett 590:3443–3455
50. Markevich NI, Hoek JB, Kholodenko BN (2004) Signaling switches and bistability arising from multisite phosphorylation in protein kinase cascades. J Cell Biol 164:353–359
51. Ortega F, Garcés JL, Mas F, Kholodenko BN, Cascante M (2006) Bistability from double phosphorylation in signal transduction. FEBS J 273:3915–3926
52. Grande MT, Sánchez-Laorden B, López-Blau C, Frutos CAD, Boutet A, Arévalo M, Rowe RG, Weiss SJ, López-Novoa JM, Nieto MA (2015) Snail1-induced partial epithelial-to-mesenchymal transition drives renal fibrosis in mice and can be targeted to reverse established disease. Nat Med 21:989–997
53. Lovisa S, LeBleu VS, Tampe BÃ et al (2015) Epithelial-to-mesenchymal transition induces cell cycle arrest and parenchymal damage in renal fibrosis. Nat Med 21:998–1009
54. Voon DC, Huang RY, Jackson RA, Thiery JP (2017) The EMT spectrum and therapeutic opportunities. Mol Oncol 11:878–891
55. Huang RY-J, Wong MK, Tan TZ et al (2013) An EMT spectrum defines an anoikis-resistant and spheroidogenic intermediate mesenchymal state that is sensitive to e-cadherin restoration by a src-kinase inhibitor saracatinib (AZD0530). Cell Death Dis 4:e915
56. Tan TZ, Miow QH, Miki Y, Noda T, Mori S, Huang RY-J, Thiery JP (2014) Epithelial-mesenchymal transition spectrum quantification and its efficacy in deciphering survival and drug responses of cancer patients. EMBO Mol Med 6:1279–1293
57. Figliuzzi M, Marinari E, Martino AD (2013) MicroRNAs as a selective channel of communication between competing RNAs: a steady-state theory. Biophys J 104:1203–1213
58. Figliuzzi M, De Martino A, Marinari E (2014) RNA-based regulation: dynamics and response to perturbations of competing RNAs. Biophys J 107:1011–1022
59. Yuan Y, Liu B, Xie P, Zhang MQ, Li Y, Xie Z, Wang X (2015) Model-guided quantitative analysis of microRNA-mediated regulation on competing endogenous RNAs using a synthetic gene circuit. Proc Natl Acad Sci U S A 112:3158–3163
60. Yuan Y, Ren X, Xie Z, Wang X (2016) A quantitative understanding of microRNA-mediated competing endogenous RNA regulation. Quant Biol 4:47–57
61. Bloom RJ, Winkler SM, Smolke CD (2015) Synthetic feedback control using an RNAi-based gene-regulatory device. J Biol Eng 9:5. https://doi.org/10.1186/s13036-015-0002-3
62. Wroblewska L, Kitada T, Endo K, Siciliano V, Stillo B, Saito H, Weiss R (2015) Mammalian synthetic circuits with RNA binding proteins

for RNA-only delivery. Nat Biotechnol 33:839–841

63. Miki K, Endo K, Takahashi S et al (2015) Efficient detection and purification of cell populations using synthetic MicroRNA switches. Cell Stem Cell 16:699–711

64. Yu M, Bardia A, Wittner BS et al (2013) Circulating breast tumor cells exhibit dynamic changes in epithelial and mesenchymal composition. Science 339:580–584

65. Ilina O, Friedl P (2009) Mechanisms of collective cell migration at a glance. J Cell Sci 122:3203–3208

66. Morel M, Shtrahman R, Rotter V, Nissim L, Bar-Ziv RH (2016) Cellular heterogeneity mediates inherent sensitivityspecificity tradeoff in cancer targeting by synthetic circuits. Proc Natl Acad Sci U S A 113:8133–8138

67. Mitchell PS, Parkin RK, Kroh EM et al (2008) Circulating microRNAs as stable blood-based markers for cancer detection. Proc Natl Acad Sci U S A 105:10513–10518

68. Fischer KR, Durrans A, Lee S et al (2015) Epithelial-to-mesenchymal transition is not required for lung metastasis but contributes to chemoresistance. Nature 527:472–476

69. Zheng X, Carstens JL, Kim J, Scheible M, Kaye J, Sugimoto H, Wu C-C, LeBleu VS, Kalluri R (2015) Epithelial-to-mesenchymal transition is dispensable for metastasis but induces chemoresistance in pancreatic cancer. Nature 527:525–530

70. Yoon J-H, Abdelmohsen K, Gorospe M (2014) Functional interactions among microRNAs and long noncoding RNAs. Semin Cell Dev Biol 34:9–14

71. Tay Y, Rinn J, Pandolfi PP (2014) The multilayered complexity of ceRNA crosstalk and competition. Nature 505:344–352

72. Hansen TB, Jensen TI, Clausen BH, Bramsen JB, Finsen B, Damgaard CK, Kjems J (2013) Natural RNA circles function as efficient microRNA sponges. Nature 495:384–388

Chapter 17

Modeling Long ncRNA-Mediated Regulation in the Mammalian Cell Cycle

Jomar F. Rabajante and Ricardo C. H. del Rosario

Abstract

Long noncoding RNAs (lncRNAs) are transcripts longer than 200 nucleotides that are not translated into proteins. They have recently gained widespread attention due to the finding that tens of thousands of lncRNAs reside in the human genome, and due to an increasing number of lncRNAs that are found to be associated with disease. Some lncRNAs, including disease-associated ones, play different roles in regulating the cell cycle. Mathematical models of the cell cycle have been useful in better understanding this biological system, such as how it could be robust to some perturbations and how the cell cycle checkpoints could act as a switch. Here, we discuss mathematical modeling techniques for studying lncRNA regulation of the mammalian cell cycle. We present examples on how modeling via network analysis and differential equations can provide novel predictions toward understanding cell cycle regulation in response to perturbations such as DNA damage.

Key words lncRNA, Cell cycle, Mathematical model, Regulation, Networks, DNA damage

1 Introduction

Long noncoding RNAs (lncRNAs) are transcribed RNAs longer than 200 nucleotides but are considered as having no potential to be translated into proteins. Although their functional evidence as a group is not yet well established, a number of individual examples have shown that they regulate gene expression during development, differentiation, and cell cycle, and that some lncRNAs have been shown to be implicated with disease [1–7]. Genome-scale identification of lncRNAs has indicated that there could be tens of thousands of lncRNA in the human genome [7–9]. Thus it is unsurprising that they have recently gained widespread attention, leading to more understanding of their biological function and mechanisms of action [5, 7, 10–14]. An example of a lncRNA with disease associations is HOTAIR (HOX transcript antisense intergenic RNA) which has been shown to be an oncogene for breast, colorectal, hepatocellular, and gastrointestinal stromal tumors [15, 16]. Examples of lncRNA implicated in non-small cell lung cancer are BANCR (BRAF-activated ncRNA), CARLo-5 (cancer-associated region lncRNA-5), MEG3 (maternally expressed 3), TUG1 (taurine upregulated 1), and MALAT1 (metastasis-associated lung adenocarcinoma transcript 1) [17]. CARLo-5 is an oncogene connected to various pathways, such as apoptosis, cell cycle regulation, and epithelial-mesenchymal transition (EMT) [17]. The enormity and diversity of lncRNA molecules provide opportunities and challenges for the discovery of new biomarkers for disease diagnosis, monitoring, prognosis, and therapy [3, 4, 17].

The cell cycle is composed of sequential events with four main phases, the G1, S, G2, and M phases. There are checkpoints in between these phases that decide if the proper conditions in the cell are met before the cycle is allowed to progress to the next phase. Accurate upregulation or downregulation of cell cycle regulatory factors, such as cyclin-dependent kinases (CDKs), are necessary for normal cell cycle progression, and dysregulation of these factors could result in disease [18–20]. DNA damage from radiation, ultraviolet light, or genotoxic substances is addressed by activating signaling pathways (e.g., initiated during G1 and G2 checkpoints) that induce cell cycle arrest and DNA repair [21–24]. If DNA damage is irreparable, the activation of programmed cell death or "suicide" called apoptosis may occur [24–26]. Irreparable DNA damage may cause apoptosis, senescence, or unregulated cell proliferation leading to cancer [4, 27]. Several lncRNAs are upregulated or downregulated by stress signals, and consequently influence the cell's DNA damage response mechanisms, e.g., through G1 arrest [15, 23]. Moreover, the involvement of lncRNAs in cell cycle control, especially in maintaining genome integrity, is not only confined with DNA damage response. The lncRNAs in cell cycle control perform different gene-regulatory roles, such as being direct transcription factor regulators, epigenetic regulators by chromatin modifier collaboration, post-transcriptional regulators by binding to target mRNA to suppress translation or modulate mRNA stability, and protein scaffolds by controlling protein-protein interactions [28]. For example, the inhibition of lncRNA TERRA (telomeric repeat-containing RNA) during the S phase can stimulate activation of telomerase that may drive telomere extension [12].

The cell cycle system is highly consistent across mammalian species. However, lncRNA types are observed to vary across mammalian species since lncRNA genes have faster evolutionary turnover (gain and lost) than protein-coding genes [29–31]. The lncRNAs perform a diversity of biological functions and it is observed that their expressions are more specific (tissue specific, spatiotemporal specific) than protein-coding genes [8].

Mathematical models have been found useful in understanding the dynamic interactions of the genes involved in regulating the cell cycle [18, 19, 22, 25, 32–34]. Thus, we propose that mathematical models of lncRNA regulation of the genes involved in the cell cycle could similarly be useful in a better understanding of the role of lncRNAs in regulating the cell cycle. Below, we present some general paths on how to set up such models and how to analyze them.

2 Pathways and Networks

To model the regulation of mammalian cell cycle by lncRNAs, one would start with a pathway or a network of genes including lncRNAs involved in a specific phase of the cell cycle. A pathway is a sequence of molecular activities resulting in changes in the cell, such as gene regulation and protein synthesis. The interaction among coding and noncoding genes, among group of molecules, or among pathways can be studied using a network. We simply denote a "network" as a list of nodes, e.g., genes, with an edge connecting two nodes if there is some interaction between them. Classical network models are static (i.e., the expression levels of the genes do not change in time) but their analysis can provide a lot of insight into the biological system being studied. In this section, we provide a brief summary of how to set up pathways and networks using available databases [35]. Note that even though we present below the use of databases for constructing a network, as many models of the cell cycle are now available, one could start with a known network of a cell cycle and then use databases to add lncRNA regulation. Biological pathways and networks can be analyzed using various computational and mathematical methods [32, 36–38], such as using techniques in network science [37, 39–42].

2.1 Databases and Networks

A number of curated databases (e.g., Gene Expression Omnibus, ENCODE, ArrayExpress, and Ensemble) contain experimental omics data (e.g., from microarray and RNA-seq experiments) and can be bioinformatically analyzed to create networks. Network construction requires finding patterns in the data using algorithms from machine learning and statistics, e.g., artificial neural networks and principal component analysis [42–47]. From these databases, we can draw networks and pathways that can be used in the dynamic modeling of gene interaction. However, networks and pathways are difficult to be inferred from a single set of experimental data alone due to combinatorial constraints and complexity of the biological processes involved. Knowledge of the mechanisms and interaction from literature, and combination of existing (e.g., from online databases) and new experimental results, is helpful in inferring and comparing networks and pathways [28, 45].

For example, datasets from DNA microarray or from RNA-seq experiments are processed and summarized to generate a gene expression matrix. This gene expression matrix is converted using algorithms to generate a gene-gene adjacency matrix, which can be visualized using networks [48–51]. In the network, nodes represent genes and edges represent significant co-expression. Gene co-expression networks illustrate correlated expression of genes in response to the stimuli introduced in the experiments [49–51]. The co-expression networks do not present mechanistic and causal interaction among the genes, but these correlation-based networks can be used as a starting point for further investigation of gene associations and interactions (e.g., gene set enrichment analysis and pathway analysis) [38, 45]. In the case of lncRNA in cell cycle, gene co-expression networks can be used to determine the cell cycle regulators that have correlated expression with lncRNAs.

Various online databases are available for investigating cell cycle regulation and lncRNAs. Functions of genes, including known lncRNAs, can be searched through the Gene Ontology Consortium database (http://geneontology.org) and GeneCards (http://www.genecards.org, a human gene database). RNAcentral (http://rnacentral.org) is one of the online public databases specific for noncoding RNAs. Cross-referencing various data sources and functional annotation of lncRNA are helpful in identifying its significantly associated biological pathways [11, 38, 46, 52]. KEGG, Reactome, WikiPathways, and PathCards (under GeneCards) are popular databases of curated pathways, which can be used for cross-referencing to infer a lncRNA-cell cycle pathway. Figure 1 shows a simple example of the diverse roles of lncRNA in pathways related to cell cycle regulation. A number of lncRNAs influence the upregulation or downregulation of cyclin and CDK (cyclin-dependent kinases), which are considered as the core regulatory factors of the cell cycle system [18, 24, 28, 53].

The lncRNAs affect cell cycle control through various pathways (Fig. 1). DNA damage induces ncRNA$_{CCND1}$ (promoter-associated ncRNA inhibiting cyclin D1) and/or gadd7 (growth-arrested DNA damage-inducible lncRNA) leading to the suppression of the corresponding cyclin-CDK complexes, which will result in G1 arrest [28]. Moreover, the depletion of ANRIL (antisense noncoding RNA in the INK4 locus) is DNA damage induced that can drive G1 arrest by promoting associated CKIs (CDK inhibitors). The activation of ANRIL plays a role in cell proliferation and inhibition of premature senescence [28]. These three pathways, via ncRNA$_{CCND1}$, gadd7, and ANRIL, suggest alternative or supplementary routes to the classical ATM/ATR pathway in inhibiting cell cycle progression as DNA damage checkpoint. The classical ATM/ATR pathway activates p53 which is the "gatekeeper" of the genome in the presence of damage in the DNA [23, 24, 28].

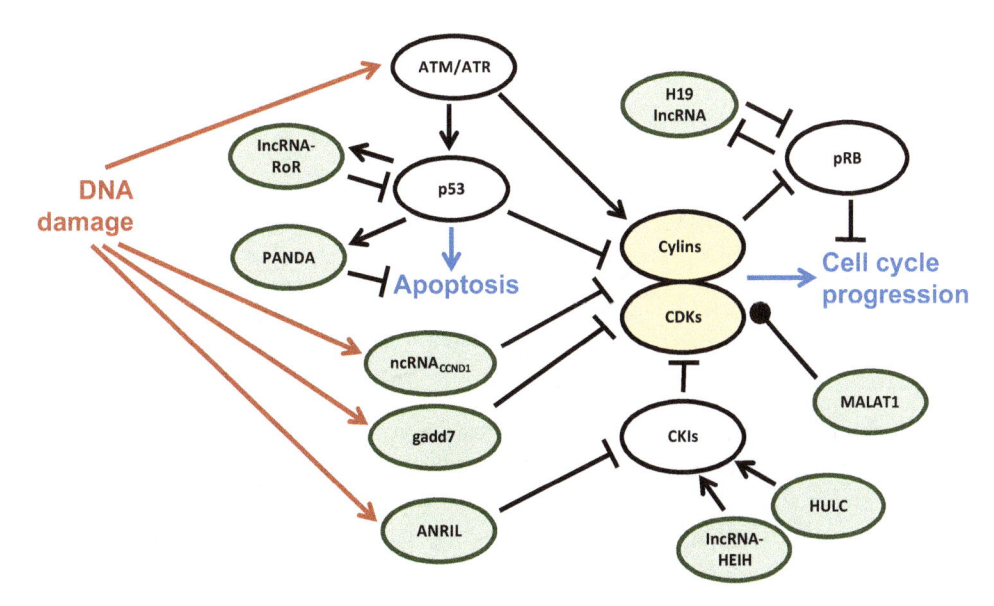

Fig. 1 Diverse roles of lncRNA (green circles) in regulating cell cycle-related pathways. Cyclins and CDKs (yellow orange circles) are core regulatory factors of the cell cycle. Black arrows indicate activation, black hammerhead lines denote inhibition, red arrows indicate external input, and black line ending in circle denotes differential action depending on the condition

Other lncRNAs affect the cell cycle in various ways (Fig. 1; [28]) through the p53 pathway (e.g., lncRNA-ROR that affects G2-M progression, and PANDA which suppresses apoptosis), through the regulation of the tumor suppressor pRB that regulates G1-S progression (e.g., H19 lncRNA affecting G1 phase), and through the regulation of CKIs (e.g., HULC affecting G1 phase, and lncRNA-HEIH affecting G0-G1 progression). Moreover, MALAT1 differentially affects cyclins and CDKs in different cell cycle phases to modulate cell cycle regulation [17, 28]. Downregulation of MALAT1 drives G1 arrest and activation of p53 pathway. Upregulation of MALAT1 may activate B-MYB leading to G2-M progression [28].

GAS5 (growth arrest specific 5) lncRNA is activated by stress signals, e.g., by nutrient deprivation and withdrawal of growth factors. This lncRNA inhibits the expression of glucocorticoid-responsive (GR) gene, which inhibits apoptosis (Fig. 2; [27]). In an alternative pathway (Fig. 2), the downregulation of GAS5 activates p53 resulting in cell cycle arrest, allowing DNA repair and cell cycle restart (rescue), which consequently avoids apoptosis. However, the activation of p53 can also initiate apoptosis if DNA damage is irreparable. There is a complex interface between apoptosis and DNA repair [54, 55]. As a result of the differential outcome of GAS5 regulation, GAS5 is considered as both a tumor suppressor and oncogene. Further studies are needed to fully understand the

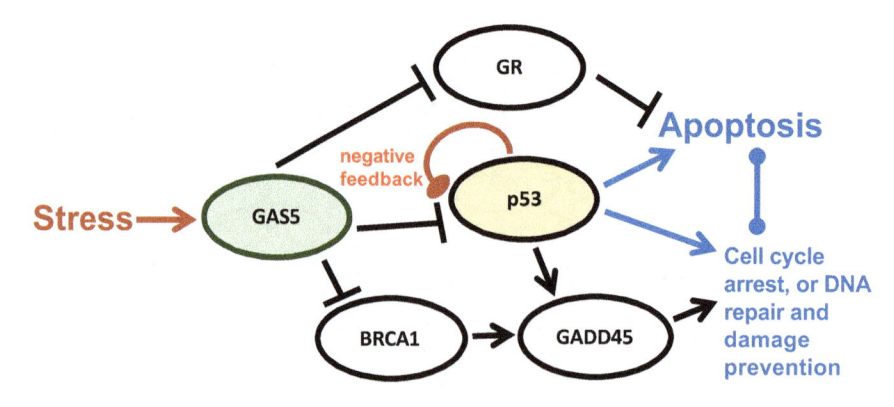

Fig. 2 GAS5 is one of the lncRNAs that has differential influence on cell cycle regulation. Line types same as in Fig. 1. Blue line ending in circle denotes complex interface between apoptosis and DNA repair

detailed function and mechanisms of GAS5 regulatory system under different conditions, especially in relation to diseases [27].

2.2 Analysis of Networks

Static measures in graph theory and network science are helpful in identifying patterns in gene-gene interaction [41, 42]. Some basic metrics are enumerated in Table 1. For example, in a gene co-expression network, a node with the largest degree characterizes the most connected gene in the network, referred to as a "hub." A hub lncRNA means that it is the most pairwise co-expressed lncRNA in the network, a potential biomarker in disease pathways. The theory behind the importance of network topology and how to analyze robustness of networks can be found in [42], while one of the most popular software for visualizing and analyzing networks is Cytoscape (http://www.cytoscape.org).

In biological networks, recurring patterns are often observed, such as network motifs that are frequently appearing subnetworks. These network motifs may have functional significance in biological processes. Examples of network motifs in gene regulation are presented in Fig. 3. Identifying the network motifs that involve lncRNAs can be helpful in determining the roles of lncRNAs in gene regulation, such as in cell cycle control.

Another valuable area in network science for studying lncRNA is the use of multipartite graphs [56, 57]. Here, we can do network projection to predict possible connections between lncRNAs [42]. We can also predict the possible association of lncRNAs, coding genes, functions, and other aspects of interest (Fig. 4). For example, if we have data on lncRNA-disease association and coding gene-disease association (diseasome), then we can hypothesize lncRNA-coding gene association [56–59]. Diseases that are found to have link to the aberration of lncRNA function are wide ranging, including neurodegeneration and cancer [3]. Similarly, if we have lncRNA-coding gene association and coding gene-

Table 1
Basic metrics in network science that are used to analyze network models

Metric	Definition
Size of the network	Number of nodes
Node degree[a]	Number of edges connected to the node (for a directed network, we can also define incoming degree and outgoing degree)
Degree distribution	Probability distribution of the degrees in the network (this can be used to determine the probability that a randomly selected node in the network has degree k)
Density of a network with N nodes	Ratio of the total number of edges in the undirected network to the total number of edges in a complete network with N nodes (total number of edges in a complete network = $N(N-1)/2$)
Path length	Number of edges in a path between two nodes
Distance	Path length of the shortest (geodesic) path between two nodes
Node eccentricity[a]	Maximum distance between the node and all other nodes in the network
Network diameter	Longest distance in the network (this is also the maximum eccentricity in the network)
Network radius	Minimum graph eccentricity in the network
Node farness	Sum of the distances between the node and all other nodes in a connected network (sometimes, the average is used rather than the summation of the distances)
Node closeness[a]	Reciprocal of the node farness (this characterizes how many steps it will take to send a signal from the node to all other nodes in a connected network)
Local clustering coefficient	Probability that two neighboring nodes of a node are directly connected to each other (this measures the local link density of the network)
Global clustering coefficient	Total number of closed triangles in a network

[a]Can be used as a centrality measure, an indicator of the significance of the node in the network

function association, then we can predict how a lncRNA influences a biological function and pathway [56, 60].

3 Dynamical Modeling of Interactions

Now that we have set up and analyzed a discrete network model, the next step is to create a mathematical model that incorporates changes in gene expression over time. We can assign corresponding expression levels to the nodes in gene regulatory networks, and interaction coefficients to the edges. This means that the network is now dynamic as interactions among genes may vary. However, analyzing a large network can be intractable, and coarse graining (simplification) may be necessary. For example, if gene X activates

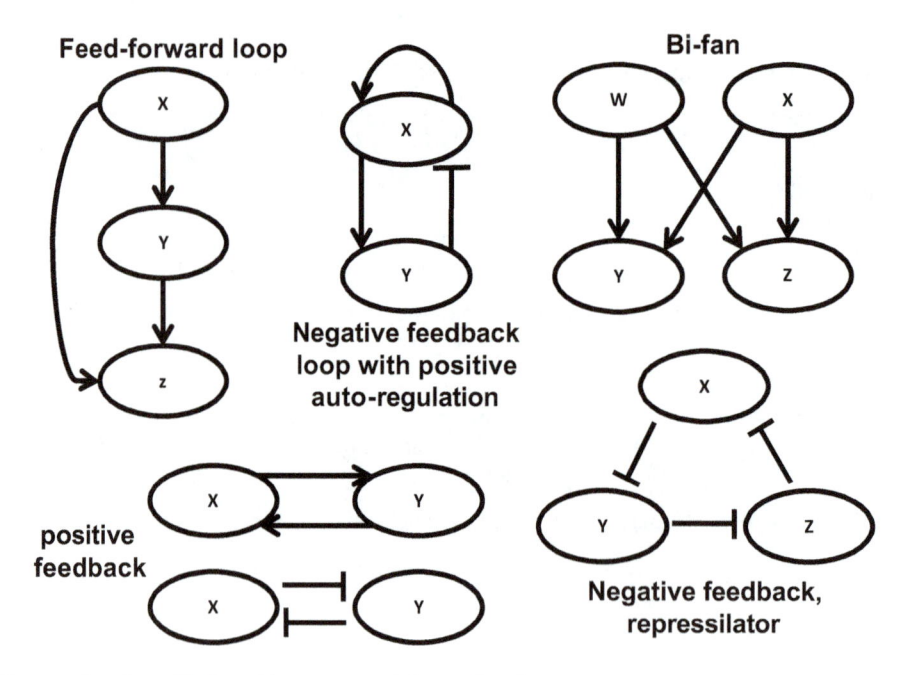

Fig. 3 Some network motifs found in gene-regulatory networks

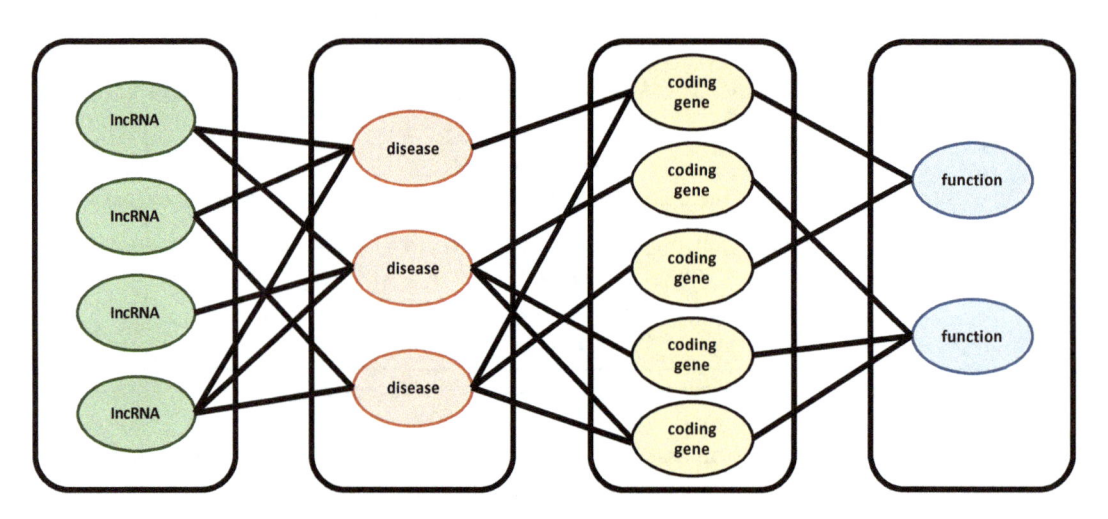

Fig. 4 Example of how a multipartite graph can be used for predicting lncRNA associations

gene Y, and Y activates gene Z, then we can simplify this as X activates Z. Nevertheless, it should be noted that coarse-grained modeling may miss details of the resulting dynamics, such as the effect of feedback delay that may cause oscillatory gene expression.

One important mathematical pattern observed by Thomas [37, 39, 40, 61–63] is the relationship between network topology and the resulting network dynamics. In a feedback cycle with more than one node, odd number of inhibitions (e.g., repressilator) can result in oscillations. An even number of inhibitions can result in equilibrium. Moreover, negative feedback loops are common ingredients of checkpoints (e.g., stabilizing self-inhibition) and also of cell cycle oscillations [19]. Positive feedback loops are common in bistable switches (e.g., in cell cycle transitions), but in some cases they can amplify oscillations driven by negative feedback [19].

There are various modeling approaches/formalisms that are available and widely used. Networks are themselves models of interaction. Other examples are differential equations and Boolean logic algorithms, which are usually formulated based on the interaction networks [32, 34, 61]. The choice of which modeling formalism to be used may depend on the applicability of the model assumptions to address the goal of the study, and on the available data. The more parameters the model has, the more experimental data are needed for parameter value fitting. Differential equation models usually require more experimental data than Boolean models [37, 61].

3.1 Differential Equations

Differential equations (DEs) are useful in modeling changes in gene expression levels. The assumptions of DE modeling include homogeneity of molecular species, continuous time, and continuous states. Homogeneity usually arises if we consider dense (not sparse) spatial distribution of molecules, and their interaction has regular macroscopic (i.e., considering population of molecules and not individual) behavior. The states here represent the concentration or expression levels of the species, such as coding and noncoding genes including lncRNA.

The general form of the DE model representing the net growth rate of the gene expression level follows: $\frac{dx}{dt} = P - D$, $x(0) = x_0$, where x is the expression level of gene X, and x_0 is the initial value (or initial condition at the start of the simulation). Here, P is the production function and D is the degradation function. The production function may include the effect of self-activation or self-inhibition of the gene, the activation and inhibition by other genes, and the residual or basal term. When formulating a model, the existence and uniqueness of its solution should be assured (*see* Lipschitz condition [32, 36, 37]). Moreover, the state variable (x) should never be negative to be biologically relevant. We also assume nonnegative parameters. The degradation function is commonly assumed to be of first order (exponential decay), $D = \gamma x$ where γ is the degradation constant. Each gene in the network will have a corresponding state variable modeled by an equation.

3.1.1 Hill Functions

There are many frameworks in DE modeling, and the choice depends on the current knowledge and assumptions on the biological system being studied [32, 64–67]. One of the classical DE modeling frameworks invokes the *law of mass action* [32, 36, 37, 68] and is commonly used in modeling gene transcription. We suppose n molecules of enhancer or repressor protein A bind to the promoter site of gene B with transcription rate k_1, resulting in complex C (i.e., $nA + B \to C$; also assume $C \to nA + B$ with rate k_2). Let the states a, b, and c be the concentration level of A, B, and C, respectively. Assuming conservation of mass (i.e., $b + c = k$ (k is constant)), then we have the following model for the change in the concentration of B and C:

$$\frac{dc}{dt} = k_1 a^n b - k_2 c \text{ and } \frac{db}{dt} = -\frac{dc}{dt} = -k_1 a^n b + k_2 c.$$

Processes may occur at different timescales, and we can approximate the "fast" state to converge to a quasi-steady state [36, 36]. Since transcription is generally slower than binding, then we set the concentration of B and C to converge to a quasi-steady state, $\frac{db}{dt} \approx 0$ and $\frac{dc}{dt} \approx 0$. Consequently, setting $k_1 a^n b - k_2 c = 0$ leads to

$$b = \frac{k\frac{k_2}{k_1}}{\frac{k_2}{k_1} + a^n} \text{ and } c = \frac{ka^n}{\frac{k_2}{k_1} + a^n}.$$

The quasi-steady-state values of b and c are of the form conforming to a Hill function ($n \geq 2$) that characterizes a sigmoid curve [36, 69]. If $n = 1$, they conform to Michaelis-Menten kinetics with hyperbolic curve. Moreover, if protein A enhances the formation of C then $c > b$; otherwise, A represses the formation of C, $b > c$.

The above discussion is one of the motivations to use the following Hill functions (commonly, $n \geq 2$ for gene regulation [37]) to represent activation and inhibition, respectively: $H_A = \frac{k_A a^n}{\phi_A^n + a^n}$ and $H_I = \frac{k_I}{\phi_I^n + a^n}$ where k_A, k_I, ϕ_A, and ϕ_I are some constants. For activation, $H_A \to k_A$ as $a \to \infty$, and $H_A \to 0$ as $a \to 0$. For inhibition, $H_I \to 0$ as $a \to \infty$, and $H_I \to k_I$ as $a \to 0$. The parameter ϕ_A (ϕ_I) is the threshold constant where $H_A = \frac{k_A}{2}$ ($H_I = \frac{k_I}{2}$) when $a = \phi_A$ ($a = \phi_I$), which is the inflection point of the sigmoid curve. For very large n, the sigmoid curve approximates a step function, where the jump happens at $a = \phi_A$ ($a = \phi_I$).

We present here three DE models; other models are also available in the literature [32, 34, 36, 37]. The interpretations of the parameters in models 1 and 2 are not necessarily the same.

3.1.2 Model 1: Additive and Independent Activation and Inhibition

Suppose the states x, y, and z are the expression level of genes X, Y, and Z, respectively. Here, lncRNA can be one of the genes. If gene Y activates gene X, and gene Z inhibits gene X, we can have the following model for the change in x:

$$\frac{dx}{dt} = R + \frac{k_A y^n}{\phi_A^n + y^n} + \frac{k_I}{\phi_I^n + z^n} + k_0 - \gamma x.$$

The first four terms are components of the production function, and the last term is the degradation term. The second term characterizes the activation of X by Y, and the third term characterizes the inhibition of X by Z. The fourth term is the residual or basal term that upregulates X even without activation. The first term reflects self-activation or self-inhibition of X (with r_i as the self-regulation rate; $r_i = 0$ if no self-regulation):

$$R = \begin{cases} \dfrac{r_A x^n}{\phi_{Ax}{}^n + x^n} & \text{self} - \text{activation} \\ \dfrac{r_I}{\phi_{Ix}{}^n + x^n} & \text{self} - \text{inhibition}. \end{cases}$$

If there are more genes influencing X, then additional terms can be added. Suppose there are m inhibitory genes, we can set the inhibitory rates to $\dfrac{k_{I,1}}{\phi_{I,1}{}^n} + \dfrac{k_{I,2}}{\phi_{I,2}{}^n} + \ldots + \dfrac{k_{I,m}}{\phi_{I,m}{}^n} + \dfrac{r_I}{\phi_{Ix}{}^n} = K$, where K is the initial gross growth rate of the expression level of X without activation, inhibition, k_0, and degradation.

3.1.3 Model 2: Multiplicative Activation and Inhibition

If gene Y activates gene X, and gene Z inhibits gene X, we can have the following model:

$$\frac{dx}{dt} = R\left(1 + \frac{k_A y^n}{\phi_A{}^n + y^n}\right)\left(\frac{k_I}{\phi_I{}^n + z^n}\right) + k_0 - \gamma x.$$

The first two terms are components of the production function, and the last term is the degradation term. The first factor in the first term characterizes the activation of X by Y, similar to model 1 except that if there is no self-regulation, we can set constant $R = r$. The second factor characterizes the inhibition of X by Z. We can set $\phi_A{}^n = k_A$ and $\phi_I{}^n = k_I$. If there are more genes influencing X, then additional factors can be introduced.

3.1.4 Model 3: Non-Hill-Type Logistic Model

If gene Y activates gene X, and gene Z inhibits gene X, we can have the following model:

$$\frac{dx}{dt} = (r_y yx - r_z zx)\left(1 - \frac{x}{x_{\max}}\right).$$

The first term in the first factor characterizes the activation of X by Y, where an increase in y contributes to a proportional increase in x. The second term in the first factor characterizes the inhibition of X by Z, where an increase in z contributes to a proportional decrease in x. The second factor limits the upregulation of x up to its maximal expression $x_{\max}$. If there are more genes influencing X, then additional terms in the first factor can be added.

The parameter values in models 1 and 2 are usually estimated by fitting time series gene expression data. Model 3 can be useful if time series dataset is not available. Moreover, models 1, 2, and 3 and other models not discussed here can be combined in a hybrid model to represent a more complex system of gene regulation.

3.1.5 *Analysis of Models*

The models that we have presented are nonlinear DEs which often do not have closed-form solutions. However, we can do qualitative analysis and numerical simulations, which are common ways to investigate the resulting dynamics of the DE system. A lncRNA may affect the transient dynamics of its associated gene regulatory network. LncRNA may also affect the steady-state expression of genes [27, 28].

In qualitative analysis, we usually investigate the long-term behavior of the solution to the DE as time approaches infinity. Examples of long-term behavior include convergence to equilibrium or steady state, and convergence to perpetual oscillation (limit cycle). Limit cycles are observed in simulations representing cell cycle oscillations and circadian rhythms [19, 20, 34]. For example, limit cycles may illustrate the periodic expression of cohesion regulator noncoding RNA (CONCR) with its maximal expression occurring during the mid-late G1 phase [70]. Chaos and heteroclinic cycles are also possible dynamics in nonlinear DEs but they are beyond the scope of this chapter; we will focus our discussions on convergence to equilibrium [32, 36].

To determine the equilibrium points of the DE model, we equate all the differential equations to zero in our model. For example, set $\frac{dx}{dt} = 0$, $\frac{dy}{dt} = 0$, and $\frac{dz}{dt} = 0$, and then find all the combinations of values of x, y, and z that satisfy these equations. Each combination is an equilibrium point, denoted by $(x^*, y^*, z^*)_{i=1, 2, \ldots, p}$ where p is the total number of combinations.

An equilibrium point can be stable or unstable (Fig. 5), which can be determined by perturbation. If the initial value of the DE model corresponds to the equilibrium, then the solution to the DE stays on the equilibrium. If we introduce arbitrary small perturbation to the initial value that is very close to the equilibrium, and the

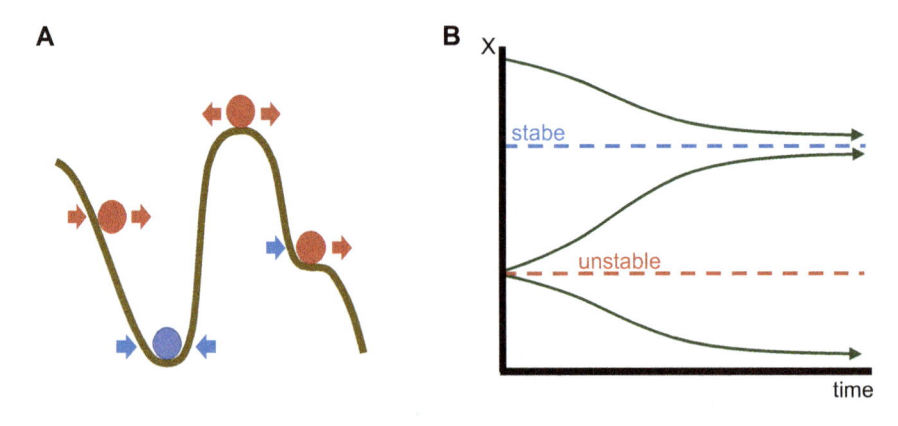

Fig. 5 Stable and unstable equilibria. (**a**) An illustration showing that a solution converging to a stable equilibrium (blue circle) will remain converging toward the same equilibrium even by introducing small perturbation. Red circles are unstable. (**b**) Solution to the DE model converges to stable equilibrium and diverges from unstable equilibrium

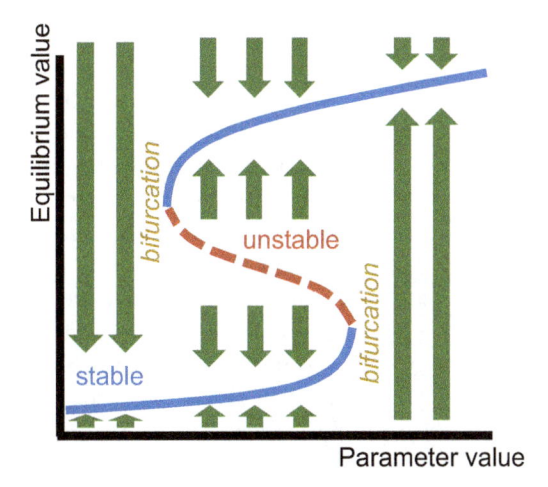

Fig. 6 An illustration of bistability in dynamical systems. The long-term fate of the solution to the DE model depends on the parameter value and initial condition

solution to the DE model converges to the same equilibrium then the equilibrium is stable. If the solution diverges, then the equilibrium is unstable. There are various mathematical techniques for stability analysis (e.g., using the Jacobian matrix); we suggest referring to available books [32, 36].

It is possible to have multiple stable equilibria (multistability). The set of all initial conditions converging to an equilibrium point is called the *basin of attraction* of that equilibrium point. Multistability, specifically bistability, plays a big role in gene regulation as switches [25, 69, 71]. Moreover, bifurcation analysis can be done to identify at what parameter values the long-term dynamics of the solution to the DE model changes its behavior [32, 36]. For example (Fig. 6), for small and high values of the parameter, there is only one stable equilibrium point. At medium values of the parameter, there are three equilibrium points where two are stable (Fig. 6). The critical parameter values between the cases with one and two stable equilibrium points are called bifurcation points. In some cases, it is also possible to find bifurcation from equilibrium toward formation of limit cycle; this is called Hopf bifurcation [19, 36, 71].

Numerical methods are used to simulate the solution to the DE models. Runge-Kutta 4 is one of the standard algorithms for simulation. There are various methods available, which have their own advantages and disadvantages [72]. Examples of software that can be utilized are Virtual Cell (http://vcell.org), Berkeley Madonna (https://www.berkeleymadonna.com), and CellDesigner (http://www.celldesigner.org). Numerical bifurcation analysis of ordinary DEs encoded in MATLAB can be implemented using the MATCONT software [73].

Validation of simulation results is indispensable in a holistic modeling of biological processes. We can determine the reliability of our simulation results by (1) identifying the scope and limitations of the applicability of our results vis-à-vis the set assumptions; (2) determining how accurate and precise our results are in comparison with available data; and (3) testing the sensitivity and robustness of our results against random noise, and against changes in assumptions, parameter values, and model structure. Parameter estimation can be done to determine how our model fits with experimental data. However, even with the number of databases that contain lncRNA information, the lack of available and appropriate data could be a challenge in cell cycle research due to the complexity and diversity of lncRNAs and other related factors.

3.2 Stochasticity

The models discussed above are deterministic. However, gene regulation contains randomness, especially because molecular interactions usually violate the homogeneity assumption of DEs. Stochastic models, such as the Gillespie algorithm and stochastic DEs (SDEs) [72, 74, 75], can be used to incorporate probability to our simulations. Here, we only discuss how to incorporate randomness to the DE models discussed above. There are many possible ways, but we discuss two techniques.

The first technique is incorporating random noise to the change in gene expression level (e.g., $\frac{dx}{dt}$). This is sometimes referred to as the demographic noise [74]. The basic form is

$$dx = (P - D)dt + \sqrt{P + D}\sigma_x dW_x.$$

In this model, it is assumed that the production function P is always nonnegative. The dW is the derivative of Wiener process (also called Brownian motion), referred to as the *white noise*. The parameter σ_x is the amplitude of white noise. For details on how to derive SDEs with demographic noise, refer to [74].

The second technique is incorporating random noise to the parameters, referred to as environmental noise [74]. Here, we use the Langevin equation, where the solution is an Ornstein-Uhlenbeck stochastic process (mean-reverting). As an example, let us use model 3: $\frac{dx}{dt} = (r_y y x - r_z z x)\left(1 - \frac{x}{x_{\max}}\right)$ with stochastic parameter r_y. Suppose the mean of r_y is $\overline{r_y}$; the corresponding SDE is

$$dr_y = \alpha(\overline{r_y} - r_y)dt + \sigma_{r_y} dW_{r_y}.$$

The term with $\overline{r_y} - r_y$ represents drift toward the mean $\overline{r_y}$. The parameter α is the drift rate and σ_{r_y} is the amplitude of white noise. The initial condition is $r_y(0) = \overline{r_y}$.

For numerical simulations of SDEs, we can use various methods, such as the Euler-Maruyama method [72]. An SDE model has many solutions, called realizations or sample paths, since we are dealing with randomness. By using the two techniques discussed

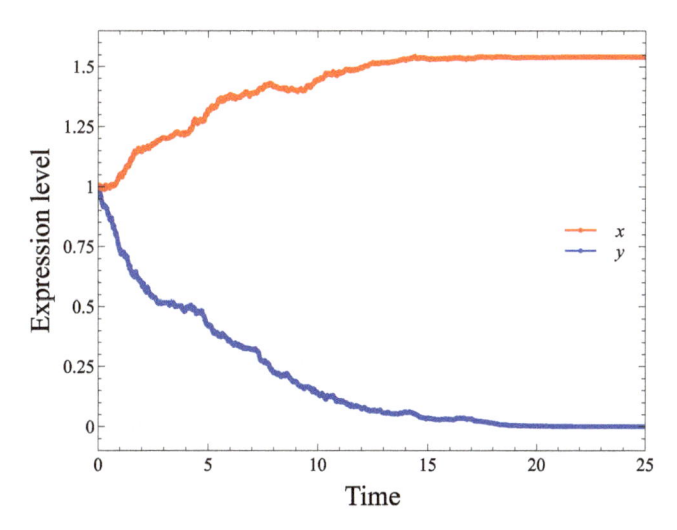

Fig. 7 Simulation of possible dynamics of lncRNA-RoR and p53 interaction. Activation of lncRNA-RoR represses p53 expression, leading to cell cycle progression

here, we can investigate the distribution of the resulting sample paths, and how the sample paths deviate from the deterministic solution.

The use of dynamic models (e.g., deterministic and stochastic DEs) is invaluable in simulating the possible dynamics of gene regulatory networks and pathways, such as those presented in Figs. 1 and 2. Differential equations can provide novel predictions toward understanding the influence of lncRNA in cell cycle regulation, especially as a response to DNA damage to maintain genome stability.

3.3 Numerical Example

Figure 7 illustrates an SDE-based simulation of the regulation of p53 by lncRNA-ROR, modeled using differential equation model 3 discussed above. The p53, which inhibits its own regulation, is repressed by lncRNA-RoR but activates the expression of lncRNA-RoR. The upregulation of lncRNA-RoR results in the downregulation of p53, which leads to cell cycle progression. The model in this example (where x and y characterize the expression levels of lncRNA-RoR and p53, respectively) is

$$dx = ryx\left(1 - \frac{x}{10}\right)dt + \sqrt{ryx\left(1 + \frac{x}{10}\right)}dW_x, x(0) = 1$$

$$dy = (-ry^2 - rxy)\left(1 - \frac{y}{10}\right)dt + \sqrt{(ry^2 + rxy)\left(1 + \frac{y}{10}\right)}dW_y, y(0) = 1$$

$$dr = (0.1 - r)dt + dW_r, r(0) = 0.1.$$

4 Conclusion

Mathematical modeling is an iterative process that includes (1) identifying the problem and the goal of the study; (2) selecting the assumptions in accordance to the set goal; (3) designing and formulating the model; (4) analyzing and simulating the solution to the model; and (5) validating the results. There are many factors related to cell cycle regulation including differential effect of lncRNAs that can be incorporated in the model, which may result in varying levels of complexity. However, we can derive or use simple models to answer the important questions that we aim to answer in the study. Our model could be as wide ranging or limited as our set assumptions. Investigating the influence of lncRNA can be studied with the lncRNA at the center, or lncRNA as just one of the factors involved in a complex interaction system.

In modeling lncRNA-cell cycle nexus, we may deal with interactions that occur in different timescales (from seconds to days) involving varying expression levels. Compared to mRNAs with a wide range of expression levels, the expression levels of lncRNAs are found to be very low to moderate [29, 76]. Proper parametrization is important when dealing with different timescales and expression levels, and also because indirect and delayed molecular signaling by lncRNA may happen. Moreover, lncRNA may influence protein-coding genes in *cis* (in close proximity) or, at a broader coverage, in *trans* (at a distance) [29, 77]. One of the reasons for the diverse dynamics of lncRNAs is that every lncRNA family may play a unique function in cell cycle regulation.

Mathematical modeling can be used as a supplementary tool to lncRNA research. Here, network analysis and differential equations are discussed, with the goal of providing tools that can aid in addressing knowledge gaps in understanding and evaluating the roles of lncRNAs in the mammalian cell cycle. Further research is needed to elucidate the mechanisms, dynamics, and implications of lncRNA-mediated regulation in the mammalian cell cycle, especially the lncRNAs linked with the initiation and progression of diseases.

Acknowledgments

This work is dedicated to the memory of Dr. Baltazar D. Aguda. JFR is supported by the PCARI-CHED IHITM 2017-018 project: Glycoproteomics of Filipino lung cancer cell lines for biomarker discovery and anti-cancer screening of natural products.

References

1. Rinn JL, Chang HY (2012) Genome regulation by long noncoding RNAs. Annu Rev Biochem 81:145–166
2. Toth KF, Hannon G (2012) Non-coding RNAs as regulators of transcription and genome organization. In: Genome organization and function in the cell nucleus. Wiley-VCH Verlag GmbH & Co., Weinheim. https://doi.org/10.1002/9783527639991.ch13
3. Wapinski O, Chang HY (2011) Long noncoding RNAs and human disease. Trends Cell Biol 21:354–361
4. Schmitt AM, Chang HY (2016) Long noncoding RNAs in cancer pathways. Cancer Cell 29:452–463
5. Wilusz JE, Sunwoo H, Spector DL (2009) Long noncoding RNAs: functional surprises from the RNA world. Genes Dev 23:1494–1504
6. Perry RB-T, Ulitsky I (2016) The functions of long noncoding RNAs in development and stem cells. Development 143:3882–3894
7. Marchese FP, Raimondi I, Huarte M (2017) The multidimensional mechanisms of long noncoding RNA function. Genome Biol 18:206
8. Derrien T, Johnson R, Bussotti G et al (2012) The GENCODE v7 catalog of human long noncoding RNAs: Analysis of their gene structure, evolution, and expression. Genome Res 22:1775–1789
9. Ziegler C, Kretz M (2017) The More the Merrier—complexity in long non-coding RNA Loci. Front Endocrinol 8:90
10. Kopp F, Mendell JT (2018) Functional classification and experimental dissection of long noncoding RNAs. Cell 172:393–407
11. Zhang J, Zhang Z, Wang Z, Liu Y, Deng L (2017) Ontological function annotation of long non-coding RNAs through hierarchical multi-label classification. Bioinformatics 34:1750–1757
12. Wang KC, Chang HY (2011) Molecular mechanisms of long noncoding RNAs. Mol Cell 43:904–914
13. St Laurent G, Wahlestedt C, Kapranov P (2015) The Landscape of long noncoding RNA classification. Trends Genet 31:239–251
14. Mattick JS, Rinn JL (2015) Discovery and annotation of long noncoding RNAs. Nat Struct Mol Biol 22:5–7
15. Ren K, Li Y, Lu H, Li Z, Li Z, Wu K, Li Z, Han X (2016) Long noncoding RNA HOTAIR controls cell cycle by functioning as a competing endogenous RNA in Esophageal squamous cell carcinoma. Transl Oncol 9:489–497
16. Zhang X, Weissman SM, Newburger PE (2014) Long intergenic non-coding RNA HOTAIRM1 regulates cell cycle progression during myeloid maturation in NB4 human promyelocytic leukemia cells. RNA Biol 11:777–787
17. Sahu A, Singhal U, Chinnaiyan AM (2015) Long noncoding RNAs in cancer: from function to translation. Trends Cancer 1:93–109
18. Aguda BD (2005) Modeling the cell division cycle. In: Lect. Notes Math. Springer-Verlag, Berlin, pp 1–22
19. Ferrell JE, Tsai TY-C, Yang Q (2011) Modeling the cell cycle: why do certain circuits oscillate? Cell 144:874–885
20. Weis MC, Avva J, Jacobberger JW, Sreenath SN (2014) A data-driven, mathematical model of mammalian cell cycle regulation. PLoS One 9: e97130
21. Heldt FS, Barr AR, Cooper S, Bakal C, Novák B (2018) A comprehensive model for the proliferation–quiescence decision in response to endogenous DNA damage in human cells. Proc Natl Acad Sci U S A 115:2532–2537
22. Aguda BD (1999) A quantitative analysis of the kinetics of the G2 DNA damage checkpoint system. Proc Natl Acad Sci U S A 96:11352–11357
23. Iwamoto K, Tashima Y, Hamada H, Eguchi Y, Okamoto M (2008) Mathematical modeling and sensitivity analysis of G1/S phase in the cell cycle including the DNA-damage signal transduction pathway. Biosystems 94:109–117
24. Iwamoto K, Hamada H, Eguchi Y, Okamoto M (2011) Mathematical modeling of cell cycle regulation in response to DNA damage: exploring mechanisms of cell-fate determination. Biosystems 103:384–391
25. Aguda BD, Algar CK (2003) A structural analysis of the qualitative networks regulating the cell cycle and apoptosis. Cell Cycle 2:538–543
26. Zhao J, Liu Y, Zhang W, Zhou Z, Wu J, Cui P, Zhang Y, Huang G (2015) Long non-coding RNA Linc00152 is involved in cell cycle arrest, apoptosis, epithelial to mesenchymal transition, cell migration and invasion in gastric cancer. Cell Cycle 14:3112–3123
27. Mazar J, Rosado A, Shelley J, Marchica J, Westmoreland TJ (2017) The long non-coding RNA GAS5 differentially regulates cell cycle arrest and apoptosis through activation of

28. Kitagawa M, Kitagawa K, Kotake Y, Niida H, Ohhata T (2013) Cell cycle regulation by long non-coding RNAs. Cell Mol Life Sci 70:4785–4794
29. Kutter C, Watt S, Stefflova K, Wilson MD, Goncalves A, Ponting CP, Odom DT, Marques AC (2012) Rapid turnover of long noncoding RNAs and the evolution of gene expression. PLoS Genet 8:e1002841
30. Szcześniak MW, Makałowska I (2016) lncRNA-RNA interactions across the human transcriptome. PLoS One 11:e0150353
31. Forouzmand E, Owens NDL, Blitz IL, Paraiso KD, Khokha MK, Gilchrist MJ, Xie X, Cho KWY (2017) Developmentally regulated long non-coding RNAs in Xenopus tropicalis. Dev Biol 426:401–408
32. Aguda B, Friedman A (2008) Models of cellular regulation. Oxford University Press, New York
33. Gauthier JH, Pohl PI (2011) A general framework for modeling growth and division of mammalian cells. BMC Syst Biol 5:3
34. Singhania R, Sramkoski RM, Jacobberger JW, Tyson JJ (2011) A hybrid model of mammalian cell cycle regulation. PLoS Comput Biol 7:e1001077
35. Aguda BD, Goryachev AB (2007) From pathways databases to network models of switching behavior. PLoS Comput Biol 3:e152
36. Müller J, Kuttler C (2015) Methods and models in mathematical biology. Springer, Berlin
37. Bernot G, Comet J-P, Richard A, Chaves M, Gouzé J-L, Dayan F (2013) Modeling and analysis of gene regulatory networks. In: Cazals F, Kornprobst P (eds) Modeling in computational biology and biomedicine. Springer, Berlin, pp 47–80
38. Jalali S, Kapoor S, Sivadas A, Bhartiya D, Scaria V (2015) Computational approaches towards understanding human long non-coding RNA biology. Bioinformatics 31:2241–2251
39. Milo R (2002) Network Motifs: simple building blocks of complex networks. Science 298:824–827
40. Alon U (2007) Network motifs: theory and experimental approaches. Nat Rev Genet 8:450–461
41. Junker BH, Schreiber F (2008) Analysis of biological networks. Wiley-Interscience, Hoboken, NJ
42. Barabasi A-L (2016) Network science. Cambridge University Press, Cambridge
43. Hecker M, Lambeck S, Toepfer S, van Someren E, Guthke R (2009) Gene regulatory network inference: data integration in dynamic models—a review. Biosystems 96:86–103
44. Marbach D, Costello JC, Küffner R et al (2012) Wisdom of crowds for robust gene network inference. Nat Methods 9:796–804
45. Creixell P, Reimand J et al (2015) Pathway and network analysis of cancer genomes. Nat Methods 12:615–621
46. Sun M, Gadad SS, Kim D-S, Kraus WL (2015) Discovery, annotation, and functional analysis of long noncoding RNAs controlling cell-cycle gene expression and proliferation in breast cancer cells. Mol Cell 59:698–711
47. Liu F, Zhang S-W, Guo W-F, Wei Z-G, Chen L (2016) Inference of gene regulatory network based on local Bayesian networks. PLoS Comput Biol 12:e1005024
48. Feng N, Ching T, Wang Y et al (2016) Analysis of microarray data on gene expression and methylation to identify long non-coding RNAs in non-small cell lung cancer. Sci Rep 6:37233
49. Filkov V (2005) Identifying gene regulatory networks from gene expression data. In: Handbook of computational molecular biology. CRC Press, Boca Raton, FL, pp 1–29
50. Leal LG, López C, López-Kleine L (2014) Construction and comparison of gene co-expression networks shows complex plant immune responses. PeerJ 2:e610
51. Zhang L, Feng XK, Ng YK, Li SC (2016) Reconstructing directed gene regulatory network by only gene expression data. BMC Genomics 17:430
52. Zhang Y, Huang H, Zhang D, Qiu J, Yang J, Wang K, Zhu L, Fan J, Yang J (2017) A review on recent computational methods for predicting noncoding RNAs. Biomed Res Int 2017:1–14
53. del Rosario RCH, Damasco JRCG, Aguda BD (2016) MicroRNA inhibition fine-tunes and provides robustness to the restriction point switch of the cell cycle. Sci Rep 6:32823
54. Foster SS, De S, Johnson LK, Petrini JHJ, Stracker TH (2012) Cell cycle- and DNA repair pathway-specific effects of apoptosis on tumor suppression. Proc Natl Acad Sci U S A 109:9953–9958
55. Nowsheen S, Yang ES (2012) The intersection between DNA damage response and cell death pathways. Exp Oncol 34:243–254
56. Ding L, Wang M, Sun D, Li A (2018) TPGLDA: Novel prediction of associations between lncRNAs and diseases via lncRNA-disease-gene tripartite graph. Sci Rep 8:1065

57. Yang X, Gao L, Guo X, Shi X, Wu H, Song F, Wang B (2014) A network based method for analysis of lncRNA-disease associations and prediction of lncRNAs implicated in diseases. PLoS One 9:e87797
58. Chen X, Yan CC, Zhang X, You Z-H (2016) Long non-coding RNAs and complex diseases: from experimental results to computational models. Brief Bioinform 18:558–576
59. Gu C, Liao B, Li X, Cai L, Li Z, Li K, Yang J (2017) Global network random walk for predicting potential human lncRNA-disease associations. Sci Rep 7:12442
60. Huang Y-A, Chan KCC, You Z-H (2018) Constructing prediction models from expression profiles for large scale lncRNA–miRNA interaction profiling. Bioinformatics 34:812–819
61. Bloomingdale P, Nguyen VA, Niu J, Mager DE (2018) Boolean network modeling in systems pharmacology. J Pharmacokinet Pharmacodyn 45:159–180
62. Thieffry D (2007) Dynamical roles of biological regulatory circuits. Brief Bioinform 8:220–225
63. Yeger-Lotem E, Sattath S, Kashtan N, Itzkovitz S, Milo R, Pinter RY, Alon U, Margalit H (2004) Network motifs in integrated cellular networks of transcription-regulation and protein-protein interaction. Proc Natl Acad Sci U S A 101:5934–5939
64. Voit EO, Radivoyevitch T (2000) Biochemical systems analysis of genome-wide expression data. Bioinformatics 16:1023–1037
65. Kikuchi S, Tominaga D, Arita M, Takahashi K, Tomita M (2003) Dynamic modeling of genetic networks using genetic algorithm and S-system. Bioinformatics 19:643–650
66. Voit EO (2013) Biochemical systems theory: a review. ISRN Biomath 2013:1–53
67. Chowdhury AR, Chetty M, Evans R (2015) Stochastic S-system modeling of gene regulatory network. Cogn Neurodyn 9:535–547
68. Voit EO, Martens HA, Omholt SW (2015) 150 Years of the mass action law. PLoS Comput Biol 11:e1004012
69. Rabajante JF, Talaue CO (2015) Equilibrium switching and mathematical properties of nonlinear interaction networks with concurrent antagonism and self-stimulation. Chaos Solitons Fractals 73:166–182
70. Marchese FP, Huarte M (2017) A long noncoding RNA in DNA replication and chromosome dynamics. Cell Cycle 16:151–152
71. Rabajante JF, Babierra AL (2015) Branching and oscillations in the epigenetic landscape of cell-fate determination. Prog Biophys Mol Biol 117:240–249
72. Sauer T (2012) Numerical analysis, 2nd edn. Pearson, Boston
73. Dhooge A, Govaerts W, Kuznetsov YA (2003) MATCONT: A MATLAB package for numerical bifurcation analysis of ODEs. ACM Trans Math Softw 29:141–164
74. Allen E (2007) Modeling with Itô stochastic differential equations. Springer, Dordrecht
75. Cardelli L, Csikász-Nagy A, Dalchau N, Tribastone M, Tschaikowski M (2016) Noise reduction in complex biological switches. Sci Rep 6:20214. https://doi.org/10.1038/srep20214
76. Ulitsky I (2016) Evolution to the rescue: using comparative genomics to understand long non-coding RNAs. Nat Rev Genet 17:601–614
77. Joung J, Engreitz JM, Konermann S et al (2017) Genome-scale activation screen identifies a lncRNA locus regulating a gene neighbourhood. Nature 548:343–346

Index

A

AGO, *see* Argonaute loading protein
Airway 8–12, 14–18
Algorithm
    Gillespie 375, 381–383, 404, 440
    K-nearest neighbor 326
    machine learning 125, 429
        deep learning 125
    random walk 306, 309, 310, 313
Alignment 56, 59, 61–62, 64, 67, 74, 93, 95, 107, 115–117, 121, 125, 161, 182, 203, 204, 235, 238
Allergen 4, 8, 11–16
Allergic 10–16
Apoptosis 16, 22, 33, 35, 39–41, 324, 346, 350, 351, 412, 421, 428, 431, 432
Archive 184, 221, 222, 224, 226, 228–230
Argonaute loading protein (AGO) 153–154, 217
Argonaute protein 313, 371
Asthma
    acute asthma 16
    chronic asthma 16

B

Back splicing 55–58, 64, 68, 69, 71
Bioinformatics
    databases
        DARNED 162
        Galaxy 116, 118, 203–205, 207–212
        KEGG 185, 189, 205, 207, 430
        miRTarBase 120, 180, 181, 183, 184, 187, 189, 226, 228, 258, 259, 269
        RADAR 162
        Reactome 43, 45, 118, 185, 189, 430
        TarBase 43, 45, 180, 181, 184, 189, 226, 228, 259, 270
        TriplexRNA 120, 205–210, 275
        UCSC 59–61, 65, 67, 80, 95, 117, 262, 271
        VANESA 185, 189
    tools
        BAM 62, 63, 95, 107
        BEDTools 80, 95, 96
        BioCatalogue Web services 202
        BioConda 116, 204

Bioinformatics Links Directory 202
bio.tools 126, 202
Bowtie 56, 62
BWA 56, 59, 60, 67
CIRI 56, 57, 59–60, 66–67
Coding Potential Calculator (CPC) 80, 95
Cufflinks 80, 95
EMBRACE 202
Fastq 59, 62, 64–66, 74, 93, 105, 106, 117
FastQC 66, 74, 80, 93, 95, 106, 117
FindCirc 56–58, 60–65, 74
Getorf 80, 95
Molecular Biology Database List 202
OMICtools 115, 126, 202
PhyloCSF 80, 95
SAM 56, 62, 64, 65, 74, 107
Segemehl 56, 57, 59, 65, 117, 203
Star 56, 117, 118, 203
TopHat 56, 80, 107
training
    de.NBI 127, 200, 201
    ELIXIR 116, 200, 202
    Goblet 200
    GTN 201
    Software Carpentry 200–201
Biomarker 56, 190, 218, 224, 258, 262, 265, 266, 304, 305, 324, 428, 432
Biomedicine 116, 219, 252, 262–267

C

Cancer
    cervical cancer 324, 325, 328, 331–336
    gynecologic cancer 323–337
    metastasis 33–49, 422
    ovarian cancer 323–326, 328–335
Cell
    macrophage
        BMDM 5, 6, 14
        PMN 7
    T-cells 5, 7, 8, 19, 21–24, 47, 220, 264
    T_H2 .... 11, 14, 15, 22

Index

Cell cycle
  checkpoint 428, 430, 435
  cyclin-CDK complex 430
  DNA damage 428, 430, 441
  G1 .... 6, 428, 430, 431, 438
  G2 428
Cell fate decision 360, 411–423
Central dogma 134, 135, 175, 341
ceRNA, *see* Competing endogenous RNAs
ChIA-PET 303, 309, 312, 317
Chromatin
  conformation 306, 307, 317
  interaction 113, 307, 309, 311–313, 317
  looping 303, 312
Chronic obstructive pulmonary disease (COPD) 8–10
Circular RNAs (circRNAs) 24,
    55–74, 112, 113, 158, 262–264, 266, 267,
    313, 314, 422, 423
Competing endogenous RNAs (ceRNA) 158,
    231, 232, 303–307, 313–316, 343, 354–356,
    360, 367–405, 415–419, 421, 423
Competition 303, 304,
    343, 354–356, 367–405
COPD, *see* Chronic obstructive pulmonary disease
Crosstalk 46, 369, 370,
    375, 377, 383–392, 394–396, 398–405
Cytoscape 331, 432
  ClueGO 331

D

Differential expression 221, 222,
    224–225, 237, 306, 317
DiGeorge syndrome critical region gene 8
    (DGCR8) 36, 46, 153, 184, 187, 217
Disease associations 190, 221,
    222, 229, 230, 252, 266, 291, 295–297,
    301–317, 427, 432
Disease ontology 325, 328, 334–335
Docker 116, 204

E

EMBRACE Data And Methods (EDAM)
    ontology 126, 202
Enhancer-promoter interaction 303
Entropy 58, 235, 326
Epithelial-to-mesenchymal transition (EMT) 37,
    39, 40, 47, 157, 188, 348–350, 359, 360, 411,
    412, 418–422, 428
Epitranscriptomics 138–140, 160, 163

G

Gene co-expression 122
Gene ontology 186, 232
Gene regulation 4, 35, 189,
    199, 216, 218, 311, 312, 314, 345, 360, 432,
    436, 437, 439
Gene regulatory circuit 4
Guilt-by-association 306, 308, 309

H

Hairpin 36, 153–155, 176–178,
    183, 184, 187, 215, 217, 222, 232, 234–236,
    238, 324
HITS-CLIP 176, 177, 180, 183
Hybridization 11, 78, 80–82,
    92, 101, 182, 222

I

ILP, *see* Integer linear programming
Immunity
  adaptive immunity 7, 21–23
  innate immunity 20–22
Infection 4, 5, 7, 12–14, 17–19,
    23–24, 218–220, 225, 229, 256, 265, 324
Inflammation
  acute inflammation 16
  chronic inflammation 8–10, 16, 17
Integer linear programming (ILP) 290,
    292–294, 298
Integrated approach 323–337
Interaction 13, 17, 18,
    20–22, 34, 35, 39–41, 46, 111, 113, 118,
    120–122, 136, 137, 140, 143, 144, 147, 151,
    152, 154, 155, 162, 175–190, 199–212,
    220–222, 226–228, 230–232, 238, 258–263,
    290–293, 296, 302–307, 309, 311–317, 329,
    341–344, 346–349, 351–356, 359, 368–371,
    376, 384, 391–394, 396, 398, 401, 404, 412,
    421, 423, 428–430, 432–442
Interaction detection 183–189
Isoform 5, 34, 35,
    40–46, 49, 95, 117, 141, 143, 147, 151, 228
  isomiR 216, 221,
    223, 224, 237, 257, 259, 273

K

Kallisto 56, 117, 118
Knowledge base 162, 212

Index

L

Ligand lipopolysaccharide (LPS) 4–6, 20, 21
Linear noise approximation 361, 375, 381
Long non-coding RNA (lncRNA) 4, 48,
    77–109, 112, 135, 230, 259, 294, 301–317,
    368, 427–442
Long range 303, 312
LPS, *see* Ligand lipopolysaccharide
Lung 4, 6–10, 12–15,
    17, 47, 264, 304, 337, 422, 428, 442

M

Mapping 56–58, 95, 107,
    117, 163, 190, 255, 263
Mathematical modeling
    differential equations
        Hill function 435–436
        mass action 378, 435
        stochasticity 440–441
    dynamics
        bistability 417–418, 439
        equilibrium 345, 435
        multistability 439
        toggle switch 348, 349
        ultrasensitivity 417–420, 422
    kinetic modeling 120, 346, 348, 367–405
    non-linear behaviour 393
MDS, *see* Minimum dominating sets
Messenger RNA (mRNA) 4, 36, 55,
    79, 111, 134, 175–190, 199–212, 216, 256,
    303, 323–337, 341, 368, 412, 428
MI, *see* Mutual information
Microarray 7, 13, 119,
    177, 222, 229, 231, 232, 308, 326, 429, 430
MicroRNA (miRNA)
    cooperativity 396, 418
    crosstalk 375, 383–390, 394, 400
    detection 176–180, 183–189, 220, 325
    functional annotation 238
    response element 18, 151–154,
        158, 230, 304, 313–315, 367
    sequencing 216, 218,
        222, 223, 234, 236–238, 257, 258
    sponge 55, 303
Minimum dominating sets (MDS) 290–298
Minimum redundancy maximum relevance 328
miRNA, *see* microRNA
miRNome 36, 41–49
Molecular mechanism 16, 150, 160, 216–217
Mutual information (MI) 315, 325–327, 398, 399

N

Nanocarrier-mediated delivery 47
ncRNA, *see* Noncoding RNAs
Network
    controllability 290, 298
    control node 292, 293, 297
    driver node 290–293, 296
    hub 37, 38, 47, 200, 202, 312, 432
    module 35, 37, 48, 309
    motif
        feedback loop 187, 188, 348–351,
            412–414
        feedforward loop 392, 394
    multi-layer 296–297
    multipartite 432, 434
    topology 37, 313, 390–392, 432, 434
Next generation sequencing (NGS) 15, 78,
    79, 91–93, 112, 114–118, 127, 161, 177, 178,
    200, 220–222, 224, 229, 232, 237, 238, 257
NFκB 4–7, 17, 20, 21
Noise
    extrinsic noise 344, 345, 387, 394–397
    intrinsic noise 345, 351, 370, 377, 380, 397
Noncoding RNAs (ncRNA)
    annotation 55–74, 80,
        93–96, 107, 121–124, 127, 162
    discovery 55, 56, 78,
        93–96, 107, 109, 113, 114, 120, 121, 123,
        133–136, 153, 161, 162
    identification 113–118
Noncoding RNAs (ncRNA)-mediated circuits .. 411–423
Nonlinearity 412
Northern blotting 126, 177, 222

P

p53 .... 33–35, 37,
    39–42, 44–49, 412, 413, 430, 431, 441
p53 mutations 34
p73 33–49
Pathway 4, 5, 11, 16,
    19, 23, 35, 37, 39–41, 43–47, 49, 113, 121,
    123, 138, 143, 150, 152–154, 157, 158, 176,
    180, 183, 185–190, 205, 207, 216, 217, 229,
    230, 263, 308, 315, 324, 325, 328, 331–334,
    345, 357–360, 370, 371, 394, 398, 404,
    428–433, 441
Patient 8–10, 12–14,
    16–18, 21, 22, 24, 34, 39, 48, 114–116, 157,
    159, 204, 232, 266, 324, 422
PCR, *see* Polymerase chain reaction

Index

Pipeline 43, 45, 56, 58–68, 73, 74, 79–80, 93, 94, 109, 161, 204, 206, 207, 211, 237, 258
Piwi-interacting RNAs (piRNAs) 112, 136, 261, 264, 265, 273, 290, 342, 422
Planemo 205, 208–210
Polymerase chain reaction (PCR) 56, 60, 68, 69, 71–73, 80, 84, 86, 87, 89, 90, 96–97, 107, 115, 126, 161, 238
PPPPY motif 41, 46
Primer
    convergent primer 68, 71
    divergent primer 56, 68, 73
Promoter–enhancer interaction 303, 312
Purification of RNA 73, 99
    RNase R 73

Q

Quality check 70, 72, 78, 82–85, 96, 99–100
    RNA loading dye 70, 78, 81, 99

R

Recycle ratio 418, 419
Regulatory modules 324–326, 328–331, 335
Relativity theory 175
RNA-induced silencing-like complex (RISC) 36, 176, 180, 217, 218, 313, 367, 371
RNA interference (RNAi) 215, 261, 274, 421
RNA isolation
    from embryonic stages 82
    from tissues 69–70, 82
    Trizol 69, 78, 82, 83
RNA modification 136–147, 163
RNA quality 70, 83, 84
RNA sequencing (RNA-seq)
    A-base addition 89–90
    end repair of cDNA fragments 88–89
    first strand cDNA preparation 87
    ligation of index adapters 89–90
    poly-A enriched RNA library 86–92
    second strand cDNA preparation 87–88

S

Signaling
    intracellular signaling 4, 356
    Juxtacrine signaling 356
Single-nucleotide polymorphisms (SNP) 9, 57, 112, 115, 119–120, 161, 226, 229, 238, 258–260, 262, 266, 305, 308
Steady state 316, 345, 351, 373–377, 380, 381, 383–391, 394–396, 401, 403, 404, 418, 438

Susceptibility 38, 385–387, 390, 391, 401, 403
    Fluctuation-dissipation relation 389
Synthetic biology 421–422

T

TAp73alpha 41–45
TAp73beta 41–45
Target mimicry 303
Target prediction 15, 17, 120, 125, 180, 182, 183, 187, 190, 228, 231, 314, 315
Taxonomy 202
Threshold 65, 93, 95, 96, 332, 345, 354, 356, 373–376, 383, 386, 391, 396–398, 401, 402, 421, 436
Tissue 4, 8, 9, 12, 13, 15–17, 24, 34, 35, 38, 47, 69–71, 78, 82–85, 96, 97, 103, 104, 109, 113, 115, 119, 121, 123, 127, 138, 141, 155, 156, 158, 177, 183, 225, 231, 232, 258, 260–262, 302, 304, 314–316, 343, 428
Toll-like receptors (TLR) 4, 6, 7, 19–21
Transcript identification 117
Transcriptome 3, 22, 56, 77, 145, 161, 186, 255, 314, 317, 405
Transfer RNA (tRNA) 81, 82, 101, 134, 138, 140, 148, 159–160, 289
Translational inhibition 36, 346

U

Untranslated region (UTR) 4, 13, 15–19, 36–38, 55, 95, 136, 144, 150–154, 157, 159, 162, 180, 184, 216, 218, 226, 228, 229, 231, 232, 313

V

Validation
    actb 71, 72
    actin B 72
    agarose 71
    BWA-MEM 67
    centrifuge 70, 73
    chloroform 60, 70
    DNase 60, 70
    DTT 71
    ethanol 60, 70, 73
    isopropanol 60, 70
    LiCl 60, 73
    $MgCl_2$ 60, 71, 72
    nuclease free water 60, 70
    random hexamer 60, 70

reverse transcriptase 70–71
RNase H 60, 71
split-mapping 67
SuperScript II Reverse Transcriptase 60, 71
TAE buffer 70
trimming 66
Trimmomatic 66
Viral miRNA 216, 223, 226, 227, 236

W

Workflow 69, 78, 85, 111–127, 200, 201, 203, 204, 210–212, 227, 229

Z

Zebrafish 77–109, 302